Advances in Wastewater Research

Series Editor

Maulin P. Shah, Environmental Microbiology Consultant, Ankleshwar, Gujarat, India

The Advances in Wastewater Research highlights the prospective future development of wastewater treatment techniques, research and field applications. The series also focuses on the limitations and challenges associated with some generally accepted wastewater strategies and evaluates the possible applications of these corrective strategies to eliminate toxic pollutants from the environment through integrated technologies in wastewater treatment. The series covers the research areas including, but not limited to, toxic pollutants removal from wastewater, methods and procedures, materials, management of wastewater, valorisation possibilities of waste resulting from the removal of toxic pollutants from wastewater, wastewater recycling, environmental impact, and wastewater policies. The series contribute to remove, reduce, or neutralize water contaminants that threaten human health, and ecosystem productivity and integrity. The series includes peer-reviewed monographs, edited volumes, textbooks, and conference proceedings.

Ali Mohd Yatoo · Surendra Suthar · Maulin P. Shah
Editors

Wastewater Monitoring and Management by Advanced Approaches

Springer

Editors
Ali Mohd Yatoo
Department of Environmental Science
University of Kashmir
Srinagar, Jammu and Kashmir, India

Maulin P. Shah
Division of Applied and Environmental Microbiology
Enviro Technology Limited
Anklesvar, Gujarat, India

Surendra Suthar
School of Environment and Natural Resources
Doon University
Dehradun, Uttarakhand, India

ISSN 3004-9725 ISSN 3004-9733 (electronic)
Advances in Wastewater Research
ISBN 978-981-95-2600-0 ISBN 978-981-95-2601-7 (eBook)
https://doi.org/10.1007/978-981-95-2601-7

This Springer imprint is published by the registered company Springer Nature Singapore Pte Ltd.
The registered company address is: 152 Beach Road, #21-01/04 Gateway East, Singapore 189721, Singapore

Preface

Wastewater Monitoring and Management by Advanced Approaches is a comprehensive guide that provides an in-depth understanding of the latest technologies and approaches for monitoring and managing wastewater. This comprehensive guide provides a deep understanding of the current state of wastewater management, the challenges faced by the industry, and the latest technological advancements in the field.

The book is divided into two parts, each focusing on a specific aspect of wastewater management. Part I is related to wastewater monitoring, which highlights and provides an overview of the latest technologies used for monitoring wastewater, including remote sensing, GIS, biosensors, electrochemical sensors, and microfluidic devices. Part II focuses on the latest approaches for treating and managing wastewater. The part discusses advanced electrochemical methods, nanotechnology applications, membrane filtration techniques, electrocoagulation, microbial fuel cells, and the potential of artificial intelligence, machine learning, etc., in wastewater management. The part also covers the potential of advanced biotechnological approaches for the removal of toxicants and enhancement of water quality.

The book is written clearly and concisely, making it accessible to both professionals and students in the field of wastewater management. With a focus on advanced approaches, this book is an essential resource for anyone looking to gain a deeper understanding of the latest technologies and approaches for monitoring and managing wastewater.

Key Features

1. Provides a comprehensive review of advanced approaches for wastewater monitoring.
2. Presents an in-depth discussion of utilizing modern and innovative technologies in wastewater management.

3. Offers applications in pollution remediation, resource recovery, and environmental sustainability.

Srinagar, India — Ali Mohd Yatoo
Dehradun, India — Surendra Suthar
Anklesvar, India — Maulin P. Shah

Contents

Part I
Approaches for Wastewater Monitoring

Chapter 1
Introduction to Wastewater Monitoring: Challenges and Opportunities

Vashisht Vishakha, Chetna Nagoda, and Meenu Chhabra

Abstract Freshwater is limited and frequently contaminated by human activity, despite its necessity. Poor sanitation and inefficient agricultural practices exacerbate the situation by putting public health and ecosystems at risk. Thus, monitoring, conservation, treatment, and sustainable management are some essential approaches that must be implemented to ensure long-term access to a clean and safe water supply. Municipal wastewater includes industrial, domestic, stormwater, and infiltration sources, and its composition depends on various factors such as population load, local climatic conditions, urban infrastructure, waste disposal practices, and industrialization. Inadequate wastewater management worsens water pollution and health risks, thus making water surveillance and monitoring an essential measure. This chapter presents an outline of wastewater monitoring processes ranging from conventional to advanced methods, opportunities, and challenges in the wastewater management sector. The conventional chemical monitoring of wastewater includes observations on organic load and toxic chemical presence, while biological monitoring detects microbial contaminants, pollution bioindicators, and the toxicity of certain water contamination on aquatic organisms. Meanwhile, advanced techniques such as membrane filtration, bioassays, and molecular biosensors enhance detection, ensuring effective wastewater management, pollution control, and environmental safety, thus offering cost-effective alternatives to traditional methods. A brief understanding of the above issues is elaborated in this chapter.

Keywords Wastewater monitoring · Biosensors · Bio-electrochemical system · Microbial fuel cell

M. Chhabra (✉)
Environmental Biotechnology Lab, Indian Institute of Technology Jodhpur, Jodhpur, Rajasthan, India
e-mail: meenuchhabra@iitj.ac.in

V. Vishakha · C. Nagoda · M. Chhabra
Department of Bioscience and Bioengineering, Indian Institute of Technology Jodhpur, Jodhpur, Rajasthan, India

A. M. Yatoo et al. (eds.), *Wastewater Monitoring and Management by Advanced Approaches*, Advances in Wastewater Research,
https://doi.org/10.1007/978-981-95-2601-7_1

1.1 Introduction

Water covers 70% of our planet, and it is reasonable to assume that it will always be abundant. Fresh water, which we use for drinking, bathing, and irrigating our fields, is extremely scarce. Fresh water makes up only 3% of the world's water supply, two-thirds of which is hidden in frozen glaciers or otherwise unusable (Ramalho 2013). Since the hydrologic cycle allows water to replenish and purify itself, it ensures the endless supply of freshwater resources. However, if the water supplies continue to be contaminated or depleted more quickly, then drinking water will be scarce in the future. Furthermore, only two-thirds of the readily available water is fit for human use; the remainder is unfit for human consumption because of pollution and other issues. Globally, there is rising worry about the depletion of water resources; therefore, access to safe and clean water is essential to maintaining a healthy population (Ramalho 2013). As part of the "2030 Agenda for Sustainable Development," the United Nations (UN) approved 17 Sustainable Development Goals (SDGs) in 2015 (United Nations, THE 17). "Ensure availability and sustainable management of water and sanitation for all" is the topic of SDG6. According to the SDG6 progress summary, two billion people, or 26% of the global population, did not have safe drinking water in 2020, leading to needless suffering from waterborne illnesses (United Nations, Summary Progress).

Sources of water pollution can be natural or anthropogenic. Chemical contamination of rivers and streams is one of the most critical environmental issues. The enormous amount of chemicals in water is discharged from industrial waste that enters rivers and streams. Water is used regularly in both our homes and businesses. However, most of it returns to these sources after being consumed and contaminating them. Around the world, 1.1 billion people do not have access to water, and 2.7 billion experience water scarcity for at least 1 month of the year. Additionally, 2.4 billion people suffer from poor sanitation, which exposes them to waterborne infections, including cholera and typhoid fever (World Wildlife Fund Report). Every year, diarrheal illnesses alone claim the lives of two million people, the majority of whom are youngsters. Reusing treated wastewater effluents from wastewater reclamation plants (WRPs) is thought to be a viable way to address the freshwater deficiency. However, since secondary effluents are considered to be a significant source of these contaminants, using reclaimed water from treated wastewater effluents may have adverse health effects linked to microbial pathogens, heavy metals, endocrine-disrupting chemicals (EDCs), pharmaceuticals and personal care products (PPCPs), and other resistant organic compounds (Tran et al. 2015).

1.2 Wastewater Composition and Characterization

Municipal wastewater consists of industrial wastewater, domestic wastewater, infiltration and inflow into sewer lines, and stormwater runoff. Domestic water is affected by population size, climate, quality of water supply, and community affluence. Moreover, the cost of water, supply pressure, and degree of industrialization also affect the composition of wastewater (Iyare et al. 2020). The types of pollutants found in wastewater vary according to the origins of the pollution (Fig. 1.1). For example, wastewater samples from petrochemical refineries and chemical processing businesses frequently contain high quantities of chemical oxygen demand (COD), sulfide, oil, grease, ammonia, organic solvents, and polycyclic aromatic hydrocarbons (PAHs). However, wastewater samples from the electrical and electroplating sectors show a low COD value and a high percentage of heavy metals (Botalova et al. 2009). The COD value is used to measure the level of contamination in wastewater. However, if the COD values are the sole criterion used, it is challenging to provide a clearly defined range to differentiate between wastewater with low, medium, and high strength. Large concentrations of wastewater constituents, such as COD, ammonia, suspended particles, or heavy metals, are typically linked to high wastewater strength (Gopala Krishna et al. 2008). The wastewater composition is also directly impacted by the architecture of sewage line systems, such as separate or combined sewer systems. Monitoring every unknown organic and inorganic component in wastewater is not feasible or cost-effective (Choi et al. 2005). It has been shown that one of the main disadvantages of using COD to assess the amount of organic matter (OM) in wastewater samples is that it includes both biodegradable and nonbiodegradable substances. Therefore, utilizing COD as a measure to assess wastewater biodegradability is quite difficult. Since the biochemical oxygen demand (BOD) test only shows the wastewater biodegradable percentage, it is also limited in its ability to evaluate the pollutant severity of wastewater based on BOD or BOD_5. Nonetheless, the BOD_5/COD ratio can be used as a gauge to evaluate the biodegradability of organic matter in wastewater (Mutamim et al. 2012).

1.3 Wastewater Monitoring

Inadequate wastewater management limits the use and makes drinking water treatment more difficult, which contributes to environmental deterioration and waterborne illnesses. However, wastewater monitoring and treatment are required to manage water reservoirs and sanitation. Wastewater monitoring is a kind of disease surveillance that provides a basis for delivering early warnings for known and unknown health risks using affordable, objective health metrics derived from random and noninvasive community-level sampling (Keshaviah et al. 2023). Monitoring of wastewater ensures adherence to environmental laws, identifies sources of pollution for appropriate treatment, guards against harm to ecosystems and public health, and

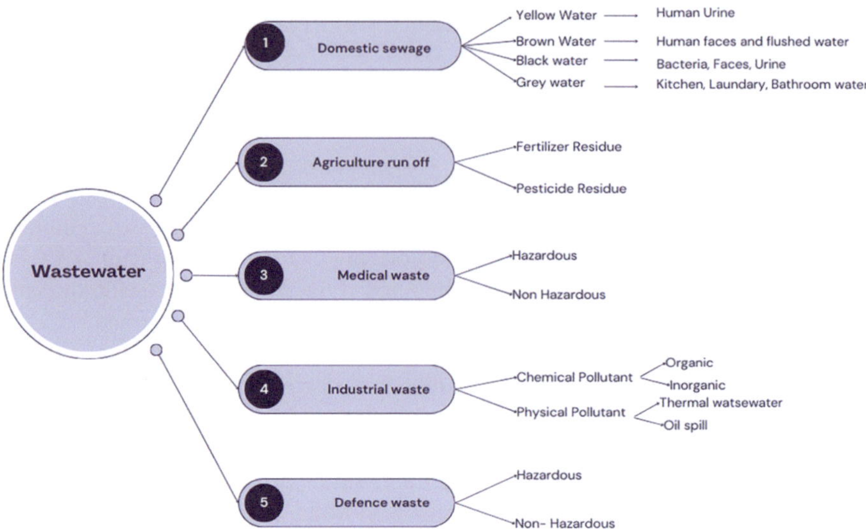

Fig. 1.1 Sources of wastewater and their chemical characteristics

aids in the optimization of wastewater treatment procedures (Vanrolleghem and Lee 2003). Four fundamental components are used in wastewater treatment plant monitoring and control: (1) sensors that provide online data, (2) insight into the process as summarized in a proper process model, (3) adequate monitoring and control strategies, and 4) actuators that implement the controller output (Vanrolleghem and Lee 2003). It is crucial to monitor wastewater to design effective treatment processes and ensure environmental compliance. Various wastewater monitoring methods are described in Fig. 1.2.

1.3.1 Physical Method

Physical monitoring of wastewater involves assessing observable properties such as temperature, pH, electrical conductivity (EC), salinity, turbidity, and suspended solids to determine water quality and treatment efficiency. Water resources like rivers and lakes are negatively impacted by two physical water quality parameters: total dissolved solids (TDS) and total suspended solids (TSS) (Adjovu et al. 2023a, b; Khan et al. 2018). TDS and TSS are fractional components of "total solids" that have been filtered out. Water bodies that have higher TDS and TSS concentrations are unable to fulfil their functions for drinking, biodiversity support, recreation, power generation, ecosystem services, regional planning, transportation routes, waste disposal, agriculture production, irrigation, energy production, and fish farming (Cheng et al. 2022; Dey and Vijay 2021). TDS are components of total solids in a water sample that, under certain circumstances, flow through nominal pore sizes of 2 μm or less. EC

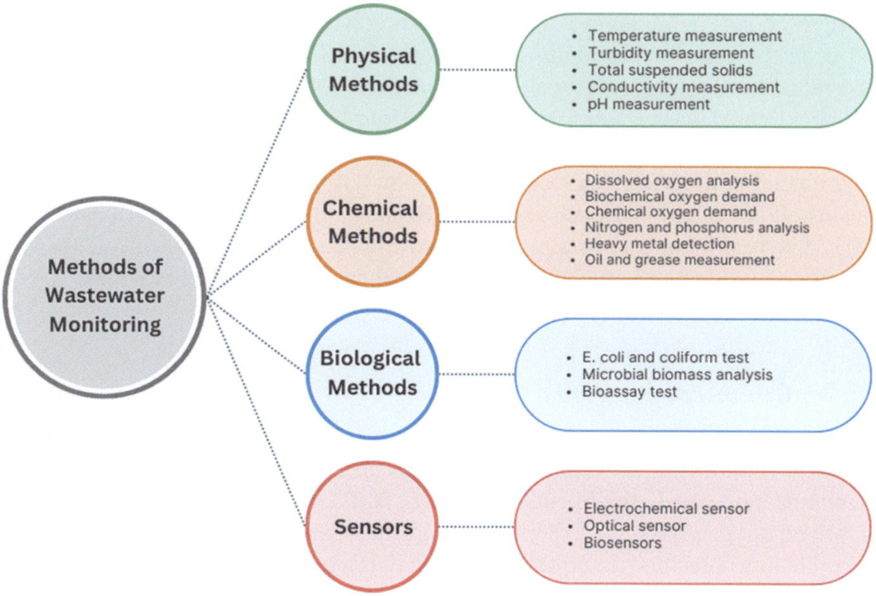

Fig. 1.2 Methods of wastewater monitoring

meters and ion chromatography are commonly used for TDS assessment (Table 1.1), with recommended limits of < 500 mg/L for drinking water and < 1500 mg/L for industrial wastewater (WHO 2020). A filter with a pore size of 2 μm or less retains the TSS constituents in any given water sample, as determined by weighing the dry residue that remains on the filter (Adjovu et al. 2023a, b). TSS constituents include particulate particles such as algae, silts, and undissolved organic and inorganic particles that affect oxygen transfer and clog treatment systems. TSS is measured via gravimetric analysis, turbidimeters, or Imhoff cones. Acceptable discharge limits for TSS are typically < 30 mg/L (Merz 2022). pH indicates acidity or alkalinity. It is typically measured using pH meters or litmus paper, with an ideal range of 6.5–8.5 for safe discharge (Merz 2022). While temperature is a critical parameter as it influences organic carbon and nitrogen in an activated sludge process, microbial activity, oxygen solubility, and chemical reaction rates in wastewater treatment (Tejaswini et al. 2020). High temperatures can reduce dissolved oxygen (DO) levels, negatively impacting aquatic ecosystems (Riffat and Husnain 2022). Monitoring is typically done using thermometers, thermistors, digital probes, and infrared sensors, with an ideal range of 10–25 °C (Tejaswini et al. 2020). Turbidity measures water clarity and the presence of suspended particles that can harbor pathogens or pollutants. Turbidity may be used to gauge the amount of particle COD in a pretreated influent and effluent because there is a linear correlation between the two variables (Mels et al. 2004). It is assessed using nephelometric turbidity units (NTU) meters, spectrophotometers, or Secchi discs in open water bodies. High turbidity levels interfere with light penetration, affecting aquatic life and photosynthesis (Spellman 2008).

However, values obtained in a single sample might differ based on the sensor model and even the operator; turbidity is therefore regarded as a rather hard parameter for evaluating the quality of water (Joannis et al. 2008). For turbidity, regulatory limits for treated wastewater generally range from 10 to 50 NTU. The color and odor analysis help detect organic and industrial pollutants. Spectrophotometry is used for color quantification, while gas chromatography (GC) identifies volatile compounds causing odors such as hydrogen sulfide (H_2S) and ammonia (Table 1.1) (Sawyer et al. 2003). Physical monitoring plays a crucial role in wastewater management, ensuring compliance with environmental regulations and protecting water resources.

1.3.2 Chemical Methods

Chemical monitoring of wastewater involves analyzing its composition to detect contaminants, assess treatment efficiency, and ensure regulatory compliance. Dissolved oxygen (DO) is a crucial factor, as it determines the behavior and activity of the heterotrophic and autotrophic bacteria living in the activated sludge (Holenda et al. 2008). BOD and COD are widely used indicators of organic pollution. BOD measures the oxygen required by microorganisms to decompose organic matter over 5 days, whereas COD quantifies both biodegradable and nonbiodegradable substances. Nutrient analysis is essential, particularly nitrate (NO_3^-) and phosphate (PO_4^{3-}) levels, as excessive concentrations lead to eutrophication. Phosphate is a necessary nutrient for plants, whereas nitrates are the by-product of the aerobic breakdown of organic nitrogenous materials (Manirakiza et al. 2022). However, they are efficient nutritional sources for algae (organic nitrogen is converted to ammonia, which is oxidized to nitrate, and organic phosphorus is transformed to inorganic phosphate). Nitrate and phosphate are particularly significant in water pollution. Heavy metal monitoring is critical for detecting toxic pollutants like lead, mercury, nickel, arsenic, chromium, copper, zinc, and cadmium, which bioaccumulate and pose severe health risks (Adekunle et al. 2012). At low quantities, organisms may experience the effects of heavy metals, which are potentially harmful trace elements. Because they are not biodegradable, heavy metals tend to bioaccumulate in aquatic life. It has been determined that the primary way that heavy metals enter the human body is through the mouth. Consuming agricultural products irrigated with untreated or inadequately treated wastewater puts people at risk for several illnesses, some of which may not show symptoms right away (Malakootian et al. 2009). Techniques such as atomic absorption spectroscopy (AAS) and inductively coupled plasma mass spectroscopy (ICP-MS) provide precise measurements of these contaminants (Sawyer et al. 2003) (Table 1.1). Additionally, oil and grease content is monitored using infrared analysis or solvent extraction methods to identify hydrocarbon pollution, which can disrupt wastewater treatment systems.

Thus, chemical monitoring is essential for evaluating wastewater quality, ensuring environmental safety, and guiding treatment processes. Regular assessment of these

Table 1.1 Parameters of wastewater monitoring and characterization

Type	Specification	Concentration	Laboratory-based test	References
Physicochemical property	pH	7–8	pH paper/strip/meter, titanium trichloride reduction, iodometric method, azide modification, permanganate modification, direct measurement, calorimetric, and titration method	Zanetti et al. (2012)
	Dissolved oxygen	5 mg/L	United States Environmental Protection Agency (USEPA) direct measurement, laboratory method, platinum cobalt standard (true and apparent). Nephelometric method	Li et al. (2022)
	Alkalinity	100–200 mg/L	EDTA trimetric method, calmagite and calorimetric method	APHA (2005)
	Conductivity	600–800 mS cm^{-1}	Visual comparison method, spectrophotometric single wavelength method, tristimulus spectrophotometric method	APHA (2005)
	Turbidity	50 NTU	Laboratory and field methods (turbidity meter)	APHA (2005)
	Total hardness	300 mg/L Transparent/colorless	Trimetric method –	APHA (2005)
	Color temperature	10–25 °C	Thermometer	APHA (2005)
Solids	TSS	200–400 mg/L	Dried at 103–105 °C, centrifuging and drying	APHA (2005)
	TDS	600 mg/L	Dried at 180 °C	APHA (2005)
	VSS	320	Ignited at 550 °C	APHA (2005)

(continued)

Table 1.1 (continued)

Type	Specification	Concentration	Laboratory-based test	References
Organic matter	COD	160–500 mg/L	Titrimetric method, colorimetric method, UV absorption method	APHA (2005)
	BOD/BOOD$_5$	80–300 mg/L	Anaerobic digestion test (5 days), respirometry test, UV absorption test	APHA (2005)
	Total organic carbon (TOC)	30–200 mg/L	High-temperature combustion method, persulfate UV/ heated persulfate oxidation method, direct HR method	APHA (2005)
Nutrients	Nitrogen	60–110 mg/L (Total Kjeldahl Nitrogen), 50–100 mg/L (NH_4-N), 0.5 mg/L (NO_3^-N), and 25 mg/L (Organic N)	The titrimetric method, the ammonia selective electrode method, the phenate method, and flow injection analysis	APHA (2005)
	Phosphate	15 mg/L (TP), 10 mg/L (Ortho-P), and 5 mg/L (Organic P)	Spectrophotometry method	APHA (2005)
Heavy metals	Cadmium Chromium Copper Mercury Nickel Aluminum Silver Zinc	0.002 mg/L 0.025 mg/L 0.07 mg/L 0.002 mg/L 0.025 0.6 0.007 0.2	The atomic absorption spectroscopy (AAS) method and inductively coupled plasma mass spectrometry/atomic emission spectrometry	APHA (2005)
Microorganisms	*Escherichia coli* *Coliphages* Roundworms *Salmonella*	10^7 colony-forming unit (CFU)/100 mL 10^5 CFU/ 100 mL 10 eggs/L 150 CFU/ 100 mL	Most probable number, multiple-tube fermentation technique, enzyme substrate coliform test	APHA (2005)

parameters helps industries and municipalities comply with regulations and minimize pollution impacts.

1.3.3 Biological Methods

Biological methods of wastewater monitoring involve assessing the presence and activity of microorganisms, bioindicators, and living organisms to evaluate water quality and pollution levels. These methods are essential for detecting organic pollution, pathogens, and toxic substances that may not be easily identified through physical or chemical analysis. Biological monitoring provides insight into the ecological impact of wastewater discharge and helps assess the effectiveness of treatment processes. One of the primary biological monitoring techniques is bacterial and microbial analysis, which involves detecting pathogenic microorganisms, such as total coliforms, *Escherichia coli, Salmonella*, and *Vibrio cholerae*. Fecal wastes contain the majority of coliforms since they are abundant in the gut flora of humans and other warm-blooded animals (Rompré et al. 2002). As a result, coliforms are found in greater quantities than pathogenic bacteria and are used as a gauge for the possible presence of enteropathogens in aquatic environments. Coliforms are described as gram-negative, rod-shaped, aerobic, and facultative anaerobic bacteria that digest lactose with gas and produce acid (Rompré et al. 2002). These bacteria are commonly detected using membrane filtration, multiple-tube fermentation, or chromogenic enzyme substrates. Depending on local regulations, the recommended limit for treated wastewater discharge is typically less than 10 colony-forming units (CFU)/100 mL for coliforms (Merz 2022). Bioassays and toxicity tests are also widely used to assess the impact of wastewater contaminants on living organisms. These tests involve exposing aquatic species such as *Daphnia* (water fleas), fish (zebrafish, trout), and algae (*Chlorella, Scenedesmus*) to wastewater samples and observing their responses. Lethal concentration (LC50) tests determine the toxicity of wastewater by measuring the concentration at which 50% of test organisms die within a specific timeframe. Such bioassays help detect the presence of heavy metals, pesticides, and industrial pollutants that may not be easily identified by chemical tests. Some of the approaches currently being used for bioassay of wastewater are illustrated in Fig. 1.3.

Biological indicators, including protozoa, algae, and benthic macroinvertebrates, serve as natural water quality monitors. Certain protozoa, such as ciliates and amoebas, thrive in well-oxygenated environments and indicate good water quality, while their absence suggests pollution (Bitton 2011). The speed at which protozoan cells divide, encyst, or excyst is extremely dynamic quantitatively. The protozoa community structure can, therefore, react swiftly to changes in the environment, indicating that it may be useful as a gauge of aquatic ecosystem shifts. The protozoan taxon responds differently to certain determinants, including temperature, salinity, nutritional availability, dissolved oxygen, organic matter, eutrophication, and pollution level (Xu et al. 2005). Protozoa possess several traits that make them effective bioindicators, which include their short life cycle, allowing rapid population shifts in response to environmental changes, and their delicate cell membranes enable quick absorption of and reaction to chemical stressors (Dias et al. 2008). While many protozoans can survive extreme conditions, their population size and biomass often

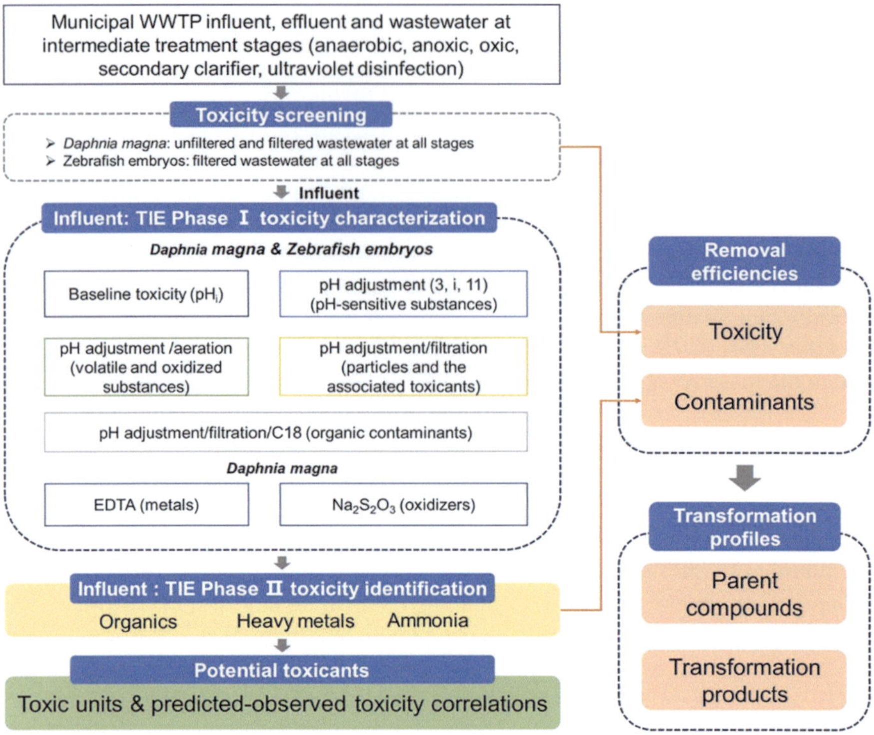

Fig. 1.3 An experimental flow chart for the municipal wastewater treatment plant (WWTP) that includes toxicity screening, toxicity identification, removal efficiency, and transformation profiles of contaminants (Liu et al. 2022)

fluctuate, reflecting the resilience of the river community to organic pollution. Additionally, protozoans are easy to quantify in laboratory settings. Numerous protozoan species have a widespread global presence, flourishing in specific environmental conditions that suit their survival, making them valuable for assessing ecological health and pollution levels (Anderson 2010). El-Tohamy et al. (2024 investigated the relationship between protozoa population with water pollution in the Nile River and concluded that the composition and structure of protozoans showed a clear influence of anthropogenic disturbance in the Nile River.

Algae, such as cyanobacteria, proliferate in nutrient-rich wastewater, signaling high levels of nitrogen and phosphorus, which contribute to eutrophication and oxygen depletion in water bodies. Rindi et al. (2011) highlighted those terrestrial algae, including green algae and diatoms, are highly responsive to climate change due to their short life cycles, rapid turnover rates, and quick adaptation to environmental shifts. Sarmaja-Korjonen et al. (2006) found that algae serve as effective indicators of environmental conditions, reflecting productivity variations during climatic fluctuations. Similarly, Hallegraeff (2010) suggested that shifts in algal communities can act as an early warning system for climate-induced uncertainties in aquatic ecosystems.

These characteristics make algae valuable bioindicators, helping researchers monitor and predict ecological changes driven by climate dynamics. Benthic macroinvertebrates, including mayflies, caddisflies, and midges, are also useful bioindicators, as their population diversity reflects water quality conditions (O'Neill and Rowan 2022).

1.3.4 Sensors for Wastewater Monitoring

Numerous studies have examined various sensor types, including biosensors (Herrera-Domínguez et al. 2023), electrochemical sensors (Zhu et al. 2024), and optical chemical sensors (Pizzoferrato 2023). These sensors are utilized to identify a variety of hazardous substances in wastewater, including inorganic (Cd^{2+}, Pb^{2+}, Zn^{2+}, etc.), organic (pesticides, pharmaceuticals), and emergent (microplastics, endocrine-disrupting chemicals) contaminants. Chemical sensors contain transducers, which are essential for converting analytical changes into signals, and they play a key role in automated systems by handling sampling, transport, and data processing (Kumar et al. 2023). These sensors are commonly used to detect contaminants such as heavy metals and nutrients (Amorim and Bento 2024; Jashari et al. 2022). Traditional monitoring methods for organic pollutants are costly, time-consuming, and require skilled personnel, which makes them impractical for on-site applications (Idris et al. 2023). Chemical sensors offer an efficient alternative and are broadly classified into optical and electrochemical sensors.

Optical sensors, including optodes, rely on chemical reactions and spectroscopic measurements. Devices such as triacetylcellulose membranes and optical fluorescent sensors use spectrophotometric methods for analyte detection, but often have limited selectivity (Robinson et al. 2022; Yang et al. 2022). The addition of selective films, such as cellulose fiber and nanofibers, improves detection, enabling high sensitivity for Zn^{2+} and Fe^{3+} ions (Kamel et al. 2024). Advanced optical sensors, such as gold-coated fiber optics with bovine serum albumin, enhance Cd^{2+} detection in water (Şolomonea et al. 2022). Optical fluorescence sensors using Santa Barbara Amorphous-15 (SBA-15) mesoporous silica offer improved sensitivity and stability (Aghili et al. 2024). However, the improper disposal of optical sensors containing chalcogenide glass or nanoparticles poses environmental risks (El-Sewify et al. 2022).

Electrochemical sensors convert electrochemical interactions into electrical signals. Modified electrodes, such as carbon electrodes with silver hexacyanoferrate nanoparticles, enhance sensitivity and performance (Borges et al. 2023). These sensors detect heavy metals and organic pollutants efficiently, though concerns exist regarding their environmental impact due to nanomaterial disposal (He et al. 2023).

Biosensors detect biological responses and convert them into electrical signals. Examples like acetylcholinesterase and urease-based biosensors have widespread applications in medicine, food safety, industrial processes, and environmental monitoring, including water quality assessment (Li et al. 2024; Rodrigues et al. 2024).

Advancements in biosensors and molecular techniques have further improved biological wastewater monitoring. Genetically engineered microorganisms (GEMs), enzyme-based biosensors, and molecular probes allow rapid and highly sensitive detection of pollutants such as heavy metals, organic toxins, and pharmaceuticals in wastewater (Calatrava et al. 2024; Pal et al. 2025). Gautam et al. (2025) have designed a paper-based sensor to monitor wastewater quality by detecting coliforms using a paper strip dip test (PSD). The PSD test optimization involved Methylene Blue (MB) dye (0.1–0.5% w/v) and ferrous sulfate heptahydrate ($FeSO_4 \cdot 7H_2O$) at concentrations of 10–100 mM, with hydrogen peroxide (H_2O_2) at concentrations of 5–100 mM. These modern techniques enhance traditional biological monitoring by providing real-time analysis and early detection of contaminants.

Biosensors offer significant advantages over conventional methods, often bulky, slow, and difficult to automate. Their portability, miniaturization, and suitability for on-site applications make them highly efficient (Kulkarni, et al., 2022). Unlike standard analytical techniques, biosensors provide faster response and recovery times, continuous monitoring, and improved precision in detecting environmental pollutants (Huang, et al., 2023). They require minimal maintenance and are particularly effective in tracking water treatment operations (Huang, et al., 2023). Identifying organic pollutants in sewage using biosensors has demonstrated higher accuracy. Bio-electrochemical sensors expand detection capabilities beyond traditional biosensors (Otero, and Magner, 2020). Additionally, integrating microbial fuel cells with biosensors enhances their performance in complex environments such as wastewater, where conventional biosensors face interference (Rojas-Villacorta, et al., 2022). However, improper disposal of biosensors containing enzymes, antibodies, or living cells may pose environmental biohazards (Williams et al., 2024).

1.4 Challenges in the Treatment of Wastewater

The presence of a wide range of pollutants in water poses a substantial challenge for policymakers and wastewater treatment facility operators. Meanwhile, the existing wastewater treatment facilities seem insufficient to completely remove these pollutants due to their variable concentration and diverse composition (Devarajan 2025). The majority of the time, high-grade contaminants must be left behind due to the potential for their removal to be either excessively costly or hazardous to the environment (Azhar, et al., 2022). Additionally, there have been reports of the retention of posttreatment contaminants that are more toxic than the compound that was removed. Operators are confronted with an additional obstacle: (1) the growing complexity and volume of effluent, (2) the contamination of aquatic bodies is more difficult to manage due to rapid urbanization and industrialization (Vasseghian, et al., 2023). This results in an increase in the pollutant load of the effluent, which ultimately reaches the environment. However, wastewater treatment systems also pose challenges in terms of energy consumption and cost of operation (Shemer, et al., 2023). Aeration, filtration, and disinfection are conventional processes that necessitate a significant amount of energy to sustain an operational cost that is steadily increasing. The majority of facilities may be forced to close or compromise the quality of their effluent to reduce production costs due to the rising cost of water treatment (N. Singh et al., 2025).

1.4.1 Emerging Pollutants in the Wastewater

Emerging pollutants represent a range of substances, including industrial chemicals, medicines, microplastics, personal care items, and pesticides that are widely detected in the discharges of wastewater, raising concerns over their health and impacts on the environment (Arman, et al., 2021). Emerging pollutants pose a significant issue due to their prevalence in wastewater and the variability in their removal efficacy with current treatment technology (Rout, et al., 2021). For instance, compounds that resist breakdown during treatment can harm aquatic life and humans, making it hard to predict their environmental effects (Luo, et al., 2014). Wastewater treatment plants struggle with these new pollutants, needing constant updates to handle them effectively. Traditional methods may not sufficiently protect against these contaminants, so new strategies like advanced oxidation and nanofiltration are required. Additionally, a risk assessment and regulatory framework should be created to manage these emerging threats (Hussain et al. 2024).

1.4.2 Conventional Methods Persist

Conventional methods for wastewater treatment encompass physical, chemical, and biological processes designed to eliminate both soluble and insoluble pollutants (Crini and Lichtfouse, 2019). The challenges and benefits of the treatment of wastewater are given in Table 1.2. Biological methods are cost-effective and efficient; however, it is ineffective for synthetic contaminants like dyes due to their resistance to aerobic biodegradation (Crini and Lichtfouse, 2019). Chemical methods generate harmful by-products and exhibit lower efficiency, whereas physical treatments are often useful (Crini and Lichtfouse, 2019).

1.4.3 Old Infrastructure

Over time and with usage, the infrastructure of water and wastewater treatment facilities undergoes corrosion and deterioration. Old infrastructure can include storage facilities, dams, tunnels, pumps, pipelines, and more equipment. The deterioration of equipment is significant due to the hazardous components of wastewater, including pathogens, toxins, chemicals, and other organic contaminants (Selvakumar and Tafuri 2012).

Table 1.2 Challenges and advantages of wastewater treatment

Treatment	Methodologies	Advantages	Challenges	References
Biosorption	Microbial immobilization on absorbent	Effective treatment and specific reduction of certain emerging pollutants	Absorbents require periodic cleaning at certain intervals	Rashid et al. (n.d.)
Screening	Coarse screening: eliminates solid substances smaller than 6 mm	Reduces disruption and obstruction of the treatment methods	Ineffective in eliminating emerging pollutants	Prabu et al. (2011)
	Fine screening: elimination of impurities ranging from 0.001 mm to 6 mm	✓ Efficiency is controlled by modifying the size of the screen spaces ✓ It is preferable to maintain the temperature in a process ✓ Lower costs and simplified procedure	The screen requires cleaning to clear out the obstruction in the small holes	
Adsorption	Methods for removing soluble chemicals using solid substrates with a highly defined surface area	✓ Effective in the precise elimination of the emerging contaminant ✓ Precise and effective elimination ✓ Can support additional treatment procedures ✓ Reduced complexity and cost, facilitating ease of adoption	✓ An accumulation of cyanotoxin in the adsorbent material ✓ The removal of unidentified types of contaminants presents challenges due to the high specificity of adsorbents ✓ The regulation of selectivity within the membranous system presents significant challenges ✓ Reverse osmosis necessitates the application of external energy	De Gisi et al. (2016)

1.4.4 Financial Challenges

Regardless of the significant initial expense on a treatment facility for wastewater, additional expenses include maintenance, operational disruptions, improper equipment, and ongoing operating expenditures. For instance, the continued use of outdated equipment may diminish treatment effectiveness and prolong wastewater processing time. Moreover, this equipment is susceptible to malfunction, leading to increased downtime and maintenance requirements (Hernández-Chover et al. 2020).

1.4.5 The Consumption of Energy

The consumption of energy has long been a key source of concern for wastewater treatment facilities since they consume roughly 60 terawatt-hours/year. Additionally, 50–60% of the energy used in municipal sewage treatment plants is used for biological treatment (Eeuroteck).

1.4.6 Maintenance of Sludge

During the treatment of wastewater through multiple stages, residues known as sludge are generated. The disposal of this collected sludge into rivers and lakes adversely affects the ecosystem and presents a considerable environmental challenge; therefore, it is essential to minimize this practice.

1.4.7 Availability of Land

The availability of land presents a significant and critical challenge in the development of wastewater treatment plants for wastewater. The treatment of wastewater and water facilities necessitates a substantial area for their installation. With the rise in population and urbanization, accessibility to land is decreasing, which poses challenges for the development of water treatment facilities.

1.5 Opportunities in the Treatment of Wastewater

1.5.1 Recovery of Resources

Wastewater consists of valuable resources, including nutrients such as phosphorus and nitrogen, energy in the form of biogas, and water, all of which can be regained and reused. Technologies such as the restoration of nutrient systems, anaerobic digestion, and membrane filtration facilitate effective resource extraction (UN World Water Development Report 2017).

1.5.2 Water Recycling and Reuse Process

Treated wastewater can be used for irrigation purposes in agriculture, industrial processes, and for potential potable water use, thereby minimizing the demand on freshwater resources (Water Reuse and Recycling). Other than agricultural applications, recycled water can be used for the irrigation of cemeteries, public parks, urban green belts, golf courses, and residential lawns. Numerous developing and developed nations, including Colombia, Syria, Pakistan, South Africa, Israel, Italy, the United States, Mexico, China, and Chile, have successfully utilized reused water for various applications. This trend is also being adopted by other countries. The system not only contributes to water conservation but also minimizes the need for fertilizers due to the presence of nutrients in the wastewater (Ofori et al. 2021).

1.5.3 Direct Potable Reuse (DPR)

Direct potable reuse involves the treatment of wastewater to a standard that allows it to be added to municipal water pipelines for drinking purposes. The initial direct potable reuse (DPR) scheme was established in Windhoek, Namibia, in 1968. However, the number of DPR projects implemented subsequently has been limited, primarily due to concerns regarding public safety. The presence of pathogens and contaminants in the DPR may pose a significant risk for diseases. However, it remains a viable option for regions experiencing significant water scarcity. (Lahnsteiner et al. 2018).

1.5.4 Indirect Potable Reuse (IPR)

IPR involves the process of discharging wastewater from treatment into the ground to facilitate the recharging of underground water levels. Compared to DPR, IPR is preferred by the public due to its ability to eliminate or deactivate pathogens, as well as enhance water quality by facilitating the diffusion of toxins through the soil. Examples of indirect potable reuse (IPR) initiatives include groundwater recharge and direct injection projects located in Orange County, California; Fairfax, Virginia; and Singapore (Hernández-Chover et al. 2020).

1.5.5 Nonpotable Domestic Use

Treated wastewater is suitable for nonpotable domestic applications, including cleaning, vehicle washing, and toilet flushing. This process contributes to the significant conservation of freshwater resources. As mentioned above, all the points are brief in Fig. 1.4.

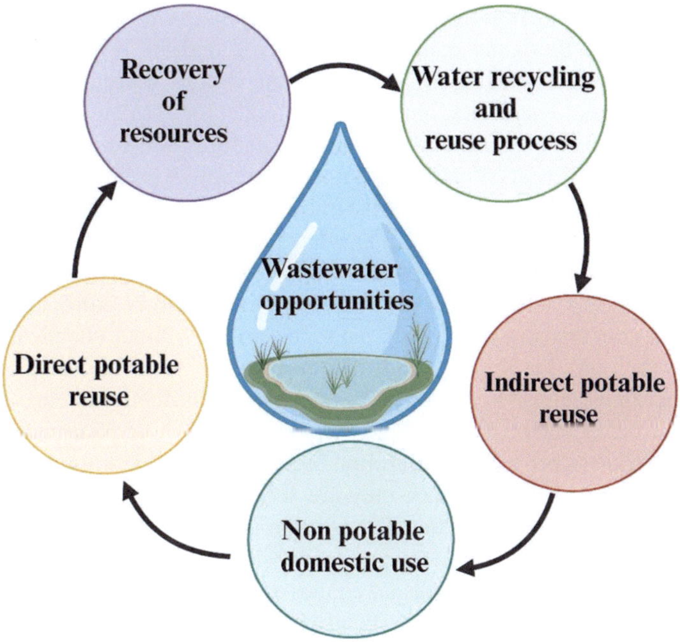

Fig. 1.4 Classification of wastewater opportunities

1.6 Assessment of Available Techniques for Wastewater Treatment

Increased industrialization and nonsustainable urbanization have led to a rapid rise in the presence of emerging contaminants in the environment. The pollutants in the examination are harmful due to their nonbiodegradable and toxic nature, encompassing detergents and personal care products (Iyare et al. 2020; Idris et al. 2023). They can potentially interfere with the functions of living organisms and result in significant harm. Consequently, the treatment of wastewater is essential for the elimination of these contaminants and the safeguarding of the environment (Li et al. 2024). Groundwater levels are declining, which makes wastewater treatment essential for fulfilling the water demand of the population. This chapter provides a detailed analysis of different wastewater treatment technologies, encompassing physical, chemical, and biological methods. The study examines the constraints associated with each methodology and highlights current developments in wastewater treatment practices. These methods are commonly employed and are selected based on the specific type of wastewater, as wastewater from various sectors contains distinct kinds of pollutants. Therefore, biological methods for wastewater treatment are considered better than physical and chemical procedures according to their environmental sustainability, cost-effectiveness, and efficiency in pollutant removal (Calatrava et al. 2024). These approaches utilize microorganisms, including bacteria, fungi, and algae, to decompose organic contaminants and nutrients by natural biochemical processes, thus reducing the production of toxic byproducts (Calatrava et al. 2024). Biological systems efficiently remove organic debris, nitrogen, and phosphorus while concurrently decreasing pathogen levels without chemical additions. In addition, they provide the benefit of resource recovery, including biogas generation via anaerobic digestion and nutrient extraction for agricultural reuse. Biological methods can be adjusted to address new pollutants, such as pharmaceuticals and microplastics, due to progress in microbial biotechnology. They are perfect for both urban and rural environments since they are scalable, energy-efficient, and appropriate for decentralized applications. Furthermore, because biological treatments are environmentally beneficial, they are more widely accepted and easily integrated into hybrid treatment systems. All things taken into account, biological wastewater treatment is a viable, adaptable, and scientifically proven way to deal with the challenges in current wastewater management. Therefore, the novel hybrid technology comes into the scientific world, which is known as microbial fuel cells (MFCs), is a bioelectrochemical system that allows for both wastewater treatment and electricity generation (Hassan et al. 2024). In an MFC, as shown in Fig. 1.5, electroactive bacteria break down organic matter in wastewater, releasing electrons as a byproduct of their metabolism. These electrons are transferred to an anode and then flow through an external circuit to a cathode, creating an electrical current. Simultaneously, protons move through a proton exchange membrane to the cathode, where they combine with oxygen in a reduction reaction. This process not only removes pollutants from wastewater but also produces bioelectricity, making MFCs a sustainable alternative

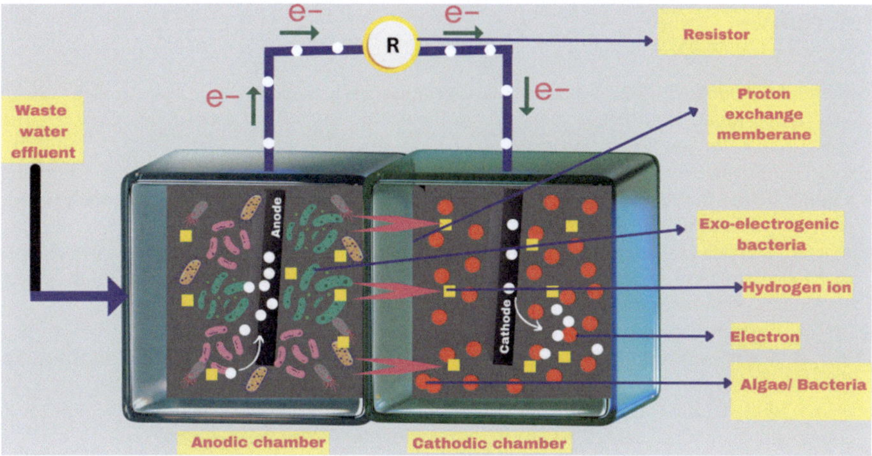

Fig. 1.5 Schematic representation of microbial fuel cells

to conventional treatment methods. The efficiency of MFCs is influenced by factors such as electrode materials, microbial species, and system configuration, offering potential improvements in energy-efficient wastewater treatment.

1.7 Conclusion and Future Perspectives

In conclusion, MFCs are a great solution for producing electricity while also bioremediating polluted groundwater sources or treating wastewater. MFCs can remove a significant amount of COD. A large number of MFCs used to treat wastewater were inoculated using anaerobic or aerobic sludge, which does not require a mediator. Similar to other microbial systems, the MFC performance is better, and also, at optimum pH and temperature, it enhances the growth of bacteria. Salinity and ionic strength are other significant factors. Higher salinity and ionic strength can improve MFC performance by increasing the conductivity of the substrate, although salt inhibits microbial growth. These are the main reasons that offer a promising approach to wastewater treatment and several distinct advantages over other methods. Overall, integrating microbial fuel cells into wastewater treatment holds great promise for enhancing sustainability and efficiency in water resource management practices.

References

Adekunle AS, Odukoya OO, Ayenimo JG, Oyekunle JAO, Doherty WO, Mamba BB, Akanni MS (2012) Removal of heavy metals from industrial effluents by water hyacinth (*Eichornia crassipes*). J Environ Chem Ecotoxicol 4(11):203–211

Adjovu GE, Stephen H, Ahmad S (2023a) Spatial and temporal dynamics of key water quality parameters in a thermal stratified lake ecosystem: the case study of Lake Mead. Earth 4(3):461–502. https://doi.org/10.3390/earth4030025

Adjovu GE, Stephen H, James D, Ahmad S (2023b) Measurement of total dissolved solids and total suspended solids in water systems: a review of the issues, conventional, and remote sensing techniques. Remote Sensing 15(14):3534. https://doi.org/10.3390/rs15143534

Aghili SA, Hajiaghababaei L, Aberoomand Azar P, Badiei A, Saber-Tehrani M (2024) Preparation of optical nano sensor for rapid detection of Fe(III) using phenyl isothiocyanate functionalized SBA-15. Chem Pap 78(9):5381–5392. https://doi.org/10.1007/s11696-024-03477-4

Amorim I, Bento F (2024) Electrochemical sensors based on transition metal materials for phenolic compound detection. Sensors 24(3):756. https://doi.org/10.3390/s24030756

Anderson OR (2010) Encyclopedia of life sciences (ELS). Wiley

APHA (2005) Standard methods for the examination of water and wastewater, 21st ed. American Public Health Association, Washington, DC

Arman NZ, Salmiati S, Aris A, Salim MR, Nazifa TH, Muhamad MS, Marpongahtun M (2021) A review on emerging pollutants in the water environment: Existences, health effects and treatment processes. Water 13(22):3258

Azhar U, Bashir MS, Babar M, Arif M, Hassan A, Riaz A, Mubashir M (2022) Template-based textural modifications of polymeric graphitic carbon nitrides towards waste water treatment. Chemosphere 302:134792

Bitton G (2011) Wastewater microbiology. Wiley

Borges PHS, Narciso LCD, de Miguel GFS, de Oliveira GS, Junior MC, da Machado AEH, Muñoz RAA, Nossol E (2023) An experimental and theoretical approach to electrochemical sensing of hydrazine at silver and copper hexacyanoferrates electrodes. Electrochimica Acta 445:142018. https://doi.org/10.1016/j.electacta.2023.142018

Botalova O, Schwarzbauer J, Frauenrath T, Dsikowitzky L (2009) Identification and chemical characterization of specific organic constituents of petrochemical effluents. Water Res 43(15):3797–3812. https://doi.org/10.1016/j.watres.2009.06.006

Calatrava V, Ballester DG, Dubini A (2024) Microalgae for bioremediation: advances, challenges, and public perception on genetic engineering. BMC Plant Biol 24(1):1261

Cheng C, Zhang F, Shi J, Kung H-T (2022) What is the relationship between land use and surface water quality? A review and prospects from remote sensing perspective. Environ Sci Pollut Res 29(38):56887–56907

Choi E-H, Klapwijk B, Mels A, Brouwer H (2005) Evaluation of wastewater characterization methods. Water Sci Technol 52(10–11):61–68. https://doi.org/10.2166/wst.2005.0679

Crini G, Lichtfouse E (2019) Advantages and disadvantages of techniques used for wastewater treatment. Environ chem lett 17(1):145–155

Dey J, Vijay R (2021) A critical and intensive review on assessment of water quality parameters through geospatial techniques. Environ Sci Pollut Res 28(31):41612–41626

Dias RJP, Wieloch AH, D'Agosto M (2008) The influence of environmental characteristics on the distribution of ciliates (Protozoa, Ciliophora) in an urban stream of southeast Brazil. Braz J Biol 68:287–295

De Gisi S, Lofrano G, Grassi M, Notarnicola M (2016) Characteristics and adsorption capacities of low-cost sorbents for wastewater treatment: a review. Sustain Materials Technol 9:10–40

Devarajan Y (2025) Nanomaterials-based wastewater treatment: addressing challenges and advancing sustainable solutions. BioNanoScience 15(1):1–14

El-Sewify IM, Radwan A, Elghazawy NH, Fritzsche W, Azzazy HME (2022) Optical chemosensors for environmental monitoring of toxic metals related to Alzheimer's disease. RSC Adv 12(50):32744–32755. https://doi.org/10.1039/D2RA05384E

El-Tohamy WS, Taher ME, Ghoneim AM, Hopcroft RR (2024) Protozoan communities serve as a strong indicator of water quality in the Nile River. Sci Rep 14(1):16382

Gautam J, Kanwar N, Sharma A, Nagoda C, Chhabra M, Mathuria M, Kanojiya R, Bhavsar SP (2025) Development of a rapid and cost-effective paper strip dip test for visual detection of coliforms and catalase-positive bacteria in water samples. J Water Process Eng 71:107163. https://doi.org/10.1016/j.jwpe.2025.107163

Gopala Krishna GVT, Kumar P, Kumar P (2008) Treatment of low strength complex wastewater using an anaerobic baffled reactor (ABR). Biores Technol 99(17):8193–8200. https://doi.org/10.1016/j.biortech.2008.03.016

Hallegraeff GM (2010) Ocean climate change, phytoplankton community responses, and harmful algal blooms: a formidable predictive challenge 1. J Phycol 46(2):220–235

Hassan M, Kanwal S, Singh RS, Muhammed Ali SA, Anwar, M, Zhao C (2024) Current challenges and future perspectives associated with configuration of microbial fuel cell for simultaneous energy generation and wastewater treatment. Int J Hydrogen Energy 323–350

Hussain S, Wang Y, Awais M, Sajjad MM, Ejaz N, Javed U, Iqbal J (2024) Integrated assessment of groundwater quality dynamics and Land use/land cover changes in rapidly urbanizing semi-arid region. Environ Res 260:119622

He Q, Wang B, Liang J, Liu J, Liang B, Li G, Long Y, Zhang G, Liu H (2023) Research on the construction of portable electrochemical sensors for environmental compounds quality monitoring. Mater Today Adv 17:100340. https://doi.org/10.1016/j.mtadv.2022.100340

Herrera-Domínguez M, Morales-Luna G, Mahlknecht J, Cheng Q, Aguilar-Hernández I, Ornelas-Soto N (2023) Optical biosensors and their applications for the detection of water pollutants. Biosensors 13(3):370. https://doi.org/10.3390/bios13030370

Hernández-Chover V, Castellet-Viciano L, Hernández-Sancho F (2020) Preventive maintenance versus cost of repairs in asset management: an efficiency analysis in wastewater treatment plants. Process Safe Environ Protection 141:215–221

Holenda B, Domokos E, Rédey Á, Fazakas J (2008) Dissolved oxygen control of the activated sludge wastewater treatment process using model predictive control. Comput Chem Eng 32(6):1270–1278. https://doi.org/10.1016/j.compchemeng.2007.06.008

Huang L, Pei Y, Jia X, Tang X, Zhang Y, Pan Y (2023) Multi-species plantation intensifies soil water competition and groundwater depletion in a water-limited desert region. For Ecol Manag 537:120953

Idris AO, Orimolade B, Dennany L, Mamba B, Azizi S, Kaviyarasu K, Maaza M (2023) A review on monitoring of organic pollutants in wastewater using electrochemical approach. Electrocatalysis 14(5):659–687. https://doi.org/10.1007/s12678-023-00834-x

Iyare PU, Ouki SK, Bond T (2020) Microplastics removal in wastewater treatment plants: a critical review. Environ Sci: Water Res Technol 6(10):2664–2675. https://doi.org/10.1039/D0EW00397B

Jashari G, Švancara I, Sýs M (2022) Characterisation of carbon paste electrodes bulk-modified with surfactants for measurements in nonaqueous media. Electrochim Acta 410:140047. https://doi.org/10.1016/j.electacta.2022.140047

Joannis C, Ruban G, Gromaire M-C, Bertrand-Krajewski J-L, Chebbo G (2008) Reproducibility and uncertainty of wastewater turbidity measurements. Water Sci Technol 57(10):1667–1673. https://doi.org/10.2166/wst.2008.292

Kamel RM, El-Sakka SS, Abbas MMA, Soliman MHA (2024) Eco-friendly fluorescent sensor for sensitive and selective detection of Zn2+ and Fe3+ ions: applications in human hair samples. J Fluoresc. https://doi.org/10.1007/s10895-024-03798-3

Keshaviah A, Diamond MB, Wade MJ, Scarpino SV, Ahmed W, Amman F, Aruna O, Badilla-Aguilar A, Bar-Or I, Bergthaler A (2023) Wastewater monitoring can anchor global disease surveillance systems. Lancet Glob Health 11(6):e976–e981

Khan I, Khan A, Khan MS, Zafar S, Hameed A, Badshah S, Rehman SU, Ullah H, Yasmeen G (2018) Impact of city effluents on water quality of Indus River: assessment of temporal and spatial variations in the southern region of Khyber Pakhtunkhwa, Pakistan. Environ Monit Assess 190:1–19

Kulkarni MB, Ayachit NH, Aminabhavi TM (2022) Recent advancements in nanobiosensors: current trends, challenges, applications, and future scope. Biosensors 12(10):892

Kumar S, Wang Z, Zhang W, Liu X, Li M, Li G, Zhang B, Singh R (2023) Optically Active nanomaterials and its biosensing applications—a review. Biosensors 13(1):85. https://doi.org/10.3390/bios13010085

Lahnsteiner J, Van Rensburg P, Esterhuizen J (2018) Direct potable reuse–a feasible water management option. J Water Reuse Desalin 8(1):14–28

Li W-L, Zhang Z-F, Kilgallon J, Sparham C, Li Y-F, Yuan Y-X (2022) Fate of household and personal care chemicals in typical urban wastewater treatment plants indicate different seasonal patterns and removal mechanisms. Environ Pollut 294:118618. https://doi.org/10.1016/j.envpol.2021.118618

Li X, Shen X, Jiang W, Xi Y, Li S (2024) Comprehensive review of emerging contaminants: detection technologies, environmental impact, and management strategies. Ecotoxicol Environ Saf 278:116420. https://doi.org/10.1016/j.ecoenv.2024.116420

Liu Y, Li F, Li H, Tong Y, Li W, Xiong J, You J (2022) Bioassay-based identification and removal of target and suspect toxicants in municipal wastewater: Impacts of chemical properties and transformation. J Hazard Mater 437:129426. https://doi.org/10.1016/j.jhazmat.2022.129426

Luo Y, Guo W, Ngo HH, Nghiem LD, Hai FI, Zhang J, Wang XC (2014) A review on the occurrence of micropollutants in the aquatic environment and their fate and removal during wastewater treatment. Sci Total environ 473:619–641

Malakootian M, Nouri J, Hossaini H (2009) Removal of heavy metals from paint industry's wastewater using Leca as an available adsorbent. Int J Environ Sci Technol 6:183–190

Manirakiza B, Gbadegesin LA, Okeke ES, Bagaragaza R, Addo FG (2022) Physico-chemical analysis of industrial wastewater pollution from Kigali special economic zone (KSEZ) and the potential impacts in the downstream regions of Kigali City in Rwanda. Sustain Water Resour Manage 8(4):95

Mels AR, Spanjers H, Klapwijk A (2004) Turbidity-based monitoring of particle concentrations and flocculant requirement in wastewater pre-treatment. Water Sci Technol 50(11):173–178. https://doi.org/10.2166/wst.2004.0685

Merz M (2022) US Environmental Protection Agency (EPA) Proposes to Reissue a National Pollutant Discharge Elimination System (NPDES) General Permit to Discharge Pollutants Pursuant to the Provisions of the Clean Water Act (CWA)

Mutamim NSA, Noor ZZ, Hassan MAA, Olsson G (2012) Application of membrane bioreactor technology in treating high strength industrial wastewater: a performance review. Desalination 305:1–11. https://doi.org/10.1016/j.desal.2012.07.033

Ofori S, Puškáčová A, Růžičková I, Wanner J (2021) Treated wastewater reuse for irrigation: Pros and cons. Sci Total Environ 760:144026

Otero F, Magner E (2020) Biosensors—recent advances and future challenges in electrode materials. Sensors 20(12):3561

O'Neill EA, Rowan NJ (2022) Microalgae as a natural ecological bioindicator for the simple real-time monitoring of aquaculture wastewater quality including provision for assessing impact of extremes in climate variance—a comparative case study from the Republic of Ireland. Sci Total Environ 802:149800. https://doi.org/10.1016/j.scitotenv.2021.149800

Pal P, Pandey B, Prakash A, Sarangi PK, Rathore SS, Agarwal V, Gupta AK, Singh AK (2025) Exploration of biosensors in innovative applications and sustainable solutions for environmental monitoring and management. In: Biotechnology for environmental sustainability. Springer, pp 561–591

Pizzoferrato R (2023) Optical chemical sensors: design and applications. Sensors 23(11):5284. https://doi.org/10.3390/s23115284

Prabu SL, Suriyaprakash TNK, Kumar JA (2011) Wastewater treatment technologies: a review. Pharma times 43(5):55–62

Ramalho RS (2013) Introduction to wastewater treatment processes. Academic

Riffat R, Husnain T (2022) Fundamentals of wastewater treatment and engineering. CRC Press. https://doi.org/10.1201/9781003134374

Rindi F, Hodkinson T, Jones M (2011) Terrestrial green algae: systematics, biogeography and expected responses to climate change. Climate Change Ecol System 78:201

Robinson KJ, Soda Y, Bakker E (2022) Recent improvements to the selectivity of extraction-based optical ion sensors. Chem Commun 58(27):4279–4287. https://doi.org/10.1039/D1CC06636F

Rodrigues JC, Bezerra CS, Lima LMA, Neves DD, Paim APS, Silva WE, Lavorante AF (2024) A multicommuted system using bacterial cellulose for urease immobilization and copper (II)-MOF colorimetric sensor for urea spectrophotometric determination in milk. Food Chem 460:140454. https://doi.org/10.1016/j.foodchem.2024.140454

Rojas-Villacorta W, Rojas-Flores S, De La Cruz-Noriega M, Chinchay Espino H, Diaz F, Gallozzo Cardenas M (2022) Microbial biosensors for wastewater monitoring: mini-review. Processes 10(10):2002

Rompré A, Servais P, Baudart J, de-Roubin M-R, Laurent P (2002) Detection and enumeration of coliforms in drinking water: current methods and emerging approaches. J Microbiol Methods 49(1):31–54. https://doi.org/10.1016/S0167-7012(01)00351-7

Rout PR, Zhang TC, Bhunia P, Surampalli RY (2021) Treatment technologies for emerging contaminants in wastewater treatment plants: a review. Sci Total Environ 753:141990

Sarmaja-Korjonen K, Seppänen A, Bennike O (2006) Pediastrum algae from the classic late glacial Bølling Sø site, Denmark: response of aquatic biota to climate change. Rev Palaeobot Palynol 138(2):95–107. https://doi.org/10.1016/j.revpalbo.2005.12.003

Sawyer CN, McCarty PL, Parkin GF (2003) Chemistry for environmental engineering and science, vol 5. McGraw-Hill, New York

Singh N, Tung S (2025) Assessment of Water Quality of Gomti River at Lucknow. Air, Soil and Water Research, 18, 11786221251328589

Shemer H, Wald S, Semiat R (2023) Challenges and solutions for global water scarcity. Membranes 13(6):612

Şolomonea B-G, Jinga L-I, Antohe V-A, Socol G, Antohe I (2022) Cadmium Ions' trace-level detection using a portable fiber optic—surface plasmon resonance sensor. Biosensors 12(8):573. https://doi.org/10.3390/bios12080573

Spellman FR (2008) Handbook of water and wastewater treatment plant operations. CRC Press. https://doi.org/10.1201/9781420075311

Selvakumar A, Tafuri AN (2012) Rehabilitation of aging water infrastructure systems: key challenges and issues. J Infrastruct Syst 18(3):202–209

Tejaswini E, Uday Bhaskar Babu G, Seshagiri Rao A (2020) Effect of temperature on effluent quality in a biological wastewater treatment process. Chem Prod Process Model 15(1). https://doi.org/10.1515/cppm-2019-0018

Tran NH, Ngo HH, Urase T, Gin KY-H (2015) A critical review on characterization strategies of organic matter for wastewater and water treatment processes. Biores Technol 193:523–533. https://doi.org/10.1016/j.biortech.2015.06.091

UN world water development report 2017: https://www.unesco.org/en/wwap/wwdr/2017

Vanrolleghem PA, Lee DS (2003) On-line monitoring equipment for wastewater treatment processes: state of the art. Water Sci Technol 47(2):1–34. https://doi.org/10.2166/wst.2003.0074

Vasseghian Y, Alimohamadi M, Dragoi EN, Sonne C (2023) A global meta-analysis of phthalate esters in drinking water sources and associated health risks. Sci Total Environ 903:166846

Williams A, Aguilar MR, Pattiya Arachchillage KG, Chandra S, Rangan S, Ghosal Gupta S, Artes Vivancos J.M (2024) Biosensors for public health and environmental monitoring: the case for sustainable biosensing. ACS Sustain Chem Eng 12(28):10296–10312

WHO Report (2020) https://www.who.int/about/accountability/results/who-results-report-2020-2021

Xu M, Cao H, Xie P, Deng D, Feng W, Xu J (2005) Use of PFU protozoan community structural and functional characteristics in assessment of water quality in a large, highly polluted freshwater lake in China. J Environ Monit 7(7):670. https://doi.org/10.1039/b504396b

Yang Y, Yi T, Liu Y, Zhao H, Liang C (2022) Design of a highly sensitive reduced graphene oxide/graphene oxide@cellulose acetate/thermoplastic polyurethane flexible sensor. Sensors 22(9):3281. https://doi.org/10.3390/s22093281

Zanetti L, Frison N, Nota E, Tomizioli M, Bolzonella D, Fatone F (2012) Progress in real-time control applied to biological nitrogen removal from wastewater. A short-review. Desalination 286:1–7. https://doi.org/10.1016/j.desal.2011.11.056

Zhu R, Liu L, Zhang G, Zhang Y, Jiang Y, Pang H (2024) Advances in electrochemistry of intrinsic conductive metal-organic frameworks and their composites: mechanisms, synthesis and applications. Nano Energy 122:109333. https://doi.org/10.1016/j.nanoen.2024.109333

Chapter 2
Aquatic Macrophyte-Based Detection and Treatment of Wastewater Pollution

Narala Gangadhara Reddy, Satya Ranjan Samal, and Kundan Samal

Abstract Aquatic macrophyte-based treatment systems (AMTS) provide an eco-friendly, sustainable, and cost-effective solution for treating domestic and industrial wastewater and for restoring polluted water bodies. In these systems, macrophytes either float directly on the water surface or grow on floating frames, with their roots suspended in the water to absorb nutrients and heavy metals. The roots release oxygen, maintaining aerobic conditions in the rhizosphere and providing surface area for microorganisms that break down contaminants. Additionally, aquatic macrophytes act as bioindicators of pollution, exhibiting stress responses such as stunted growth, leaf discoloration, or reduced biomass when exposed to toxins, heavy metals, or excess nutrients, thereby helping to monitor water quality and eutrophication. The effectiveness of AMTS depends on factors such as water depth, vegetation coverage, pollutant concentration, residence time, and plant species used. Regular harvesting is essential to prevent the release of accumulated pollutants back into the water. Further research is needed to optimize the system's performance and promote its large-scale application.

Keywords Aquatic macrophyte · Depth · Aquatic macrophyte-based treatment system (AMTS) · Root system · Nutrients

2.1 Introduction

Water pollution from agricultural runoff, domestic sewage, and industrial effluents poses a serious threat to aquatic ecosystems by introducing a wide range of contaminants, including toxic chemicals, heavy metals, pathogens, nutrients, and organic matter (Qiu et al. 2023; Kurniawan et al. 2025; Wu et al. 2024). Additionally,

N. Gangadhara Reddy
School of Building and Civil Engineering, Fiji National University, Samabula, Suva, Fiji

S. R. Samal · K. Samal (✉)
School of Civil Engineering, KIIT University, Bhubaneswar, Odisha, India
e-mail: kundan.samalfce@kiit.ac.in

A. M. Yatoo et al. (eds.), *Wastewater Monitoring and Management by Advanced Approaches*, Advances in Wastewater Research,
https://doi.org/10.1007/978-981-95-2601-7_2

nutrient-rich leachate from landfills and illegal sewage discharges into surface waters further deteriorate water quality (Yatoo et al. 2024a). The excessive influx of nutrients promotes the rapid growth of algae and plankton, leading to eutrophication, which can result in complete surface coverage of the water body (Cao et al. 2024; Tang et al. 2024). This prevents atmospheric oxygen from diffusing into the water, causing a sharp decline in dissolved oxygen levels and eventually leading to the death of fish and other aquatic organisms (Yatoo et al. 2025). Therefore, it is essential to both prevent pollutants from entering water bodies and remediate those already present (Zhang et al. 2018; Luca et al. 2019). Various remediation strategies, such as sludge and sediment dredging, oxygenation, cultivating oxygen-releasing aquatic plants, and installing constructed floating wetlands (CFWs), have been applied worldwide. Biological wastewater treatment technologies, including rotating biological contactors (RBC), oxidation ponds, lagoons, constructed wetlands (CW), trickling filters (TF), vermifilters (VF), and others, offer additional solutions for managing water pollution (Aurangzeb et al. 2014; Kar and Samal 2022; Zhu et al. 2011).

Aquatic macrophyte-based treatment systems (AMTS) were first introduced by Pallis in 1915 and later applied in Arcata, California, in 1986. Over time, this technology has been implemented worldwide under various names, including ecological floating beds (EFB), planting floating beds, artificial floating beds, and vegetated floating beds. AMTS represents a sustainable, nature-based solution for water purification, leveraging the natural self-cleansing ability of ecosystems. It is simple to construct, requires no specialized skills for operation, and generates no harmful by-products during the treatment process (Abdallah 2012; Mahapatra et al. 2022). The system consists of aquatic plants anchored on buoyant structures, with their shoots extending above the floating platform and roots submerged in the water. These roots not only absorb nutrients from the water but also offer a surface for beneficial microbial communities that aid in the breakdown of pollutants. Key pollutant removal mechanisms within AMTS include trapping suspended particles, sedimentation, mineralization, organic matter degradation, nitrification–denitrification, oxidation, and ammonification (Headley and Tanner 2012; Samal and Trivedi 2020). This chapter focuses on the design aspects of AMTS, including nutrient uptake efficiency, biomass harvesting techniques, practical applications, and limitations, providing a comprehensive understanding of the system's potential for wastewater management.

2.2 Aquatic Macrophyte as Bioindicator of Wastewater Pollution

Aquatic macrophyte as bioindicators of wastewater pollution refers to the use of submerged, emergent, or floating plants to monitor the health and pollution levels of water bodies, leveraging their sensitivity to contaminants and changes in water quality (Bai et al. 2025). Macrophytes such as *Eichhornia crassipes* (water hyacinth), *Lemna* spp. (duckweed), and *Ceratophyllum demersum* (coontail) exhibit specific physiological and morphological responses to pollutants like heavy metals, excessive nutrients, organic toxins, and changes in pH (Gao et al. 2017; Samal and Dash 2021). These plants respond to water pollution through visible signs like stunted growth, chlorosis (yellowing of leaves), and biomass reduction, or biochemical changes such as the accumulation of stress enzymes and alterations in chlorophyll content. Macrophytes can also accumulate toxic substances, including cadmium, mercury, and lead, making them effective indicators of metal contamination (Colares et al. 2020; Xu et al. 2024). The presence or absence of certain species can signal nutrient overload (eutrophication) or toxicity, with sensitive species often disappearing under extreme conditions, while more tolerant ones proliferate. Monitoring macrophytes thus offers a cost-effective, non-invasive, and continuous method to assess the extent and nature of pollution over time. Since these plants reflect the cumulative effects of multiple stressors, they are particularly valuable in assessing complex wastewater discharges from industrial, agricultural, and urban sources (Zhu et al. 2023; Samal et al. 2024). In ecological monitoring programs, the health of macrophyte populations provides critical insights into the overall quality of the aquatic environment, complementing chemical analyses and enhancing the effectiveness of water management strategies.

2.3 Aquatic Macrophyte Classification

There are four primary types of plants used in the design of constructed floating wetlands (CFWs) for wastewater treatment: submerged macrophytes, emergent macrophytes, floating-leaved macrophytes, and free-floating macrophytes. Submerged plants grow entirely underwater, with or without root systems. Their leaves often have a waxy coating that helps prevent damage. These plants are less suitable for treating raw sewage but are effective in tertiary treatment systems for polishing treated effluents (Hwang et al. 2016; Lynch et al. 2015). Common examples include *Cladophora* spp., *Enteromorpha* spp., *Potamogeton* spp., *Ceratophyllum* spp., *Myriophyllum* spp., *Elodea canadensis, Elodea nuttallii, Ulva lactuca*, and *Egeria densa*. Emergent macrophytes grow along the edges and banks of water bodies, with their roots anchored in soil and stems, leaves, and flowers extending above the water surface. These plants rely on aerial reproduction and absorb nutrients from both water and soil. The oxygen released from their roots maintains aerobic conditions in the rhizosphere, encouraging bacterial growth for pollutant degradation (Samal et al. 2021; Cao and Zhang 2014). Common emergent species include *Phragmites australis, Glyceria maxima, Scirpus validus*, and *Iris pseudacorus* (Brix and Sorrell 1996; Borne 2014). Floating-leaved macrophytes are rooted in the substrate, with their leaves floating on the water's surface, fully exposed to sunlight. These leaves, often leathery, circular, and hydrophobic, are connected to roots via long petioles. The elongation of petioles is regulated by ethylene loss, which triggers the production of gibberellic acid, a growth hormone. Examples include *Nymphaea odorata, Nuphar lutea,* and *Hydrocotyle vulgaris* (Ayaz and Saygin 1996). Free-floating macrophytes have their photosynthetic structures above the water while roots hang below the surface. These plants cover the pond's surface, inhibiting algal growth and creating stable chemical and physical conditions. Due to their high capacity for nutrient and heavy metal uptake, they are suitable for treating raw sewage and both primary and secondary effluents. Frequently used free-floating species include *Lemna* spp., *Pistia stratiotes*, *Hydrocotyle* spp., and *Salvinia* spp. (Jasper et al. 2013; Samal et al. 2019).

2.4 Design Factors of Vegetated Floating System

2.4.1 *Type of Vegetation*

Selecting appropriate plant species is crucial for the effective design of AMTS, as plants play a direct role in pollutant removal and facilitate microbial growth. Ideal plant species should exhibit a high capacity for absorbing nutrients, heavy metals, and other contaminants (Cao and Zhang 2014). They must also adapt well to local climatic conditions, thrive in aquatic environments, be noninvasive, and possess robust oxygen diffusion from leaves to roots, along with a dense, submerged root system (Syukor et al. 2016). Additionally, aesthetic appeal can be a desirable factor. While emergent macrophytes are commonly used, the choice of species depends on both the regional climate and the specific pollutants targeted for treatment (Billore et al. 2009). Utilizing a mixture of plant species in AMTS often yields better results than relying on a single species, as it promotes biodiversity and enhances pollutant uptake. Regular harvesting is essential to prevent the rerelease of accumulated pollutants into the water, maintain aerobic conditions, and create space for new plant growth. As plant density increases, pollutant removal efficiency improves, but beyond a certain point, chlorophyll content may decline, impairing plant metabolism and respiration, which can negatively impact the treatment process (Gao et al. 2017).

2.4.2 *Buoyancy*

In natural floating wetlands, buoyancy is achieved through the oxygen stored within the roots and leaves of macrophytes, which allows the plants to remain afloat on the water's surface. Additionally, methane (CH_4) gas generated during the anaerobic decomposition of organic matter in the sediment becomes trapped within the plant roots, further enhancing their ability to float (Kumari and Tripathi 2015). To improve stability and facilitate plant growth, low-cost floating materials can also be integrated into the system. A variety of materials, including polyvinyl chloride (PVC) pipes, foam blocks, bamboo, coconut peat, and polystyrene sheets, can be used to create buoyant mats that support the macrophytes (Xu et al. 2024). In artificial AMTS setups, floating rafts or other buoyant structures are often used to provide initial

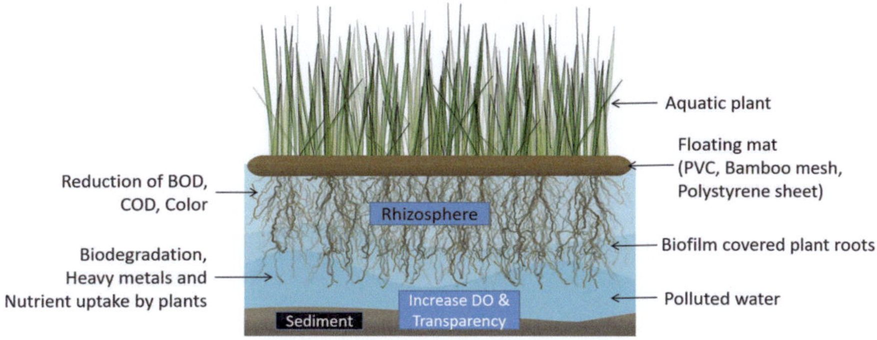

Fig. 2.1 Design of an aquatic macrophyte-based treatment system (AMTS)

support for plant establishment until the macrophytes develop sufficient root systems and achieve natural buoyancy on their own (Basile et al. 2012; Samal et al. 2021). These floating frameworks not only anchor the plants but also ensure they remain evenly distributed, maximizing contact between the roots and the water for effective pollutant absorption. Figure 2.1 presents a basic schematic of an AMTS, illustrating the use of floating structures to enhance plant growth and pollutant removal efficiency.

2.4.3 Water Depth

In an AMTS, plant roots grow underwater, anchored to sediments. However, fluctuations in water levels can pose challenges, as plants may become fully submerged or suffer damage. To prevent this, it is recommended to maintain a minimum water depth of 0.8–1 m (Saeed et al. 2016). In nutrient-rich wastewater, a shallower depth of around 0.2 m can suffice since the nutrients are readily available near the surface. However, for nutrient-deficient wastewater, such as stormwater, a depth of 0.8–3 m is advised, allowing the roots to extend to the benthic zone where nutrients are more concentrated. This deeper rooting helps plants absorb essential nutrients efficiently (McAndrew and Ahn 2017). AMTS is not suitable for stormwater treatment in areas with low rainfall, as it requires a permanent water pool to remain functional. Shallow depths enhance the removal of fine suspended solids by trapping them within the root mat, while deeper depths promote the sedimentation of larger particles, as a free-water zone forms beneath the floating roots (Swarnalatha and Radhakrishnan 2015). The optimal depth of the AMTS depends on the macrophyte species used, as different plants have varying root structures and depth requirements. It is crucial to ensure that the roots remain suspended within the water column to maximize pollutant uptake and maintain effective treatment performance (Benvenuti et al. 2018).

2.4.4 Coverage and Shading

Vegetation coverage plays a critical role in determining the oxygen diffusion into the water, which directly impacts the treatment performance of an AMTS. If the vegetation coverage exceeds 50%, there is a significant risk of creating anoxic conditions, as dense plant canopies obstruct the diffusion of atmospheric oxygen into the water. Limited sunlight penetration further disrupts photosynthesis, leading to a decline in photosynthetic algal species (Zhang et al. 2018; Samal et al. 2021). The breakdown of the natural algal–bacterial symbiosis also reduces the population of aerobic bacteria, which rely on oxygen for their survival. As a result, the aerobic decomposition of organic matter slows, especially at the lower layers of the AMTS, altering the entire physicochemical balance of the system. Without sufficient aerobic activity, the degradation of pollutants becomes less efficient, and the accumulation of organic matter may increase, potentially leading to foul odors or eutrophication (Wibowo et al. 2023). Studies have shown that maintaining vegetation coverage above 50% can negatively impact treatment efficiency by reducing oxygen levels and impairing microbial processes essential for water purification (Zhao et al. 2012). For optimal performance, plant density must be carefully managed to maintain a balance between pollutant uptake by macrophytes and oxygen availability for aerobic microorganisms. Proper monitoring of vegetation coverage can prevent these negative outcomes and sustain the ecological balance needed for effective water treatment.

2.4.5 Biofilm

In an AMTS, biofilms develop on the surfaces of plant roots and other submerged structures, playing a crucial role in enhancing the treatment process. These biofilms consist of sticky communities of bacteria, protozoa, and fungi that adhere to both living (e.g., roots) and nonliving surfaces. The formation of biofilm is strengthened by root exudates and microbial metabolites, which help bind the microorganisms into a cohesive layer (Chaudhuri et al. 2014). Biofilms are composed primarily of proteins, extracellular enzymes, polysaccharides, and nucleic acids, creating a matrix that supports microbial life and contributes to the overall treatment process (Syukor et al. 2016). Biofilms significantly improve system efficiency by trapping suspended solids, increasing nutrient and heavy metal retention, and enhancing nitrification–denitrification processes. They also boost water retention capacity and exhibit antimicrobial activity, helping control pathogenic microorganisms (Saeed et al. 2016). To further enhance performance, artificial biofilm carriers, such as bamboo fibers, polystyrene-based carriers, plastic media, or magnetic particles, can be introduced into the system, providing additional surfaces for biofilm growth. Plant roots serve as natural scaffolds for microbial colonization, offering an extensive surface area that supports the formation and development of biofilms (Samal et al. 2019). This synergy between plant roots and biofilms plays a pivotal role in pollutant removal, transforming nutrients and contaminants into less harmful forms through microbial activity, making AMTS a highly efficient and sustainable solution for wastewater treatment.

2.5 Pollutant Removal Mechanism of AMTS

2.5.1 Organics Removal

Aquatic macrophytes are essential contributors to wastewater treatment, particularly in systems such as ecological floating beds and vegetated floating systems. These plants have evolved a variety of mechanisms to engage with, assimilate, and eliminate contaminants, including organic materials and solids, from wastewater. They directly absorb organic pollutants through their roots, stems, and leaves. Once these compounds enter the plant, they may be stored, transformed, or integrated into metabolic processes. As wastewater passes through a system filled with aquatic macrophytes, the dense network of roots and rhizomes serves as a physical barrier that traps and settles suspended particles (Gao et al. 2018; Zhu et al. 2023; Samal et al. 2017). The surfaces of the roots are often negatively charged, enabling them to attract and adsorb positively charged particles and molecules, including various organic compounds. The rhizosphere, the area surrounding the plant roots, is rich in microbial life. These microorganisms establish symbiotic relationships with the plants, obtaining nutrients released from the roots while simultaneously degrading organic compounds present in the water. Certain complex organic molecules, such as hydrocarbons, can be broken down into simpler, less toxic forms by these microbes. Additionally, through the process of photosynthesis, aquatic macrophytes release oxygen into their surrounding environment. This increase in oxygen concentration supports the growth of aerobic bacteria, which further aids in the decomposition of organic materials in wastewater (Li et al. 2010; Groot et al. 2005). Some contaminants can be oxidized directly, rendering them neutral or less harmful when oxygen is present.

Moreover, aquatic macrophytes help create calm zones in the water through their physical structure, which promotes sedimentation. When water movement is minimized, suspended particles, including organic solids, settle more easily. Over time, this process leads to the formation of a sediment layer at the bottom, effectively removing these solids from the water column. The surfaces of macrophyte roots provide ideal environments for biofilm development. These biofilms consist of matrices formed by microbial communities embedded in a slimy extracellular matrix. They can metabolize and break down a wide variety of organic compounds, significantly reducing the organic load in wastewater (Kumari and Tripathi 2015). Table 2.1 illustrates various mechanisms by which pollutants are removed in AMTS.

Table 2.1 Pollutants removal mechanism in aquatic macrophyte-based treatment system (AMTS) (Samal et al. 2019; Luca et al. 2019; Lu et al. 2015; Verma and Suthar 2015)

Pollutant type	Removal mechanism	Effective plant species
Nitrogen (N)	Direct plant uptake, nitrification, denitrification	*Typha angustifolia, Phragmites australis, Lemna minor*
Phosphorus (P)	Direct plant uptake, precipitation by biofilm	*Eichhornia crassipes, Lemna gibba, Pistia stratiotes*
Heavy metals	Phytoextraction, adsorption to roots, and biofilm	*Salvinia minima, Eichhornia crassipes*
Organic matter	Direct uptake, microbial degradation in biofilm	*Pistia stratiotes, Typha angustifolia*
Pathogens	UV disinfection, predation, natural die-off	Any dense canopy species
Suspended solids	Sedimentation, filtration by roots	*Typha angustifolia, Phragmites australis*

2.5.2 *Nutrient Removal*

Plants absorb nutrients from the water column to support their growth. As they develop, they assimilate and incorporate essential nutrients, especially nitrogen (N) and phosphorus (P), into their biomass. Nitrogen is primarily taken up in the forms of ammonium (NH_4^+) and nitrate (NO_3^-), which are then converted into amino acids, proteins, and other organic compounds within the plant. Phosphorus is absorbed as phosphate (PO_4^{3-}) and is utilized in the synthesis of ATP, DNA, RNA, and phospholipids within plant cells (Abed et al. 2017; Wibowo et al. 2023).

Nitrogen is also removed from the water through the nitrification–denitrification process. In nitrification, ammonium is oxidized by aerobic bacteria, such as *Nitrosomonas* and *Nitrobacter*, to produce nitrate. This process occurs in the aerobic zones surrounding plant roots and on the floating mat. Denitrification takes place in anaerobic or anoxic zones, where certain bacteria can reduce nitrates and nitrites to nitrogen gas (N_2), which then volatilizes into the atmosphere, effectively removing nitrogen from the system (Chan et al. 2022). Phosphate-accumulating organisms (PAOs) can store excess phosphate beyond their metabolic requirements (Fig. 2.2). These organisms absorb and accumulate phosphate from the water when it is readily available (Lu et al. 2015). As water flows through the AMTS, particulate matter, including organic material rich in nutrients, can settle and adhere to the floating mat or become trapped in the root matrix. Some AMTS incorporate specific materials, such as certain clays or biochar, which can physically or chemically adsorb nutrients, enhancing nutrient removal (Kyambadde et al. 2004).

Overall, nutrient removal in an AMTS occurs through a combination of plant uptake, microbial transformations, physical processes, and trophic interactions. Table 2.2 illustrates the applications of AMTS in treating various types of wastewater.

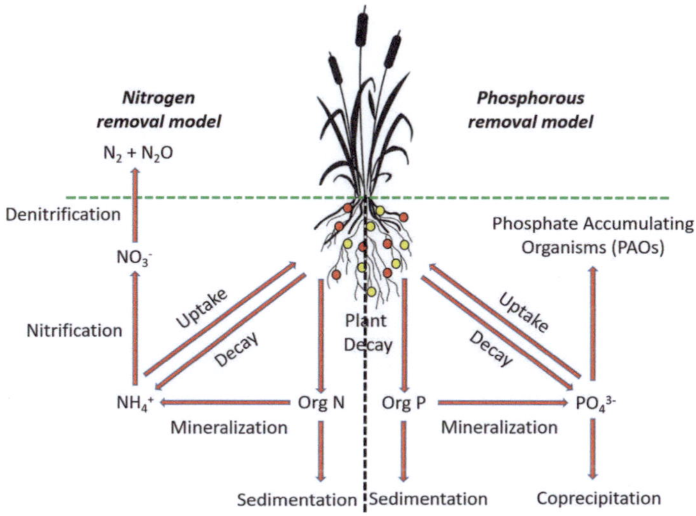

Fig. 2.2 Nutrient removal mechanism in an aquatic macrophyte-based treatment system (AMTS)

2.5.3 Heavy Metals Removal

Aquatic plants within AMTS can directly absorb heavy metals from the water column, leading to their accumulation in plant tissues. This uptake process is species-specific, meaning that certain plants are more efficient at accumulating specific metals than others. Hyperaccumulator plants, which can concentrate significant amounts of metals in their biomass without exhibiting toxicity symptoms, are particularly noteworthy in this regard. The roots of floating plants, along with the biofilm composed of microorganisms and their extracellular products developing around them, can adsorb and immobilize heavy metals (Miretzky et al. 2006; Swarnalatha and Radhakrishnan 2015). The biofilm and root exudates contain organic compounds capable of chelating or binding to metals, thereby effectively reducing their concentrations in the surrounding water (Verma and Suthar 2015; Samal and Dash 2024). Certain microorganisms within the biofilm can facilitate the transformation of heavy metals from soluble to insoluble forms through processes such as precipitation, complexation, or reduction (Xie et al. 2022). For instance, sulfate-reducing bacteria can convert soluble metals into insoluble metal sulfides, effectively removing them from the water column.

Additionally, some heavy metals may adhere to particulate matter in the water column. The presence of AMTS can enhance the sedimentation of these particles, leading to the settling and subsequent removal of metals from the water. Furthermore, microorganisms within the AMTS can convert certain heavy metals, like mercury, into volatile compounds. These volatile compounds can evaporate from the water surface, thus removing the metal from the system (Lima et al. 2013; Samal et al.

Table 2.2 The use of Aquatic macrophyte-based treatment system (AMTS) for the treatment of various types of wastewater

Types of wastewater	Plant species	Organics removal (%)	Nutrient removal (%)	Operation period (days)	References
Landscape water	*H. vulgaris, I. A. calamus*	–	TN 59.8, NH_4^+-N 52.4, NO_3^--N 54.72, TP 89, PO_4^{3-}-P 93.9	133	Zhu et al. (2023)
Urban water body	C. *indica, L. salicaria*	COD 13.7	NH_4^+-N 41.8, NO_3^--N 100, NO_2^--N 80.6	180	Qiu et al. (2023)
Municipal wastewater	*M. aquaticum*	COD 96.1 ± 12.3	TN 99.2 ± 11.2, NH_4^+-N 99.82 ± 16.4, TP 98.3 ± 14.3	–	Kumwimba et al. (2022)
Stormwater	*A. subcordatum*	–	TN 20	8	McAndrew and Ahn (2017)
Simulated wastewater	*C. indica*	COD 88.3	TN 80.7, NO_3^--N 98.1, NH_4^+-N 59.4, TP 74	–	Zhang et al. (2018)
Pond water + NH_4NO_3 + KH_2PO_4	*T. domingensis*	–	TP 90, NH_4^+ 96	43	Luca et al. (2019)
Secondary effluent	*C. papyrus*	BOD 83.1	TN 66, TP 61	3	Kyambadde et al. (2004)
Simulated wastewater	*S. japonicus*	–	TP 74-98	10	Geng et al. (2017)
Lake water	I. *aquatica*	–	TN 76.5, TP 61	–	Li et al., (2010)
Polluted river water	*P. australis*	BOD 51, COD 40	NH_4^+-N 44, TP 23.4	–	Saeed et al. (2016)
Lake water	*T. orientalis*	SS 80	TN 66, TP 74	7	Lu et al. (2015)
Synthetic greywater	*P. australis*	BOD 33, COD 27	PO_4^{3-}-P 19	7	Abed et al. (2017)
Secondary effluent	*I. pseudacorus*	–	TN 74-88	3	Gao et al. (2017)
River water	*F. karka*	BOD 50	TN 50, NH_4 55	–	Billore et al. (2009)

(continued)

Table 2.2 (continued)

Types of wastewater	Plant species	Organics removal (%)	Nutrient removal (%)	Operation period (days)	References
Domestic sewage	*T. domingensis*	BOD 56, COD 55, TSS 78	TKN 41, TP 37	11	Benvenuti et al. (2018)
River water + KNO_3 + NH_4Cl	*I. pseudacorus*	–	TN 90	3	Gao et al. (2018)

2020). Figure 2.3 illustrates the mechanisms by which aquatic plants uptake heavy metals.

Bio-concentration factor (BCF) is the ability of a plant to accumulate heavy metals in the plant tissue, and it can be calculated using the following equation (Ndimele et al. 2014):

$$\text{BCF} = \frac{\text{Heavy metals conc. in plant tissue}}{\text{Initial metal conc. in solution}} \tag{2.1}$$

Translocation factor (TF) is the ability of the plant to translocate heavy metals from its roots to the shoot, and it can be calculated using the following equation (Das et al. 2016):

$$\text{TF} = \frac{\text{Metals conc. in leaf}}{\text{Metals conc. in roots}} \tag{2.2}$$

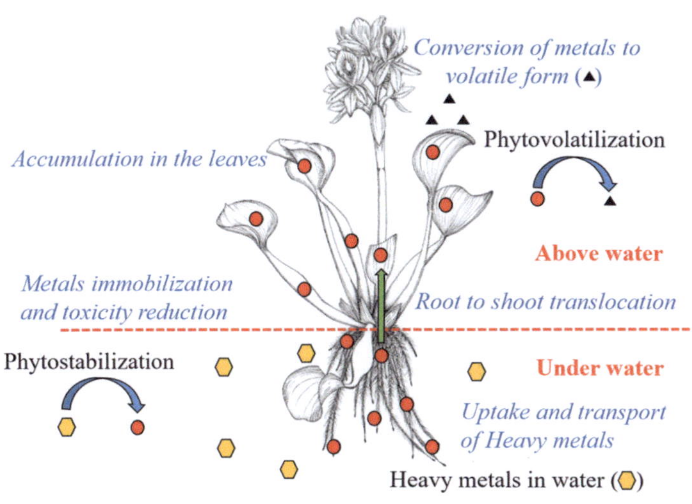

Fig. 2.3 Heavy metals uptake in an aquatic plant

It's worth noting that while AMTS can be effective in removing heavy metals, the efficiency varies based on factors like the specific heavy metal in question, the type of plants used, the age and health of the plants, water chemistry, residence time, and overall design and management of the system. The integration of multiple remediation strategies, such as combining AMTS with other treatments or using diverse plant species, can improve overall heavy metal removal efficiency. Treatment of heavy metals-rich wastewater using AMTS is shown in Table 2.3.

Table 2.3 The use of aquatic macrophyte-based treatment system (AMTS) for the treatment of heavy metals rich wastewater

Heavy metals	Plants	Conditions	Removal efficiency (%)	References
Fe, Mn	*P. stratiotes*, *E. crassipes*	Acid mine drainage + coal mining industry + domestic wastewater	Fe 87, Mn 72 Fe 89, Mn 74	Wibowo et al. (2023)
Cd, Cu, Pb, Ni, Cr	P. *stratiotes* L. *gibba*	Industrial polluted water	Cd 84.4, Cu 74, Pb 72, Ni 74, Cr 88 Cd 95, Cu 93, Pb 82, Ni 93, Cr 92	Xie et al. (2022)
Cd, Fe, Cr, Pb,	*T. angustifolia*	Waste stabilization pond	Cd 61.07, Pb 62.07, Cr 69.17, Fe 81.17	Syukor et al. (2016)
Pb, Fe	*P. stratiotes*	Synthetic leachate	Pb 59.86, Fe 56.56	Chan et al. (2022)
Cu, Zn, Cd	*P. australis*	Sewage + industrial effluents	Cd 60, Cu 78, Zn 61	Kumari and Tripathi (2015)
Cd, Cu, Pb	*Lemna* spp.	Simulated wastewater	Cd 95, Pb 93, Cu 86.5	Basile et al. (2012)
Cr, Zn	*E. crassipes*	Synthetic solutions	Cr 80, Zn 96	Swarnalatha and Radhakrishnan (2015)
Cd	*L. minor*	Concentrations of 0.5, 1.0, 1.5, 2.0, 2.5, and 3.0 mg/L	42–78	Chaudhuri et al. (2014)
Pb, Cu	*P. stratiotes*	Steel industry effluent	Pb 70, Cu 66	Aurangzeb et al. (2014)
Cd, Pb	*L. gibba*	Concentration of 2, 5, and 10 mg/L	Cd 84.8 (2 mg/L at pH 7), Pb 98.1 (10 mg/L at pH 7)	Verma and Suthar (2015)
Cr, Pb	*C. demersum*	Simulated wastewater	Pb 95, Cr 84	Lima et al. (2013)

2.6 Biomass Harvesting and Utilization

Biomass harvesting is a crucial component of the AMTS used for wastewater treatment. The biomass accumulated in AMTS results from the uptake of nutrients, organic pollutants, and, in some cases, heavy metals from the wastewater. Regular harvesting of biomass is essential to maintain the health and productivity of the plants. This practice not only promotes efficient nutrient uptake but also prevents the plants from dying and releasing the accumulated pollutants back into the water. The frequency of harvesting varies based on plant species, wastewater characteristics, climatic conditions, and the specific objectives of the treatment system (Ayaz and Saygin 1996). Typically, workers manually cut and collect the plant biomass, a labor-intensive method suitable for smaller systems. In contrast, larger systems may utilize machinery, similar to agricultural equipment, for cutting, collecting, and transporting the biomass (Samal et al. 2019). The harvested biomass can be vermicomposted to produce organic fertilizer (Yatoo et al. 2024b; Yatoo et al. 2025). However, if heavy metals are present, it is crucial to test the compost and limit its use to nonfood crops. Additionally, biomass can be harnessed for bioenergy production through various methods. For example, it can be anaerobically decomposed to generate biogas, a methane-rich gas that serves as a valuable energy source (Gao et al. 2018). Biomass can also be converted into syngas, which can be burned directly or processed into liquid fuels. Furthermore, dry biomass can be incinerated to produce heat or steam for electricity generation. Some aquatic plants, after proper treatment, may even be used as feed for livestock or fish (Saeed et al. 2016). However, it is essential to ensure that any pollutants in the biomass do not enter the food chain. Certain aquatic plants, such as reeds, have traditionally been utilized for making thatched roofs or other construction materials. The cellulose-rich biomass can also be processed into paper or cardboard. Moreover, aquatic plants that contain bioactive compounds can be extracted for medicinal or health-related uses (Zhu et al. 2011).

Overall, the biomass harvested from AMTS in wastewater treatment applications represents a valuable resource. With proper management and utilization, this waste product can be transformed into a variety of beneficial commodities, effectively turning a treatment process into a resource recovery operation.

2.7 Conclusion and Future Prospects

Aquatic macrophytes play a crucial role in the bioremediation of wastewater. They utilize various mechanisms, such as direct nutrient uptake, rhizofiltration, and the promotion of biofilm development, to effectively reduce pollutants, including nutrients and heavy metals, from wastewater. Additionally, the collaboration between macrophytes and the microbial communities associated with AMTS enhances the overall effectiveness of the treatment process. The sustainable nature of AMTS, highlighted by biomass harvesting and the potential for resource recovery, emphasizes their viability as a cost-effective and environmentally friendly solution for wastewater treatment. This approach is particularly beneficial in areas where conventional wastewater infrastructure may be insufficient or impractical to implement. Future research can focus on optimizing plant selection, improving root zone microbial interactions, and integrating AMTS with advanced treatment technologies to enhance nutrient removal efficiency. Moreover, their application in mitigating the impacts of climate change, such as reducing greenhouse gas emissions and restoring aquatic habitats, makes AMTS a promising solution for achieving sustainable water management and ecological balance.

References

Abdallah MA (2012) Phytoremediation of heavy metals from aqueous solutions by two aquatic macrophytes, *Ceratophyllum demersum* and *Lemna gibba* L. Environ Technol 33:1609–1614. https://doi.org/10.1080/09593330.2011.640354

Abed SN, Almuktar SA, Scholz M (2017) Remediation of synthetic greywater in mesocosm-Scale floating treatment wetlands. Ecol Eng 102:303–319. https://doi.org/10.1016/j.ecoleng.2017.01.043

Aurangzeb N, Nisa S, Bibi Y, Javed F, Hussain F (2014) Phytoremediation potential of aquatic herbs from steel foundry effluent. Braz J Chem Eng 31:881–886. https://doi.org/10.1590/0104-6632.20140314s00002734

Ayaz SC, Saygin Ö (1996) Hydroponic tertiary treatment. Water Res 5:295–1298. https://doi.org/10.1016/0043-1354(95)00284-7

Bai J, Qin Y, Zhao J, Song Y (2025) Investigation of agricultural nutrient removal by ecological ditches using meta-analysis. Agr Ecosyst Environ 380:109401. https://doi.org/10.1016/j.agee.2024.109401

Basile A, Sorbo S, Conte B, Cobianchi RC, Trinchella F, Capasso C, Carginale V (2012) Toxicity, accumulation, and removal of heavy metals by three aquatic macrophytes. Int J Phytoremediat 14:374–387. https://doi.org/10.1080/15226514.2011.620653

Benvenuti T, Hamerski F, Giacobbo A, Bernardes AM, Ferreira JZ, Rodrigues MAS (2018) Constructed floating wetland for the treatment of domestic sewage: a real scale study, J Environ Chem Eng 6:5706–5711. https://doi.org/10.1016/j.jece.2018.08.067

Billore SK, Prashant Sharma JK (2009) Treatment performance of artificial floating reed beds in an experimental mesocosm to improve the water quality of river Kshipra. Water Sci Technol 60:2851–2859. https://doi.org/10.2166/wst.2009.731

Borne KE (2014) Floating treatment wetland influences on the fate and removal performance of phosphorus in stormwater retention ponds. Ecol Eng 69:76–82. https://doi.org/10.1016/j.ecoleng.2014.03.062

Brix H, Sorrell BK (1996) Oxygen stress in wetland plants: comparison of de-oxygenated and reducing root environments. Funct Ecol 10:521–526. https://doi.org/10.2307/2389945

Cao W, Zhang Y (2014) Removal of nitrogen (N) from hypereutrophic waters by ecological floating beds (EFBs) with various substrates. Ecol Eng 62:148–152. https://doi.org/10.1016/j.ecoleng.2013.10.018

Cao M, Sy ND, Yu CP, Gan J (2024) Removal of neonicotinoid insecticides in a large-scale constructed wetland system. Environ Pollut 344:123303. https://doi.org/10.1016/j.envpol.2024.123303

Chan MY, Tee CS, Chai TT, Sim YL, Beh WL (2022) Evaluation of electro-assisted phytoremediation (EAPR) system for heavy metal removal from synthetic leachate using *Pistia stratiotes*. Int J Phytorem 24:1376–1384. https://doi.org/10.1080/15226514.2022.2031863

Chaudhuri D, Majumder A, Misra AK, Bandyopadhyay K (2014) Cadmium removal by *Lemna minor* and *Spirodela polyrhiza*. Int J Phytoremediat 16:1119–1132. https://doi.org/10.1080/15226514.2013.821446

Colares GS, Dell'Osbel N, Wiesel PG, Oliveira, GA, Lemos PHZ, da Silva FP et al (2020) Floating treatment wetlands: a review and bibliometric analysis. Sci Total Environ 714:136776. https://doi.org/10.1016/j.scitotenv.2020.136776

Das S, Goswami S, Talukdar AD (2016) Physiological responses of water hyacinth, *Eichhornia crassipes* (Mart.) Solms, to cadmium and its phytoremediation potential. Turk J Biol 40:84–94

Gao L, Zhou W, Huang J, He S, Yan Y, Zhu W, Wu S, Zhang X (2017) Nitrogen removal by the enhanced floating treatment wetlands from the secondary effluent. Bioresourc. Technol. 234:243–252. https://doi.org/10.1016/j.biortech.2017.03.036

Gao L, Zhou W, Wu S, He S, Huang J, Zhang X (2018) Nitrogen removal by thiosulfate-driven denitrification and plant uptake in enhanced floating treatment wetland. Sci Total Environ 621:1550–1558. https://doi.org/10.1016/j.scitotenv.2017.10.073

Geng Y, Han W, Yu C, Jiang Q, Wu J, Chang J, Ge Y (2017) Effect of plant diversity on phosphorus removal in hydroponic microcosms simulating floating constructed wetlands. Ecol Eng 107:110–119. https://doi.org/10.1016/j.ecoleng.2017.06.061

Groot TT, van Bodegom PM, Meijer HAJ, Harren FJM, (2005) Gas transport through the root-shoot transition zone of rice tillers. Plant Soil 277:107–116. https://doi.org/10.1007/s11104-005-0435-4

Headley TR, Tanner CC (2012) Constructed wetlands with floating emergent macrophytes: an innovative stormwater treatment technology. Crit Rev Environ Sci Technol 42:2261–2310. https://doi.org/10.1080/10643389.2011.574108

Hwang HM, Fiala MJ, Park D, Wade TL (2016) Review of pollutants in urban road dust and stormwater runoff: Part 1. Heavy metals released from vehicles. Int J Urban Sci 20:334–360. https://doi.org/10.1080/12265934.2016.1193041

Jasper JT, Nguyen MT, Jones ZL, Ismail NS, Sedlak DL, Sharp JO, Luthy RG, Horne AJ, Nelson KL (2013) Unit process wetlands for removal of trace organic contaminants and pathogens from municipal wastewater effluents. Environ Eng Sci 30:421–436. https://doi.org/10.1089/ees.2012.0239

Kar S, Samal K (2022) Hydro economy: environmental sustainability of water and wastewater resources and infrastructure. In: Recent developments in sustainable infrastructure (ICRDSI-2020)—GEO-TRA-ENV-WRM. Lecture notes in civil engineering, vol 207. Springer, Singapore. https://doi.org/10.1007/978-981-16-7509-6_15

Kumari M, Tripathi BD (2015) Efficiency of *Phragmites australis* and *Typha latifolia* for heavy metal removal from wastewater. Ecotox Environ Saf 112:80–86. https://doi.org/10.1016/j.ecoenv.2014.10.034

Kumwimba MN, Li X, Huang J, Muyembe DK, Dzakpasu M, Sanganyado E (2022) Performance of various fillers in ecological floating beds planted with *Myriophyllum aquaticum* treating municipal wastewater. Sci Tot Environ 842:156827. https://doi.org/10.1016/j.scitotenv.2022.156827

Kurniawan SB, Imron MF, Mustofa RR, Najiya D, Said NSM, Buhari J et al (2025) Phytotreatment of tofu effluent using water lettuce (Pistia stratiotes) and potential of biogas production from resultant biomass. J Water Process Eng 69:106672. https://doi.org/10.1016/j.jwpe.2024.106672

Kyambadde J, Kansiime F, Gumaelius L, Dalhammar G (2004) A comparative study of *Cyperus papyrus* and *Miscanthidium violaceum* based constructed wetlands for wastewater treatment in a tropical climate. Water Res 38:475–485. https://doi.org/10.1016/j.watres.2003.10.008

Li XN, Song HL, Li W, Lu XW, Nishimura O (2010) An integrated ecological floating-bed employing plant, freshwater clam and biofilm carrier for purification of eutrophic water. Ecol Eng 36:382–390. https://doi.org/10.1016/j.ecoleng.2009.11.004

Lima LKS, Pelosi BT, Silva MGC, Vieira MGA (2013) Lead and chromium biosorption by *Pistia stratiotes* biomass. Chem Eng Trans 32:1045–1050

Lu HL, Ku CR, Chang YH (2015) Water quality improvement with artificial floating islands. Ecol Eng 74:371–375. https://doi.org/10.1016/j.ecoleng.2014.11.013

Luca GAD, Mufarrege MM, Hadad HR, Maine MA (2019) Nitrogen and phosphorus removal and *Typha domingensis* tolerance in a floating treatment wetland. Sci Tot Environ 650:233–240. https://doi.org/10.1016/j.scitotenv.2018.09.042

Lynch J, Fox LJ, Owen JS, Sample DJ (2015) Evaluation of commercial floating treatment wetland technologies for nutrient remediation of stormwater. Ecol Eng 75:61–69. https://doi.org/10.1016/j.ecoleng.2014.11.001

Mahapatra S, Samal K, Dash RR (2022) Waste stabilization pond (WSP) for wastewater treatment: a review on factors, modelling and cost analysis. J Environ Manage 308:114668. https://doi.org/10.1016/j.jenvman.2022.114668

McAndrew B, Ahn C (2017) Developing an ecosystem model of a floating wetland for water quality improvement on a stormwater pond. J Environ Mange 202:198–207. https://doi.org/10.1016/j.jenvman.2017.07.035

Miretzky P, Saralegui A, Fernández CA (2006) Simultaneous heavy metal removal mechanism by dead macrophytes. Chemosphe 62:247–254. https://doi.org/10.1016/j.chemosphere.2005.05.010

Ndimele PE, Kumolu-Johnson CA, Chukwuka KS, Ndimele CC, Ayorinde OA, Adaramoye OR (2014) Phytoremediation of iron (Fe) and copper (Cu) by water hyacinth (*Eichhornia crassipes* (Mart.) Solms). Trends Appl Sci Res 9:485–493. https://scialert.net/abstract/?doi=tasr.2014.485.493

Qiu Y, Ji Y, Tian Y, Li H, Li J, Li Z, Liao M, Liu G, Feng Y (2023) Engineering demonstration of the remediation of urban water using a novel MES enhanced ecological floating bed: from construction to long-term performance. Chem Eng J 454:140024. https://doi.org/10.1016/j.cej.2022.140024

Saeed T, Paul B, Afrin R, Muyeed AA, Sun G (2016) Floating constructed wetland for the treatment of polluted river water: a pilot scale study on seasonal variation and shock load. Chem Eng J 287:62–73. https://doi.org/10.1016/j.cej.2015.10.118

Samal K, Dash RR (2021) Modelling of pollutants removal in integrated vermifilter (IVmF) using response surface methodology. Clean. Eng. Technol. 2:100060. https://doi.org/10.1016/j.clet.2021.100060

Samal K, Dash RR (2024) Experiments and modeling to develop a *Pistia stratiotes* based floating vegetated system (FVS) for the removal of heavy metals (Pb, Zn, Cr, Cu, Ni). Sci Tot Environ 926:171981. https://doi.org/10.1016/j.scitotenv.2024.171981

Samal K, Dash RR, Bhunia P (2017) Treatment of wastewater by vermifiltration integrated with macrophyte filter: a review. J Environ Chem Eng 5:2274–2289. https://doi.org/10.1016/j.jece.2017.04.026

Samal K, Kar S, Trivedi S (2019) Ecological floating bed (EFB) for decontamination of polluted water bodies: design, mechanism and performance. J. Environ. Managem. 251:109550. https://doi.org/10.1016/j.jenvman.2019.109550

Samal K, Kar S, Trivedi S, Upadhaya S (2021) Assessing the impact of vegetation coverage ratio in a floating water treatment bed of *Pistia stratiotes*. SN App Sci 3. https://doi.org/10.1007/s42452-020-04139-2

Samal K, Naushin Y, Kumari P (2020) Challenges in the implementation of phyto fuel system (PFS) for wastewater treatment and harnessing bio-energy. J Environ Chem Eng 8:104388. https://doi.org/10.1016/j.jece.2020.104388

Samal K, Samal SR, Mishra S, Nayak JK (2024) Sources, transport, and accumulation of synthetic microfiber wastes in aquatic and terrestrial environments. Water 16(16):2238. https://doi.org/10.3390/w16162238

Samal K, Trivedi S (2020) A statistical and kinetic approach to develop a floating bed for the treatment of wastewater. J Environ Chem Eng 8:104102. https://doi.org/10.1016/j.jece.2020.104102

Swarnalatha K, Radhakrishnan B (2015) Studies on removal of Zinc and Chromium from aqueous solutions using water Hyacinth. Pollut 1:193–202. https://doi.org/10.7508/pj.2015.02.007

Syukor AR, Sulaiman S, Siddique MNI, Zularisam AW, Said MIM (2016) Integration of phytogreen for heavy metal removal from wastewater. J Clean Prod 112:3124–3131. https://doi.org/10.1016/j.jclepro.2015.10.103

Tang X, Chen Y, Zheng W, Chen L, Liu H, Li M, Yang Y (2024) Enhancing neonicotinoid removal in recirculating constructed wetlands: The impact of Fe/Mn biochar and microbial interactions. J Hazard Mater 476:135139. https://doi.org/10.1016/j.jhazmat.2024.135139

Verma R, Suthar S (2015) Lead and cadmium removal from water using duckweed-*Lemna gibba* L.: impact of pH and initial metal load. Alex Eng J 54:1297–1304. https://doi.org/10.1016/j.aej.2015.09.014

Wibowo YG, Nugraha AT, Rohman A (2023) Phytoremediation of several wastewater sources using *Pistia stratiotes* and *Eichhornia crassipes* in Indonesia. Environ Nanotechnol Monitor Manage 20:100781. https://doi.org/10.1016/j.enmm.2023.100781

Wu W, Hu X, Xie H, Hu Z, Liang S, Guo Z (2024) Treatment of total petroleum hydrocarbon by free water surface constructed wetland: effects of different combinations of submerged and free-floating plants. Desalination Water Treat 100908. https://doi.org/10.1016/j.dwt.2024.100908

Xie P, Zahoor F, Iqbal SS, Zahoor US, Noman M, Din ZU, Yang W (2022) Elimination of toxic heavy metals from industrial polluted water by using hydrophytes. J Clean Prod 352:131358. https://doi.org/10.1016/j.jclepro.2022.131358

Xu F, Peng Y, Gu X, Sun S, Li A, He S (2024) Revealing sulfur-iron coupling mechanism for enhanced autotrophic denitrification in ecological floating beds. Biores Technol 402:130800. https://doi.org/10.1016/j.biortech.2024.130800

Yatoo AM, Hamid B, Sheikh TA, Ali S, Bhat SA, Ramola S et al (2024a) Global perspective of municipal solid waste and landfill leachate: generation, composition, eco-toxicity, and sustainable management strategies. Environ Sci Pollution Res 31(16):23363–23392

Yatoo AM, Ali MN, Baba ZA, Alsohim AS, Muthukumaran M, Sayyed RZ (2024b) Effect of macrophyte biomass-based vermicompost and vermicompost tea on plant growth, productivity, and biocontrol of Fusarium wilt disease in tomato. Biocatal Agric Biotechnol 103320

Yatoo AM, Suthar S, Elrys AS, Ali MN, Hussain S (2025) Management of aquatic macrophytes by vermicomposting: a green technology for organic resource management, nutrient recovery, and agricultural sustainability. J Soil Sci Plant Nutrition 1–18

Zhang L, Sun Z, Xie J, Wub J, Cheng S (2018) Nutrient removal, biomass accumulation and nitrogen-transformation functional gene response to different nitrogen forms in enhanced floating treatment wetlands. Ecol Eng 112:21–25. https://doi.org/10.1016/j.ecoleng.2017.12.021

Zhao F, Xi S, Yang X, Yang W, Li J, Gu B, He Z (2012) Purifying eutrophic river waters with integrated floating island systems. Ecol Eng 40:53–60. https://doi.org/10.1016/j.ecoleng.2011.12.012

Zhu LD, Li ZH, Ketola T (2011) Biomass accumulations and nutrient uptake of plants cultivated on artificial floating beds in China's rural area. Ecol Eng 37:1460–1466. https://doi.org/10.1016/j.ecoleng.2011.03.010

Zhu Z, Wang Y, Han XY, Wang WH, Li HM, Yue ZQ, Chen W, Xue FR (2023) Strengthen the purification of eutrophic water and improve the characteristics of sediment by functional ecological floating bed suspended calcium peroxide and sponge iron jointly. J Environ Mange 325:116610. https://doi.org/10.1016/j.jenvman.2022.116610

Chapter 3
Remote Sensing: Utilizing Satellite Imagery to Monitor Wastewater Pollution

Prem Shankar and Pragati Singh

Abstract The aquatic environment, water resources, and human health face a significant threat from water pollution. To address this issue effectively, comprehensive, dynamic, and precise monitoring of water quality is essential. Remote sensing technologies are commonly utilized for gathering water quality data due to their temporal and spatial advantages. Unmanned aerial vehicles (UAVs), multispectral sensors, and hyperspectral sensors have all enhanced data access and retrieval techniques, which have significantly boosted remote sensing potential for water quality monitoring. This chapter explores the application of remote sensing to water quality evaluation, emphasizing novel developments in data sources and retrieval methods. This chapter discusses methods for retrieving water quality measurements, including total suspended solids (TSM), chlorophyll-a (Chl-a), colored dissolved organic matter (CDOM), chemical oxygen demand (COD), total nitrogen (TN), and total phosphorus (TP). The chapter additionally examines key concerns with atmospheric correction, remote sensing data resolution, and the applicability of retrieval models across spatial, temporal, and water complexity domains. Potential solutions to these challenges are proposed. This chapter may serve as a comprehensive reference for upcoming innovations and research in water quality evaluation.

Keywords Wastewater pollution · Remote sensing · Water quality indicators · Satellite data · Geographical information system (GIS) · Modelling

---

P. Shankar (✉)
Department of Forest and Ecology, Indian Institute of Remote Sensing (ISRO), Dehradun, Uttarakhand 248001, India
e-mail: premmathuria1@gmail.com

P. Singh
Department of Environment Management, Indian Institute of Forest Management, Bhopal, Madhya Pradesh 462003, India
e-mail: psinghfri21@iifmbhopal.edu.in

A. M. Yatoo et al. (eds.), *Wastewater Monitoring and Management by Advanced Approaches*, Advances in Wastewater Research,
https://doi.org/10.1007/978-981-95-2601-7_3

3.1 Introduction

Water-related concerns, including pollution, water scarcity, and other related problems, have been recognized in the Global Risks Report 2020 as the sixth most significant worldwide issue. In particular, wastewater pollution has become a challenging and complex issue that poses serious risks to the scientific and economic foundations of many nations (Han et al. 2016). Effective monitoring of wastewater pollution is necessary in order to conduct thorough evaluations of the quality of the water at the regional and national levels. Therefore, the effective resolution of these issues depends on the deployment of accurate and timely water quality monitoring technology (Syeed et al. 2023; Swain and Sahoo 2017).

Although accurate, traditional methods for measuring water quality parameters are typically time-consuming, expensive, and labor-intensive, which limits their applicability for large-scale monitoring (Baltodano et al. 2024). In contrast, remote sensing technology offers significant advantages in terms of both spatial and temporal coverage, which makes it useful for tracking the dynamic fluctuations and spatial distribution of water quality (Anand et al. 2024; Kallio et al. 2001; Song et al. 2011; Malahlela 2019). Water quality studies have made use of a variety of spaceborne sensors operating in visible (VIS), infrared (IR), and microwave wavelengths because of the high frequency of data collection and the capacity to cover enormous areas. Sea surface temperature and salinity can only be estimated with limited success using devices like radiometers and synthetic aperture radars operating in the microwave spectrum (Zhang et al. 2003). Meanwhile, a wide variety of spectral remote sensing data has been successfully utilized for retrieving water quality in the VIS and IR spectra (Becker et al. 2019; Pyo et al. 2019; Shen et al. 2020). Since its September 2021 release, Landsat-9 has provided academics all over the world with widespread access to data. Due to their flexibility, mobility, and capacity to carry a variety of spectrum sensors, unmanned aerial vehicles (UAVs) have become a popular airborne instrument for water quality monitoring in addition to satellite-based platforms (Ying et al. 2021; Cheng et al. 2020; Mohsan et al. 2023).

Understanding the connection between the amounts of water constituents and the scattering signals, such as water-leaving radiance, that are picked up by remote sensors is essential to retrieving water quality. Retrieval models can be created for various optically active components (OACs) by utilizing the connection between remote sensing reflectance and inherent optical characteristics (IOPs). IOPs are inherent optical characteristics of water that are only influenced by the makeup of the water body and are not impacted by external factors. This makes it possible to directly retrieve parameters related to water quality, including suspended matter (SM), colored dissolved organic matter (CDOM) (Huang et al. 2024), and chlorophyll-a (Chl-a) (Palmer et al. 2015; Wang et al. 2018; Li et al. 2018). Chemical oxygen demand (COD), dissolved oxygen (DO), total nitrogen (TN), total phosphorus (TP), and other nonoptically active chemicals do not display direct optical properties. On the other hand, these can be approximated using artificial intelligence (AI) methods or by analyzing the connections between various water components (Sun et al. 2014; El

Din et al. 2017). Since interactions between the atmosphere, water surface, and water body affect radiation at different wavelengths, accurately obtaining multiple data points related to water quality necessitates the intricate calculation of water-leaving reflectance. There are four main types of remote sensing-based water quality retrieval techniques: physical, semi-empirical, empirical, and intelligent models. Notably, one kind of empirical model is AI-based models, which depend on statistical methods.

Significant progress in remote sensing-based water quality retrieval has been documented in several reviews, with particular attention paid to the utilization of multi-source remotely sensed data and different retrieval techniques (Wang et al. 2019a, b; Dörnhöfer and Oppelt 2016; Sagan et al. 2020; Chawla et al. 2020). Despite recent advancements, a comprehensive evaluation remains crucial to fully grasp the current state of water quality research, particularly given the pressing demand for sustainable development. This study provides an in-depth examination most recent progress in optical remote sensing techniques for extracting water quality parameters, encompassing data origins, retrieval frameworks, and computational methods. It also addresses critical challenges, such as spectral interference among water quality indicators, the spatial and temporal applicability of inversion models, and atmospheric correction. This chapter proposes remedies to these obstacles in remote sensing methods to evaluate the quality of the water.

3.2 Spatial and Temporal Databases for Remote Sensing-Based Water Quality Assessment

3.2.1 Remote Sensing Data from Satellites

3.2.1.1 Multispectral Data

Water quality retrieval by remote sensing makes extensive use of multispectral data, which usually consists of three to ten bands. This comprises information from sensors like National Oceanic and Atmospheric Administration's (NOAA) Advanced Very High Resolution Radiometer (AVHRR) (Batur and Maktav 2018; Odermatt et al. 2018), European Space Agency's (ESA) Sentinel-2, Thematic Mapper (TM), Enhanced Thematic Mapper Plus (ETM+), Operational Land Imager (OLI), and France's Satellite Pour l'Observation de la Terre (SPOT) satellites. Researchers have also used China's Gaofen (GF) and Huanjing (HJ) series data more recently (Shang and Shen 2016; Li et al. 2015). The most common source of data for monitoring water quality parameters is the Landsat series, due to its accessibility and availability with regard to geographic, temporal, and spectral resolution. These data are particularly useful for assessing total suspended matter (TSM), chemical oxygen demand (COD), and total phosphorus (TP) concentrations (Vakili and Amanollahi 2020; Wang et al. 2019a, b).

3.2.1.2 Hyperspectral Data

Multiband data with high spectral resolution (usually around 0.01 μm) is provided by hyperspectral satellites. Hyperspectral datasets have been utilized in assessing water quality, including Hyperion data from the United States and Hyperspectral Imager System (HIS) data from China's HJ-1 satellite (Cao et al. 2018). Comparing these datasets to multispectral data, their high spectral resolution provides a multitude of bands that may be possibly and selectively exploited to build water quality retrieval algorithms, hence enabling a more precise differentiation of spectral mixing. This significantly enhances the retrieval precision of algorithms for water quality parameters, as well as shows great promise for wider application (Kudela et al. 2015; Sudduth et al. 2015; Hestir et al. 2015).

3.2.2 Data from Nonsatellite Remote Sensing

In recent years, nonsatellite remote sensing data has been increasingly utilized for assessing water quality. For these reasons, a variety of aerial remote sensing data and ground-based spectral measurements are useful sources. The speedy advancement of unmanned aerial vehicle (UAV) technology has resulted in the creation of lightweight UAV systems that are very useful and practical instruments for controlling aquatic ecosystems. These systems come with multispectral cameras, hyperspectral sensors, infrared detectors, and light detection and ranging (LiDAR) (Becker et al. 2019; Ouma et al. 2018). The Compact Airborne Spectrographic Imager (CASI) in Canada and the Airborne Visible/Infrared Imaging Spectrometer (AVIRIS) in the USA both have 48 channels and are widely used for monitoring aquatic ecosystems. In a similar vein, data from China's Chinese Imaging Spectrometer (CIS) on aircraft have been used to track aquatic ecosystems. However, there are some significant drawbacks to using data from aircraft, including high operating costs and possible hazards. Portable spectrometers provide an adaptable substitute for airborne spectrometers, enabling the observation of water surface reflectance within tunable spectral bands. The portable field spec radiometer Spectra Vista Corporation High-Resolution 1024 (SVC HR-1024), for example, has 1024 bands that span a spectral range of 350–2500 nm (Li et al. 2015). Other popular portable spectrometers used in water quality control include the L1-1800, Analytical Spectral Devices (ASD) Field Spec Pro, and Portable Spectroradiometer (PSR)-1100. Higher spectral and spatial resolutions are easily achieved by these nonsatellite spectrometers, which also produce continuous spectral curves of ground objects. They are also less impacted by air interference than data from satellites. Despite these benefits, nonsatellite data collecting methods, such as flying or taking measurements on the ground, are sometimes expensive and have a limited capability for large-scale monitoring of vast quantities of water, such as rivers and lakes. Table 3.1 lists a few widely used satellites with wide spectral ranges for retrieving water quality. While near infrared (NIR) is typically utilized for clearer waters and shortwave infrared (SWIR) for turbid waters, Liu's research indicates that

Table 3.1 Satellites available to retrieve water quality via remote sensing

	Satellite sensor	Spatial resolution (m)	Temporal resolution (Day)	Bands	Data availability
Multispectral	Sentinel-2A	10/20	5	13	2015–Present
	Sentinel-2B	10/20	5	13	2017–Present
	Sentinel-3 OLCI	500–1000	27	20	2016 – Present
	Landsat-9	30	16	7	2021–Present
	Landsat-8 OLI	30	16	7	2013–Present
	ASTER	15–90 m		3	1999–Present
	SPOT 6	1.5 m	26	4	2012–Present
	MODIS	250–500–1000	0.5	9	1999–Present
	WorldView-2/3	1.85/1.24	1.1	8	2009/2014–Present
	NOAA-16 AVHRR	1100–4000	9	6	2000–Present
	MERIS	300–1200	1	15	2002–Present
Hyper-spectral	PRISMA	30		66/173	2019–Present
	ZY1-02D	30	3	166	2019–Present
	OHS	10	2	32	2018–present
	GF5-AHSI	30	3	330	2018–present
	VIIRS	375–750	0.5	22	2011–Present
	PROBA CHRIS	18–36	7	19	2001–Present
	Hyperion	30	16	42	2000–Present
UAV sensors	GaiaSky-mini	0.04	–	176	–
	ZK-VNR-FPG480	0.09	–	270	–

the combination of NIR and SWIR can enhance the effectiveness of water quality monitoring (Liu et al. 2019).

3.3 Mode of Water Quality Parameters Retrieval

3.3.1 Empirical Mode

The empirical mode (EM), a type of correlation-based statistical regression model, was constructed using ground-measured water quality indicators and reflectance from optimal spectral bands or combinations of bands (Ouma et al. 2018). Its main benefit is that it's simple and easy to use. By choosing more exact spectral channels, it can potentially boost the precision of retrieving water quality parameters (Li et al.

2017). The empirical form does, however, have three major disadvantages. First, simple empirical models usually fall short of the precision requirements for determining water quality parameter amounts due to the complex interactions of variables, including phytoplankton pigments, coarse dissolved and suspended matter (CDSM), and colored dissolved organic matter (CDOM), which introduce significant uncertainty in the inversion process (Palmer et al. 2015). Second, models created for one body of water could not apply to other areas (Gao et al. 2015); empirical models have weak generalizability across many geographies and historical periods. Lastly, to maintain a relatively high level of accuracy, the empirical model necessitates a large field survey of water quality data.

3.3.2 Analytical Mode

The analytical mode (AM) establishes the connection between water quality characteristics and the radiance or reflectance of radiation released from water surfaces by simulating the flow of light through the atmosphere and water using bio-optical and radiative transfer models. A broad framework explaining the connection between apparent optical reflectance and water body parameters was presented by Gordon et al. in 1975 (Gordon et al. 1975). An essential starting point for the future advancement of analytical models in water quality retrieval is this quasi-single-scattering diffuse reflectance model.

$$\text{Rrs} = f\left[\frac{bb(\lambda)}{a(\lambda) + bb(\lambda)}\right]$$

Rrs stands for reflectance, λ for wavelength, a for water's total absorption coefficient, bb for water's total backscattering coefficient, and f for the functional relation.

Based on sample data, Gilerson calculated the chlorophyll-a (Chl-a) content in lakes in eastern Nebraska using the bio-optical inversion model (Gao et al. 2015). Setting up these models is difficult, though, because a large number of variables must be measured, such as the water's intrinsic optical qualities, surface reflectance features, and water quality metrics. Further complicating the model-building process are discrepancies in the spectral resolutions of measurements made on the ground and by satellite sensors. This makes it challenging to achieve precise quantitative remote sensing inversion in analysis mode, and its practical applicability is still restricted (Palmer et al. 2015; Le et al. 2009a, b).

3.3.3 Semi-empirical Mode

Both the analytical and empirical modes' components are integrated in the semi-empirical mode (SEM). Researchers can determine water quality variables using a minimal amount of on-site data and reflectance or radiance measurements by using a framework based on robust analytical models. Scientists have successfully utilized this method to accurately assess various water quality components, including those related to visibility and turbidity, suspended matter (SM), chlorophyll-a (Chl-a), and CDOM (Hunter et al. 2010; Zhou et al. 2013). For instance, Lathrop (Lathrop and Lillesand 1989) observed that the SPOT-1 R band and NIR band demonstrated heightened sensitivity to suspended matter concentration. This discovery enabled the calculation of SM concentrations within the SEM framework by using the R–G band ratio and NIR band. The semi-empirical approach offers enhanced transferability compared to purely empirical methods by considering the optical properties of water quality parameters.

3.3.4 Artificial Intelligence (AI) Mode

Using implicit algorithms that produce explicit equations, the AI mode (AIM) is distinct from the empirical, analytical, and semi-empirical modes discussed above. AIM demonstrates proficiency in capturing both linear and nonlinear associations due to the complex interrelationships among sediment deposits, water quality indicators, and surface phenomena. This approach proves more adept at managing the intricate and diverse background influences compared to conventional statistical or physical models. To retrieve water quality, researchers have successfully used a variety of AI techniques, including neural networks (NN), random forest (RF), and support vector machines (SVM), with largely good results (Becker et al. 2019; Chebud et al. 2012). One of them that is frequently utilized is the artificial neural network (ANN) model. The artificial neural network (ANN) model utilizes input data from various bands of remote sensing, including gray values, radiance values, or reflectance, to generate water quality information. The neural network approach's adaptability as a nonlinear approximation technique makes it especially useful for obtaining parameters related to water quality. This is most valuable since it is often challenging to establish a simple linear correlation between the concentrations of components in water bodies and their spectral reflectance. Numerous training samples and a great deal of expertise are necessary when building a neural network for water quality retrieval; otherwise, problems like underfitting or overfitting may occur, severely reducing the inversion accuracy of the model (Schmidhuber 2015). Support vector machine (SVM) models, on the other hand, perform better with smaller sample sizes. SVMs ensure global optimality by solving convex optimization issues and addressing the computing difficulties posed by high-dimensional spaces through the application of kernel functions.

SVMs are especially useful in solving problems with high dimensionality, nonlinearity, short sample sizes, and local minima (Singh et al. 2011). Nevertheless, adding more training samples can make the SVM algorithm implementation more difficult and time-consuming. Convolutional neural network (CNN) models have been used recently to retrieve water quality parameters. For example, Yang used CNN and a matrix extension approach to generate a chlorophyll-a (Chl-a) inversion model that produces 5×5 grid pixel blocks. In (Yang et al. 2022), the matrix expansion method and the CNN's structure are explained in detail. CNN use has aided in lowering the amount of network parameters, reducing overfitting problems, and speeding up computation.

3.4 Development in Querying Water Quality Parameters

3.4.1 Total Suspended Matter

A critical factor in determining the health of water is the concentration of total suspended matter (TSM). Increased TSM levels in inland lakes and estuaries can alter the visual characteristics of the water, such as the euphotic zone's intensity, which can impact aquatic plant development and the water bodies' primary productivity (Havens 2003; Hou et al. 2017). Numerous techniques to estimate TSM from satellite data have been devised, and successful real-world implementations have also been documented (Chen et al. 2015; Caballero et al. 2014). Table 3.2 lists several models for TSM inversion. Dekker, for instance, found that the green (G) and red (R) bands (TM, HRV) show a strong association with TSM concentration, following an exponential relationship ($r^2 = 0.99$) (Dekker et al. 2002). This was noted in lakes that were turbid and had TSM concentrations between 5 and 50 g/m^3. According to Dörnhöfer and Oppelt (2016), Wang used Landsat data to do a semi-empirical analysis of variations in SM concentration in the Zhuhai estuary between 1987 and 2015. The findings revealed spatiotemporal swings in SM levels throughout the estuary coast. Furthermore, positive findings have been obtained when empirical approaches using SPOT-6 data have been utilized to determine SM concentrations in narrow, mountainous rivers (Isidro et al. 2018).

Reflectance in the 700–900 nm and 580–680 nm spectral regions is ideal for retrieving TSM from remote sensing data because it is extremely sensitive to variations in TSM concentration. By utilizing these features, Kallio studied lakes in southern Finland using Airborne Imaging Spectrometer for Applications (AISA) imagery data, and he created a useful technique for calculating TSM concentration using single-band reflectance in the range of 705–714 nm (Kallio et al. 2001). Miller used a statistical method to quantify the concentration of suspended matter in the northern Gulf of Mexico in offshore habitats using the Moderate Resolution Imaging Spectroradiometer (MODIS) data's channel 1 and channel 2 bands (Miller and McKee 2004). Furthermore, at the mouth of the Garonne River in France,

Table 3.2 Inversion models of water quality parameter (TSM, Chl-a, CDOM, COD, TN, and TP)

	Data	Mode	Models	R^2	References
TSM	Sentinel-2	EM	Neural network	0.80	Saberioon et al. (2020)
	Landsat TM and OLI	SEM	$\log_{(CTSM)} = \frac{-b \mp \sqrt{b^2 - 4ac - \frac{\log R(865)}{\log R(655)}}}{2a}$	–	Wang et al. (2018)
	MODIS	EM	$C_{TSM} = 1.91 * 1140.25 R(645)$	0.89	Miller and McKee (2004)
	MODIS	IAM	Neural network (MBP)	0.72	Chen et al. (2015)
	SPOT 6	EM	$C_{TSM} = 947.579(R(660))^{4.07}$	0.65	Isidro et al. (2018)
Chl-a	MERIS	SEM	$C_{Chl\text{-}a} = -(16.2 \pm 1.8) + (136.2 \pm 3.2) * [R^{-1}(667) * R(748)$	0.93	Dall'Olmo and Gitelson (2005)
	MERIS	EM	$C_{Chl\text{-}a} = 61.32 R^{-1}(665) * R(708) - 37.94$	0.97	Moses et al. (2009)
	Airborne	AIM	Convolution neural networks	0.92	Pyo et al. (2019)
	Sentinel MSI	EM	Neural network	0.85	Saberioon et al. (2020)
	TM &OLI	EM	$C_{Chl\text{-}a} = -11.64 * NDVI 26.89 * GNDVI + 8.01$	0.97	Rodríguez-López et al. (2020)
CDOM	Sentinel-2	EM	$C_{Chl\text{-}a} = 61.32 R^{-1}(665) * R(708) - 37.94$	0.53	Shang et al. (2021)
	GOCI	EM	$C_{CDOM}(400) = 0.2355(R(412)/R(555))^{-1.34}$	0.85	Moon et al. (2012)
	VIIRS	EM	$C_{CDOM}(412) = 3.18(R(551)/R(671))^{-1.046}$	0.87	Joshi et al. (2017)
	MODIS	EM	$C_{CDOM}(412) = 0.507(R(469) + R(645))/R(555)$	0.50	Zhang et al. (2021)
	OLI	EM	$C_{CDOM}(400) = 40.75 e^{-2.463(R(563)/R(655))}$	0.83	Chen et al. (2017)
COD	TM	EM	$C_{COD}(400) = e^{(0.367+1.2454 \ln R(560)/R(660))}$	0.66	Yang et al. (2011)
	TM	EM	$C_{COD} = 2.76 - 17.27 R(475) + 72.15 R(560) - 12.11 R(660)$	0.79	Wang et al. (2004)
	OLI	IAM	Neural network	0.92	El Din et al. (2017)
TN and TP	OLI	EM	$C_{TN} = 0.11 + 2.07(R(655)/(R(562 + R(865)))^{1.264}$	0.84	Li et al. (2017)

(continued)

Table 3.2 (continued)

	Data	Mode	Models	R^2	References
	OLI	EM	$C_{TP} = -0.2 + 0.005e^{(R(443)+R(562)+R(655))/R(583)}$	0.8	Li et al. (2017)
	MODIS	IAM	Genetic programming (C_{TP})	0.58	Chang et al. (2013)
	TM	IAM	Neural network (C_{TP})	0.95	Chebud et al. (2012)

Doxaran discovered a strong correlation between the reflectivity ratio of 850–550 nm and higher suspended matter content for extremely turbid waters (< 985 g/m^3). This finding highlights the possibility of the near-infrared band for estimating suspended matter concentration (Doxaran et al. 2002). Table 3.2 lists different TSM inversion models, where MBP stands for matrix backpropagation neural network, R (λ) stands for remote sensing reflectance at wavelength λ, and C_{TSM} is for total suspended matter concentration. The studies' reported TSM content ranges from 0.5–55 μg/L, and their R^2 values vary from 0.89 to above 0.65.

3.4.2 Chlorophyll-a

The distribution of plankton biomass is mostly determined by chlorophyll-a (Chl-a). This parameter is essential for evaluating eutrophication and primary productivity in water bodies. It has been successful to estimate the amount of Chl-a using both hyperspectral and multispectral data (Pyo et al. 2019; Feng et al. 2015). Reflectance in the red (R) and green (G) bands usually increases with increasing Chl-a concentration, while reflection in the blue band usually decreases. As a result, the concentration of Chl-a determines which band is best for estimation (Pulliainen et al. 2001). Gitelson observed that when the concentration of Chl-a increases, the peak reflectance shifts from roughly 680–715 nm, with an increase in peak amplitude. He underlined the significance of the 700 nm peak in figuring out the amount of Chl-a present in inland waters (Wang and Chen 2024). Palmer used Medium Resolution Imaging Spectrometer (MERIS) data from 2007 to 2012 to develop a method for determining the concentration of Chl-a and evaluating its geographical and temporal changes. Palmer concentrated on large, turbid, and complex shallow lakes (Chen et al. 2013). Furthermore, Boucher used Landsat images from 2013 to 2015 to assess several algorithms that were already in use across 192 lakes. According to his research, at the regional level, remote sensing can be a practical and cost-effective method for monitoring Chl-a in lakes (Boucher et al. 2018). We have summarized four types of models for retrieving Chl-a concentration.

1. Band ratio model: The band ratio model estimates Chl-a concentration using the reflectance ratios between bands. For instance, Kevin developed a quantitative inversion model using the reflectance ratio of 672–704 nm (Ruddick et al. 2001). Sampas utilized aerial hyperspectral and MERIS imagery data to analyze various channel ratios, identifying an optimal band ratio algorithm for Chl-a retrieval with a classification accuracy of 78% (Koponen et al. 2002).
2. First-order differential model: The first-order differential model, introduced by Rundquist, mitigates noise influence by analyzing spectral inflection points and reflectance extremes (Wang and Chen 2024). The ideal wavelength for differential estimation of high Chl-a concentrations was determined by Luo to be 690 nm. (Han and Rundquist 1997).
3. 3 Three-band model: The three-band model utilizes three spectral bands to calculate Chl-a content, offering improved precision by reducing the influence of water components' interactions (Gitelson et al. 2003). A model for turbid waters that is semi-analytical in nature showed high correlation between simulated and measured Chl-a values using MERIS data (Gitelson et al. 2008). Gitelson's three-band model, based on eastern Nebraska lake data, explained over 89% of Chl-a variation (Gitelson et al. 2009). Le's enhancement included a fourth band to address absorption and backscattering effects (Le et al. 2009a, b).
4. Artificial intelligence model: Zhang et al. (2002) used a neural network algorithm for Chl-a retrieval, achieving high accuracy (Zhang et al. 2002). Singh's support vector machine models showed superior performance in Chl-a estimation compared to classification models (Singh et al. 2011). Song showed that neural network models performed better in spatial transferability than three-band models using a genetic algorithm in conjunction with spectral and observational information (Song et al. 2013). The convolutional neural network algorithm provided reliable Chl-a inversion results with an $R^2 > 0.81$ (Yang et al. 2022).

Table 3.2 provides specific information on Chl-a concentration inversion models. The concentration of Chl-a is represented by $C_{\text{Chl-a}}$, while the remote sensing reflectance at wavelength λ is shown by $R(\lambda)$. The geographical application of empirical and semi-analytical techniques, as well as the difficulties in accurately correcting for atmospheric correction, which can bring errors and complexities, often pose limitations. These restrictions are addressed by developments in artificial intelligence (AI) algorithms, which lessen the requirement for substantial prior knowledge. Table 3.2 lists the inversion models together with their corresponding Chl-a concentrations, which range from 0 to 409 mg/m^3. The models' R^2 values, which show how well they work in different scenarios, range from 0.84 to 0.97.

3.4.3 Colored Dissolved Organic Matter

The complex composition of CDOM has a major impact on chemical reactions, underwater light fields, and water color. As CDOM concentration rises, it may absorb more visible and ultraviolet light, expanding its absorption into the blue band at 440 nm. Kowalczuk used the remote sensing reflectance ratio between 490 and 590 nm as the individual variable (Kowalczuk et al. 2005) to construct an inversion model for the Baltic Sea's CDOM concentration. An empirical model for the Mississippi River was proposed by Eurico and Richard, who found a high correlation between reflectance ratios (510/555) (D'Sa and Miller 2003) and CDOM absorption at 412 nm. Kutser used quasi-synchronous Advanced Land Imager (ALI) data for lakes in southern Sweden and Finland to study CDOM absorption at 420 nm. Based on this analysis, he developed a retrieval algorithm that uses bands ranging from 525–605 nm to 630–690 nm (Kutser et al. 2005). A regression approach for retrieving yellow matter was developed by testing several remote sensing spectral reflectance ratios. Moon et al. used bands at 412 nm and 555 nm from Geostationary Ocean Color Imager (GOCI) images (Moon et al. 2012) to create a CDOM absorption coefficient model. In Taihu Lake, Duan looked at spectral slopes at 443 nm and CDOM absorption coefficients to study black water bloom events (Duan et al. 2014).

According to recent research, CDOM absorption coefficients for inland waterways can be successfully reversed using Landsat-8 data (Herrault et al. 2016; Olmanson et al. 2016; Chen et al. 2017). Table 3.2 shows a variety of CDOM concentration inversion models, where $R(\lambda)$ is the wavelength-dependent reflectance for remote sensing, C_{CDOM} stands for SM concentration, and MBP is for matrix back propagation neural networks. According to the models, R^2 values vary from 0.63 to 0.97, while CDOM concentrations range from 0.08 to 11.3/m.

3.4.4 Chemical Oxygen Demand

Since nothing is known about the optical characteristics of COD in aquatic bodies, bio-optical models are inappropriate for assessing COD. Wang used an empirical method that combined Landsat TM satellite images with field-measured data to recover COD in Shenzhen's reservoir. Using satellite data for COD estimate is achievable, according to the results, at a significance level of $p = 0.01$ (Wang et al. 2004). The value of the R^2 was 0.791. Tao built on this empirical approach by developing an enhanced CODM index that used Landsat-8 OLI bands B1, B7, and the ratio of B4/B3 to quantitatively estimate the CODM content in Taihu Lake with a relative error of 7.8% (Tao et al. 2014). Din used a Support Vector Machine (SVM) model and a Back-Propagation Neural Network (BPMN) model for COD retrieval based on Landsat-8 data because of the intricate nonlinear relationship between COD concentration and remote sensing reflectance, which is impacted by the interactions among different water quality parameters. According to Yang et al. (2011), the accuracy of the BPMN model was superior to that of the SVM model ($R^2 = 0.918$). Table 3.2 displays several COD concentration inversion models, where CCOD is the COD concentration, and $R(\lambda)$ is the wavelength-dependent remote sensing reflectance. The COD concentrations in these models vary from 1.6 to 4.8 mg/L, and their R^2 values fall between 0.66 and 0.92.

3.4.5 *TP and TN Concentrations*

Total phosphorus (TP) and total nitrogen (TN) are vital markers of eutrophication in lakes and rivers, indicating serious problems with water pollution that undermine biological processes and threaten global water security (Sherwood and Qualls 2001; Savage et al. 2010). The imprecise spectral response mechanisms present difficulties for precise remote sensing retrieval of TN and TP (Gao et al. 2015), despite their significance. Using statistical techniques and Landsat TM data, Wu created a TP inversion model for the Qiantang River (Wu et al. 2010). To evaluate the nutrient distribution in Lake Champlain (Isenstein and Park 2014), Isenstein used in situ measurements in conjunction with Landsat ETM + data to develop a multivariate linear regression model. Regression models for TN and TP retrieval were successfully established by Liangjiang et al. using the ratio technique, yielding high R^2 values of 0.839 and 0.934, respectively (Liangjiang et al. 2013). Chang et al. examined the TP distribution in Tampa Bay, western Florida (Chang et al. 2013), using MODIS imagery and a genetic algorithm. Furthermore, Chebud et al. evaluated the TP distribution in the Kissimmee River basin (Chebud et al. 2012) using neural network techniques. Table 3.2 shows several inversion models for concentrations of TP and TN, where CTP and CTN stand for TP and TN concentrations, respectively, and $R(\lambda)$ is the wavelength-resolved remote sensing reflectance. In these models, the concentration ranges for TP and TN are 0.02–0.96 mg/L and 0.18–1.78 mg/L, respectively. The R^2 values for TP and TN are 0.58–0.99 and 0.78–0.92, respectively.

3.5 Water Quality Retrieval: Challenges and Possible Solutions

3.5.1 *Atmospheric Correction*

In water quality retrieval, air scattering has a major impact on the water-leaving radiance measured at the top of the atmosphere by a sensor, in addition to bottom absorption and water component parameters (Barnes et al. 2021; Manzo et al. 2018; Sun et al. 2022). Since air molecules scatter most of the signal, aerosol scattering in the atmosphere can account for up to 80–90% of the signal seen by sensors in most ocean remote sensing spectral ranges (Chen et al. 2018). The water-leaving radiance is significantly reduced by this air interference; this makes it challenging to conduct an appropriate water quality analysis. As a result, atmospheric correction is essential for increasing retrieval precision and lowering radiation errors brought on by atmospheric factors. The spatially adaptive fast atmospheric correction (ATCOR) (Sriwongsitanon et al. 2011; Tian et al. 2010), MODerate resolution atmospheric TRANsmission (MODTRAN), and 6S are examples of common atmospheric correction models. Nevertheless, a lot of atmospheric algorithms depend on fixed atmospheric profiles or preestablished average values that ignore temporal and geographical variability. The low flexibility and considerable uncertainty of atmospheric correction algorithms present serious problems to this constraint. According to Warren et al. (2019), there is a need to improve six atmospheric correction algorithms: C2RCC v1.0, Polymer v4.6, Sen2Cor v2.4.0, Acolite v20170718, iCOR v1.0, and l2gen version 7.5.1. For the inversion of water quality metrics to be reliable, atmospheric correction model accuracy is essential.

Since inland waterways present difficulties with atmospheric correction, developing high-precision algorithms is necessary to improve the precision of water quality monitoring (Mao et al. 2013). To boost the retrieval accuracy of water quality (Sterckx et al. 2015), a sensor-generic adjacency preprocessing approach has been developed to address adjacency effects in near-infrared bands induced by bright land and dark water. Effectiveness has also been demonstrated by merging algorithms that mix different bands from multiple remote sensing data sources (Bi et al. 2018). High-resolution sensors on UAV platforms can offer water quality retrievals with less atmospheric influence (Cillero Castro et al. 2020) in addition to satellite data. Thus, a satellite and UAV image reflectance association study is necessary to develop a universal atmospheric correction model.

3.5.2 Remotely Sensed Data Resolution

Remote sensing of water quality and ocean color can be accomplished efficiently using satellite data from sensors like GOCI, MERIS, and MODIS. (Giardino et al. 2019). However, getting data with high spectral, geographical, and temporal precision for inland waters at the same time is quite difficult, particularly for small lakes and narrow rivers (Meng et al. 2015; Shen et al. 2016). Low temporal resolution in high spatial resolution data frequently results in a lack of long-term or dynamic monitoring capabilities, while high temporal resolution in high spatial resolution data cannot offer enough spatial detail (Holderness et al. 2013). Moreover, even though continuous spectral bands are provided for water quality analysis by hyperspectral sensors such as Ocean and Land Color Instrument (OLCI), Huanjing-1A Hyperspectral Imager System (HJ-1A HIS), Hyperspectral Imager for the Coastal Ocean (HICO), and Visible Infrared Imaging Radiometer Suite (VIIRS), they usually lack spatial resolution (Toming et al. 2017). Furthermore, weather has a significant influence on the quality of remote sensing images, particularly during the monsoon season when cloud cover and fog are common.

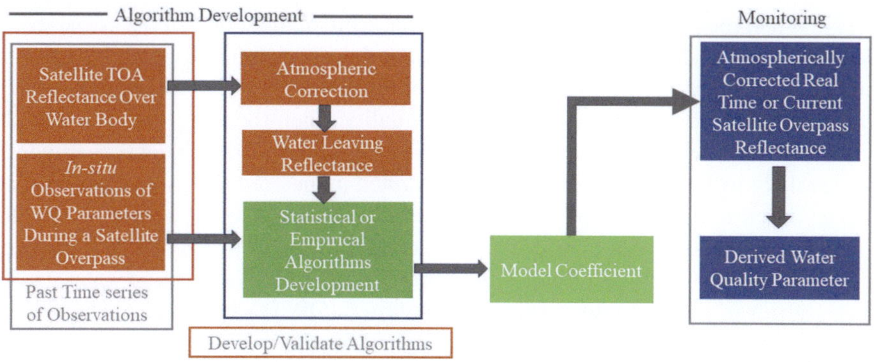

Fig. 3.1 Flow diagram of quantitative techniques for water quality estimation through remote sensing

Progress in data fusion technologies has made it possible to synthesize high-resolution spatiotemporal data from MODIS and Landsat imagery. Several noteworthy methods are the Flexible Spatiotemporal Data Fusion Method (FSDAF) (Wu et al. 2015), the Spatiotemporal Integrated Temperature Fusion Model (STITFM), and the Spatial and Temporal Adaptive Reflectance Fusion Model (STARFM). According to Zhu et al. (2016) and Wang and Atkinson (2018), these methods satisfy the demand for thorough, long-term regional investigations by combining multisource remote sensing data with high geographical and temporal resolutions. To generate high-resolution image sequences that especially meet the monitoring requirements of small lakes and rivers (Chang et al. 2014), spatiotemporal fusion techniques integrate data from several sources. In addition, a method called spatio-temporal-spectral fusion has been put out to improve the quality of data by merging high spectral, temporal, and spatial resolutions from different sources (Huang et al. 2013). Figure 3.1 depicts the flow diagram of quantitative methods for remote sensing-based water quality estimation.

3.5.3 Retrieval Model Applicability

Models of water quality created using remote sensing data show a notable spatial dependency, frequently customized to particular places by utilizing local water quality factors and spectral reflectance. Because of geographical differences in inherent optical properties (IOPs), these models may work well in one location but not well enough in another. As an illustration, Huang's research on three rivers in Qingyang City showed the limitations of empirical models in various geographical areas (Matthews 2011). These models are limited in their seasonal applicability temporally by short-term in situ data. Chl-a and dissolved oxygen levels in water can be affected by variations in climatic variables, such as temperature (Kim et al.

2017). When model optical parameters deviate from actual conditions, errors occur (Sayers et al. 2019). Retrieving water quality is made more difficult by spectral interactions, especially in complex inland waters. It is difficult to distinguish contributions from distinct components of water quality when there are several optically active constituents present because overlapping signals result from these constituents. Retrieval algorithms' accuracy is thus constrained in these environments (Zhou et al. 2020). Given the intricate nature of the elements that make up water quality and how they change over time, improving the applicability of retrieval models requires addressing changes in water types and applying new approaches and datasets. To lower retrieval uncertainties, optical water types (OWTs) have been adapted for usage in inland lakes after initially being created for the world's oceans. The established OWT typology can be applied to a variety of aquatic systems (Spyrakos et al. 2018). Compared to conventional techniques like OC3M, the use of generalized additive models (GAMs) has increased the accuracy of satellite-based Chl-a algorithms, especially in seasonal and magnitude predictions (Wang et al. 2019a, b). The application of machine learning techniques, which can capture intricate nonlinear interactions, is growing. Environmental variables such as water surface temperature (WST) derived from MODIS data are incorporated into models (Li et al. 2021; Reichstein et al. 2019) and are frequently utilized. An OWT framework that integrates many techniques, such as empirical, neural network, and semi-analytical models, has dynamically improved Chl-a retrieval in both space and time (Neil et al. 2019).

High spectral resolution provided by airborne hyperspectral data suggests that optically passive compounds can be monitored by comparing them to active water quality constituents like phosphorus and nitrogen (Niu et al. 2021). Calibration and validation of satellite data by extensive in situ observations are crucial for long-term monitoring. This strategy is demonstrated by the global DYNAQUA project, which tracks water quality over time (Staehr et al. 2022; Yang et al. 2019) by integrating multisensor systems with meteorological and sampling data. The combination of satellite and aerial remote sensing has provided a comprehensive approach for the evaluation of inland water quality (Gholizadeh et al. 2016).

3.6 Conclusion and Future Perspectives

This study examines developments in remote sensing for the retrieval of water quality, alongside an emphasis on data sources, different types of algorithms, models for different water quality metrics, and a discussion of existing issues along with suggested fixes. UAV-based spectrometers and multispectral and hyperspectral data provide high-resolution monitoring in temporal, geographic, and spectral domains. The article highlights four main retrieval modes: analytical, semi-empirical, empirical, and intelligent algorithms. While analytical methods are more precise but more expensive, empirical methods are simpler but have a physical base that limits them. Artificial intelligence (AI) techniques have become popular for recovering nonoptically active compounds. Modern satellite technology and machine learning have made it possible to expand remote sensing capabilities beyond seas to inland waters by advancing models for parameters like SM, Chl-a, S, COD, TN, and TP from empirical to theoretical methodologies. Using bibliometric analysis, prominent authors and works on the topic were found. Sensor performance, atmospheric correction, and spatiotemporal variability are major obstacles that need to be overcome in order to advance both current and future research endeavors.

References

Anand V, Oinam B, Wieprecht S (2024) Machine learning approach for water quality predictions based on multispectral satellite imageries. Ecol Inform 102868

Baltodano A, Agramont A, Lekarkar K, Spyrakos E, Reusen I, van Griensven A (2024) Exploring global remote sensing products for water quality assessment: Lake Nicaragua case study. Remote Sens Appl Soci Environ 36:101331

Barnes BB, Hu C, Bailey SW, Pahlevan N, Franz BA (2021) Cross-calibration of MODIS and VIIRS long near infrared bands for ocean color science and applications. Remote Sens Environ 260:112439

Batur E, Maktav D (2018) Assessment of surface water quality by using satellite images fusion based on PCA method in the Lake Gala, Turkey. IEEE Trans Geosci Remote Sens 57(5):2983–2989

Becker RH, Sayers M, Dehm D, Shuchman R, Quintero K, Bosse K, Sawtell R (2019) Unmanned aerial system based spectroradiometer for monitoring harmful algal blooms: a new paradigm in water quality monitoring. J Great Lakes Res 45(3):444–453

Bi S, Li Y, Wang Q, Lyu H, Liu G, Zheng Z et al (2018) Inland water atmospheric correction based on turbidity classification using OLCI and SLSTR synergistic observations. Remote Sens 10(7):1002

Boucher J, Weathers KC, Norouzi H, Steele B (2018) Assessing the effectiveness of Landsat 8 chlorophyll a retrieval algorithms for regional freshwater monitoring. Ecol Appl 28(4):1044–1054

Caballero I, Morris EP, Ruiz J, Navarro G (2014) Assessment of suspended solids in the Guadalquivir estuary using new DEIMOS-1 medium spatial resolution imagery. Remote Sens Environ 146:148–158

Cao Y, Ye Y, Zhao H, Jiang Y, Wang H, Shang Y, Wang J (2018) Remote sensing of water quality based on HJ-1A HSI imagery with modified discrete binary particle swarm optimization-partial least squares (MDBPSO-PLS) in inland waters: a case in Weishan Lake. Eco Inform 44:21–32

Chang NB, Xuan Z, Yang YJ (2013) Exploring spatiotemporal patterns of phosphorus concentrations in a coastal bay with MODIS images and machine learning models. Remote Sens Environ 134:100–110

Chang NB, Vannah BW, Yang YJ, Elovitz M (2014) Integrated data fusion and mining techniques for monitoring total organic carbon concentrations in a lake. Int J Remote Sens 35(3):1064–1093

Chawla I, Karthikeyan L, Mishra AK (2020) A review of remote sensing applications for water security: quantity, quality, and extremes. J Hydrol 585:124826

Chebud Y, Naja GM, Rivero RG, Melesse AM (2012) Water quality monitoring using remote sensing and an artificial neural network. Water Air Soil Pollut 223:4875–4887

Chen J, Zhang M, Cui T, Wen Z (2013) A review of some important technical problems in respect of satellite remote sensing of chlorophyll-a concentration in coastal waters. IEEE J Selected Top Appl Earth Observations and Remote Sens 6(5):2275–2289

Chen J, Quan W, Cui T, Song Q (2015) Estimation of total suspended matter concentration from MODIS data using a neural network model in the China eastern coastal zone. Estuar Coast Shelf Sci 155:104–113

Chen J, Zhu WN, Tian YQ, Yu Q (2017) Estimation of colored dissolved organic matter from Landsat-8 imagery for complex inland water: case study of Lake Huron. IEEE Trans Geosci Remote Sens 55(4):2201–2212

Chen X, Feng L (2018) Remote sensing of lakes' water environment. In: Comprehensive remote sensing; Liang S, (ed) pp 249–277 Elsevier: Amsterdam, The Netherlands

Cheng K, Chan S, Lee JH (2020) Remote sensing of coastal algal blooms using unmanned aerial vehicles (UAVs). Mar Pollut Bull 152:110889

Cillero Castro C, Domínguez Gómez JA, Delgado Martín J, Hinojo Sánchez BA, Cereijo Arango JL, Cheda Tuya FA, Díaz-Varela R (2020) An UAV and satellite multispectral data approach to monitor water quality in small reservoirs. Remote Sens 12(9):1514

Dall'Olmo G, Gitelson AA (2005) Effect of bio-optical parameter variability on the remote estimation of chlorophyll-a concentration in turbid productive waters: experimental results. Appl Opt 44(3):412–422.

Dekker AG, Vos RJ, Peters SWM (2002) Analytical algorithms for lake water TSM estimation for retrospective analyses of TM and SPOT sensor data. Int J Remote Sens 23(1):15–35

Dörnhöfer K, Oppelt N (2016) Remote sensing for lake research and monitoring—recent advances. Ecol Ind 64:105–122

Doxaran D, Froidefond JM, Castaing P (2002) A reflectance band ratio used to estimate suspended matter concentrations in sediment-dominated coastal waters. Int J Remote Sens 23(23):5079–5085

D'Sa EJ, Miller RL (2003) Bio-optical properties in waters influenced by the Mississippi River during low flow conditions. Remote Sens Environ 84(4):538–549

Duan H, Ma R, Loiselle SA, Shen Q, Yin H, Zhang Y (2014) Optical characterization of black water blooms in eutrophic waters. Sci Total Environ 482:174–183

El Din E, Zhang Y, Suliman A (2017) Mapping concentrations of surface water quality parameters using a novel remote sensing and artificial intelligence framework. Int J Remote Sens 38(4):1023–1042

Feng L, Hu C, Han X, Chen X, Qi L (2015) Long-term distribution patterns of chlorophyll-a concentration in China's largest freshwater lake: MERIS full-resolution observations with a practical approach. Remote Sens 7(1):275–299

Gao Y, Gao J, Yin H, Liu C, Xia T, Wang J, Huang Q (2015) Remote sensing estimation of the total phosphorus concentration in a large lake using band combinations and regional multivariate statistical modeling techniques. J Environ Manage 151:33–43

Gholizadeh MH, Melesse AM, Reddi L (2016) A comprehensive review on water quality parameters estimation using remote sensing techniques. Sensors 16(8):1298

Giardino C, Brando VE, Gege P, Pinnel N, Hochberg E, Knaeps E et al (2019) Imaging spectrometry of inland and coastal waters: state of the art, achievements and perspectives. Surv Geophys 40:401–429

Gitelson AA, Gritz Y, Merzlyak MN (2003) Relationships between leaf chlorophyll content and spectral reflectance and algorithms for non-destructive chlorophyll assessment in higher plant leaves. J Plant Physiol 160(3):271–282

Gitelson AA, Dall'Olmo G, Moses W, Rundquist DC, Barrow T, Fisher TR et al (2008) A simple semi-analytical model for remote estimation of chlorophyll-a in turbid waters: validation. Remote Sens Environ 112(9):3582–3593

Gitelson AA, Gurlin D, Moses WJ, Barrow T (2009) A bio-optical algorithm for the remote estimation of the chlorophyll-a concentration in case 2 waters. Environ Res Lett 4(4):045003

Gordon HR, Brown OB, Jacobs MM (1975) Computed relationships between the inherent and apparent optical properties of a flat homogeneous ocean. Appl Opt 14(2):417–427

Han L, Rundquist DC (1997) Comparison of NIR/RED ratio and first derivative of reflectance in estimating algal-chlorophyll concentration: a case study in a turbid reservoir. Remote Sens Environ 62(3):253–261

Han D, Currell MJ, Cao G (2016) Deep challenges for China's war on water pollution. Environ Pollut 218:1222–1233

Havens KE (2003) Submerged aquatic vegetation correlations with depth and light attenuating materials in a shallow subtropical lake. Hydrobiologia 493:173–186

Herrault PA, Gandois L, Gascoin S, Tananaev N, Le Dantec T, Teisserenc R (2016) Using high spatio-temporal optical remote sensing to monitor dissolved organic carbon in the Arctic River Yenisei. Remote Sens 8(10):803

Hestir EL, Brando VE, Bresciani M, Giardino C, Matta E, Villa P, Dekker AG (2015) Measuring freshwater aquatic ecosystems: the need for a hyperspectral global mapping satellite mission. Remote Sens Environ 167:181–195

Holderness T, Barr S, Dawson R, Hall J (2013) An evaluation of thermal Earth observation for characterizing urban heatwave event dynamics using the urban heat island intensity metric. Int J Remote Sens 34(3):864–884

Hou X, Feng L, Duan H, Chen X, Sun D, Shi K (2017) Fifteen-year monitoring of the turbidity dynamics in large lakes and reservoirs in the middle and lower basin of the Yangtze River, China. Remote Sens Environ 190:107–121

Huang B, Zhang H, Song H, Wang J, Song C (2013) Unified fusion of remote-sensing imagery: Generating simultaneously high-resolution synthetic spatial–temporal–spectral earth observations. Remote Sens Lett 4(6):561–569

Huang J, Chen J, Mu Y, Cao C, Shen H (2024) Remote-sensing monitoring of colored dissolved organic matter in the Arctic Ocean. Mar Pollut Bull 204:116529

Hunter PD, Tyler AN, Carvalho L, Codd GA, Maberly SC (2010) Hyperspectral remote sensing of cyanobacterial pigments as indicators for cell populations and toxins in eutrophic lakes. Remote Sens Environ 114(11):2705–2718

Isenstein EM, Park M-H (2014) Assessment of nutrient distributions in Lake Champlain using satellite remote sensing. J Environ Sci 26:1831–1836

Isidro CM, McIntyre N, Lechner AM, Callow I (2018) Quantifying suspended solids in small rivers using satellite data. Sci Total Environ 634:1554–1562

Joshi ID, D'Sa EJ, Osburn CL, Bianchi TS, Ko DS, Oviedo-Vargas D et al (2017) Assessing chromophoric dissolved organic matter (CDOM) distribution, stocks, and fluxes in Apalachicola Bay using combined field, VIIRS ocean color, and model observations. Remote Sens Environ 191:359–372

Kallio K, Kutser T, Hannonen T, Koponen S, Pulliainen J, Vepsäläinen J, Pyhälahti T (2001) Retrieval of water quality from airborne imaging spectrometry of various lake types in different seasons. Sci Total Environ 268(1–3):59–77

Kim HC, Son S, Kim YH, Khim JS, Nam J, Chang WK et al (2017) Remote sensing and water quality indicators in the Korean West coast: Spatio-temporal structures of MODIS-derived chlorophyll-a and total suspended solids. Marine Pollution Bull 121(1–2):425–434

Koponen S, Pulliainen J, Kallio K, Hallikainen M (2002) Lake water quality classification with airborne hyperspectral spectrometer and simulated MERIS data. Remote Sens Environ 79(1):51–59

Kowalczuk P, Olszewski J, Darecki M, Kaczmarek S (2005) Empirical relationships between coloured dissolved organic matter (CDOM) absorption and apparent optical properties in Baltic Sea waters. Int J Remote Sens 26(2):345–370

Kudela RM, Palacios SL, Austerberry DC, Accorsi EK, Guild LS, Torres-Perez J (2015) Application of hyperspectral remote sensing to cyanobacterial blooms in inland waters. Remote Sens Environ 167:196–205

Kutser T, Pierson DC, Kallio KY, Reinart A, Sobek S (2005) Mapping lake CDOM by satellite remote sensing. Remote Sens Environ 94(4):535–540

Lathrop RG, Lillesand TM (1989) Monitoring water quality and river plume transport in Green Bay, Lake Michigan with SPOT-1 imagery. Photogramm Eng Remote Sens 55(3):349–354

Le CF, Li YM, Zha Y, Sun D, Yin B (2009a) Validation of a quasi-analytical algorithm for highly turbid eutrophic water of Meiliang Bay in Taihu Lake, China. IEEE Trans Geosci Remote Sens 47(8):2492–2500

Le C, Li Y, Zha Y, Sun D, Huang C, Lu H (2009b) A four-band semi-analytical model for estimating chlorophyll a in highly turbid lakes: the case of Taihu Lake, China. Remote Sens Environ 113(6):1175–1182

Li J, Chen X, Tian L, Huang J, Feng L (2015) Improved capabilities of the Chinese high-resolution remote sensing satellite GF-1 for monitoring suspended particulate matter (SPM) in inland waters: Radiometric and spatial considerations. ISPRS J Photogramm Remote Sens 106:145–156

Li Y, Zhang Y, Shi K, Zhu G, Zhou Y, Zhang Y, Guo Y (2017) Monitoring spatiotemporal variations in nutrients in a large drinking water reservoir and their relationships with hydrological and meteorological conditions based on Landsat 8 imagery. Sci Total Environ 599:1705–1717

Li J, Yu Q, Tian YQ, Becker BL, Siqueira P, Torbick N (2018) Spatio-temporal variations of CDOM in shallow inland waters from a semi-analytical inversion of Landsat-8. Remote Sens Environ 218:189–200

Li S, Song K, Wang S, Liu G, Wen Z, Shang Y et al (2021). Quantification of chlorophyll-a in typical lakes across China using Sentinel-2 MSI imagery with machine learning algorithm. Sci Total Environ 778:146271

Liangjiang X, Changchun H, Yunmei L, Xia C (2013) Deriving concentration of TN, TP based on hyper spectral reflectivity. Remote Sens Technol Appl 28(4):681–688

Liu H, Zhou Q, Li Q, Hu S, Shi T, Wu G (2019) Determining switching threshold for NIR-SWIR combined atmospheric correction algorithm of ocean color remote sensing. ISPRS J Photogramm Remote Sens 153:59–73

Malahlela OE (2019) Spatio-temporal assessment of inland surface water quality using remote sensing data in the wake of changing climate. In: IOP conference series: earth and environmental science, vol 227, No. 6. IOP Publishing, p 062012

Manzo C, Braga F, Zaggia L, Brando VE, Giardino C, Bresciani M, Bassani C (2018) Spatio-temporal analysis of prodelta dynamics by means of new satellite generation: the case of Po river by Landsat-8 data. Int J Appl Earth Obs Geoinf 66:210–225

Mao Z, Chen J, Hao Z, Pan D, Tao B, Zhu Q (2013) A new approach to estimate the aerosol scattering ratios for the atmos-pheric correction of satellite remote sensing data in coastal regions. Remote Sens Environ 132:186–194

Matthews MW (2011) A current review of empirical procedures of remote sensing in inland and near-coastal transitional waters. Int J Remote Sens 32(21):6855–6899

Meng X, Shen H, Zhang L, Yuan Q, Li H (2015) A unified framework for spatio-temporal-spectral fusion of remote sensing images. In: 2015 IEEE International geoscience and remote sensing symposium (IGARSS). IEEE, pp 2584–2587

Miller RL, McKee BA (2004) Using MODIS Terra 250 m imagery to map concentrations of total suspended matter in coastal waters. Remote Sens Environ 93(1–2):259–266

Mohsan SAH, Othman NQH, Li Y, Alsharif MH, Khan MA (2023) Unmanned aerial vehicles (UAVs): Practical aspects, applications, open challenges, security issues, and future trends. Intel Serv Robot 16(1):109–137

Moon JE, Park YJ, Ryu JH, Choi JK, Ahn JH, Min JE et al (2012) Initial validation of GOCI water products against in situ data collected around Korean peninsula for 2010–2011. Ocean Sci J 47:261–277

Moses WJ, Gitelson AA, Berdnikov S, Povazhnyy V (2009) Satellite estimation of chlorophyll-$ a $ concentration using the red and NIR bands of MERIS—the Azov sea case study. IEEE Geosci Remote Sens Lett 6(4):845–849

Neil C, Spyrakos E, Hunter PD, Tyler AN (2019) A global approach for chlorophyll-a retrieval across optically complex inland waters based on optical water types. Remote Sens Environ 229:159–178

Niu C, Tan K, Jia X, Wang X (2021) Deep learning based regression for optically inactive inland water quality parameter estimation using airborne hyperspectral imagery. Environ Pollut 286:117534

Odermatt D, Danne O, Philipson P, Brockmann C (2018) Diversity II water quality parameters from ENVISAT (2002–2012): a new global information source for lakes. Earth Syst Sci Data 10(3):1527–1549

Olmanson LG, Brezonik PL, Finlay JC, Bauer ME (2016) Comparison of Landsat 8 and Landsat 7 for regional measurements of CDOM and water clarity in lakes. Remote Sens Environ 185:119–128

Ouma YO, Waga J, Okech M, Lavisa O, Mbuthia D (2018) Estimation of reservoir bio-optical water quality parameters using smartphone sensor apps and landsat ETM+: review and comparative experimental results. J Sens 2018(1):3490757

Palmer SC, Hunter PD, Lankester T, Hubbard S, Spyrakos E, Tyler AN et al (2015) Validation of Envisat MERIS algorithms for chlorophyll retrieval in a large, turbid and optically-complex shallow lake. Remote Sens Environ 157:158–169

Pulliainen J, Kallio K, Eloheimo K, Koponen S, Servomaa H, Hannonen T et al (2001) A semi-operative approach to lake water quality retrieval from remote sensing data. Sci Total Environ 268(1–3):79–93

Pyo J, Duan H, Baek S, Kim MS, Jeon T, Kwon YS et al (2019) A convolutional neural network regression for quantifying cyanobacteria using hyperspectral imagery. Remote Sens Environ 233:111350

Reichstein M, Camps-Valls G, Stevens B, Jung M, Denzler J, Carvalhais N, Prabhat F (2019) Deep learning and process understanding for data-driven earth system science. Nature 566(7743):195–204

Rodríguez-López L, Duran-Llacer I, González-Rodríguez L, Abarca-del-Rio R, Cárdenas R, Parra O et al (2020) Spectral analysis using LANDSAT images to monitor the chlorophyll-a concentration in Lake Laja in Chile. Ecol Inform 60:10118

Ruddick KG, Gons HJ, Rijkeboer M, Tilstone G (2001) Optical remote sensing of chlorophyll a in case 2 waters by use of an adaptive two-band algorithm with optimal error properties. Appl Opt 40(21):3575–3585

Saberioon M, Brom J, Nedbal V, Souček P, Císař P (2020) Chlorophyll-a and total suspended solids retrieval and mapping using Sentinel-2A and machine learning for inland waters. Ecol Indicators 113:106236

Sagan V, Peterson KT, Maimaitijiang M, Sidike P, Sloan J, Greeling BA et al (2020) Monitoring inland water quality using remote sensing: potential and limitations of spectral indices, bio-optical simulations, machine learning, and cloud computing. Earth-Sci Rev 205:103187

Savage C, Leavitt PR, Elmgren R (2010) Effects of land use, urbanization, and climate variability on coastal eutrophication in the Baltic Sea. Limnol Oceanogr 55(3):1033–1046

Sayers MJ., Bosse KR, Shuchman RA, Ruberg SA, Fahnenstiel GL, Leshkevich GA et al (2019) Spatial and temporal variability of inherent and apparent optical properties in western Lake Erie: Implications for water quality remote sensing. J Great Lakes Res 45(3):490–507

Schmidhuber J (2015) Deep learning in neural networks: an overview. Neural Netw 61:85–117

Shang P, Shen F (2016) Atmospheric correction of satellite GF-1/WFV imagery and quantitative estimation of suspended particulate matter in the Yangtze estuary. Sensors 16(12):1997

Shang Y, Liu G, Wen Z, Jacinthe PA, Song K, Zhang B et al (2021) Remote estimates of CDOM using Sentinel-2 remote sensing data in reservoirs with different trophic states across China. J Environ Manage 286:112275

Shen H, Meng X, Zhang L (2016) An integrated framework for the spatio–temporal–spectral fusion of remote sensing images. IEEE Trans Geosci Remote Sens 54(12):7135–7148

Shen M, Duan H, Cao Z, Xue K, Qi T, Ma J et al (2020) Sentinel-3 OLCI observations of water clarity in large lakes in eastern China: implications for SDG 6.3. 2 evaluation. Remote Sens Environ 247:111950

Sherwood LJ, Qualls RG (2001) Stability of phosphorus within a wetland soil following ferric chloride treatment to control eutrophication. Environ Sci Technol 35(20):4126–4131

Singh KP, Basant N, Gupta S (2011) Support vector machines in water quality management. Anal Chim Acta 703(2):152–162

Song K, Wang Z, Blackwell J, Zhang B, Li F, Zhang Y, Jiang G (2011) Water quality monitoring using Landsat Themate Mapper data with empirical algorithms in Chagan Lake, China. J Appl Remote Sens 5(1):053506–053506

Song K, Li L, Tedesco LP, Li S, Duan H, Liu D et al (2013) Remote estimation of chlorophyll-a in turbid inland waters: three-band model versus GA-PLS model. Remote Sens Environ 136:342–357

Spyrakos E, O'donnell R, Hunter PD, Miller C, Scott M, Simis SG et al (2018) Optical types of inland and coastal waters. Limnol Oceanogr 63(2):846–870

Sriwongsitanon N, Surakit K, Thianpopirug S (2011) Influence of atmospheric correction and number of sampling points on the accuracy of water clarity assessment using remote sensing application. J Hydrol 401(3–4):203–220

Staehr SU, Van der Zande D, Staehr PA, Markager S (2022) Suitability of multisensory satellites for long-term chlorophyll assessment in coastal waters: a case study in optically-complex waters of the temperate region. Ecol Ind 134:108479

Sterckx S, Knaeps S, Kratzer S, Ruddick K (2015) SIMilarity Environment Correction (SIMEC) applied to MERIS data over inland and coastal waters. Remote Sens Environ 157:96–110

Sudduth KA, Jang GS, Lerch RN, Sadler EJ (2015) Long-term agroecosystem research in the Central Mississippi River Basin: hyperspectral remote sensing of reservoir water quality. J Environ Qual 44(1):71–83

Sun D, Qiu Z, Li Y, Shi K, Gong S (2014) Detection of total phosphorus concentrations of turbid inland waters using a remote sensing method. Water Air Soil Pollut 225:1–17

Sun X, Zhang Y, Shi K, Zhang Y, Li N, Wang W et al (2022) Monitoring water quality using proximal remote sensing technology. Sci Total Environ 803:149805

Swain R, Sahoo B (2017) Improving river water quality monitoring using satellite data products and a genetic algorithm processing approach. Sustain Water Qual Ecol 9:88–114

Syeed MM, Hossain MS, Karim MR, Uddin MF, Hasan M, Khan RH (2023) Surface water quality profiling using the water quality index, pollution index and statistical methods: a critical review. Environ Sustain Indicators 18:100247

Tao, Y., Xu, M., & Ma, J. (2014, July). Estimation of COD Mn in Tai lake basin using Landsat-8 satellite. In 2014 IEEE geoscience and remote sensing symposium. IEEE, pp 3866–3869

Tian L, Lu J, Chen X, Yu Z, Xiao J, Qiu F, Zhao X (2010) Atmospheric correction of HJ-1A/B CCD images over Chinese coastal waters using MODIS-Terra aerosol data. SCIENCE CHINA Technol Sci 53:191–195

Toming K, Kutser T, Uiboupin R, Arikas A, Vahter K, Paavel B (2017) Mapping water quality parameters with sentinel-3 ocean and land colour instrument imagery in the Baltic Sea. Remote Sens 9(10):1070

Vakili T, Amanollahi J (2020) Determination of optically inactive water quality variables using Landsat 8 data: a case study in Geshlagh reservoir affected by agricultural land use. J Clean Prod 247:119134

Wang Q, Atkinson PM (2018) Spatio-temporal fusion for daily Sentinel-2 images. Remote Sens Environ 204:31–42

Wang J, Chen X (2024) A new approach to quantify chlorophyll-a over inland water targets based on multi-source remote sensing data. Sci Total Environ 906:167631

Wang C, Li W, Chen S, Li D, Wang D, Liu J (2018) The spatial and temporal variation of total suspended solid concentration in Pearl River Estuary during 1987–2015 based on remote sensing. Sci Total Environ 618:1125–1138

Wang S, Garcia M, Bauer-Gottwein P, Jakobsen J, Zarco-Tejada PJ, Bandini F et al (2019a). High spatial resolution monitoring land surface energy, water and CO_2 fluxes from an unmanned aerial system. Remote Sens Environ 229:14–31

Wang Y, Liu D, Wang Y, Gao Z, Keesing JK (2019b) Evaluation of standard and regional satellite chlorophyll-a algorithms for moderate-resolution imaging spectroradiometer (MODIS) in the Bohai and Yellow Seas, China: a comparison of chlorophyll-a magnitude and seasonality. Int J Remote Sens 40(13):4980–4995

Wang Y, Xia H, Fu J, Sheng G (2004) Water quality change in reservoirs of Shenzhen, China: detection using LANDSAT/TM data. Sci Total Environ 328(1–3):195–206

Warren MA, Simis SG, Martinez-Vicente V, Poser K, Bresciani M, Alikas K et al (2019) Assessment of atmospheric correction algorithms for the Sentinel-2A MultiSpectral Imager over coastal and inland waters. Remote Sens Environ 225:267–289

Wu C, Wu J, Qi J, Zhang L, Huang H, Lou L, Chen Y (2010) Empirical estimation of total phosphorus concentration in the mainstream of the Qiantang River in China using Landsat TM data. Int J Remote Sens 31(9):2309–2324

Wu P, Shen H, Zhang L, Göttsche FM (2015) Integrated fusion of multi-scale polar-orbiting and geostationary satellite observations for the mapping of high spatial and temporal resolution land surface temperature. Remote Sens Environ 156:169–181

Yang B, Liu Y, Ou F, Yuan M (2011) Temporal and spatial analysis of COD concentration in East Dongting Lake by using of remotely sensed data. Procedia Environ Sci 10:2703–2708

Yang J, Holbach A, Wilhelms A, Qin Y, Zheng B, Zou H et al (2019) Highly time-resolved analysis of seasonal water dynamics and algal kinetics based on in-situ multi-sensor-system monitoring data in Lake Taihu, China. Sci Total Environ 660:329–339

Yang H, Du Y, Zhao H, Chen F (2022) Water quality Chl-a inversion based on spatio-temporal fusion and convolutional neural network. Remote Sens 14(5):1267

Ying H, Xia K, Huang X, Feng H, Yang Y, Du X, Huang L (2021) Evaluation of water quality based on UAV images and the IMP-MPP algorithm. Eco Inform 61:101239

Zhang H, Yao B, Wang S, Wang G (2021) Remote sensing estimation of the concentration and sources of coloured dissolved organic matter based on MODIS: a case study of Erhai lake. Ecol Ind 131:108180

Zhang Y, Pulliainen J, Koponen S, Hallikainen M (2002) Application of an empirical neural network to surface water quality estimation in the Gulf of Finland using combined optical data and microwave data. Remote Sens Environ 81(2–3):327–336

Zhang Y, Pulliainen JT, Koponen SS, Hallikainen MT (2003) Water quality retrievals from combined Landsat TM data and ERS-2 SAR data in the Gulf of Finland. IEEE Trans Geosci Remote Sens 41(3):622–629

Zhou G, Tian G, Chen J, Li J, Gong A (2020) Research of coupling effects among various water quality components. Spectrosc Spect Anal 30:470–475

Zhou L, Roberts DA, Ma W, Zhang H, Tang L (2013) Estimation of higher chlorophylla concentrations using field spectral measurement and HJ-1A hyperspectral satellite data in Dianshan Lake, China. ISPRS J Photogramm Remote Sens 88:41–47

Zhu X, Helmer EH, Gao F, Liu D, Chen J, Lefsky MA (2016) A flexible spatiotemporal method for fusing satellite images with different resolutions. Remote Sens Environ 172:165–177

Chapter 4
Biosensors: Potential of Microbes and Microbial Enzymes in Detecting and Managing Emerging Pollutants in Wastewater

Sayantan Biswas, Trisheema Sen, Linkon Bharali, and Anjana Singha

Abstract Biosensors are now highly effective instruments for the quick and accurate identification of contaminants in water, providing a number of benefits over traditional analytical techniques. Because of their special qualities—such as high selectivity, adaptability to a variety of conditions, and simplicity of manipulation—microbes and microbial enzymes have attracted increasing attention as sensing elements for biosensor systems in recent times. Among various classes of biosensors available, microbially generated biosensors are affordable, small, lightweight, and portable. They are highly suitable for the identification and tracking of a wide range of contaminants due to their superior structural and functional properties. A variety of microbial enzymes, such as oxidoreductases, hydrolases, and lyases, can be utilized as potential biosensors. In this chapter, we have discussed the potential of microbial enzymes to be utilized as biosensors and also the varied advantages of microbial-based biosensors in wastewater remediation. Also, we have discussed the challenges and hurdles related to harnessing as well as utilizing various microbial-based biosensors for water pollution detection and management.

Keywords Biosensor · Wastewater treatment · Detection and management of pollutants · Microbes · Microbial enzymes

S. Biswas · T. Sen · L. Bharali (✉)
Department of Chemistry, National Institute of Technology Silchar, Cachar, Assam, India
e-mail: bharalilinkon@gmail.com

A. Singha
Department of Chemistry, Indian Institute of Technology Guwahati, Assam 781039, India

Department of Chemistry, Karimganj Polytechnic, Assam 788712, India

A. M. Yatoo et al. (eds.), *Wastewater Monitoring and Management by Advanced Approaches*, Advances in Wastewater Research,
https://doi.org/10.1007/978-981-95-2601-7_4

4.1 Introduction

The development of devices that can detect different compounds using microbes is depicted in the intriguing history of biosensors, which follows the convergence of microbiology, analytical chemistry, and engineering. Leland C. Clark developed enzyme electrodes in 1962, which sparked the idea of biosensors and served as the basis for later biosensor technologies (Popenda et al. 2024b). The preliminary microbial biosensors were created in the 1980s, and they used yeast and bacterial cells to detect different kinds of analytes. Advances in molecular biology and biotechnology in the 1990s made it possible to genetically modify microbes to enhance their specificity and sensing capabilities. Since then, The development of synthetic biology has transformed the field, enabling the creation of custom microbes with tailored sensory abilities.

Real-time and remote monitoring is now possible because of the development of wireless as well as smartphone-based biosensor technologies (Nikoleli et al. 2016; Newman and Turner 2007; Williams et al. 2024). The term "biosensor" denotes that the apparatus consists of two components: a bio element plus a sensor element. Figure 4.1 illustrates the fundamental ideas behind how a biosensor works. An enzyme, antibodies, living cells, tissues, or any other particular bio element can identify a particular analyte, and the sensor element—which could be an electric current as well as electrode potential—transduces the biomolecule's change into an electrical signal. The bio element exhibits high specificity for its target analyte, remaining unresponsive to other analytes (Lianos et al. 2014; Turner 2013).

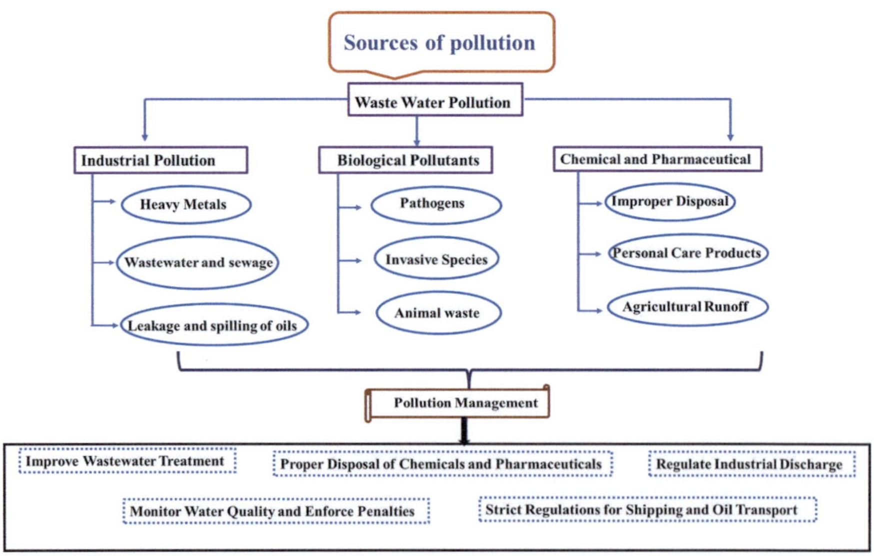

Fig. 4.1 Sources of different types of pollutants

Utilizing biosensors' ability to monitor and regulate different treatment stages is a key component of using these devices for wastewater treatment as well as management. Biosensors provide important benefits for guaranteeing effective and efficient wastewater management by continuously monitoring and detecting contaminants in real-time. An outline of their work in this area is provided in the next section.

4.1.1 Continuous Monitoring of Pollutants

(a) **Heavy Metals Detection**

Due to human activity and natural processes, heavy metals can now be found almost anywhere in soil, water, and air, along with food in numerous forms of hydroxides, oxides, sulphates, phosphates, silicates, and organic compounds. Enzymes, bacteria, fungi, as well as plant and animal cells, are examples of biological materials that may detect heavy metals (Popenda et al. 2024b; Zhou et al. 2024). To reduce the harmful effects of heavy metals, microbes use a variety of processes, including biosorption, bioaccumulation, bioleaching, bio-transformation, bio-volatilization, and bio-mineralization (Tang et al. 2024). Figure 4.2 describes the mechanism involved in the removal of heavy metals from the wastewater.

(b) **Organic Pollutants Detection**

The use of pesticides, industrial solvents, and pharmaceuticals are examples of organic pollutants that biosensors may detect and quantify. These sensors work best when they use microbial or enzymatic elements. This is essential for determining how well treatment procedures, such as biological degradation, work (Luo et al. 2024).

(c) **Nutrient Level Detection**

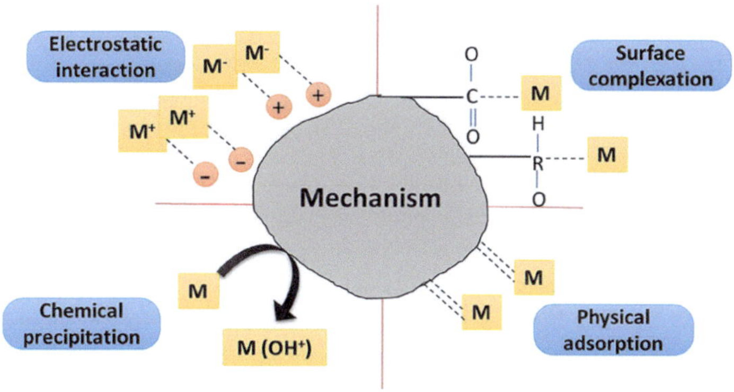

Fig. 4.2 The mechanism involved in the removal of heavy metals from the wastewater (Carolin et al., 2023)

Water bodies may become eutrophic due to an excess of nutrients such as phosphorus, nitrogen, and potassium. Through the use of biosensors, wastewater treatment procedures can be altered to preserve safe levels of these nutrients. In natural waterways, phosphorus can be found as both organic and inorganic phosphates. Nitrogen is present in the form of ammonia, nitrate, or nitrite (Korkut Uru 2024). Nitrate and nitrite are converted into quantifiable compounds by enzymes such as nitrate reductase or nitrite reductase, which catalyze the reaction, usually yielding a quantifiable electrochemical signal. By modifying metabolic byproducts, phosphate-accumulating microorganisms might be used in biosensors to identify orthophosphate (Hossain et al. 2024; Korkut Uru 2024).

4.1.2 Process Management

(a) **Biochemical oxygen demand (BOD)**

The amount of biodegradable organic carbon and pollution levels are linked by the biochemical oxygen demand (BOD), which is essential for evaluating the quality of water (Preininger et al. 1994; Sonawane et al. 2020).

(b) **Management of sludge**

Biosensors that monitor the microbial activity in sludge and ensure conditions are correct for organic waste to be broken down. Accordingly, the treatment of activated sludge processes may be carried out more efficiently (Abdullah et al. 2017).

(c) **Identification** of toxic substances

Some materials in wastewater interfere with the already existing biological process used to remediate it. Operators then have the opportunity to act before the treatment process in the impacted (Farré and Barceló 2003).

4.1.3 Microbial Biosensors and Their Components

In recent years, microbial biosensors as an evolving analytical tool, have drawn wide attention to detect specific chemical and biological materials in different conditions, using microorganisms such as bacteria, fungi, algae, and yeast. These biosensors are highly attractive in the area of real-time, sensitive, and personalized detection for industrial operations (pollutants), health care (saliva analysis), or environmental monitoring. The basic components of microbial biosensors include a transducer, immobilization matrix, and signal processing unit, as well as a biorecognition element. The biorecognition element has an impact on the target analyte, resulting in a biological reaction, usually with a microorganism. This reaction is converted to an electrical signal by the transducer, which is processed by a signal processing unit for useful data (values) to be derived. What is guaranteed, however, are the stability and activity of microorganisms on the surface due to the immobilization matrix (Inda and Lu 2020; Singh and Singh 2013; Xu and Ying 2011).

4.2 Waste Water Pollution

4.2.1 Origin of Water Pollution

The genesis of water contamination can be related to several natural as well as real factors (Mitiku 2020). Human activity is mostly responsible for the water pollution, but some pollutants arise naturally. There can be the following reasons for water pollution:

Industrial activities cause water contamination due to the discharge of hazardous chemicals into bodies of water. When industrial pollutants are released into the environment, it has a very negative effect on ecosystems and water quality, which is detrimental to both the natural world as well as human health. Water pollution from industries is an important method of contamination (Mokarram et al. 2020).

Industries often discharge their toxic chemicals, which include solvents and heavy metals (such as lead, mercury, and cadmium), along with synthetic compounds, into nearby water bodies. Some of these substances do represent residues from chemical factories, mining operations, or industrial activities. Dumping acidic or alkaline industrial waste into water can disturb the natural pH of it, and is dangerous for aquatic life (Issakhov et al. 2021).

Mining activities also lead to the diffusion of heavy metals (e.g., arsenic, cadmium, and mercury) in water bodies. Their high toxicity and bio-accumulative nature make them a serious threat to human health, among other risks, such as to aquatic life. Because heavy metals are present in the recycling of batteries and electronics, if waste is poorly disposed of, it has a good chance of contaminating water sources (Sonone et al. 2021; Vardhan et al. 2019) Fig. 4.3 displays various biosensors for the detection of heavy metals in wastewater.

The oil extraction and refining processes, both used to produce diesel fuel, cause spills and leaks, which are pending at offshore drilling rigs that can be spilled in long-haul pipelines as well. These happen due to the growing prevalence of ocean, freshwater river, and shore contamination, which threatens ecosystems as well as marine life. Moreover, hydrocarbons can accidentally escape or be discharged from crude oil storage tanks and refineries into water bodies (Purwendah et al. 2019).

Otherwise, the business of manufacturing active pharmaceutical ingredients (APIs) openly discards its leftovers into freshwater bodies. These criticisms include the possibility of endocrine disruption, antibiotic resistance due to possible persistence in water, and consequent influence on aquatic life (Fig. 4.4 displays Biosensors for organic pollutants and pharmaceuticals in wastewater). Chemical production

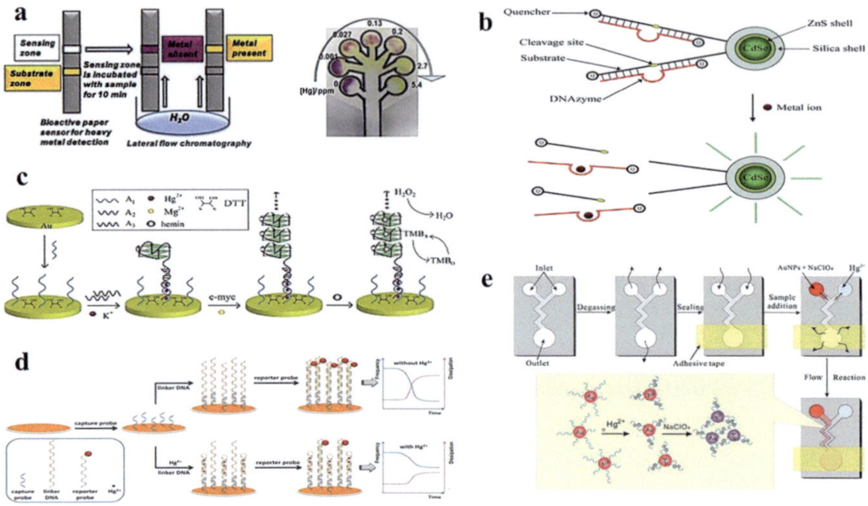

Fig. 4.3 Biosensors for the detection of heavy metals (Ejeian et al., 2018)

plants release insecticides, solvents, and several other fertilizers into water bodies, which are the source of numerous hazardous pollutants (Wilkinson et al. 2022).

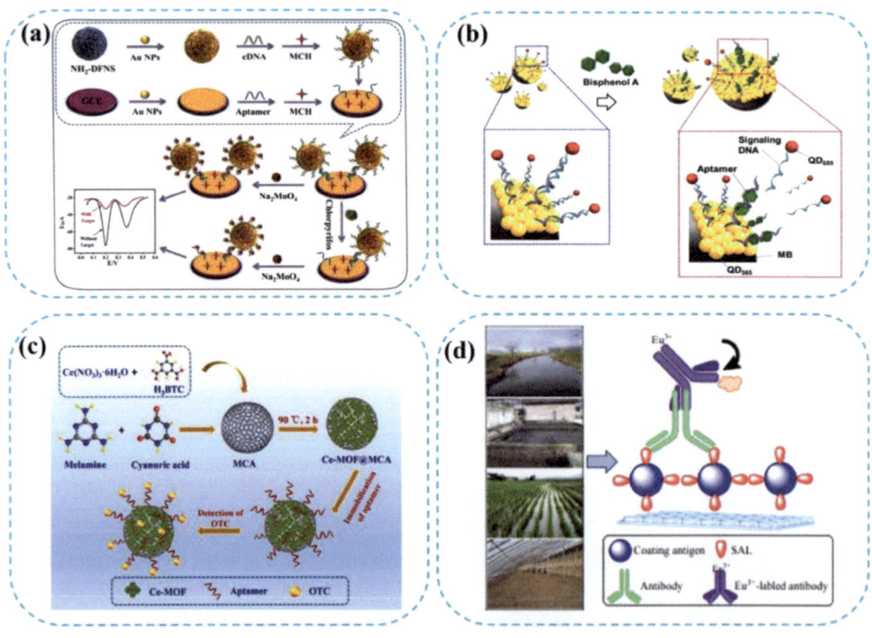

Fig. 4.4 Biosensors for organic pollutants and pharmaceuticals in wastewater (Mao et al., 2021)

Harmful organisms, including bacteria, viruses, parasites, and invasive species, can contaminate water bodies and threaten aquatic life, human health, and ecosystems. This phenomenon is known as biological pollution. The primary biological contaminants that cause water contamination are listed below:

- Runoff from agriculture and discharge of sewage and animal wastes contribute to the contamination of water sources by many pathogenic bacteria; for example, *Salmonella* species, *Vibrio cholerae*, *E. coli* or *Escherichia coli*. Bacteria like those that cause dysentery, gastroenteritis, and cholera can be transmitted as waterborne human illnesses. Pathogens emitted from human and animal feces can bring about waterborne viruses like rotavirus, norovirus, as well as hepatitis A in bodies of water (Some et al. 2021). Drinking contaminated water can cause viral infections in the liver and intestines. In contaminated water, protozoan parasites like Giardia lamblia, along with Cryptosporidium, are frequently discovered. These microbes can lead to serious gastrointestinal disorders, especially in people with weakened immune systems.
- Water bodies are intentionally or accidentally invaded by introduced aquatic species and can be hosted by Asian carp, zebra mussel, and water hyacinth due to introduction and ballast water disc tally. By changing habitat structures and nutrient dynamics, this species can degrade water quality and is notorious for its ability to shape aquatic ecosystems with repercussions on native biota through a combination of ecosystem alterations (e.g., changes in physical structure and the disturbance regime) or by loose competition among other strategies. Invasive microbial species can also impact water quality and ecosystem health, for example, through new pathogens or an altered balance of microbes coating surfaces in conduit systems (Amoatey and Baawain 2019).
- The hormone residue level in animal faces is high, and it contains many kinds of diseases and nutrients, including organic matter. It is produced in appreciable amounts. This litter can contribute to biological pollution and eutrophication, which occurs as it enters water systems.

Biological pollution: Wildlife populations, particularly large mammals and many bird species that frequent water bodies, can contribute to emerging waterborne diseases (usually through urine contamination) and may occasionally form components of biological pollution over the long term (Swarup and Patra 2005).

- A major source of water pollution that impacts aquatic ecosystems and water quality is household garbage. Improper residential waste disposal releases a range of contaminants into water bodies, whether it be by runoff, sewage systems, or direct dumping (Melati et al. 2020). These pollutants, which endanger both human health and the environment, might be physical, chemical, or organic in nature. Some of the main ways that household waste pollutes water are listed below:

    (a) In fact, many regions, including developing countries, dump untreated or inadequately treated sewage directly into lakes, rivers, and marine waters. Such pathogens, organic matter, and hazardous chemicals in this sewage are among the sources of environmental degradation as well as waterborne diseases. Household gray water, which includes wastewater generated by activities such as washing clothes, dishwashing, and showering, also contains chemicals besides heavy detergents and soaps, which, if left untreated, can pollute water (Sutrisno 2017).

(b) Because most wastewater treatment plants are not constructed to completely eliminate pharmaceuticals and drug waste holds the potential of making its way into water supply systems when medications are poured down the sewage or sink. Consequently, in the water bodies present may be active pharmaceutical ingredients (APIs) that can impact aquatic life and potentially lead to antibiotic resistance. Personal care products, like shampoos, soaps, and lotions, usually contain synthetic chemicals such as microbeads, which put aquatic life in danger with their capacity to bioaccumulate (Rivera-Utrilla et al. 2013).

(c) Microbial biosensors provide a sensitive, economical, and environmentally friendly way to identify and track contaminants, making them a possible solution to the problem of water pollution. These biosensors use the innate ability of microbes to adapt to changes in their surroundings to rapidly detect pollutants such as heavy metals, organic pollutants, and pathogens in water. When used in conjunction with real-time water quality monitoring, they allow for the early identification of pollution, prompt response, and efficient water management. By adding microbial biosensors to current water treatment systems, pollution control can be improved, and water resources can be made safer and cleaner (Waldrop 2021).

4.3 Microbial Biosensors

Due to their ability to monitor and identify pollutants, pathogens, and additional dangerous compounds in real time, microbial biosensors are an important part of water treatment processes. These biosensors are very helpful for determining pollutants, evaluating water quality, and making sure water treatment procedures are operating efficiently.

Microbial biosensors are analytical tools that identify the presence of particular chemical substances or environmental factors using microorganisms, usually bacteria (Rodriguez-Mozaz et al. 2006). These biosensors work by combining a physical transducer and a biological component (the microbe) to generate a detectable signal in response to a target material (Hashemi Goradel et al. 2018). An outline of how microbial biosensors function is provided below:

To ensure that the resulting microbial component will interact adequately with the target analyte. Compounds like pollutants, nutrients, or toxins can trigger the reactions of some kinds of bacteria, fungi, and other microorganisms. The Microbes fuse during biorecognition. When it encounters a target analyte, this metabolic product can induce specific biological responses (e.g., changes in gene expression or the production of certain enzymes) (Lei et al. 2006; Xu and Ying 2011).

The transducer transforms the microorganism's biological response into a detectable signal. Biological reactions can be transformed into electrical, optical, or electrochemical signals by the transducer (Thévenot et al. 2001; Xu and Ying 2011).

- Changes in conductivity, voltage, or current brought on by microbial activity are measured by electrochemical transducers.
- Variations in light absorption, fluorescence, or luminescence associated with microbial responses are detected by optical transducers.
- Piezoelectric transducers quantify mass changes or mechanical vibrations brought on by the metabolism or development of microorganisms.

The target analyte-containing sample is exposed to the biosensor. A biological reaction results from the microorganism's interaction with the target. A biological reaction, including metabolic activity, growth inhibition, or the synthesis of certain metabolites, is triggered by the microorganism's interaction with the analyte. The transducer detects a change in the qualities (mechanical, optical, or electrical) as a result of this biological activity.

The biological response is transformed into a quantifiable signal by the transducer. The electronics of the biosensor subsequently evaluate and interpret this signal to provide data regarding the target analyte's concentration (Su et al. 2011; Xu and Ying 2011).

The final signal is delivered to a data logging system or shown on a screen, where the analyte concentration is measured. For precise quantification, the results could need to be calibrated, or they might be instantaneous.

4.3.1 Types of Microbial Biosensors

Classifications based on Transducer Types

- Electrochemical microbial biosensors

These biosensors track variations in electrical characteristics (such as conductivity, voltage, or current) brought on by biochemical processes or microbial metabolism, namely (Bilal and Iqbal 2019; Su et al. 2011):

- Amperometry: Calculate variations in current brought on by redox processes, such as the consumption of oxygen during microbial respiration.
- Potentiometric: Measure voltage variations (e.g., pH variations brought on by microbial activity).
- Conductometric: Assess variations in electrical conductivity (such as the generation of ions during the metabolism of microorganisms) (Lagarde and Jaffrezic-Renault 2011; Thévenot et al. 2001; Xu and Ying 2011).

Applicative uses involve the identification of heavy metals, organic contaminants, and wastewater's biochemical oxygen demand (BOD).

- Optical microbial biosensors

These biosensors identify modifications in light characteristics brought about by microbial interactions with the target analyte, such as absorption, fluorescence, or luminescence, namely (Bilal and Iqbal 2019; Su et al. 2011):

- Fluorescence-based: Utilize microbes that have been genetically modified to generate fluorescent proteins in response to particular analytes (Lei et al. 2006; Su et al. 2011).
- Bioluminescence-based: Make use of microorganisms that naturally release light (like *Vibrio fischeri*) or when subjected to specific chemicals.
- Surface plasmon resonance (SPR): Assesses variations in light reflection at a sensor surface brought on by microbial activity (Gabriel et al. 2014; Su et al. 2011).

Applications commonly involve the identification of environmental contaminants, pathogens, and poisons in water and food samples (Bilal and Iqbal 2019).

Classifications Environmental microbial biosensors:

These types of microbial biosensors majorly involve target analytes such as toxins, insecticides, heavy metals, and organic chemicals are examples of pollutants. One example is the use of *Pseudomonas putida* to identify organic contaminants such as phenols (Xu and Ying 2011).

The major applications use air, water, and soil quality monitoring in the environment.

- Medical and clinical microbial biosensors:

In these types of microbial biosensors, the target analytes are pathogens and poisons, as well as useful metabolites with respect to human health. Examples are the utilization of *Lactobacillus* species to distinguish particular contaminations in clinical settings. Keeping with this approach, and the generation of biosensors that can realize lactate, urea, or glucose in physiological fluids (Haleem et al. 2021).

Applications commonly deal with the detection of diseases, infections, and prosthetics (Kaur et al. 2022).

- Food safety microbial biosensors:

These include contaminants, spoilage organisms, and foodborne pathogens. Predictive microbiology researches to quantify the behavior of microorganisms in food with mathematical models that predict how bacteria grow contingent upon time (Yasmin et al. 2016). The major examples range from biosensors that detect *E. coli* or *Salmonella* strains in their food products. *Saccharomyces cerevisiae* has been used in the food and beverage industries to control fermentation processes. Applications for testing food safety, preserving quality in the production of food, and watching over spoilage (Xu and Ying 2011) (Fig 4.5 represents the types of microbial biosensors based on transducer types and target analytes)

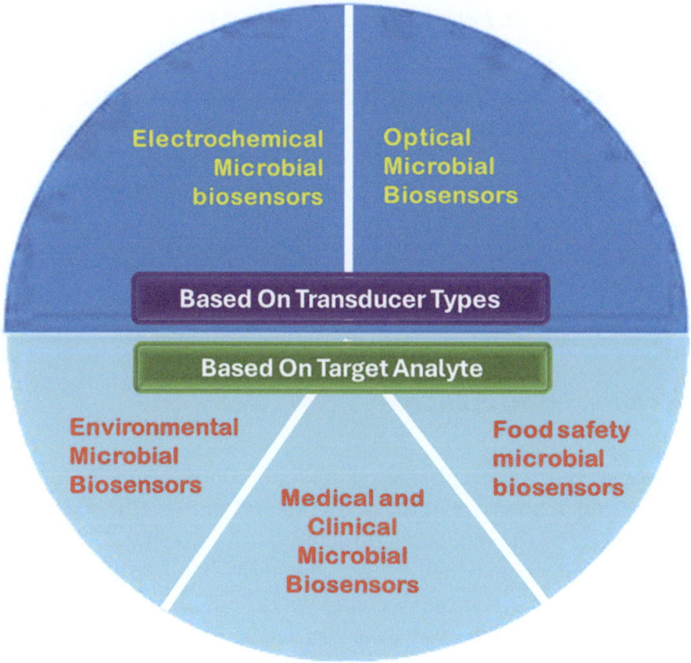

Fig. 4.5 Classification of the types of microbial biosensors: based on transducer types and target analyte

4.4 Microbial Enzymes in Biosensors

Because of their sensitivity to particular substrates and catalytic qualities, microbial enzymes are commonly used as biorecognition elements in biosensors. These enzymes are essential for turning a biochemical reaction into a detectable signal (Hashemi Goradel et al. 2018). They are obtained from microorganisms, including yeast, fungi, and bacteria.

4.4.1 Role of Microbial Enzymes as Biorecognition Elements

The biological recognition elements used in microbial biosensors, by which they can selectively sense the target analytes, are due to the enzymatic activity of microorganisms. Enzymes can convert the presence of an analyte into a measurable signal due to their versatility—they are highly selective and catalytically efficient. This form of biorecognition is particularly evident at the level of microbial enzymes, as seen in microbial sugar-specific biosensors, for example (Nguyen et al. 2019).

4.4.2 Biorecognition Function

Because their active sites are clearly defined, microbial enzymes are extremely selective for their substrates. Because of its specificity, the biosensor can reliably identify the presence of a given analyte even in intricate combinations.

Biochemical reactions are facilitated by enzymes, which frequently transform the target analyte into a product that the sensor can detect. The biosensor's sensitivity is increased by this catalytic activity, making it possible for it to identify analytes at low concentrations (Kaur et al. 2019).

4.4.3 Mechanisms of Action

Substrate binding: When the enzyme comes into contact with its target molecule or substrate, it forms a complex at the active site of the metallic ions. This specific binding is controlled by the molecular structure of the enzyme.

Catalysis is a process in which an enzyme turns a substrate into one or more products. This shift typically causes a change in the environment that can be quantified through pH changes, production of electroactive molecules, or bioluminescence (Newman and Turner 2007; Amine et al. 2006).

Altering signal transduction: The signal transduction of the biosensor is adjusted to reflect the change by enzymatic reaction. The transducer converts this biological response into a measurable signal, an electrical current, an optical event, or a temperature change (Amine et al. 2006).

4.4.4 Commonly Used Microbial Enzymes and Their Specific Applications in Pollutant Detection

Since microbial enzymes may catalyze reactions with a wide range of environmental pollutants, they are frequently used in biosensors for pollution detection because of their specificity (Rebollar-Pérez et al. 2016). The following is a summary of typical microbial enzymes and how they are specifically utilized to identify pollutants:

- Laccase:

Sources include certain bacteria and fungi, including *Pleurotus*, *Aspergillus*, and *Trametes versicolor*. The function is to oxidize a variety of phenolic and aromatic molecules by reducing oxygen to water.

Its various uses can be as follows:

- Phenolic compounds: Phenolic pollutants are frequently found in industrial wastewater from activities like paper production and petroleum refining. Laccase is used to identify these contaminants.

- Synthetic chemicals and dyes: It can identify and break down additional xenobiotics in wastewater as well as synthetic dyes.
- Pesticides: Finding phenolic pesticides in environmental samples, such as chlorophenols (Casero et al. 2013; Rodríguez-Delgado et al. 2015).

• Peroxidase:

These varieties of microbial enzymes can have the following uses:

- Aromatic hydrocarbons: To identify a variety of aromatic hydrocarbons in contaminated air and water, such as toluene and benzene.
- Phenolic chemicals: Peroxidase is used to identify and break down phenolic chemicals in sources such as plants, microbes, and fungi (such as *Aspergillus* and *horseradish peroxidase*) (Ahammad 2012).

The function involves the catalysis of the hydrogen peroxide-induced oxidation of substrates. They are primarily utilized in wastewater, just like laccase does (Ahammad 2012; Castillo et al. 2003).

- Pesticides: Identifying pesticides such as atrazine and herbicides in agricultural runoff.

• Urease:

The principal sources of urease are plants, fungi, and bacteria (*Klebsiella*, *Proteus*, and *Helicobacter pylori*). Its function is to hydrolyze urea to produce carbon dioxide and ammonia.

The main uses involve:

- Nitrogen pollution: Ammonia and urea, which are frequently present in wastewater from businesses like fertilizer manufacture and agricultural runoff, are detected using biosensors (Singh et al. 2008; Dindar et al. 2011).
- Monitoring water quality involves looking for contaminants with a nitrogen base that cause eutrophication in water bodies.

4.5 Applications of Microbial Biosensors in Detecting Emerging Pollutants

Microbial biosensors are ideal for identifying contaminants that are frequently present in low amounts because they use innate or engineered abilities that allow microbes to perceive and react to particular chemical substances (Soni et al. 2022). The sensitivity, selectivity, and adaptability of microbial biosensors render them particularly promising for detecting emerging contaminants in complex matrices. Some of the key applications of microbial biosensors are:

• Pharmaceutical and personal care product detection (PPCPs):

Microbial biosensors are able to identify residual personal care items like odors and preservatives, as well as medications like hormones and antibiotics in wastewater and surface waters. For example, biosensors that use bacteria modified to respond to particular antibiotics could be used to track the degree of pollution in water bodies (Narayanan et al. 2022).

- Pesticide and herbicide monitoring:

Agriculture-related chemicals like pesticides and herbicides can be found in soil and water, and their quantities can be measured using biosensors. They can be used to continuously monitor the regions around agricultural fields to make sure that contamination levels stay below acceptable bounds (Aynalem and Muleta 2021).

- Detection of heavy metals

The detection of heavy metals such as mercury, lead, and cadmium in contaminated water and soil can be accomplished by the engineering of microbial biosensors. In the event of the presence of particular heavy metals, certain bacteria's metal-binding proteins can be used to generate a detectable signal (Verma and Singh 2005).

- Monitoring endocrine disrupting chemicals (EDCs):

Through the use of microbial biosensors, endocrine disruptors such as phthalates and bisphenol A (BPA) can be identified. These sensors can help identify the presence and concentration of EDCs in different habitats, which can be useful in determining how they might affect human health and wildlife populations (Wang et al. 2023).

- Monitoring of microplastics:

Microplastics and artificial nanomaterials are developing contaminants that are alarming due to their persistent nature and possible toxicity in the environment. Though more difficult, microbial biosensors are being designed to detect by-products of microplastic degradation or the presence of particular nanomaterials. This entails designing microorganisms to interact with these chemicals or their breakdown products. This application is critical for assessing the environmental impact of microplastics in aquatic and terrestrial ecosystems (Popenda et al. 2024a).

- Bioremediation monitoring:

Using microbial biosensors for bioremediation monitoring, one may optimize and track the degradation of contaminants in polluted environments by using the natural or synthetic bacteria's ability for detection. Microorganisms are used in bioremediation to break down or detoxify contaminants, changing them into less hazardous forms. The effectiveness and efficiency of the bioremediation process are greatly dependent on microbial biosensors (Divyasorubini et al. 2022).

- Detection of antibiotic resistance genes (ARGs):

A promising technique for tracking the emergence of antibiotic resistance in many situations, including wastewater, agricultural runoff, and clinical settings, is detecting the presence of antibiotic resistance genes (ARGs) using microbial biosensors.

Antimicrobial resistance (AMR), a developing issue that can make drugs ineffective in fighting bacterial infections, is exacerbated by ARGs. Microbial biosensors provide a sensitive and targeted way to identify ARGs in real time, enabling improved monitoring and management of this worldwide health hazard (Magnano San Lio et al. 2023)

- Water quality control:

A novel method for recognizing and measuring pollutants, poisons, and other dangerous materials in water systems, such as drinking water, wastewater, and natural bodies of water, is water quality monitoring, making use of microbial biosensors. Microbial biosensors are designed to detect particular pollutants and contaminants, providing sensitive, real-time, and reasonably priced monitoring options. By addressing issues with the environment and public health, these gadgets can aid in ensuring the safety and purity of water (Lagarde and Jaffrezic-Renault 2011).

4.6 Case Studies and Examples of Microbial Biosensors Detecting Emerging Pollutants

The application of microbial biosensors to identify emerging pollutants—those recently recognized as potential threats to human health and the environment—has increased. Endocrine-disrupting chemicals (EDCs), microplastics, pharmaceuticals and personal care products (PPCPs), and antibiotics constitute a variety of these pollutants.

4.6.1 Case Study 1. Detection of Endocrine-Disrupting Chemicals (EDCs) in Water

Synthetic hormones, phthalates, and bisphenol A (BPA) are examples of endocrine-disrupting chemicals (EDCs) that can interfere with an animal's or human's hormone systems, possibly resulting in problems with development and reproduction (Wang et al. 2023; Shah et al. 2023).

- Design of microbial biosensor:

To identify estrogenic chemicals in water, a microbial biosensor using genetically modified *E. coli* was created. This biosensor has been developed with a reporter gene that is activated in the presence of estrogenic contaminants like BPA and artificial hormones. The reporter gene is regulated by an estrogen-responsive element. The biosensor emits a quantifiable light signal upon detecting these chemicals.

During testing in wastewater treatment facilities and river samples, the biosensor was able to identify synthetic estrogens and BPA at low amounts, even at parts-per-billion (ppb) levels. EDCs in treated water could be continuously assessed thanks to the real-time monitoring abilities, which enabled to guarantee that these chemicals were eliminated before being released into streams and lakes. A system for early detection of EDC contamination is made feasible by the microbial biosensor. It is more rapid and economical than traditional analytical methods such as high-performance liquid chromatography (HPLC) (Scognamiglio et al. 2016).

4.6.2 Case Study 2: Monitoring Pharmaceuticals and Antibiotics in Wastewater

The inappropriate disposal of pharmaceuticals and antibiotics has led to their widespread utilization, making them a substantial source of emerging pollution. Aquatic habitats are impacted by these contaminants, which can also result in resistance to antibiotics (Adegoke et al. 2019).

- Design of microbial biosensors:

Using genetically modified *Pseudomonas putida*, a microbial biosensor has been developed that can identify the widely used antibiotic sulfamethoxazole. The biosensor was engineered to initiate a reporter gene regulated by the sulfonamide resistance element, thus producing a fluorescent signal upon encountering sulfamethoxazole. To track the effectiveness of sulfamethoxazole elimination in real-time, a biosensor has been installed in wastewater treatment facilities. Additionally, samples from nearby rivers and groundwater that are being treated for wastewater are tested for it. At values less than 10 ppb, the biosensor accurately found residual sulfamethoxazole in treated water. It made constant tracking possible without requiring laborious sample preparation, which made it incredibly useful for in-situ testing. The data revealed how well the current treatment methods removed medications from wastewater (Campaña et al. 2019).

4.6.3 Case Study 3: A Microbial Biosensor for Detecting Microplastics

Microplastics (particles smaller than 5 mm) are an increasing threat to the environment owing to their longstanding presence in aquatic ecosystems and ability to absorb hazardous contaminants (Popenda et al. 2024a; Behera et al. 2023).

- Design of microbial biosensors:

Using cultivated bacteria that transmit a signal when exposed to substances generated by microplastics that break down plastic, researchers devised a microbial biosensor. Phthalates and bisphenol A, which seep out of microplastics, are among the broken-down products that the biosensor detects. The biosensor presented an invaluable instrument for evaluating the disintegration of microplastics in the environment since it could identify the byproducts of microplastic degradation at exceedingly low quantities. This approach offers a way to track the effects of microplastic pollution on water quality in real-time, as well as aids in mapping their distribution (Tang et al. 2023).

4.6.4 *Case Study 4: Personal Care Product (PPCP) Identification in Urban Runoff*

Cosmetics, scents, and preservatives are examples of pharmaceutical and personal care products (PPCPs) that may contaminate water sources and endanger the environment when they get into wastewater systems (Berberich et al. 2019).

- Design of microbial biosensors:

A yeast called *S. cerevisiae* has been used to develop a microbial biosensor that could identify parabens, a common preservative present in personal care goods. The biosensor's reaction to parabens was intended to be a colorimetric signal generated by activating a reporter gene. In locations with high human densities and industrial activity, the biosensor was used to monitor the levels of parabens in urban sewage and stormwater drainage systems. To guarantee that PPCPs have been effectively eliminated from the water before it is consumed, it is also utilized in drinking water treatment facilities. The biosensor enabled the early detection of pollution in urban runoff by detecting parabens at barely detectable concentrations (ppb levels). The real-time monitoring gave information that might help develop enhanced water treatment solutions to handle emerging contaminants in metropolitan areas (Yang et al. 2020).

4.6.5 *Case Study 5: Perfluoroalkyl Substance (PFAS) Monitoring*

Synthetic chemicals called perfluoroalkyl substances (PFAS) are found in water-repellent textiles, nonstick cookware, and firefighting foams. Due to their environmental persistence, PFAS are hazardous to the environment and human health (Ahmadi Tabar et al. 2023).

- Design of microbial sensors:

Using genetically engineered *E. coli*, researchers created a microbial biosensor that is sensitive to PFAS exposure. By causing the production of a fluorescent protein in response to PFAS, the biosensor facilitates the visual detection of these chemicals. In comparison to conventional methods for analysis, these biosensors are more affordable, have excellent sensitivity, and allow for real-time monitoring. Microbial biosensors offer an invaluable instrument for tackling environmental challenges and preserving public health, as pollution from these newly discovered pollutants spreads (Moro et al. 2022).

4.7 Conclusion and Future Prospects

In summary, microbial biosensors represent a potential and novel strategy for identifying and regulating emerging contaminants in wastewater. These biosensors give advantages such as high specificity, cost-effectiveness, and real-time monitoring capabilities by leveraging microbes' and enzymes' innate capabilities. They can efficiently target pollutants like medicines, industrial chemicals, and microplastics, which are frequently resistant to traditional treatment procedures. Despite their potential, they nonetheless face various challenges. Maintaining microbial viability and activity in changing environmental circumstances, as well as obtaining regulatory approvals, might complicate their implementation. Furthermore, the diverse composition of wastewater can interfere with sensor performance.

However, the advancement and setup of modern technology present promising opportunities for boosting data interpretation and pollutant detection accuracy. The development of portable, multifunctional biosensors that combine detection and treatment capabilities has the potential to transform wastewater management. The short operational life of biosensors, which results from the gradual decrease of enzyme or microbial activity, is one issue with their current application. The creation of self-regenerating microbial systems, either by genetic manipulation or by introducing symbiotic connections within microbial populations, may be the main focus of future developments. Although small-scale wastewater treatment facilities and pilot studies are currently using microbial biosensors, one of the main objectives for the future is to scale up these kinds of devices for industrial use. To guarantee that microbial biosensors which can manage the quantity and complexity of contaminants found in massive operations of industrial wastewater treatment, developments in bioprocessing as well as engineering will be required. Overall, microbial biosensors are a cutting-edge solution that can make a substantial contribution to sustainable wastewater treatment. They serve an important role in maintaining environmental health and public safety by addressing both monitoring and remediation, ultimately contributing to provide a cleaner and safer water supply.

References

Abdullah NA, Ramli S, Mamat H, Khan S, Gomes C (2017) Chemical and biosensor technologies for wastewater quality management. Int J Adv Res Publ 1(6):1–10. www.ijarp.org

Adegoke AA, Singh G, Stenström TA (2019) Biosensors for monitoring pharmaceutical nanocontaminants and drug resistant bacteria in surface water, subsurface water and wastewater effluent for reuse. In: Nanoparticles pharmacotherapy. Elsevier, pp 525–559

Ahammad AJS (2012) Hydrogen peroxide biosensors based on horseradish peroxidase and hemoglobin. J Biosens Bioelectron s9. https://doi.org/10.4172/2155-6210.s9-001

Ahmadi Tabar F, Lowdon JW, Bakhshi Sichani S, Khorshid M, Cleij TJ, Diliën H, Eersels K, Wagner P, van Grinsven B (2023) An overview on recent advances in biomimetic sensors for the detection of perfluoroalkyl substances. Sensors 24(1):130

Amine A, Mohammadi H, Bourais I, Palleschi G (2006) Enzyme inhibition-based biosensors for food safety and environmental monitoring. Biosens Bioelectron 21(8):1405–1423. https://doi.org/10.1016/j.bios.2005.07.012

Amoatey P, Baawain MS (2019) Effects of pollution on freshwater aquatic organisms. Water Environ Res 91(10):1272–1287. https://doi.org/10.1002/wer.1221

Aynalem B, Muleta D (2021) Microbial biosensors as pesticide detector: an overview. J Sens 2021(1):5538857

Behera A, Mahapatra SR, Majhi S, Misra N, Sharma R, Singh J, Singh RP, Pandey SS, Singh KRB, Kerry RG (2023) Gold nanoparticle assisted colorimetric biosensors for rapid polyethylene terephthalate (PET) sensing for sustainable environment to monitor microplastics. Environ Res 234:116556

Berberich J, Li T, Sahle-Demessie E (2019) Biosensors for monitoring water pollutants: a case study with arsenic in groundwater. In: Separation science and technology, vol 11. Elsevier, Amsterdam, pp 285–328

Bilal M, Iqbal HMN (2019) Microbial-derived biosensors for monitoring environmental contaminants: Recent advances and future outlook. Process Saf Environ Prot 124:8–17. https://doi.org/10.1016/j.psep.2019.01.032

Carolin CF, Kamalesh T, Senthil Kumar P, Rangasamy G (2023) A critical review on the sustainable approaches for the removal of toxic heavy metals from water systems. Ind Eng Chem Res 62(22):8575–8601

Campaña AL, Florez SL, Noguera MJ, Fuentes OP, Ruiz Puentes P, Cruz JC, Osma JF (2019) Enzyme-based electrochemical biosensors for microfluidic platforms to detect pharmaceutical residues in wastewater. Biosensors 9(1):41

Casero E, Petit-Domínguez MD, Vázquez L, Ramírez-Asperilla I, Parra-Alfambra AM, Pariente F, Lorenzo E (2013) Laccase biosensors based on different enzyme immobilization strategies for phenolic compounds determination. Talanta 115:401–408. https://doi.org/10.1016/j.talanta.2013.05.045

Castillo J, Gáspár S, Sakharov I, Csöregi E (2003) Bienzyme biosensors for glucose, ethanol and putrescine built on oxidase and sweet potato peroxidase. Biosens Bioelectron 18(5–6):705–714. https://doi.org/10.1016/S0956-5663(03)00011-3

Dindar B, Karakuş E, Abasıyanık F (2011) New urea biosensor based on urease enzyme obtained from helicobacter pylori. Appl Biochem Biotechnol 165(5–6):1308–1321. https://doi.org/10.1007/s12010-011-9348-2

Divyasorubini S, Kandage SM, Liyanage S, Rajapakse C, Silva GN (2022) Microbial biosensors for real-time monitoring of the bioremediation processes. In: Bioremediation of environmental pollutants: emerging trends and strategies, pp 111–144

Ejeian F, Etedali P, Mansouri-Tehrani HA, Soozanipour A, Low ZX, Asadnia M, Taheri-Kafrani A, Razmjou A (2018) Biosensors for wastewater monitoring: a review. Biosens Bioelectron 118:66–79

Farré M, Barceló D (2003) Toxicity testing of wastewater and sewage sludge by biosensors, bioassays and chemical analysis. TrAC Trends Anal Chem 22(5):299–310. https://doi.org/10.1016/S0165-9936(03)00504-1

Gabriel GVM, Lopes PS, Viviani VR (2014) Suitability of Macrolampis firefly and Pyrearinus click beetle luciferases for bacterial light off toxicity biosensor. Anal Biochem 445(1):73–79. https://doi.org/10.1016/j.ab.2013.09.017

Haleem A, Javaid M, Singh RP, Suman R, Rab S (2021) Biosensors applications in medical field: a brief review. Sens Int 2:100100. https://doi.org/10.1016/j.sintl.2021.100100

Hashemi Goradel N, Mirzaei H, Sahebkar A, Poursadeghiyan M, Masoudifar A, Malekshahi ZV, Negahdari B (2018) Biosensors for the detection of environmental and urban pollutions. J Cell Biochem 119(1):207–212. https://doi.org/10.1002/jcb.26030

Hossain MI, Khaleque MA, Ali MR, Bacchu MS, Hossain MS, Shahed SMF, Saad Aly MA, Khan MZH (2024) Development of electrochemical sensors for quick detection of environmental (soil and water) NPK ions. RSC Adv 14(13):9137–9158. https://doi.org/10.1039/d4ra00034j

Inda ME, Lu TK (2020) Microbes as biosensors. Annu Rev Microbiol 74:337–359. https://doi.org/10.1146/annurev-micro-022620-081059

Issakhov A, Alimbek A, Zhandaulet Y (2021) The assessment of water pollution by chemical reaction products from the activities of industrial facilities: numerical study. J Clean Prod 282:125239. https://doi.org/10.1016/j.jclepro.2020.125239

Kaur B, Kumar S, Kaushik BK (2022) Recent advancements in optical biosensors for cancer detection. Biosens Bioelectron 197:113805. https://doi.org/10.1016/j.bios.2021.113805

Kaur J, Choudhary S, Chaudhari R, Jayant RD, Joshi A (2019) Enzyme-based biosensors. In: Bioelectronics and medical devices: from materials to devices—fabrication, applications and reliability. Elsevier, Amsterdam. https://doi.org/10.1016/B978-0-08-102420-1.00013-3

Korkut Uru S (2024) A critical perspective on the applied potential in amperometric phosphate biosensors. J Water Process Eng 58(January):104886. https://doi.org/10.1016/j.jwpe.2024.104886

Lagarde F, Jaffrezic-Renault N (2011) Cell-based electrochemical biosensors for water quality assessment. Anal Bioanal Chem 400(4):947–964. https://doi.org/10.1007/s00216-011-4816-7

Lei Y, Chen W, Mulchandani A (2006) Microbial biosensors. Anal Chim Acta 568(1–2):200–210. https://doi.org/10.1016/j.aca.2005.11.065

Lianos GD, Glantzounis GK, Mangano A, Rausei S, Roukos DH (2014) Colorectal liver metastases guidelines, tumor heterogeneity and clonal evolution: can this be translated to patient benefit? Future Oncol 10(10):1723–1726. https://doi.org/10.2217/fon.14.91

Luo Y, Sun Y, Wei X, He Y, Wang H, Cui Z, Ma J, Liu X, Shu R, Lin H, Xu D (2024) Detection methods for antibiotics in wastewater: a review. Bioprocess Biosyst Eng 47(9):1433–1451. https://doi.org/10.1007/s00449-024-03033-0

Mao K, Zhang H, Pan Y, & Yang Z (2021) Biosensors for wastewater-based epidemiology for monitoring public health. Water Res 191: 116787

Magnano San Lio R, Barchitta M, Maugeri A, La Rosa MC, Favara G, Agodi A (2023) Updates on developing and applying biosensors for the detection of microorganisms, antimicrobial resistance genes and antibiotics: a scoping review. Front Public Health 11:1240584

Melati I, Maghfiroh M, Jasalesmana T (2020) Public perception and contribution of domestic waste to water pollution of Lake Matano, South Sulawesi-Indonesia. IOP Conf Ser Earth Environ Sci 535(1). https://doi.org/10.1088/1755-1315/535/1/012065

Mitiku AA (2020) Water pollution: causes and prevention. October

Mokarram M, Saber A, Sheykhi V (2020) Effects of heavy metal contamination on river water quality due to release of industrial effluents. J Clean Prod 277:123380. https://doi.org/10.1016/j.jclepro.2020.123380

Moro G, Chiavaioli F, Zubiate P, Del Villar I, Baldini F, De Wael K, Moretto LM, Giannetti A (2022) Albumin-based optical and electrochemical biosensors for PFAS detection: a comparison. In: AISEM Annual Conference on sensors and microsystems, pp 1–15

Narayanan M, El-Sheekh M, Ma Y, Pugazhendhi A, Natarajan D, Kandasamy G, Raja R, Kumar RMS, Kumarasamy S, Sathiyan G (2022) Current status of microbes involved in the degradation of pharmaceutical and personal care products (PPCPs) pollutants in the aquatic ecosystem. Environ Pollut 300:118922

Newman JD, Turner APF (2007) Historical perspective of biosensor and biochip development. Handbook Biosens Biochips. https://doi.org/10.1002/9780470061565.hbb002

Nguyen HH, Lee SH, Lee UJ, Fermin CD, Kim M (2019) Immobilized enzymes in biosensor applications. Materials12(1):1–34. https://doi.org/10.3390/ma12010121

Nikoleli G, Karapetis S, Bratakou S, Nikolelis DP, Tzamtzis N, Psychoyios VN, Psaroudakis N (2016) Biosensors for security and bioterrorism applications. Biosensors for security and bioterrorism applications, March, pp 431–442. https://doi.org/10.1007/978-3-319-28926-7

Popenda A, Wiśniowska E, Manuel C (2024a) Biosensors in environmental analysis of microplastics and heavy metal compounds–a review on current status and challenges. Desalination Water Treat 100456

Popenda A, Wiśniowska E, Manuel C (2024b) Biosensors in environmental analysis of microplastics and heavy metal compounds—a review on current status and challenges. Desalin Water Treat 319(May):100456. https://doi.org/10.1016/j.dwt.2024.100456

Preininger C, Klimant I, Wolfbeis OS (1994) Optical Fiber sensor for biological oxygen demand. Anal Chem 66(11):1841–1846. https://doi.org/10.1021/ac00083a011

Purwendah EK, Mangku DGS, Periani A (2019) Dispute settlements of oil spills in the sea towards sea environment pollution. 317 (IConProCS), pp 245–248. https://doi.org/10.2991/iconprocs-19.2019.51

Rebollar-Pérez G, Campos-Terán J, Ornelas-Soto N, Méndez-Albores A, Torres E (2016) Biosensors based on oxidative enzymes for detection of environmental pollutants. Biocatalysis 1(1):118–129. https://doi.org/10.1515/boca-2015-0010

Rivera-Utrilla J, Sánchez-Polo M, Ferro-García MÁ, Prados-Joya G, Ocampo-Pérez R (2013) Pharmaceuticals as emerging contaminants and their removal from water. a review. Chemosphere 93(7):1268–1287. https://doi.org/10.1016/j.chemosphere.2013.07.059

Rodríguez-Delgado MM, Alemán-Nava GS, Rodríguez-Delgado JM, Dieck-Assad G, Martínez-Chapa SO, Barceló D, Parra R (2015) Laccase-based biosensors for detection of phenolic compounds. TrAC Trends Anal Chem 74:21–45. https://doi.org/10.1016/j.trac.2015.05.008

Rodriguez-Mozaz S, Lopez De Alda MJ, Barceló D (2006) Biosensors as useful tools for environmental analysis and monitoring. Anal Bioanal Chem 386(4):1025–1041. https://doi.org/10.1007/s00216-006-0574-3

Scognamiglio V, Antonacci A, Patrolecco L, Lambreva MD, Litescu SC, Ghuge SA, Rea G (2016) Analytical tools monitoring endocrine disrupting chemicals. TrAC, Trends Anal Chem 80:555–567

Shah MM, Ahmad K, Boota S, Jensen T, La Frano MR, Irudayaraj J (2023) Sensor technologies for the detection and monitoring of endocrine-disrupting chemicals. Front Bioeng Biotechnol 11:1141523

Singh A, Singh R (2013) Microbial biosensors. Polymer Prod Chem Process Techn Anal Appl 16:183–202. https://doi.org/10.1201/b15539

Singh M, Verma N, Garg AK, Redhu N (2008) Urea biosensors. Sens Actuators B Chem 134(1):345–351. https://doi.org/10.1016/j.snb.2008.04.025

Some S, Mondal R, Mitra D, Jain D, Verma D, Das S (2021) Microbial pollution of water with special reference to coliform bacteria and their nexus with environment. Energy Nexus 1:100008. https://doi.org/10.1016/j.nexus.2021.100008

Sonawane JM, Ezugwu CI, Ghosh PC (2020) Microbial fuel cell-based biological oxygen demand sensors for monitoring wastewater: state-of-the-art and practical applications. ACS Sens 5(8):2297–2316. https://doi.org/10.1021/acssensors.0c01299

Soni T, Vivekanand V, Pareek N (2022) Biosensors as an effective tool for detection of emerging water and wastewater pollutants. In: Biodegradation and detoxification of micropollutants in industrial wastewater. Elsevier, Amsterdam, pp 39–54

Sonone SS, Jadhav S, Sankhla MS, Kumar R (2021) Water Contamination by heavy metals and their toxic effect on aquaculture and human health through food chain. Lett Appl NanoBioScience 10(2):2148–2166. https://doi.org/10.33263/LIANBS102.21482166

Su L, Jia W, Hou C, Lei Y (2011) Microbial biosensors: a review. Biosens Bioelectron 26(5):1788–1799. https://doi.org/10.1016/j.bios.2010.09.005

Sutrisno E (2017) The study of river pollution related to domestic water in the perspective of community legal culture. South East Asia J Contemp Business Econ Law 12(2):156–162

Swarup D, Patra RC (2005) Environmental pollution and its impact on domestic animals and wildlife. Indian J Animal Sci 75(2):231–240

Tang H, Xiang G, Xiao W, Yang Z, Zhao B (2024) Microbial mediated remediation of heavy metals toxicity: mechanisms and future prospects. Front Plant Sci 15(July):1–23. https://doi.org/10.3389/fpls.2024.1420408

Tang Y, Hardy TJ, Yoon J-Y (2023) Receptor-based detection of microplastics and nanoplastics: current and future. Biosens Bioelectron 234:115361

Thévenot DR, Toth K, Durst RA, Wilson GS (2001) Electrochemical biosensors: recommended definitions and classification. Biosens Bioelectron 16(1–2):121–131. https://doi.org/10.1016/S0956-5663(01)00115-4

Turner APF (2013) Biosensors: sense and sensibility. Chem Soc Rev 42(8):3184–3196. https://doi.org/10.1039/c3cs35528d

Vardhan KH, Kumar PS, Panda RC (2019) A review on heavy metal pollution, toxicity and remedial measures: current trends and future perspectives. J Mol Liq 290:111197. https://doi.org/10.1016/j.molliq.2019.111197

Verma N, Singh M (2005) Biosensors for heavy metals. Biometals 18:121–129

Waldrop MM (2021) Microbes for better sewage treatment. Proc Natl Acad Sci USA 118(32):1–5. https://doi.org/10.1073/pnas.2112863118

Wang L, Wang H, Tizaoui C, Yang Y, Ali J, Zhang W (2023) Endocrine disrupting chemicals in water and recent advances on their detection using electrochemical biosensors. Sens Diagnostics 2(1):46–77

Wilkinson JL, Boxall ABA, Kolpin DW, Leung KMY, Lai RWS, Wong D, Ntchantcho R, Pizarro J, Mart J, Echeverr S, Garric J, Chaumot A, Gibba P, Kunchulia I, Seidensticker S, Lyberatos G, Morales-salda JM, Kang H (2022) Pharmaceutical Pollution World's Rivers 119(8):1–10. https://doi.org/10.1073/pnas.2113947119/-/DCSupplemental.Published

Williams A, Aguilar MR, Pattiya Arachchillage KGG, Chandra S, Rangan S, Ghosal Gupta S, Artes Vivancos JM (2024) Biosensors for public health and environmental monitoring: the case for sustainable biosensing. ACS Sustain Chem Eng 12(28):10296–10312. https://doi.org/10.1021/acssuschemeng.3c06112

Xu X, Ying Y (2011) Microbial biosensors for environmental monitoring and food analysis. Food Rev Intl 27(3):300–329. https://doi.org/10.1080/87559129.2011.563393

Yang L, Zhou Y, Shi B, Meng J, He B, Yang H, Yoon SJ, Kim T, Kwon B-O, Khim JS (2020) Anthropogenic impacts on the contamination of pharmaceuticals and personal care products (PPCPs) in the coastal environments of the Yellow and Bohai seas. Environ Int 135:105306

Yasmin J, Ahmed MR, Cho B-K (2016) Biosensors and their applications in food safety: a review. J Biosyst Eng 41(3):240–254. https://doi.org/10.5307/jbe.2016.41.3.240

Zhou Y, Zheng S, Qin W (2024) Electrochemical biochemical oxygen demand biosensors and their applications in aquatic environmental monitoring. Sens Bio-Sensing Res 44(March):100642. https://doi.org/10.1016/j.sbsr.2024.100642

Chapter 5
Application of GIS to Analyze Wastewater Pollution and Its Impact on Surrounding Environment

Shelu Kumari and Vinay Bhushan Kumar

Abstract Water pollution has recently emerged as one of the big global issues due to its impact on people's lives and the environment, posing substantial human health risks. Geospatial methods, strategies, and technologies to collect or analyze spatial data are very important tools used for purposes in fields like water quality monitoring, wastewater management, hydrology applications in general (for example, rainfall), urban planning, and public health. Common methods for this type of approach include remote sensing, geospatial information systems (GIS), and spatial statistics. By leveraging such technologies, we gain valuable information on where and how pollution sources are distributed in space and time; the levels or concentrations of pollutants recorded by these devices also help us estimate incidences that may be detrimental to human health as well as the surrounding ecosystem. For example, remote sensing can be used to collect data on wastewater quality and pollution sources. Remote sensing can map the spatial distribution characteristics of water pollution, which can be realized by satellite and aerial imagery techniques for locating sources to determine the location and extent. Detection of the most polluted areas is an essential knowledge base for preventing some scenarios and minimizing others. Moreover, it may help in identifying sources of high pollution levels and evaluating the efficiency of any existing measures to control pollution. In conjunction with geospatial modelling, several frameworks of different types have been developed for pollution modeling. So those frameworks give you a way to build your models in a much more controlled way that should ensure they are all as consistent and accurate as possible. Working in conjunction with such frameworks, the use of geospatial approaches has been identified as a potentially useful method for evaluating pollution impacts on human health and the environment.

S. Kumari
Department of Botany, J. D. Women's College (Patliputra University), Patna, Bihar, India

V. B. Kumar (✉)
Department of Botany, T.P.S. College (Patliputra University), Patna, Bihar, India
e-mail: vinaykumar10121976@gmail.com

A. M. Yatoo et al. (eds.), *Wastewater Monitoring and Management by Advanced Approaches*, Advances in Wastewater Research,
https://doi.org/10.1007/978-981-95-2601-7_5

Keywords Wastewater pollution · Satellite data · Geographical information system (GIS) · Geospatial techniques

5.1 Introduction

Industrialization, over time, has had a major impact on both urbanization and economic growth. Unfortunately, it has also caused a significant decline in the state of the environment, primarily due to the discharge of dangerous chemical and biological pollutants into the air, water, and soil in both gaseous and solid forms (Kaplan et al. 2024). Bănăduc et al. (2024) assert that human and natural activity-induced environmental contamination has emerged as a significant issue in modern civilization. Uneven water distribution stresses the globe due to growing pollution and consumption (Boyd 2019). Water makes up 70% of the Earth's surface and 50–95% of the mass of all living organisms (Dargaville and Hutmacher 2022). Problems, including droughts in certain regions, flooding in others, and a growing demand for clean water, led to pollution, scarcity, and ecological dangers (Altıok et al. 2023). Water pollution is the untreated discharge of contaminants into natural water bodies, such as lakes, rivers, seas, aquifers, reservoirs, and groundwater, either directly or indirectly. This pollution has a negative impact on ecosystems, particularly on aquatic and plant life (Bănăduc et al. 2024). Water can become contaminated in two primary ways: (1) through point sources, which are easily identifiable sources of contaminants like pipes or ditches; and (2) through nonpoint sources, which are sources of cumulative low-concentration pollution releases from wide areas. Nonpoint source contamination is frequently caused by fertilizer and pesticide leaching, which includes xenobiotic compounds and nitrogen from agricultural areas, as well as nutrient and phosphate runoff that eventually reaches stormwater and water bodies (Ingrao et al. 2023). The four primary types of contaminants found in water are microbiological, inorganic, organic, and radioactive. Some of these compounds have the potential to be extremely harmful, even at very low quantities (Kumar et al. 2023).

Water pollution can also be categorized based on the kind of contaminants present, such as chemical, physical, or biological. Chemical pollution encompasses many contaminants, such as industrial chemicals, pesticides, and heavy metals. Physical pollution includes things like variations in water temperature, turbidity, and sediment concentration. The presence of potentially harmful microorganisms like bacteria and viruses is referred to as "biological pollution." Water contamination can result from a variety of sources, including industrial discharges, agricultural runoff, urban stormwater runoff, and untreated sewage. It is crucial to locate and monitor these sources to manage water resources and minimize pollution (Kaplan et al. 2024). Standards for managing water quality have been established to avoid contaminating water bodies and to offer guidance when selecting water sources for various uses. The evaluation of water quality is crucial to many businesses, including residential, commercial, and agricultural water supplies, fisheries and aquaculture, outdoor leisure, and the overall health of ecosystems (Boyd 2019). The water's quality is determined by its biological, physical, and chemical properties, each of which includes a range of specific metrics (Akhtar et al. 2021). While temperature, turbidity, color, taste, odor, particles, and electrical conductivity (EC) are examples of physical qualities, bacteria, algae, viruses, and protozoa are examples of biological markers. pH levels, heavy metal concentrations, inorganic and organic component concentrations, water hardness, dissolved oxygen (DO), chemical oxygen demand (COD), and biochemical oxygen demand (BOD) are examples of indicators of chemical composition. However, these assessment methods can be complex and may not always provide a trustworthy image of the water's quality. Furthermore, due to their high costs, labor-intensive procedures, extensive time commitments, and potential risks, these approaches might not be able to identify every contaminant in the water. The results of these tests are typically compared with established rules to assess and classify the water quality; however, this conventional approach may be outdated and subjective (Alvizuri-Tintaya et al. 2022). Therefore, to improve the resilience and reliability of traditional in-situ water quality monitoring, it is imperative to combine biological and physicochemical research with distant assessment methodologies. Wastewater management is especially crucial in developing countries due to the strain that growing urbanization, industrialization, and population growth are putting on the infrastructure. Improper wastewater treatment has led to the pollution of water bodies, posing a major hazard to public health and the environment. Untreated wastewater is discharged into rivers, lakes, and coastal areas because many of these cities lack adequate sewage infrastructure. The main effects of inadequate wastewater management, such as the rise in waterborne illnesses including cholera, dysentery, and typhoid, have a detrimental impact on urban populations' health. Another factor contributing to environmental degradation is polluted streams, which destroy aquatic life and decrease biodiversity. Additionally, groundwater contamination is a possibility (Saranya et al. 2024). In many cities, groundwater is a major source of drinking water. Investments in sewage treatment infrastructure, such as sewage treatment facilities, are essential to ensure that wastewater is sufficiently cleaned before being released into the environment. In urban areas, innovative methods can improve risk reduction, sustainable management, and quality of life. Decentralized wastewater treatment and wastewater reuse and recycling are two examples of these strategies (Singh et al. 2023).

An integral component of urban infrastructure is wastewater management, which comprises the collection, handling, and recycling of wastewater generated by residences, businesses, and industries. Effective wastewater management is essential for maintaining environmental quality, promoting sustainable urban development, and safeguarding public health. Because of their dense population, extensive industrialization, and economic activity, urban areas generate a significant volume of wastewater (Singh et al. 2023). Reusing and recycling treated wastewater is essential to conserving water and lessening the demand on freshwater resources.

Wastewater management, which includes the gathering, treating, and recycling of wastewater produced by homes, companies, and industries, is a crucial part of urban infrastructure. According to Čampulová et al. (2018), sustaining environmental quality, encouraging sustainable urban development, and protecting public health all depend on effective wastewater management. Due to their dense population, considerable industrialization, and economic activity, urban areas produce a large amount of wastewater. Water conservation and reducing the demand on freshwater resources depend on the reuse and recycling of treated wastewater (Gao et al. 2018). When combined with methods such as the analytic hierarchy process (AHP), geographic information systems (GIS) provide a geographical framework for the integration, analysis, and presentation of wastewater management data. Numerous studies have used AHP and GIS-based multi-criteria decision analysis (MCDA) to identify study sites, particularly in stormwater recharge zones (Saragih et al. 2020; Roy et al. 2022). GIS is useful for tracking pollution plumes in aquatic bodies, locating and tracking the origins of wastewater outflow, and spotting land use changes that can affect wastewater systems. It can also be used by stakeholders to map the wastewater infrastructure, assess treatment facility capacity and distribution, and pinpoint the locations of contamination sources. GIS-based analysis can be used to identify sites at risk of flooding or sewage overflow, which can help with planning infrastructure upgrades or expansions (Abdessamed and Abderrazak 2019; Saranya et al. 2024). Machine learning improves geographic information systems (GIS) by automating data processing and deriving conclusions from large datasets (Akouemo and Povinelli 2017). Machine learning algorithms can analyze complex relationships between variables such as weather patterns, land use changes, and population expansion to predict future demands on wastewater infrastructure. Effective wastewater treatment is essential for protecting the environment and public health (Velimirović et al. 2021).

5.2 A Summary of Remote Sensing and Water Quality Evaluation

Traditional in situ methods provide only point estimates of the water conditions at specific times and locations, making it nearly impossible to monitor broader trends. This makes it difficult to collect spatial and temporal variations of water quality indices in large water bodies (Loaiza et al. 2023). One of the main problems with in situ approaches is that it can be hard to record broad temporal and spatial fluctuations and trends in water quality, especially in vast bodies of water. Topographic and accessibility issues frequently impede the monitoring and management of entire water systems, and in situ sampling is costly, time-consuming, and labor-intensive. Furthermore, mistakes made during field sampling and laboratory analysis may have an impact on the precision and accuracy of the data that was gathered (Gholizadeh et al. 2016). These constraints have been addressed by remote sensing techniques, which have over 40 years of demonstrated efficacy in monitoring and evaluating inland water quality. Researchers often use the visible and near-infrared (NIR) portions of the solar spectrum to create relationships between the reflectance of the water column and physical and biogeochemical parameters, like the concentration of phytoplankton (chlorophyll-a [Chl-a]), the water transparency, and the suspended sediments (both mineral and organic) in various water bodies (Zheng and DiGiacomo 2017).

Although remote sensing is a useful tool for evaluating the quality of water, it needs to be integrated with field surveys and conventional sample methods because it is insufficiently precise when used alone. Combining computer-based water quality modeling, in situ measurements, and remote sensing can lead to a more thorough understanding of aquatic system water quality. Furthermore, data sharing and cooperation across public, federal, and commercial organizations can greatly increase the amount and caliber of data needed for regional research. Four main benefits of combining remote sensing with other water quality monitoring systems were listed by Pahlevan et al. (2022). The energy spectrum of reflected solar radiation from aquatic bodies can be altered by the interaction of optically active components with light, which can be observed using remote sensing (Bukata et al. 2018). These optically active components make up many important water quality metrics in surface waters. However, other factors—like viruses, chemicals, and acidity—cannot be identified by remote sensing because they have no direct impact on the spectral properties of reflected light. However, these characteristics can also be inferred or calculated from strong correlations with markers of water quality that can be found using remote sensing (Ritchie et al. 2003; Kallio 2000).

5.3 Airborne and Space-Based Sensors for Water Quality Research

Based on their platforms, two categories of observing sensors can be generally classified. Within the Earth's atmosphere, airborne sensors are fixed on platforms like boats, balloons, helicopters, or aircraft, whereas spaceborne sensors are carried by spacecraft or satellites outside the atmosphere. It is necessary to have a solid understanding of the features of various sensors to select the right one for a study. Due to this, a variety of satellite and aircraft-based remote sensing technologies that are frequently used in water quality evaluations are showcased, along with their respective spectral attributes, including bands, spatial resolution, and revisit intervals. When doing water quality tests, this tabular data is helpful in choosing the best sensors from the many that are on the market. Other popular sensor types used in oceanographic remote sensing are microwave radiometers (MWR) and synthetic aperture radar (SAR). Passive microwave radiometers measure microwave radiation, which is emitted between 1 and 1000 GHz at wavelengths ranging from submillimeter to centimeter. Sea surface temperature (SST) and sea surface salinity (SSS), two important indicators of water quality, may be determined by oceanographers thanks to the physical processes underlying energy emission at these wavelengths (Adjovu et al. 2023). Microwave radiometers have also been used for monitoring atmospheric and terrestrial radiation, as well as for zenith-pointing surface instruments that observe the Earth's atmosphere immediately above the stationary sensor in meteorological research. Due to their unique properties, microwave radiometers are frequently used in oceanographic and water quality research. To create two- or three-dimensional photographs of objects, synthetic aperture radar (SAR), a form of radar that may be installed on aircraft or spacecraft, is utilized (Hupton 2009; Minvielle et al. 2024). SAR is not often used for direct water quality research or parameter measurement, despite being frequently used for identifying water contamination (such as oil spills), ocean topography, sea surface wind speeds, and ice monitoring. However, as studies like Zhang et al. (2003) have shown, SAR data can improve the evaluation of water quality indicators when paired with data from other sensors.

5.4 Assessing the Quality of Water Through Remote Sensing Techniques

Water quality studies require determining the chemical, physical, and biological characteristics of water bodies in addition to identifying possible sources of contamination that lower water quality (Usali and Ismail 2010). Nevertheless, several important aspects of water quality, such as pH, total nitrogen (TN), ammonia nitrogen (NH_3–N), nitrate nitrogen (NO_3–N), and dissolved phosphorus (DP), are frequently ignored in the literature that is currently available. Their weak optical characteristics and low signal-to-noise ratios are the main causes of this absence. However, these parameters are essential parts of water quality indices and constitute a difficult field of research in water quality evaluations aided by remote sensing, which ought to motivate researchers to carry out more work. The most popular techniques for determining the concentrations of eleven water quality measures are then reviewed in the publication (Chidiac et al. 2023). Chl-a, colored dissolved organic matter (CDOM), Secchi disk depth (SDD), turbidity, total suspended sediments (TSS), water temperature (WT), total phosphorus (TP), dissolved oxygen (DO), biochemical oxygen demand (BOD), and chemical oxygen demand (COD) are the water quality indicators that are included in this analysis.

5.4.1 Chlorophyll-a

The concentration of chlorophyll-a (Chl-a), which is necessary for photosynthesis, is directly correlated with algal blooms, which are frequently brought on by freshwater eutrophication processes (Lim and Choi 2015). Chl-a is a vital component of oxygenic photosynthesis and is present in algae, plants, and cyanobacteria. Chl-a modulates the link between nutrient concentration, especially phosphorus, and algal productivity, acting as a major trophic status indicator. Although chlorophyll mainly reflects green light, most of the energy from violet-blue and orange-red light wavelengths is absorbed by it, giving it its characteristic green color. When Chl-b is present in addition to Chl-a, the absorption spectrum gets wider. The ratio of produced Chl-b to chlorophyll-a molecules is generally larger under lower light circumstances, which increases photosynthetic production (Kume et al. 2018). The absorption spectra of the pigments Chl-a and Chl-b are shown in Fig. 5.1. Several investigations have demonstrated that the spectral sensitivity at shorter wavelengths, especially in the blue band, decreases when the concentration of Chl-a increases (Binh et al. 2022). Several studies that investigate the use of remote sensing to measure chlorophyll-a concentration are included in this chapter.

Narrow bands of pictures are necessary to evaluate the concentrations of chlorophyll-a and how they change over time and space in a body of water (Kolluru et al. 2023). There is some uncertainty, according to a survey of the literature on the application of multispectral sensors in empirical methodologies. Numerous studies show that the broad wavelength spectral data available from current satellites (such as Landsat and Satellite Pour l'Observation de la Terre [SPOT]) do not effectively differentiate chlorophyll in waters with high suspended sediments, even though many authors report satisfactory results using broadband sensors for chlorophyll estimation (Kolluru et al. 2023). According to Östlund et al. (2001), this restriction is most noticeable in extremely turbid and eutrophic settings where the spectral signal from suspended sediments predominates. The spectral bands ranging from blue to green are useful in precisely estimating the amounts of Chl-a present in Case 1 waters. But because the components of Case 2 waters are so diverse, identifying Chl-a becomes

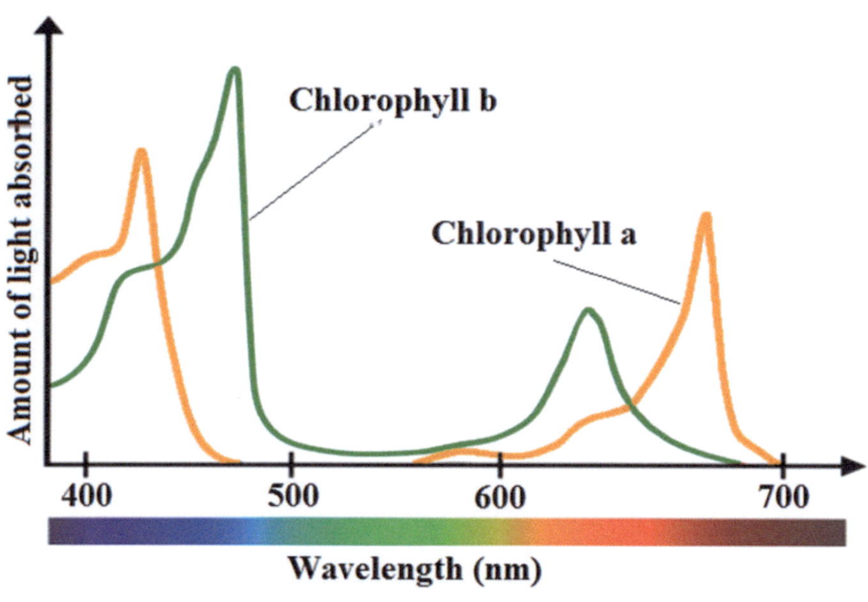

Fig. 5.1 The absorption spectrum of both the Chl-*a* and the Chl-*b* pigments

a more difficult task requiring sophisticated methodologies and techniques (Mobley et al. 2004). In addition to phytoplankton, a combination of dead particle organic matter, inorganic particulate matter, and terrestrial dissolved organic matter affects the optical characteristics of Case 2 waters (both inland and coastal). Due to this lack of statistical correlation, measuring the concentration of chlorophyll-a is more difficult and less accurate. Moreover, the blue-green zone in Case 2 waters is frequently obscured by Gelbstoff absorption, suggesting that chlorophyll-a algorithms designed for Case 1 waters are not appropriate for Case 2 waters (Gemperli 2004). Since spectral band ratios in the remotely detected signal can help minimize the impacts of the air–water interface, atmosphere, and irradiance, a range of visible spectral bands and their ratios are often used to estimate Chl-a (Lillesand et al. 2014). With a peak reflectance at 550 nm (green) and around 700 nm (near-infrared, or NIR), Chl-a displays strong scattering and absorption properties between 450 and 475 nm (blue) and at 670 nm (red). Various techniques for determining chlorophyll-a concentrations in turbid waters have been developed using the reflectance peak at 700 nm and its ratio to the reflectance at 670 nm (Gitelson et al. 2008). Potential wavelengths for determining chlorophyll concentrations using first derivatives were determined by Han and Jordan (2005) to be the spectral bands at 630–645, 660–670, 680–687, and 700–735 nm. Multiple spectral bands can be used to successfully study the scattering and absorption features of Chl-a, as noted by Barraza-Moraga et al. (2022). According to He et al. (2020), a combination of the Advanced Visible-Infrared Imaging Spectrometer (AVIRIS) band centered around 713 nm and the 667 nm band yields the best sensitivity ratio for retrieving chlorophyll in inland waterways. It's also been shown that the ideal ratio (R674/R705) can be used to evaluate the amounts of chlorophyll in inland lakes and rivers (Thiemann and Kaufmann 2002). With estimated R values of -0.66, -0.70, -0.64, and -0.64 for Bands 2, 3, 4, and the band ratio (Band 5/Band 3), Lim and Choi (2015) found a strong correlation between chlorophyll-a (Chl-a) concentration and various Operational Land Imager (OLI) bands as well as band ratios using Landsat-8/OLI. All these results were obtained at a significance level of $p < 0.01$. Using Band 2, Band 5, and the ratio of Band 2/Band 4, Kim et al. (2014) also used Landsat-8/OLI to measure the concentration of Chl-a. Mannheim et al. (2004) found that the maximum sensitivity to variations in Chl-a concentration was observed in the baseline and reflectance curve between 672 and 742 nm (CHRIS spectral bands 8–12). Using data from the Moderate Resolution Imaging Spectroradiometer (MODIS), Sea-viewing Wide Field-of-view Sensor (SeaWiFS), Medium Resolution Imaging Spectrometer (MERIS), and RapidEye, Choe et al. (2011) estimated the concentrations of Chl-a in turbid waters using two-band and three-band models based on band ratios, including the Red and NIR bands. Furthermore, the third-largest freshwater lake in China, Taihu Lake, has Chl-a concentrations that can be estimated according to a method developed by Qi et al. (2014) based on empirical orthogonal function (EOF) analysis. This EOF approach examined the spectrum variation of normalized Rayleigh-corrected reflectance (Rrc) data at 469, 555, 645, and 859 nm using 28 concurrent MODIS and field observations. It then connected that variance to Chl-a concentrations. A novel empirical algorithm for estimating the concentration of Chl-a in Poyang Lake, China's largest freshwater lake, was developed by Feng et al. (2014) using atmospherically corrected data from the Medium Resolution Imaging Spectrometer (MERIS) and a normalized green–red difference index (NGRDI). Many Chl-a concentration calculation algorithms, according to a review of the literature, rely on wavelengths in the vicinity of 675 and 700 nm. For instance, the spectral bands of Landsat/Enhanced Thematic Mapper Plus (ETM+) are neither broad enough nor ideally positioned to detect Chl-a. Similarly, it can be argued that multispectral sensors like Advanced Spaceborne Thermal Emission and Reflection Radiometer (ASTER), Indian Remote Sensing–Linear Imaging Self-Scanning Sensor III (IRS-LISS III), and Satellite Pour l'Observation de la Terre–High-Resolution Visible (SPOT-HRV) are also not very suitable for Chl-a detection because their spectral locations are equivalent in the Red/NIR range.

5.4.2 Colored Dissolved Organic Matters (CDOM)

Colored dissolved organic matter (CDOM) comprises naturally occurring, water-soluble, biogenic, and heterogeneous organic molecules that have a yellow to brown hue and can be found in both freshwater and saltwater environments (Marinho et al. 2024). When these brown compounds are present in large enough quantities, they can give water a yellow–brown color, which is why they are called yellow matter, or CDOM. Total suspended solids (TSS) and Chl-a have an impact on water color. The absorbance spectra of CDOM can overlap with the absorption of chlorophyll at a wavelength of 443 nm, where chlorophyll concentrations are often measured, accounting for more than 50% of the overall absorption (Ruescas et al. 2017). The reflectance values in the blue and green parts of the spectrum, particularly below roughly 500 nm, are greatly influenced by the absorbance of colored dissolved organic matter (CDOM), which grows exponentially as the wavelength decreases. The utilization of models for phytoplankton production and methods for retrieving Chl-a based on remotely sensed ocean color may become more challenging due to this effect (Miller et al. 2002). Furthermore, with large quantities of CDOM, Strömbeck and Pierson (2001) shows that absorption in the red-light spectrum can be significant. Carbon dynamics and aquatic ecology studies require the use of remote sensing of CDOM (Mannino et al. 2008). When colored dissolved organic matter (CDOM) is present, several satellite remote sensing instruments, including MODIS and SeaWiFS, may identify changes in the color of the water in rivers, lakes, and seas (Antoine et al. 2008). Additionally, CDOM has an impact on the underwater light field and the intrinsic optical properties (IOP) of water, which in turn changes the reflectance of the water that is detected by remote sensors. Consequently, a useful technique for estimating CDOM concentration over wide geographic and temporal scales is to invert remote sensing data (Aurin et al. 2018). The absorption properties of CDOM, its absorption coefficients at 440 nm, are frequently utilized as surrogates for CDOM content in ocean color research (Nelson and Siegel 2002). Chl-a concentrations in algorithms obtained from sensors like Coastal Zone Color

Scanner (CZCS) are empirically inverted, enabling the detection of CDOM under the supposition that it covaries with chlorophyll (O'Reilly and Werdell 2019). Hyperspectral data and recently created remote sensing reflectance models, such as EO-1 Hyperion with the Matrix Inversion Method (MIM), have been used to estimate CDOM as one of the components of ocean color (Qiang et al. 2023). Kutser et al. (2005) used the band ratio of EO-1/ALI Bands 2 and 3 to evaluate the CDOM content in the lakes of Southern Finland. Current developments in colored dissolved organic matter (CDOM) estimation seek to improve the precision of CDOM inversion through the incorporation of new elements like bottom effects, the integration of hyperspectral remote sensing data with semi-analytical models, and the application of sophisticated computational methods (Zhu et al. 2011). CDOM absorption is commonly reported in terms of color and platinum-cobalt units (PCU) in many water quality monitoring studies. There is a substantial link between CDOM absorption and PCU color, according to recent studies. Light absorption coefficients at wavelengths are a frequent way to report CDOM absorption (Corbett 2007). The absorption coefficients of CDOM for both MODIS (Moderate Resolution Imaging Spectroradiometer) and SeaWiFS (Sea-viewing Wide Field-of-view Sensor) can be separately inverted from remote sensing reflectance (Rrs) in semi-analytical models (Zhu et al. 2011). Radiance and the underwater light field have been simplified, and radiative transfer equations constitute the foundation of these semi-analytical models (Qiang et al. 2023). Nonetheless, spectral signals of suspended sediments and chlorophyll frequently overlap with those of CDOM, making remote sensing of CDOM in riverine and coastal waters more difficult and error-prone than in oceanic waters (Pan et al. 2008). Due to its broad range of narrow bands and spectrum reactions to the inherent optical characteristics (IOP) of water, hyperspectral remote sensing proves helpful in these difficult scenarios. Research has shown that the evaluation of water's IOP in coastal environments can be enhanced by high spectral resolution (10 nm or higher) (Doxaran et al. 2006). Still, as was already indicated, the spatial and temporal variability of riverine and coastal waters, as well as interference from spectral signals of suspended sediments and chlorophyll, may cause the ideal bands for detecting CDOM to not always be situated at the same wavelengths.

A major problem is locating relevant wavelengths within the hundreds of small bands of hyperspectral reflectance (Karaska et al. 2004). To lower the dimensionality of the hyperspectral data, methods such as band selection, derivative analysis, spectral indices, or hyperspectral transformation should be used initially (Xu and Gong 2007). A shipboard data collection technique is necessary not just for high spatial resolution underwater CDOM observation but also for calibrating and validating remotely sensed CDOM concentrations. Additionally, Qiang et al. (2023) point out that a few studies have demonstrated a link between CDOM and dissolved organic carbon (DOC) (Vignudelli et al. 2004), suggesting that CDOM is frequently utilized as a vital indicator of the dynamics of dissolved organic carbon (DOC) in freshwater and coastal marine environments. According to a survey of the literature, most research focuses on four primary areas: underwater CDOM measurements, water-surface reflected radiance measured by remote sensors on aircraft or satellites, in situ hyperspectral observations, and functional data processing.

5.4.3 Secchi Disk Depth

One visual characteristic of water that is closely related to the different elements found in water bodies is the Secchi depth. The Secchi depth and the total suspended solids (TSS) concentration of water bodies have an inverse connection (Gholizadeh et al. 2016). The most widely used technique for determining water transparency is based on the ideas of light attenuation, according to Bigham Stephens et al. (2015). The most widely used operational measure of water transparency was created in 1865 by Pietro Angelo Secchi SJ and is a circular disk used to evaluate the quality of lakes and seas. Once the disk's design is completely submerged, it is lowered into the water while still attached to a line. The Secchi disk depth (SDD) is the term used to describe this measurement, which is also connected to water turbidity. Apart from the lakes with the most vibrant colors and low levels of chlorophyll-a and nonalgal turbidity (such as clay or calcium carbonate), SDD is a good indicator of trophic circumstances (algal abundance). There is a negative correlation between Secchi depth and TSS levels in aquatic environments. As a result, measuring SDD and keeping an eye on water transparency can be accomplished with the use of remote sensing. Lee et al. (2015) recently proposed a model for Secchi disk depth (SDD) estimation that is based only on the diffuse attenuation coefficient at a wavelength corresponding to maximum transparency for such interpretations, in contrast to the classical model, which heavily relies on the beam attenuation coefficient. As such, several researchers have utilized remote sensing techniques, exhibiting a robust association between sensor data and SDD values (Gholizadeh et al. 2016). SDD and atmospherically corrected satellite radiance have a strong correlation. When considering the ratio of backscattering to total scattering coefficients for suspended particles, SDD corresponds with reflectance slightly below the surface in the atmospherically corrected Multi-Spectral Scanner (MSS) Green band (Wen et al. 2024). This correlation is fairly accurate for SDD values <16 m (Liu et al. 2003). Several remote sensing data sets, including Thematic Mapper (TM) (Álvarez-Robles et al. 2006), MSI (Yang et al. 2022), IKONOS, and even video data, have been used to build significant algorithms for calculating SDD. Landsat-TM is one of the most popular sensors among them for estimating SDD. Strong correlations between SDD and TM data were observed by Zhang et al. (2022), particularly during high tide. Numerous relationships between Secchi depth and total suspended solids (TSS), colored dissolved organic matter (CDOM), total phosphorus, and Chl-a have been documented in the literature. Prior research has shown that certain spectral bands and band ratios are useful for measuring Secchi depth. For example, Bhatti et al. (2010) used Advanced Land Observing Satellite–Advanced Visible and Near Infrared Radiometer type 2 (ALOS-AVNIR-2) data and discovered a high association between Secchi depth and the reflectance ratio of R750/R560 (NIR/Green). Thiemann and Kaufmann (2002) used Compact Airborne Spectrographic Imager (CASI) and Hyperspectral Mapper (HyMap) data to evaluate Secchi disk transparency and chlorophyll-a levels in the Mecklenburg Lake District of Germany. The spectrum between a baseline and the 400–750 nm range revealed a substantial correlation with Secchi disk transparency

(SDT) measured in situ. A Secchi depth measurement method based on Band 1 (445–530 nm), Band 2 (520–610 nm), and Band 3 (640–720 nm) from IKONOS data was developed by Ekercin (2007). A formula utilizing TM1 and the ratios of TM3/TM2, TM1/TM2, and TM2/TM1 was developed by Mancino et al. (2009). Powell et al. (2008) have presented a regression equation connecting in situ Secchi disk transparency measurements to the blue, green, and red bands of TM. Additionally, based on the results of the Kloiber et al. (2002) study and analyses of TM and MSS images, recommendations were given for a Landsat-based technique of evaluating water clarity.

5.4.4 Turbidity and Total Suspended Sediments

An optical property of water called turbidity makes light absorb and scatter instead of moving in straight lines. While colored dissolved or particulate matter and chlorophyll-a govern absorption, suspended sediments account for most of this scattering (Myint and Walker 2002). Since suspended matter is the main cause of water turbidity, turbidity measurement—often regarded as the inverse of clarity—is widely employed to evaluate the quantities of suspended silt in rivers (Sehgal et al. 2022). The number of suspended particles in the water determines its turbidity; the more suspended particles there are, the murkier or more turbid the water, and the more difficult it is for light to pass through. The intricate makeup of the suspended particles modifies the water body's reflectance, giving rise to color changes. As a result, using water color alone to evaluate remotely sensed data is frequently inadequate and erroneous. Furthermore, turbidity and total suspended matter are regarded as significant in much research because they are related to incoming sunlight, which affects photosynthesis and the growth of algae and plankton. These features are also closely correlated with the Secchi disk depth. It is routine practice to evaluate turbidity and suspended particle concentrations, together with their temporal and spatial fluctuations, using remote sensing techniques. In theory, a robust algorithm sensitive to total suspended matter (TSM) can be produced by carefully choosing and applying a single band (as found that compared to other sensors, Landsat/TM was employed substantially more frequently. CASI and AISA hyperspectral pictures on airborne sensors for rivers and other case studies needing enhanced spectral and spatial resolution, as well as ALOS/AVNIR-2 and IKONOS on spaceborne sensors, were used to analyze turbidity and suspended materials.

5.4.5 Total Phosphorus

All forms of phosphorus, including dissolved, inorganic, and organic forms, are measured in studies on total phosphorus (TP). When phosphorus levels are high, plants and algae can grow quickly. Phosphates are vital nutrients for plants. Total phosphorus and Chl-a concentration are correlated either directly or indirectly; this relationship is frequently mediated by Secchi depth, which is used to measure the transparency or clarity of water (Swanson and Zurawell 2006). Rivers that pass through different land-use zones can carry a variety of substances and materials, such as fertilizers, runoff from residential areas, suspended sediments, and more. For example, when a river or creek flows through an area used for agriculture, the amount of phosphorus in the surface water may be higher than in other materials. The main causes of the high phosphorus and nitrogen levels in surface waterways, which pose a threat to many ecosystems worldwide, are fertilizer-rich agricultural runoff and vast amounts of wastewater treatment plant effluents. According to Zhao et al. (2022), there is a strong exponential association between total suspended materials and Secchi disk transparency. These materials are frequently used as carriers for TP. Numerous investigations have demonstrated a general trend wherein waterbodies' concentrations of Chl-a increase when TP levels rise (Müller-Navarra et al. 2004). Additionally, Heiskary and Wilson (2005) discovered that a certain amount of phosphorus can adhere to soil erosion-related suspended particles, which rivers then carry downslope. These studies showed a significant relationship between Secchi disk depth and chlorophyll-a concentration, indicating that these variables may be indirectly utilized to estimate TP concentration in theory.

5.4.6 *Water Temperature*

Water temperature is a crucial factor that impacts air–water interactions and the many activities taking place in the water, which in turn impacts physical, chemical, and biological processes in aquatic habitats. Water temperature affects many chemical constituents' solubility and availability, which has a major effect on dissolved oxygen levels. The solubility of oxygen diminishes with increasing water temperature, which is particularly significant for aquatic life. Analysis of the temporal variations in water temperature brought on by seasonal variations is also crucial. Furthermore, there is a direct correlation between the temperature of the water column and the amount, flow, and interaction of some pollutants, such as fertilizers. Because rivers are smaller than other bodies of water, it is more difficult to remotely sense the temperature of the water in them because of the challenges involved in recording temperature swings at the resolution (pixel size) of thermal infrared (TIR) data (Handcock et al. 2006). As endangered fish species are extremely susceptible to rising water temperatures, this factor is especially crucial for regulating their numbers. A network of in-stream gauges that track temporal changes at sites is usually used in traditional techniques of determining water temperature; however, this approach is frequently constrained by sparse spatial and temporal sampling. The gauges' sites, which are mostly in large streams and rivers, restrict the geographical spread of river temperatures. Using remote sensing techniques is a promising substitute for tracking and measuring stream temperatures with known accuracies and uncertainties. Temperature distribution in streams and rivers may be accurately measured using remotely detected thermoluminescence images. Thermal radiation emitted from the top 0.1 mm of the water surface within the 3–5 and 8–14 μm wavebands can be detected by sensors to assess water temperature remotely (Torgersen et al. 2001).

The method of detecting TIR radiation emission (3–14 μm) from satellites to determine regional and global sea-surface temperature (SST) is a well-established practice in oceanography (Parkinson 2003). At first, TIR remote sensing was mostly used for applications related to lakes and the shore, including tracking thermal pollution from cooling water that power plants released into the environment (Chen et al. 2003). But starting in the 1990s, government organizations started using aerial TIR remote sensing over large river segments to track the quality of the water, locate cold-water influx sources, and create spatially referenced river temperature models. Because TIR image sensors have different spectral bands across a range of wavelengths, there are currently many of them that can be used to measure water temperature. It is crucial to carefully consider the minimum instrument noise and environmental effects when choosing the proper band or bands for accurate water temperature estimations. Averaging various bands can yield a more accurate approximation of the real temperature by reducing noise from ambient or sensor changes (Handcock et al. 2006).

5.5 Overview of Water Indicators Retrieval Through Remote Sensing

Many satellite images can be used to assess the quality of water, but Landsat TM (Thematic Mapper) images have gained popularity due to their temporal coverage, geographical resolution, and affordability (Adjovu et al. 2023). One of the earliest sensors still in use for assessing water quality today, the Landsat-5 sensor acquired the TM data between 1984 and November 2011 (Markogianni et al. 2013). Thematic Mapper (TM), Multi-Spectral Scanner (MSS), Enhanced Thematic Mapper (ETM), and Operational Land Imager (OLI) are among the Landsat sensors that have been successful in measuring a wide range of significant water quality parameters, including total phosphorus, total suspended solids, turbidity, dissolved oxygen, biochemical oxygen demand, and chemical oxygen demand (Markogianni et al. 2013; Schaeffer et al. 2013). Nevertheless, evaluating water quality metrics with Landsat data has several significant drawbacks. Significant limitations are placed on intraseasonal monitoring by the 16-day repeat cycle, especially in regions with high cloud cover. Furthermore, according to Gholizadeh et al. (2016), the characteristics pertaining to water quality must be associated with an "inherent optical property" (IOP) that is measurable by the satellite sensor. Examining the relationship between Secchi disk transparency (SDT) and the radiance seen by Landsat TM and MSS in various spectral bands, numerous possible sources of error associated with fluctuating atmospheric conditions were identified. There are currently few space-based hyperspectral sensors with sufficient spatial resolution to monitor smaller lakes and ponds, even though spaceborne hyperspectral sensors with low spatial resolutions, like MERIS and MODIS, have been used for years to monitor coastal and large inland water bodies. With its 30 m spatial resolution, the Hyperion hyperspectral image spectrometer, which was launched on the EO-1 platform in November 2000, represented a breakthrough for operational monitoring (USGS 2024).

5.6 Conclusion and Future Prospects

Water quality has been severely deteriorated by anthropogenic activities and increasing industrial growth. One of the most dependable, economical, and useful ways to monitor water quality parameters in different water bodies, such as lakes, rivers, and groundwater, is to combine conventional in situ sampling with remote sensing and GIS approaches. According to recent research, a range of satellite and aircraft sensors can precisely measure water quality markers. More efficient technology for determining the features of water quality has recently created hyperspectral satellite photography, including the Hyperspectral Imager for the Coastal Ocean (HICO), which can capture up to 200 spectral channels simultaneously. Additionally, the concurrent acquisition of more narrowly spaced bands by hyperspectral aerial sensors expands the possibility of evaluating and tracking many water quality parameters. Thus, using data from remote sensing to monitor and assess problems with water quality can help to improve the management of water resources. Remote sensing assessments of water quality, however, are rarely used as the basis for administrative decisions. Instead, the majority of current techniques center on models of continuous (ship- or buoy-based) or periodic (boat-based) monitoring. It can be difficult for managers and policymakers to comprehend the technical capabilities, constraints, and justifications of remote sensing techniques if they lack technical competence. Although it is necessary to make managers' impressions of water quality clearer, doing so can be difficult at times. It is highly recommended that optical and remote sensing researchers get in closer contact with water resource management organizations—beyond just publishing papers in peer-reviewed journals—to explore the most effective ways to address crucial monitoring demands using the instruments that are currently available. This chapter has shown how inland water quality can be assessed using satellite and aerial remote sensing methods. Because they have a higher spectral and spatial resolution as well as a larger variety of spectral bands, which enable more accurate retrieval of water quality data, airborne sensors are more versatile than spaceborne sensors. Airborne sensors are ideal for monitoring smaller bodies of water, like rivers, ponds, estuaries, and tributaries, while satellite sensors perform better for observing larger bodies of water. Additionally, tabulated data on the spectrum, spatial, and temporal properties of both airborne and spaceborne sensors are included in this chapter. These tabulated data can be used as a guide when choosing sensors for pertinent investigations.

References

Abdessamed D, Abderrazak B (2019) Coupling HEC-RAS and HEC-HMS in rainfall–runoff modeling and evaluating floodplain inundation maps in arid environments: case study of Ain Sefra city, Ksour Mountain, SW of Algeria. Environ Earth Sci 78(19):586

Adjovu GE, Stephen H, James D, Ahmad S (2023) Overview of the application of remote sensing in effective monitoring of water quality parameters. Remote Sens 15(7):1938

Akhtar N, Ishak MIS, Ahmad MI, Umar K, Md-Yusuff MS, Anees MT et al (2021) Modification of the water quality index (WQI) process for simple calculation using the multi-criteria decision-making (MCDM) method: a review. Water 13(7):905

Akouemo HN, Povinelli RJ (2017) Data improving in time series using ARX and ANN models. IEEE Trans Power Syst 32(5):3352–3359

Alparslan E, Aydöner C, Tufekci V, Tüfekci H (2007) Water quality assessment at ömerli dam using remote sensing techniques. Environ Monit Assess 135:391–398

Altıok E, Şen F, Wolska J, Cyganowski P, Bryjak M, Kabay N et al (2023) Separation of boron and arsenic from geothermal water with novel gel-type chelating ion exchange resins: batch and column sorption-elution studies. Molecules 28(23):7708

Álvarez-Robles JA, Zarazaga-Soria FJ, Ángel M, Latre RB, Muro-Medrano PR (2006) Water quality monitoring based on sediment distribution using satellite imagery. In: Proceedings of the 9th AGILE conference on geographic information science, Visegrad, Hungary, 20–22 April 2006, pp 144–150

Alvizuri-Tintaya PA, Villena-Martínez EM, Avendaño-Acosta N, Lo-Iacono-Ferreira VG, Torregrosa-López JI, Lora-García J (2022) Contamination of water supply sources by heavy metals: the price of development in Bolivia, a Latin American reality. Water 14(21):3470

Antoine D, d'Ortenzio F, Hooker SB, Bécu G, Gentili B, Tailliez D, Scott AJ (2008) Assessment of uncertainty in the ocean reflectance determined by three satellite ocean color sensors (MERIS, SeaWiFS and MODIS-A) at an offshore site in the Mediterranean Sea (BOUSSOLE project). J Geophys Res Oceans 20:113

Aurin D, Mannino A, Lary DJ (2018) Remote sensing of CDOM, CDOM spectral slope, and dissolved organic carbon in the global ocean. Appl Sci 8(12):2687

Bănăduc D, Curtean-Bănăduc A, Barinova S, Lozano VL, Afanasyev S, Leite T et al (2024) Multi-interacting natural and anthropogenic stressors on freshwater ecosystems: their current status and future prospects for 21st century. Water 16(11):1483

Barraza-Moraga F, Alcayaga H, Pizarro A, Félez-Bernal J, Urrutia R (2022) Estimation of chlorophyll-a concentrations in Lanalhue Lake using Sentinel-2 MSI satellite images. Remote Sens 14(22):5647

Bhatti A, Rundquist D, Schalles J, Ramirez L (2010) Application of hyperspectral remotely sensed data for water quality monitoring: accuracy and limitation. In: Proceedings of the accuracy symposium, Leicester, UK, 20–23 July 2010

Bigham Stephens DL, Carlson RE, Horsburgh CA, Hoyer MV, Bachmann RW, Canfield DE (2015) Regional distribution of Secchi disk transparency in waters of the United States. Lake Reser Manag 31(1):55–63

Binh NA, Hoa PV, Thao GTP, Duan HD, Thu PM (2022) Evaluation of Chlorophyll-a estimation using Sentinel 3 based on various algorithms in southern coastal Vietnam. Int J Appl Earth Observ Geoinform 112:102951

Boyd CE (2019) Water quality: an introduction. Springer, Berlin

Bukata RP, Jerome JH, Kondratyev AS, Pozdnyakov DV (2018) Optical properties and remote sensing of inland and coastal waters. CRC Press, New York

Čampulová M, Michálek J, Mikuška P, Bokal D (2018) Nonparametric algorithm for identification of outliers in environmental data. J Chemom 32(5):e2997

Chen C, Shi P, Mao Q (2003) Application of remote sensing techniques for monitoring the thermal pollution of cooling-water discharge from nuclear power plant. J Environ Sci Health A 38:1659–1668

Chidiac S, El Najjar P, Ouaini N, El Rayess Y, El Azzi D (2023) A comprehensive review of water quality indices (WQIs): history, models, attempts and perspectives. Rev Environ Sci Bio/technol 22(2):349–395

Choe E-Y, Lee J-W, Lee J-K (2011) Estimation of chlorophyll-a concentrations in the nakdong river using high-resolution satellite image. Korean J Remote Sens 27:613–623

Corbett CA (2007) Colored dissolved organic matter (CDOM) workshop summary. University of South Florida, Tampa

Dargaville BL, Hutmacher DW (2022) Water as the often neglected medium at the interface between materials and biology. Nat Commun 13(1):4222

Doxaran D, Cherukuru N, Lavender SJ (2006) Apparent and inherent optical properties of turbid estuarine waters: measurements, empirical quantification relationships, and modeling. Appl Opt 45:2310–2324

Ekercin S (2007) Water quality retrievals from high resolution IKONOS multispectral imagery: a case study in Istanbul. Turkey. Water Air Soil Pollut 183:239–251

Feng L, Hu C, Han X, Chen X, Qi L (2014) Long-term distribution patterns of chlorophyll-a concentration in China's largest freshwater lake: MERIS full-resolution observations with a practical approach. Remote Sens 7:275–299

Gao W, Shen Q, Zhou Y, Li X (2018) Analysis of flood inundation in ungauged basins based on multi-source remote sensing data. Environ Monit Assess 190:1–13

Gemperli C (2004) Determination of water quality parameters in Indian ponds using remote sensing methods. University of Zurich, Zurich

Gholizadeh MH, Melesse AM, Reddi L (2016) A comprehensive review on water quality parameters estimation using remote sensing techniques. Sensors 16(8):1298

Gitelson AA, Dall'Olmo G, Moses W, Rundquist DC, Barrow T, Fisher TR, Gurlin D, Holz J (2008) A simple semi-analytical model for remote estimation of chlorophyll-a in turbid waters: validation. Remote Sens Environ 112:3582–3593

Han L, Jordan KJ (2005) Estimating and mapping chlorophyll-a concentration in Pensacola Bay, Florida using Landsat ETM+ data. Int J Remote Sens 26:5245–5254

Handcock R, Gillespie A, Cherkauer K, Kay J, Burges S, Kampf S (2006) Accuracy and uncertainty of thermal-infrared remote sensing of stream temperatures at multiple spatial scales. Remote Sens Environ 100:427–440

He W, Chen S, Liu X, Chen J (2008) Water quality monitoring in a slightly-polluted inland water body through remote sensing—case study of the guanting reservoir in Beijing. China. Front Environ Sci Eng China 2:163–171

He J, Chen Y, Wu J, Stow DA, Christakos G (2020) Space-time chlorophyll-a retrieval in optically complex waters that accounts for remote sensing and modeling uncertainties and improves remote estimation accuracy. Water Res 171:115403

Heiskary S, Wilson B (2005) Minnesota lake water quality: developing nutrient criteria, 3rd edn. Minnesota Pollution Control Agency, St. Paul

Hupton JR (2009) Three-dimensional target modeling with synthetic aperture radar. Master's Thesis, California Polytechnic State University, San Luis Obispo

Ingrao C, Strippoli R, Lagioia G, Huisingh D (2023) Water scarcity in agriculture: an overview of causes, impacts and approaches for reducing the risks. Heliyon 14:542

Kallio K (2000) Remote sensing as a tool for monitoring lake water quality. In: Hydrological and limnological aspects of lake monitoring. Wiley, New York, p 237

Kaplan G, Yalcinkaya F, Altıok E, Pietrelli A, Nastro RA, Lovecchio N et al (2024) The role of remote sensing in the evolution of water pollution detection and monitoring: a comprehensive review. Phys Chem Earth a/b/c 54:103712

Karaska MA, Huguenin RL, Beacham JL, Wang M-H, Jensen JR, Kaufmann RS (2004) AVIRIS measurements of chlorophyll, suspended minerals, dissolved organic carbon, and turbidity in the Neuse River, North Carolina. Photogramm Eng Remote Sens 70:125–133

Kim S-I, Kim H-C, Hyun C-U (2014) High resolution ocean color products estimation in Fjord of Svalbard, arctic sea using Landsat-8 oli. Korean J Remote Sens 30:11542

Kloiber SM, Brezonik PL, Bauer ME (2002) Application of landsat imagery to regional-scale assessments of lake clarity. Water Res 36:4330–4340

Kolluru S, Gedam SS, Chander S, Sahay A (2023) Development of chlorophyll-a concentration estimation algorithm for turbid productive inland waters in India. Geocarto Int 15:2171143

Kumar V, Sharma M, Sondhi S, Kaur K, Sharma D, Sharma S, Utreja D (2023) Removal of inorganic pollutants from wastewater: innovative technologies and toxicity assessment. Sustainability 15(23):16376

Kume A, Akitsu T, Nasahara KN (2018) Why is chlorophyll b only used in light-harvesting systems? J Plant Res 131:961–972

Kutser T, Pierson DC, Kallio KY, Reinart A, Sobek S (2005) Mapping lake CDOM by satellite remote sensing. Remote Sens Environ 94:535–540

Lee Z, Shang S, Hu C, Du K, Weidemann A, Hou W, Lin J, Lin G (2015) Secchi disk depth: a new theory and mechanistic model for underwater visibility. Remote Sens Environ 169:139–149

Lillesand T, Kiefer RW, Chipman J (2014) Remote sensing and image interpretation. Wiley, New York

Lim J, Choi M (2015) Assessment of water quality based on Landsat 8 operational land imager associated with human activities in Korea. Environ Monit Assess 187:1–17

Liu Y, Islam MA, Gao J (2003) Quantification of shallow water quality parameters by means of remote sensing. Prog Phys Geogr 27:24–43

Loaiza JG, Rangel-Peraza JG, Monjardín-Armenta SA, Bustos-Terrones YA, Bandala ER, Sanhouse-García AJ, Rentería-Guevara SA (2023) Surface water quality assessment through remote sensing based on the box–cox transformation and linear regression. Water 15(14):2606

Mancino G, Nolè A, Urbano V, Amato M, Ferrara A (2009) Assessing water quality by remote sensing in small lakes: the case study of monticchio lakes in southern Italy. iFor Biogeosci for 2:154–161

Mannheim S, Segl K, Heim B, Kaufmann H (2004) Monitoring of lake water quality using hyperspectral chris-proba data. In: Proceedings of the 2nd CHRIS/PROBA workshop, ESA/ESRIN, Frascati, Italy, 28–30 April 2004, pp 28–30

Mannino A, Russ ME, Hooker SB (2008) Algorithm development and validation for satellite-derived distributions of DOC and CDOM in the us middle Atlantic bight. J Geophys Res Oceans 113:542

Marinho RR, Martinez JM, de Oliveira TCS, Moreira WP, de Carvalho LAS, Moreira-Turcq P, Harmel T (2024) Estimating the colored dissolved organic matter in the Negro River, Amazon Basin, with in situ remote sensing data. Remote Sens 16(4):613

Markogianni V, Dimitriou E, Tzortziou M (2013) Monitoring of chlorophyll-a and turbidity in Evros River (Greece) using Landsat imagery. In: Proceeding of the 1st international conference on remote sensing and geoinformation of environment, Paphos, Cyprus

Miller RL, DelCastillo CE, Powell RT, DSa E, Spiering B (2002) Mapping CDOM concentration in waters influenced by the Mississippi River Plume; NASA technical reports server: Cleveland

Minvielle P, Massaloux P, Giovannelli J-F (2024) Fast 3D synthetic aperture radar imaging from polarization-diverse measurements. http://arxiv.org/abs/1506.07459. Accessed 26 July 2024

Mobley C, Stramski D, Bissett W, Boss E (2004) Optical modeling of ocean waters: is the case 1–case 2 classification still useful? Oceanography 17:60–67

Müller-Navarra DC, Brett MT, Park S, Chandra S, Ballantyne AP, Zorita E, Goldman CR (2004) Unsaturated fatty acid content in seston and tropho-dynamic coupling in lakes. Nature 427:69–72

Myint S, Walker N (2002) Quantification of surface suspended sediments along a river dominated coast with NOAA AVHRR and SeaWiFS measurements: Louisiana, USA. Int J Remote Sens 23:3229–3249

Nechad B, Ruddick K, Park Y (2010) Calibration and validation of a generic multisensor algorithm for mapping of total suspended matter in turbid waters. Remote Sens Environ 114:854–866

Nelson NB, Siegel DA (2002) Chromophoric DOM in the open ocean. Biogeochemistry of marine dissolved organic matter. Academic Press, San Diego, pp 547–578

O'Reilly JE, Werdell PJ (2019) Chlorophyll algorithms for ocean color sensors-OC4, OC5 and OC6. Remote Sens Environ 229:32–47

Östlund C, Flink P, Strömbeck N, Pierson D, Lindell T (2001) Mapping of the water quality of Lake Erken, Sweden, from imaging spectrometry and Landsat Thematic Mapper. Sci Total Environ 268:139–154

Pahlevan N, Smith B, Alikas K, Anstee J, Barbosa C, Binding C et al (2022) Simultaneous retrieval of selected optical water quality indicators from Landsat-8, Sentinel-2, and Sentinel-3. Remote Sens Environ 270:112860

Pan X, Mannino A, Russ ME, Hooker SB (2008) Remote sensing of the absorption coefficients and chlorophyll a concentration in the United States Southern middle Atlantic Bight from SeaWiFS and MODIS-Aqua. J Geophys Res Oceans 54:113

Parkinson CL (2003) Aqua: an earth-observing satellite mission to examine water and other climate variables. IEEE Trans Geosci Remote Sens 41:173–183

Powell R, Brooks C, French N, Shuchman R (2008) Remote sensing of lake clarity. Michigan Tech Research Institute, Ann Arbor

Qi L, Hu C, Duan H, Barnes BB, Ma R (2014) An EOF-based algorithm to estimate chlorophyll a concentrations in Taihu Lake from MODIS land-band measurements: implications for near real-time applications and forecasting models. Remote Sens 6:10694–10715

Qiang S, Song K, Shang Y, Lai F, Wen Z, Liu G et al (2023) Remote sensing estimation of CDOM and DOC with the environmental implications for lake Khanka. Remote Sens 15(24):5707

Ritchie JC, Zimba PV, Everitt JH (2003) Remote sensing techniques to assess water quality. Photogramm Eng Remote Sens 69:695–704

Roy PD, Selvam S, Gopinath S, Logesh N, Sánchez-Zavala JL, Lakshumanan C (2022) Geochemical evolution and seasonality of groundwater recharge at water-scarce southeast margin of the Chihuahuan Desert in Mexico. Environ Res 203:111847

Ruescas AB, Hieronymi M, Koponen S, Kallio K, Camps-Valls G (2017) Retrieval of coloured dissolved organic matter with machine learning methods. In: Proceedings of the 2017 IEEE international geoscience and remote sensing symposium (IGARSS). IEEE, pp 2187–2190

Saragih DF, Ahamad MSS, Abdullah R (2020) Application of geographic information systems (GIS) in the multi criteria site selection of retention pond for urban rainwater management. In: Proceedings of AICCE'19: transforming the nation for a sustainable tomorrow 4. Springer, Berlin, pp 317–330

Saranya A, Al-Mazroa A, Maashi M, Nithya TM, Priya V (2024) Remote sensing and machine learning approach for zoning of wastewater drainage system. Desal Water Treat 319:100549

Schaeffer BA, Schaeffer KG, Keith D, Lunetta RS, Conmy R, Gould RW (2013) Barriers to adopting satellite remote sensing for water quality management. Int J Remote Sens 34:7534–7544

Sehgal D, Martínez-Carreras N, Hissler C, Bense VF, Hoitink AJF (2022) A generic relation between turbidity, suspended particulate matter concentration, and sediment characteristics. J Geophys Res Earth Surf 127(12):e2022JF006838

Singh BJ, Chakraborty A, Sehgal R (2023) A systematic review of industrial wastewater management: evaluating challenges and enablers. J Environ Manag 348:119230

Strömbeck N, Pierson DC (2001) The effects of variability in the inherent optical properties on estimations of chlorophyll a by remote sensing in Swedish freshwaters. Sci Total Environ 268:123–137

Swanson H, Zurawell R (2006) Steele lake water quality monitoring report; monitoring and evaluation branch, environmental assurance division. Alberta Environment, Edmonton

Thiemann S, Kaufmann H (2002) Lake water quality monitoring using hyperspectral airborne data: a semiempirical multisensor and multitemporal approach for the Mecklenburg Lake District. Germany. Remote Sens Environ 81:228–237

Torgersen CE, Faux RN, McIntosh BA, Poage NJ, Norton DJ (2001) Airborne thermal remote sensing for water temperature assessment in rivers and streams. Remote Sens Environ 76(3):386–398

Usali N, Ismail MH (2010) Use of remote sensing and gis in monitoring water quality. J Sustain Dev 3:228–238

USGS (2024) Earth observing 1 (EO-1). http://eo1.usgs.gov/. Accessed 25 Feb 2024

Velimirović LZ, Janković R, Velimirović JD, Janjić A (2021) Wastewater plant reliability prediction using the machine learning classification algorithms. Symmetry 13(8):1518

Vignudelli S, Santinelli C, Murru E, Nannicini L, Seritti A (2004) Distributions of dissolved organic carbon (DOC) and chromophoric dissolved organic matter (CDOM) in coastal waters of the northern Tyrrhenian Sea (Italy). Estuar Coast Shelf Sci 60:133–149

Wen Z, Wang Q, Ma Y, Jacinthe PA, Liu G, Li S et al (2024) Remote estimates of suspended particulate matter in global lakes using machine learning models. Int Soil Water Conserv Res 12(1):200–216

Xu B, Gong P (2007) Land-use/land-cover classification with multispectral and hyperspectral EO-1 data. Photogramm Eng Remote Sens 73:955–965

Yang L, Yu D, Yao H, Gao H, Zhou Y, Gai Y et al (2022) Capturing Secchi disk depth by using Sentinel-2 MSI imagery in Jiaozhou Bay, China from 2017 to 2021. Mar Pollut Bull 185:114304

Zhang Y, Pulliainen JT, Koponen SS, Hallikainen MT (2003) Water quality retrievals from combined Landsat TM data and ERS-2 SAR data in the gulf of finland. IEEE Trans Geosci Remote Sens 41:622–629

Zhang Y, Shi K, Sun X, Zhang Y, Li N, Wang W et al (2022) Improving remote sensing estimation of Secchi disk depth for global lakes and reservoirs using machine learning methods. Gisci Remote Sens 59(1):1367–1383

Zhao H, Ma Y, Fang J, Hu L, Li X (2022) Particle size distribution and total suspended solid concentrations in urban surface runoff. Sci Total Environ 815:152533

Zheng G, DiGiacomo PM (2017) Uncertainties and applications of satellite-derived coastal water quality products. Progr Oceanogr 159:45–72

Zhu W, Yu Q, Tian YQ, Chen RF, Gardner GB (2011) Estimation of chromophoric dissolved organic matter in the Mississippi and Atchafalaya river plume regions using above-surface hyperspectral remote sensing. J Geophys Res Oceans 15:116

Chapter 6
Microfluidic Devices: Harnessing the Power of Tiny Devices in Analysing Wastewater Quality

Zeba Sadia, Syed Sadiq Abbas, Saimah Khan, and Jamal Akhtar Ansari

Abstract The reduction in aquatic biodiversity and the growing water shortage are caused by habitat degradation and water pollution. Over the previous few decades, freshwater availability has decreased. Monitoring contaminants in wastewater effluents is essential to identifying areas of water contamination that require treatment. Multiple hazardous components in wastewater can be difficult to trace using conventional detection methods since they vary over time and come from different sources. There is increasing interest in making microfluidic apparatuses because of their phenomenal responsiveness, ease, dynamic response, negligible cost, constancy, and selectivity. Depicted by their ability to control small volumes of fluids through microchannels, they are emerging as moderate devices in normal checking, particularly in analysing wastewater quality. These miniature apparatuses furnish significant opportunities over the conventional approach, including reduced sample volume, faster processing times, high sensitivity, and the potential for on-site real-time analysis. Additionally, these systems can combine several analytical tasks into a single, small instrument, including sample preparation, segregation, identification, and quantification. Microfluidic devices can accurately measure contaminants like heavy metals, organic pollutants, pathogens, and nutrient levels. This chapter explores the applications of microfluidic technology in wastewater analysis, emphasizing its impact on improving water quality management and environmental protection.

Z. Sadia · S. S. Abbas
Department of Zoology, Shia PG College, University of Lucknow, Lucknow, India
e-mail: zebasadia46@gmail.com

S. S. Abbas
e-mail: drsadiqabbas@rediffmail.com

S. Khan · J. A. Ansari (✉)
Department of Chemistry, Integral University, Lucknow, India
e-mail: jamalakhtar@iul.ac.in

S. Khan
e-mail: saimah@iul.ac.in

A. M. Yatoo et al. (eds.), *Wastewater Monitoring and Management by Advanced Approaches*, Advances in Wastewater Research,
https://doi.org/10.1007/978-981-95-2601-7_6

Keywords Microfluidic devices · Pollutant measurement · Water contamination · Faster processing times · Fabrication process · Wireless sensor networks · Water quality monitoring

Abbreviations

LOD	Limit of detection
LOQ	Limit of quantification
NSET	Nanometal surface energy transfer
CRET	Chemiluminescence resonance energy transfer
IET	Interfacial electron transfer
IPET	Intramolecular photoinduced electron transfer
PEC	Photoelectrochemical
LoC	Lab-on-a-chip
LoD	Lab-on-a-disc
CAD	Computer-aided design
PSA	Pressure-sensitive adhesive
PDMS	Polydimethylsiloxane
PMMA	Polymethylmethacrylate
COC	Cyclic olefin copolymer
μPAD	Paper-based analytical device
IO	Ionogel
NOM	Natural organic matter
RE	Reference electrode
RGB	Red green blue
WE	Working electrode
GO	Graphene oxide
CE	Counter electrode
μm	Micrometre
mm	Millimetre
PMB	Phosphomolybdenum
nM	Nanomolar

6.1 Introduction

Due to water contamination and limited freshwater supplies, the world is currently experiencing a water crisis. Contamination of water can lead to serious illnesses that endanger human life. The main cause of water contamination is the unregulated and unpermitted discharge of organic and inorganic waste from domestic and commercial wastes into surface waterways. These point sources are responsible for the influx of viruses, fertilizers, and heavy metals into the surface water, among other hazardous materials (Ibrahim et al. 2020). In addition, non-point sources like sewage and industrial by-products, land run-off from melting snow and rain, leakage from subterranean waste disposal sites, and agricultural waste significantly contribute to water contamination. Heavy metals, including Pb (lead), Cd (cadmium), and As (arsenic), naturally exist in the environment at higher concentrations, including the aquatic environment, and are hazardous to human health. Because heavy metals can cause mutations or cancer, their concentrations beyond the permissible limit can have a detrimental impact on public health (Chen et al. 2019a, b; Sharma et al. 2021; Bhattacharya et al. 2018). Ammonia (NH_3) is a chemical that is produced and used extensively in many different industries. It is the nitrogenous contaminant that appears in wastewater most frequently (Lee et al. 2009). Depending on the pH of the solution, ammonia can exist in aqueous solutions as either gaseous NH_3 or ammonium ion (NH_4^+). Because ammonia is poisonous, its presence in wastewater is considered a serious hazard. Ammonia toxicity poses a threat to fish and various other aquatic organisms, leading to eutrophication and loss of dissolved oxygen (Nxumalo et al. 2020). As a result, monitoring the quality of the water has gained prominence globally. Conventional techniques for environmental contaminant monitoring, including mass spectrometry (MS), atomic absorption spectroscopy (AAS), fluorescence spectroscopy, ultraviolet–visible (UV–vis) spectroscopy, inductively coupled plasma optical emission spectroscopy (ICP-OES), and high-performance liquid chromatography (HPLC), provide high sensitivity and precision even at low analyte concentrations (Mukherjee et al. 2021; Sekar et al. 2021). The majority of the currently used monitoring approaches involve costly equipment, lengthy processing times, laborious procedures, and highly qualified lab staff, making them insufficient. There is currently a need for a portable, sensitive, and selective sensor that can check water quality in situ and continuously (Jaywant and Arif 2019). The ability to analyse several samples with great accuracy and precision, as well as to use automated equipment to reach the low limit of detection (LOD) and the low limit of quantification (LOQ), is a benefit of laboratory-based approaches. The primary drawback of this equipment is that it is typically not portable, which means that in situ analysis is neglected and sample stability may be jeopardized. In terms of economics, this equipment is costly, and handling and preparing the samples require skilled staff. But this kind of technology is always developing, leading to more affordable, lightweight, and easily controlled equipment that may be used in traditional lab settings. This is the situation with portable kits, which enable on-site testing and have proven to be particularly helpful in resource-constrained situations due to their speed, simplicity, and affordability. The drawbacks of portable devices are their lack of accuracy and the need for skilled staff to operate them. They frequently do not entirely satisfy user needs because they simply offer estimates of the composition of water (Santana et al. 2020, Yuan et al. 2023).

On the other hand, water quality analysis probes are an alternative to portable kits that have been developed and brought to market. Along with their portability, the key advantage of these devices is their great precision. The latest iterations incorporate wireless sensor networks, enabling the collection and transmission of data via the

Internet (Gilmore et al. 2016). However, because of their high costs and the requirement for physical work during data collection, they are not always affordable. As this technology develops, humans will be able to monitor water quality precisely at the point of need by using efficient, transportable, or self-sufficient sensors in rivers, lakes, and water supply networks; they can also be linked to industrial and municipal wastewater treatment facilities. Additionally, the Internet will be used to make information available to citizens. These sensors might eventually be worn, enabling the user to keep an eye on the water environment around them. They can detect very low concentrations of contaminants, making them ideal for early detection of pollutants or pathogens. Microfluidic systems often require less reagent and sample volume, reducing the overall cost of analysis. Many microfluidic devices are small and portable, making them suitable for on-site environmental monitoring. The miniaturization of the process leads to faster reaction times, which is crucial for real-time or near-real-time detection (Suqi et al. 2024; Zhang et al. 2024). The paper highlights the growing need for quick, cost-effective, and on-site detection methods for bacterial contaminants in water quality monitoring. This is crucial for ensuring safe drinking water and environmental health. CIP-functionalized microwires (CIP-MWs) are designed specifically to target bacteria like *Escherichia coli*. These microwires are integrated into a low-cost microfluidic sensor that detects bacterial binding events using conductometric transduction. This integration of CIPs into a microfluidic sensor directly addresses the growing need for rapid, cost-effective, and on-site bacterial detection solutions (Akhtarian et al. 2023). A paper-based microfluidic device has been developed for wastewater surveillance, specifically to detect viral nucleic acid biomarkers such as SARS-CoV-2 and influenza. The device uses reverse transcription loop-mediated isothermal amplification (RT-LAMP) technology, which first converts viral RNA into complementary DNA (cDNA) and then amplifies it for detection. This process is faster than the traditional laboratory-based quantitative reverse transcription polymerase chain reaction (RT-qPCR) method, completing the analysis in 1.5 h compared to 4 h for RT-qPCR (Pan et al. 2020). A comprehensive analysis of commonly used detection techniques in microfluidic devices is presented, highlighting methods such as fluorescence, electrochemical detection, and mass spectrometry, which are crucial for identifying specific contaminants. The book chapter further discusses how microfluidic sensors can effectively detect harmful substances that interfere with hormonal systems (Zhang et al. 2024).

The growing use of microfluidic devices to assess water quality is covered in this chapter. The harmful consequences of water pollution on the health of humans are discussed in this book chapter to highlight the significance of routine water quality monitoring. The signal transduction methods used in the microfluidic water quality monitoring devices will depend on various detection techniques that are summarized. The chapter also discusses fabrication techniques used to create microfluidic devices and highlights the various microfluidic instruments that have been looked into for monitoring water contamination. The evaluation further examines the present obstacles linked to the utilization of microfluidic apparatus in water surveillance, as well as upcoming outlooks (Figure 6.1).

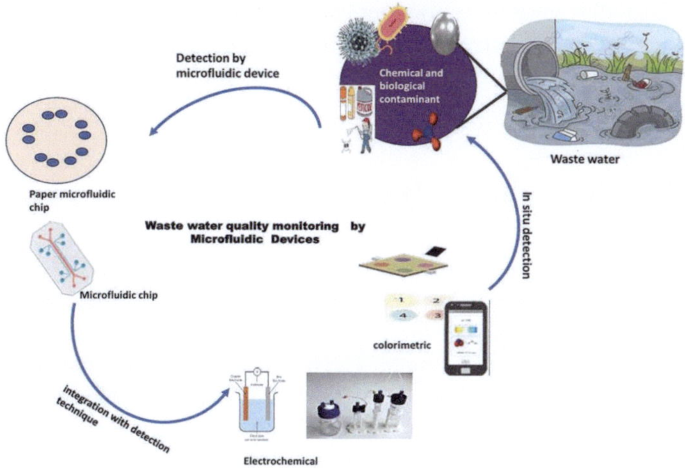

Fig. 6.1 Diagrammatic illustration of the utilization of microfluidics in conjunction with various techniques for identifying prevalent pollutants in water

6.2 Detection Mechanisms

The most popular detection methods that have been researched over the past 5 years for environmental quality monitoring systems based on microfluidics include fluorescence detection, electrochemical processes, and colorimetric testing. Other techniques include chemiluminescence, photoelectrochemistry, absorbance-based techniques, quartz crystal microbalance, as well as surface plasmon resonance (SPR). We concentrate on the most common approaches in the last few years, which consist of colorimetric, chemiluminescence, fluorescence, and electrochemical techniques in this article.

6.2.1 *Electrochemical Analysis*

An electrochemical cell with three electrodes—a working electrode, a reference electrode, and a counter electrode—is typically utilized for the detection procedure (Aryal et al. 2024a; b). Electrochemical analysis stands as a potent method of detection, primarily reliant on an analyte's electrochemical interaction with an electrode's surface. Redox (reduction–oxidation) reactions are the primary means of achieving this interaction in electrochemistry, and electrical signals like current, potential, and charge can be used to measure them. High-sensitivity electrochemical analysis enables detection at extremely low levels, down to the picomole range. Disposable paper electrodes, disposable glass electrodes, carbon paste electrodes, and screen-printed electrodes are only a few of the electrode types that have been studied.

Furthermore, studies have been conducted on portable set-ups such as smartphone-based systems, handheld potentiostats, and handmade electronics (Baranwal et al. 2022).

6.2.2 *Fluorescence Detection*

Fluorescence is a widely used optical detection method for monitoring water contamination. It entails emitting light with a longer wavelength after absorbing light with a specific wavelength (Fu and Wang 2018). The apparatus consists of a detector, a focusing lens, filters, and a light source. The fluorescent detection method consists of three stages: emission of fluorescence, excited-state lifetime, and excitation. During the processing step, an illumination source with the proper wavelength is employed to cause the fluorophores to glow. The resulting light is screened to detach the excitation and emission photons, and the intensity of the emission photons is measured and utilized as a target analyte concentration indicator (Fu and Wang 2018, Aryal et al. 2024a, b). The analyte's signal in a fluorescence measurement is dependent on both the analyte's concentration and the intensity of the incident excitation light. Receptors must function as chelating agents to engage with the analytes and recognize heavy metal ions. To find heavy metal ions, fluorescent sensors use mechanisms of energy or electron transfer. Chemiluminescence resonance energy transfer (CRET), nanometal surface energy transfer (NSET), and Förster resonance energy transfer (FRET) are a few examples of fluorescent microfluidic systems based on energy transfer processes. Electron transfer methods encompass intramolecular photoinduced electron transfer (IPET) as well as interfacial electron transfer (IET).

6.2.3 *Colorimetric Detection*

Colorimetry in conjunction with microfluidic devices, especially paper-based ones, is commonly utilized as the most effective detection method because of its simplicity of use, outstanding stability, visual outputs, compatibility with affordable reporting systems, and capacity to operate in isolated areas. Figure 6.2 illustrates the stages of colorimetric analysis to identify contaminants in wastewater. Colorimetry relies on how a substance's colour changes in the presence of the target. A chemical reaction between the material and a reactive agent causes this colour change. These substances cause colour development on paper; examples include enzymatic assays, redox indicators, pH markers, and different nanoparticles (carbon, silver, and gold dots). Occasionally, the reaction modifies the pH or redox characteristics rather than changing colour. In these cases, a colour change in response to the chemical stimulus is displayed using a redox or pH indicator. For accurate colorimetric readings in microfluidic devices, colour constancy is essential since it allows for more accurate measurements and increases the efficacy of the analytical procedure. Colour

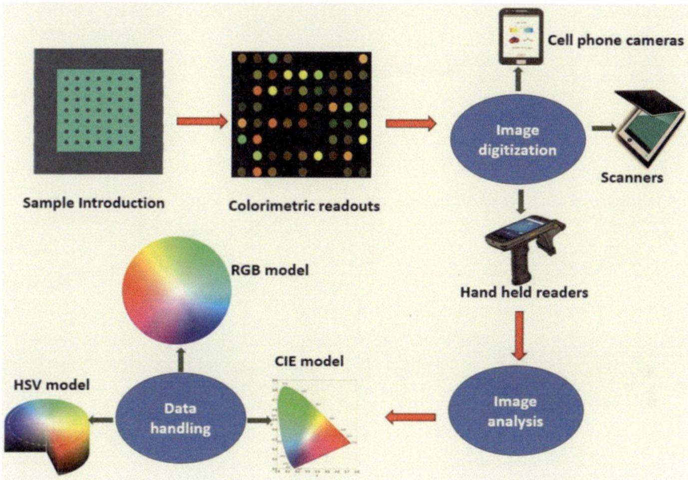

Fig. 6.2 Steps engaged in colorimetric identification within polluted water samples

consistency in readings is affected by several parameters, including substrate alterations, digitalization and data processing choices, assay execution, and device design. Usually, an optical set-up records the colour variation for further examination (Kudr et al. 2017; Morbioli et al. 2017).

6.2.4 Photoelectrochemical

Photoelectrochemical (PEC) sensing is a crucial field within electrochemical detection, which has experienced significant advancements in recent times. The PEC sensor offers heightened sensitivity, uncomplicated instrumentation, cost-effectiveness, exceptional selectivity, and the ability to detect various types of analytes. Photoelectrochemistry refers to systems in which light is utilized to influence electrochemical processes. The PEC system consists of an electrode workstation, a data gathering system, a monochromator, an electrochemical workstation, and lastly an irradiation source. The reference electrode (RE), counter electrode (CE), and working electrode (WE) make up the three-electrode system. The working electrode material needs to be a semiconductor to allow an electron–hole pair to form when higher-energy photons are absorbed on the photoactive surface. In the electrolyte solution, the photogenerated electron–hole charge carrier then takes part in the redox process and generates an electrical signal (Shu and Tang 2017; Zhao et al. 2015) (Fig. 6.3).

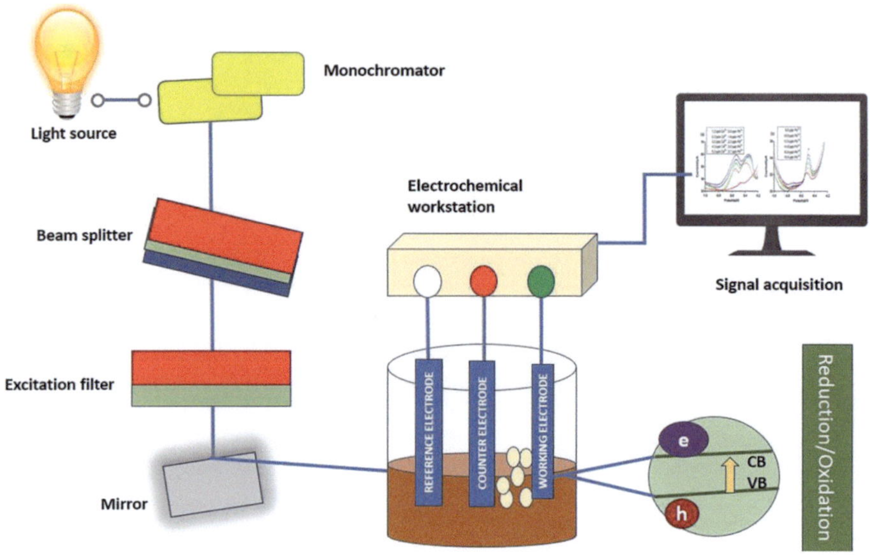

Fig. 6.3 Diagrammatic illustration of photoelectrochemical sensors' operating principle

6.3 Fabrication Methods

The cellulosic paper serves as a substrate with hydrophilic properties. To construct microfluidic channels, specific methodologies must be employed to adjust the hydrophobic nature of the cellulosic substrate. Two approaches have been utilized for the patterning of microfluidic channels. One method involves the selective creation of a hydrophobic surface on the cellulosic film. The alternative approach entails transforming the cellulosic film into a hydrophobic substrate initially, followed by the selective modification of a portion of the surface to regain its hydrophilic properties. Various techniques have demonstrated feasibility in establishing microfluidic channels on the cellulosic substrate, including flexographic printing, plasma treatments, inkjet printing, photolithography, screen printing, laser cutting, wax printing, etc.

Fabrication of Microfluidic Devices Used for Water Contamination Detection

6.3.1 *Wax Printing*

The simplest method for producing analytical equipment on paper is wax printing. This approach involves designing the hydrophobic barrier's layout using a computer. Programs like Adobe Photoshop, Microsoft Word, Paint, and others can be used for this. Depending on the sample type, the matrix in which the analyte will be identified, or the kind of reagents employed to produce colour or electrochemical potential, the pattern design may differ among studies. The chosen pattern can then be printed using wax printers onto Whatman filter paper (Liu et al. 2019; Nogueira et al. 2019). After that, the paper is heated to roughly 40 °C, allowing the wax ink to fully enter the paper. As a result, a hydrophobic barrier is formed across the thickness of the paper. The result is an impermeable barrier that contains a hydrophilic area following the wax printing design. A transparent tape or laminated sheet is applied to one of the paper's surfaces to guarantee that it stays dry on that side and to stop reagents and/or samples from leaking to the opposite side by permeation (Das et al. 2022).

6.3.2 Centrifugal Microfluidics

The two main platforms on which microfluidic devices operate are lab-on-a-chip (LoC) and lab-on-a-disc (LoD). While liquid flow in the LoD platform is mainly regulated by centrifugal forces, external micropumps are necessary in the LoC platform to move the liquid through microchannels. Centrifugal, Coriolis, Euler, and capillary forces are employed by the centrifugal microfluidic rotating platform to move and work with liquids. These platforms allow for fluid manipulation without the need for external pumping, in contrast to LoC devices. They have several benefits, such as being simple to multiplex, managing particles, having actuation forces that can be scaled, being independent of the physiochemical parameters of fluids (such as pH and viscosity), and not requiring a high-voltage power supply. Environmental monitoring frequently makes use of centrifugal platforms (Maguire et al. 2018).

6.3.3 Xurography

The word 'xurography' comes from the combination of the Greek words '*xuron*', which means razor, and '*graphe*', which means writing. The technique of xurography involves the patterning of sheets using a blade, as implied by its name. In the past few years, xurography has gained prominence as a cost-effective, straightforward, and rapid method for fabricating both 2D and 3D microfluidic devices. Its inception can be credited to Bartholomeusz, who initially utilized this methodology to generate microfeatures on various films, which include hard and soft films, as well as thick and thin polymer films, with 6 μm width and 25 μm thickness. To build structures that are both positive and negative, various cutting techniques, including emulated tangential, true tangential, and drag knife, were used in conjunction with blade angle orientations of 30°, 45°, and 60°. Using computer-aided design (CAD) software, microfluidic channels are created in a typical manufacturing process. A cutting plotter machine is then utilized to print a polymer and/or two-sided pressure-sensitive adhesive (PSA) sheet. The microfluidic device is then completed by carefully aligning and integrating the polymer with patterns and/or PSA films with a substrate along with a polymer cover layer. The smallest dimension that can be obtained with this technology is about 200 μm (Islam et al. 2015).

6.3.4 Thread-Based Analytical Devices (MTADs)

In terms of environmental monitoring and general analytical examination, the μTADs show how porous materials can be successfully used (Agustini et al. 2021; Tan et al. 2021). Like paper-based analytical devices (μPADs), these devices are great options for budget-conscious uses. Additionally, the global market encourages the use of applications that do not require complicated alterations to materials (Farajikhah et al. 2019; Weng et al. 2019). Nowadays, there are many different types of threads available for different uses. These include synthetic threads like polyester, polyether, polyurea, and acrylic, as well as natural threads like silk, wool, and linen (Oliveira et al. 2022). Because they are both porous, threads' flow properties along with detecting techniques are comparable to those of paper (Tan et al. 2021). However, thread-based tools are believed to be more appropriate for applications involving clothing than paper tools. This is due to the ease with which threads can be used to sew clothes directly or imprint walls onto fabric (Xiao et al. 2019; Tan et al. 2021; Xia et al. 2021). Furthermore, two other potential low-cost detection strategies are distance- and barcode-based detection (Tan et al. 2021). The idea behind distance-based detection is that distinct analytes might cause variable wetting capabilities, which in turn can lead to recognition (Shimazu et al. 2022; Jarujamrus et al. 2018; Alsaeed and Mansour 2020).

6.3.5 Paper-Based Analytical Devices (MPADs)

Paper is a cheap and widely available material that has been utilized in a variety of ways for many years (Gupta et al. 2019). The characteristics of the material, such as chemical composition, wetting behaviour, and porosity, can be easily customized for various applications (Gao et al. 2019). The porous quality of paper and its high proportion of surface to volume make it similar to other porous materials in that they allow for passive fluid manipulation and regulation. Modifying the chemical make-up of the fibres, such as the degree of polarity, can improve interactions with samples and is crucial in the design and functioning of devices (Ma et al. 2019; Soum et al. 2019; Lim et al. 2019). For instance, nitrocellulose paper is an excellent choice for covalently immobilizing biomolecules because of its strong ability to bind to proteins through the nitrate groups present on their surfaces When compared to other forms of paper, filter paper and chromatography paper excel in possessing consistent pore size and breadth. The devices are based on the paper material itself, which is subjected to analytical analysis through the integration of several sensing or detection techniques (Nishat et al. 2021). According to discussions, Raman sensing, fluorescent, and

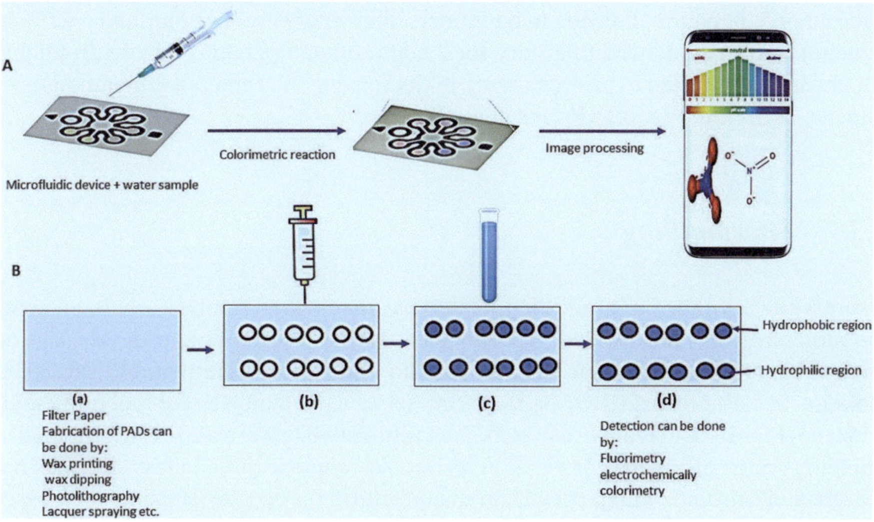

Fig. 6.4 **A** A simultaneous pH and nitrite colorimetric measurement system based on a smartphone, a microfluidic apparatus in which each sensing region is filled with reagents; **B** (a) unfabricated Whatman filter paper, (b) preloading of reacting reagents, (c) sample loading on the PAD, and (d) evolution of colour in the PAD fabricated using unique hydrophobic barriers and distinct hydrophilic zones

electrochemical, as well as colorimetric, are some of the several types of detection technologies that are currently available (Zheng et al. 2021; Li et al. 2021; Fu and Wang 2018). Review papers offering in-depth analyses of advancements in μPADs can be accessed through the following resources (Lim et al. 2019; Soum et al. 2019; Kaneta et al. 2019; Carrell et al. 2019) (Fig. 6.4).

6.3.6 Inkjet Printing Technique

Inkjet printing is a widely employed procedure in the realm of printing manufacturing, delineated into two primary classifications: photopolymer- and powder-based (Khorsandi et al. 2021; Aladese and Jeong 2021; Loo et al. 2019; Olmos et al. 2019). Through the utilization of this method, hydrophobic inks are deployed to form the boundaries of channels based on paper apparatuses (Fan 2018; Lim et al. 2019; Weng et al. 2019). It is noteworthy that printing using inkjets can be utilized on various paper types, whereas laser printers offer swift and high-capacity printing procedures (Bamshad and Cho 2021; Nishat et al. 2021). Abe and associates developed a method for producing structured paper and microfluidic multianalyte chemical sensing instruments using the inkjet printing process. Using inkjet printing on filter paper, they produced 3D hydrophilic microfluidic designs with a sensing zone of 1.5 mm × 1.5 mm and flow channels of 550 μm width. To make the paper substrate completely hydrophobic in nature, it needs to be immersed in a polystyrene solution (1 wt.% in toluene). It is then allowed to air dry for 2 h at room temperature. The hydrophobic paper can be subjected to printing using inkjet sensing inks and picolitre amounts of toluene, as seen in Fig. 6.5 (Prince et al. 2019).

6.3.7 Micromilling

Using a process called micromilling, materials are removed from their bulk to create the structures that are needed. The finished enclosed microfluidic device can be created by attaching the manufactured parts to a substrate (Rahim and Ehsan 2021; Hossain and Rahman 2018), or they can serve as a template for polydimethylsiloxane (PDMS) (Javidanbardan et al. 2021; Jiménez-Díaz et al. 2019). Compared to numerous other affordable fabrication techniques, micromilling is fast and does not require a cleanroom, which speeds up manufacturing processes considerably, especially for prototype reviews (Nguyen et al. 2019). Various materials have currently been investigated for the production of microfluidic devices through micromilling, with polymethylmethacrylate (PMMA) and aluminium emerging as the two most favoured materials (Jiménez-Díaz et al. 2019; Nguyen et al. 2019; Behroodi et al. 2020; Javidanbardan et al. 2021; Saptaji et al. 2021). The aluminium micromilled templates can be utilized for multiple casting cycles, consequently leading to a decrease in the final device's expenses (Guckenberger et al. 2015; Nguyen et al. 2019). On the contrary, micromilling is accompanied by several constraints that necessitate attention. For instance, the milling tools employed in micromilling are susceptible to fracture, particularly in scenarios demanding high precision (e.g. 25 μm) (Charles et al. 2018). Furthermore, intricate 3D characteristics and designs may not work for micromilling, notwithstanding the potential ability of customized milling tools to generate structures with predetermined shapes (Ku et al. 2018; Javidanbardan et al. 2021). Micromilling exclusively eliminates substances originating from outside surfaces, thereby making them microchannels. This fusion can be achieved through mechanical means (e.g. using screws), thermal methods (e.g. bonding two PMMA plates through heat), or by employing surface modifications and adhesives like tapes (Gonçalves et al. 2021; Madureira et al. 2019; Owens and Hart 2018; Kosoff et al. 2018).

6.3.8 Laser Micromachining

The use of laser micromachining was applied in the realm of cost-effective microfluidics, as indicated by previous studies (Persson et al. 2022). An illustration of this is the prevalent use of CO_2 laser in the process of microfabrication (Chen et al. 2019a, b; Buchroithner et al. 2021; Nishat et al. 2021). Throughout the manufacturing phase, the laser's energy is concentrated on specific areas of the workpieces, leading to

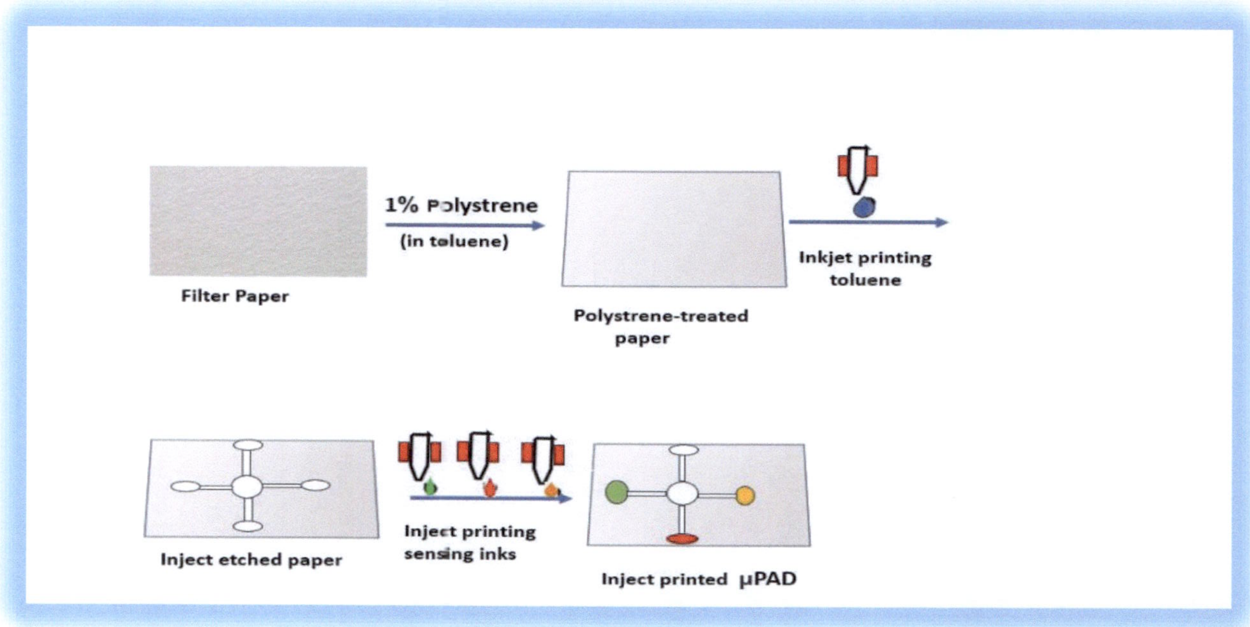

Fig. 6.5 Illustration showing the steps involved in creating the microfluidic multianalyte chemical sensing paper, made using inkjet printing with microfluidic channels

melting and vaporization of the materials. Typically, CO_2 lasers with a wavelength of 10.6 μm are utilized in these procedures (Persson et al. 2022). Certainly, advanced laser systems or shorter wavelengths (e.g. femtosecond lasers) could be implemented to enhance cutting precision; however, these techniques are unsuitable for economical microfluidics due to the associated additional expenses (Elgohary et al. 2020; Andriukaitis et al. 2022). In the context of materials employed in laser micromachining, a variety of substances including rigid and malleable materials, like glass, PMMA, cyclic olefin copolymer (COC), and even paper, are feasible options (Islam et al. 2018; Lin et al. 2019). It is important to note that to prevent thermal stress-induced cracks, surface coatings can be administered on the glass slides. Furthermore, laser micromachining is capable of producing both moulds and final products after the bonding process (Mahmud et al. 2018; Gao et al. 2019; Ma et al. 2019). The assembling and bonding processes applied to laser-cut devices are comparable to those used on micromilled electronics (Nguyen et al. 2019; Persson et al. 2022).

6.4 Microfluidic Devices for the Detection of Water Contaminants

6.4.1 Detection of Nutrients

To keep aquatic habitats healthy and in balance, nutrients are essential. However, an abundance of nutrients in water due to industrial effluents, wastewater discharge, or agricultural run-off can have detrimental effects on the environment. Eutrophication is one such consequence that causes aquatic species to have lower oxygen levels and eventually die from nutrient overload. Furthermore, cyanobacteria in algal blooms can create dangerous substances that endanger people, domestic animals, and aquatic life. As a result, to be able to understand the overall water quality and the sources of contamination, it is essential to do comprehensive monitoring of the nutrients in water systems. Provided a comprehensive examination of microfluidic instruments used for nutrient detection in water, including nitrite, nitrate, ammonia nitrogen, phosphate, and silicate detection devices. Owing to the comprehensive character of this review (Li et al. 2023). papers published following Li et al.'s publication will be emphasized. An ionogel (IO)-based hybrid polymethylmethacrylate (PMMA) paper handheld instrument for NO^{3-} as well as nitrite detection in water was introduced by Catalan-Carrio et al. ZnO is trapped on the paper surface of this apparatus-based microfluidic section, which causes nitrate to reduce to NO^{2-} against proximity. The solution then flows onto the device's IO segment, where the sample's nitrite reacts with Griess reagent to produce a visible change in colour in the IO. The apparatus demonstrated nitrite detection thresholds of 0.47 mg L^{-1} and nitrate detection levels of 2.3 mg L^{-1}. It is crucial to remember that colour stabilization could take up to 50 min, making it less than ideal for field applications (Catalan-Carrio et al. 2022).

Developed a colorimetric-based instrument for the underwater detection of phosphate and nitrate. This device is made of PMMA and has two layers with inlets and flow channels that enable dual chemistry to occur simultaneously. For colour assessment, two optical cells were used in parallel; the phosphomolybdenum (PMB) assay was used for phosphate detection, while the Griess test was used for nitrate detection. The instrument displayed its usefulness in real-world scenarios by delivering 592 measurements throughout the course of an 8-day field test. It also displayed nitrate detection limits of 97 nM and phosphate detection limits of 15 nM (Luy et al. 2022). Similar to this, developed a colorimetric tool for real-time nitrate assessment in samples of surface water. UV absorption was utilized to detect nitrates; a light-emitting diode (LED) was used as the source of light, and a photodiode was employed to quantify the signal that was produced. The scientists used a tiny capacitive deionization cell for separating natural organic matter (NOM) from nitrate ions, which can cause problems with accurate nitrate detection by absorbing UV light. The nitrate limit of quantification (LOQ) was 0.12 mgL^{-1}, and the limit of detection (LOD) was 0.03 mgL^{-1} (Li et al. 2023).

6.4.1.1 Detection of Ammonia in Industrial Wastewater

Ammonia (NH_3) is extensively utilized and manufactured across various industries. Within wastewater, ammonia stands out as the predominant nitrogenous pollutant. Due to its harmful nature, wastewater containing ammonia is extremely dangerous. Its toxicity can have detrimental effects on aquatic life, causing oxygen depletion and eutrophication. Studies have shown that aquatic organisms are at risk of death even at quantities as low as 0.2 ppm (mgL^{-1}). Presented a method for measuring and visually detecting ammonia found in industrial wastewater using μPAD. Utilizing wax printing, the μPAD was made. An inexpensive monitoring system for detecting ammonia in wastewater has been created by this research. It is an updated colorimetric approach that combines microfluidic technologies and the Nessler reagent. Ammonia concentrations of 0–5 ppm can be detected by this method, with a detection limit of 3.34 ppm (Nxumalo et al. 2020).

6.4.1.2 Detection of Nitrate in Wastewater

Nitrate is a naturally occurring ion in the nitrogen cycle. It plays a crucial role in plant growth, yet excessive amounts can lead to significant water nutrient pollution (Charbaji et al. 2021). Various sources contribute to nitrate in water, such as fertilizers, manure run-off, air deposition, and wastewater treatment plants. Due to its stability in oxygenated conditions, nitrate becomes the main form of nitrogen in water (Silva et al. 2018). Drinking nitrate-rich water increases the risk of thyroid issues, central nervous system birth abnormalities, and colorectal cancer. There are several techniques for determining the nitrate content in water, but they frequently need pricey tools and labour-intensive analysis. Additionally, these methods involve specific sample handling and trained personnel. Introduced a microfluidic nitrate detecting device based on paper, created through wax printing. The design, featuring a folding architecture, enhanced the quality, uniformity, and detection limit of the device. The nitrate detection device proposed by the study has limits for both detection and quantification in water of 0.53 and 1.18 ppm, respectively (Charbaji et al. 2021).

6.4.2 Detection of Phosphate in Water

Water holds immense significance in our lives as one of the most crucial elements. Groundwater gets polluted in several ways, with high concentrations of pollutants like phosphates endangering human health. To tackle this issue, different approaches have been suggested, like the integration of microelectrode with a microfluidics system. However, this method proves to be overly complex, demanding large sample sizes, and posing challenges in monitoring. Introduced a novel technique utilizing μPAD for quantifying phosphate concentrations in water using a colorimetric test. The value of RGB revealed a colour that serves to measure the phosphate concentration, detecting the presence of phosphates through a reagent embedded in the paper. Rubber-stamp PDMS material is used to construct the μPAD; it is processed in an oven to establish a water-resistant barrier, followed by the application of chemicals on the paper. The test involves adding a single water droplet to yield results, which are then analysed. The accuracy of the results assessed through image is notably precise. This method stands out due to its minimal fluid requirement, high specificity, and cost-effectiveness. It can detect phosphate concentration within the range of 120 $mg{\cdot}mL^{-1}$ using just 1 mL of water, making it suitable for routine water analysis (Waghwani et al. 2019).

6.4.3 Heavy Metal Detection

Heavy metals are naturally occurring metals that have densities at least five times that of water (Al Osman et al. 2019). Water pollution from heavy metals has become a growing concern in recent decades (Kinuthia et al. 2020; Almeida et al. 2018; Santangelo et al. 2019). In light of escalating worries, inexpensive microfluidic devices offer a cost-effective solution for ongoing water monitoring concerning global heavy metal pollution. The rapid progress is evident through consistent innovations achieved in recent years, with numerous uses stemming from paper-based microfluidics (Almeida et al. 2018). A tiny PAD designed for lead ions (Pb^{2+}) present in drinking water has remarkable selectivity and accuracy, according to Wang et al. (2022a; b). This tool allows quick visual quantitative detection by measuring the particle dam's colour bar extension length. The nanofluidic paper-based analytical device (μPAD) was created for the detection of mercury (Hg^{2+}) in different water samples. This device can promptly provide on-site Hg_2^+ concentration results through a smart Android. The smartphone app for analysis is exceptionally receptive and easy to use, allowing inexperienced users to perform sample testing with this device (Jarujamrus et al. 2018). Similar to this, a device was developed to quantify the amount of pollution caused by silver (Ag^+) in freshwater. It achieved remarkable selectivity, a recovery rate of 96.8%, and a detection threshold of 453.7 nM (Wang et al. 2019). Analytical devices based on microfluidic paper have been employed in similar applications to detect Cu^{2+} (Quinn et al. 2018; Sharifi et al. 2020). Effectively carried out an analysis of heavy metals using a microfluidic device based on a sponge, which was combined using ion-selective electrodes to collect ions from heavy metals (Cd^{2+} and Pb^{2+}) as well as clinically relevant non-metallic chemical ions such as K^+, Cl^+, and Na^+ (Ding et al. 2021) (Table 6.1).

Table 6.1 Determination of heavy metals by nanofluidic technique (Mesquita et al. 2022)

Applications	Substrates and fabrication	Detection technique	Contributions and results	References
Detection of Hg^{2+} concentration in water samples	μPAD	Colorimetric	Higher sensitivity; lower LOD (0.003 mg/L, 3 SD blank/slope of the calibration curve); small sample volume uptake; short analysis duration	Jarujamrus et al. (2018)
Detection of Cu^{2+} concentration in water samples	μPAD	Colorimetric	Adaptable to measuring a broad variety of Cu concentrations (within roughly 20–500,000 ppb)	Quinn et al. (2018); Sharifi et al. (2020)
Detection of Ag^+ and Hg^{2+} in water samples	μPAD	Fluorescence recovery changes from reaction with GO surface	1.33 and 1.01 p.m. LODs; quick analysis	Khoshbin et al. (2020)
Detection of Cu^{2+}, Co^{2+}, Ni^{2+}, Hg^{2+}, and Mn^{2+} in the water sample	μPAD, wax printing	Colorimetric	Cu, Co, Ni, Hg, and Mn, with the lowest detectable amounts of 0.32, 0.59, 5.87, 0.20, and 0.11 mg/L, respectively	Kamnoet et al. (2021)
Detection of Pb^{2+} and Cd^{2+} in the water sample	3D printed chip	Epitaxial graphene conductivity	Excellent sensitivity for detecting Pb at low concentrations	Santangelo et al. (2019)
Recovery of metal ions (Cu^{2+}, Zn^{2+}, Ni^{2+}, Ag^+ and Zn^{2+}/Cu^{2+} mixtures) from water sample	Carbon cloth	Electrooxidation process	Regeneration of rare earth trace pollutants; controlled decorating of materials using metal nanoparticle patterns	(Allioux et al. 2018; Santangelo et al. 2019)

6.4.4 Non-metallic Pollutants

Non-metallic substances, which are inherently highly complex, are prevalent pollutants in water. In today's world, portable microfluidic devices are crucial in analysing the quality of water for a wide range of toxins, including residues from pharmaceuticals, owing to their numerous benefits (Barocio et al. 2021). Scala-Benuzzi et al. (2018) created an electrochemical paper-based immunocapture assay (EPIA) to measure ethinylestradiol in water samples quantitatively. The study documented that the assay successfully reached the lowest possible limit of detection of 0.1 ng/L and a linear range spanning from 0.5 to 120 ng/L (Scala-Benuzzi et al. 2018). A lipase-embedded paper-based device was used in another study to identify the chemical chlorpyrifos. It was found that the quantification limit was 0.198 mg/L and the detection limit was 0.065 mg/L (Scala-Benuzzi et al. 2018). It is important to remember that wash water from bitter guard, brinjal, grapes, cauliflower, and coriander leaves can serve as samples for tests (Sankar et al. 2020). An exceedingly sensitive electrochemical paper sensor for detecting bisphenol A (BPA) in drinking water was developed and a μPAD to track the amount of nitrate in drinking water was unveiled (Jemmeli et al. 2020; Mako et al. 2020). Caetano et al. (2018) created a microfluidic device using a textile thread, which was combined with an electrochemical biosensor, for detecting phenol levels in tap water. Microplastics are one type of non-metallic pollution that has attracted a lot of study interest lately. Microfluidic devices can enhance research related to microplastics in various ways, including identification and separation. Nevertheless, there is a scarcity of inexpensive microfluidic devices designed for these purposes. A 3D-printed microfluidic resistive pulse sensor with cost-effective and excellent throughput was developed by Pollard et al. to characterize microplastics from tea bags and algae with spherical and rod shapes. When microplastics and algae are present, this apparatus can promptly filter liquids at a rate of 1 L/min (Pollard et al. 2020). Researchers are advised to utilize cost-effective nanofluidic equipment to fully realize the potential of achieving consistent and trustworthy long-term evaluation of environmental microplastics.

6.4.5 Determination of Pesticides in Water

There is a greater chance of pesticide deposition in soil, agricultural goods, and water due to the increased usage of pesticides in agriculture, which can be dangerous to human well-being and other living organisms. Even in small quantities, pesticide residues can significantly impact the neurological systems of animals if they enter the food chain (Yan et al. 2018). In addition, pesticide build-ups generally collect in streams and lakes because of overflow from downpours or water systems in regions with far and wide pesticide use, adding to water contamination. Consequently, a helpful, lightweight, and easy-to-use scientific instrument for pesticide identification is still required. Utilizing chlorpyrifos and carbaryl as specific illustrations,

Table 6.2 Some devices and detection techniques for pollutants in wastewater (Rai et al. 2022)

Pollutants	Mode of determination	Substrate	Method of fabrication	References
2,4-D pesticide	Fluorescence	Cellulose paper	Wax printing	Zhang et al. (2019)
E. coli	Polymerase chain reaction	PMMA	Lithography	Dharmasiri et al. (2010)
E. coli	Light scattering	Chromatography paper	Photolithography	San & Yoon (2014)
Organophosphate and carbamate pesticides	Acetylcholinesterase (AChE) inhibition	Whatman paper	Wax printing	Jin et al. (2020)
Organophosphorus pesticides	Surface-enhanced Raman scattering	Polymer	Laser printing	Guselnikova et al. (2019)
Carbofuran ($C_{12}H_{15}NO_3$) and alachlor ($C_{12}H_{18}ClNO_2$) pesticide	Surface-enhanced Raman scattering	Silicon wafer	Photolithography	Parisi et al. (2015)
NO_3, NO_2, NH_3	Colorimetric	Polycarbonate plate	CNC milling	Hwang et al. (2013)

showed a bioactive nanofluidic paper-based tool made by laser slicing that might be utilized to examine organophosphorus and carbamate pesticides in water. The sensor stage uses a compound immobilized creative paper plan for discovery. The chemical restraint triggers signal transduction by hydrolyzing the substrate, prompting varieties in the caught pointer's pH, bringing about a variety change. A visual camera then, at that point, records this variety shift, and variety facilitates are utilized as the scientific boundary to gauge the pesticide. This proposed technique effectively identified carbaryl and chlorpyrifos with a location breaking point of 0.24 and 2.00 $g{\cdot}L^{-1}$, as well as consistency paces of 4.2 and 5.5%, respectively (Fernández-Ramos et al. 2020) (Table 6.2).

6.4.6 *Detection of Bacterial Samples in Water*

Nitrate is a naturally occurring ion in the nitrogen cycle. It plays a significant role in plant growth, yet excessive amounts can lead to significant water nutrient pollution (Charbaji et al. 2021). Different origins contribute to nitrate in water, for instance, fertilizers, manure run-off, air deposition, and wastewater treatment plants. Due to its stability in oxygenated conditions, nitrate becomes the main form of nitrogen in water (Silva et al. 2018). Drinking nitrate-rich water increases the risk of thyroid issues, central nervous system birth abnormalities, and colorectal cancer. There are several techniques for determining the nitrate content in water, but they frequently need pricey tools and labour-intensive analysis. Additionally, these methods involve specific sample handling and trained personnel. Introduced a microfluidic nitrate detecting device based on paper, created through wax printing. The design, featuring a folding architecture, enhanced the quality, uniformity, and detection limit of the device. The nitrate detection device proposed by the study has limits for both detection and quantification in water of 0.53 and 1.18 ppm, respectively (Charbaji et al. 2021).

6.5 Challenges and Future Aspects

Without a doubt, routine drug quality monitoring in the pharmaceutical industry cannot be replaced by current μPAD technologies with standard instrumental analytical processes. The obstacle faced by μPAD applications in the pharmaceutical industry is automating the platform to handle various samples and the lack of method validation to compare their analytical performance with conventional methods. μPAD often encounters issues with low sensitivity and selectivity, posing additional obstacles to detecting medicinal compounds in complex sample matrices. One potential tactic to think about is using on-device sample preparation to improve and selectively extract the analyte. This would streamline user procedures and increase sensitivity and selectivity. The detection limits for drug trace analysis may be improved by creating a reliable and portable detector and switching from colorimetry to more sensitive optical detection methods like chemiluminescence and its derivatives (e.g. biochemiluminescence [BCL] and electrochemiluminescence [ECL]). One more difficulty in drug analysis is the selective detection of drug enantiomers. Adding chiral recognition components to μPADs is one way to increase the potential applications of paper microfluidics in drug analysis in the future. One major disadvantage of paper-based microfluidic systems is that every time a paper-based device is used, unanticipated errors are introduced due to the intrinsic unpredictability of paper. These discrepancies can lead to variations in the results of paper-based tests, fluid wicking rates in the devices, and the accuracy of hydrophobic barriers on paper. Improving the paper-making processes' uniformity could significantly increase the accuracy of paper-based microfluidic devices. The uneven distribution of analytes, reagents, or reaction products inside the test zones is caused by the coffee-ring effect, which occurs when solutions dry on paper and causes solutes to concentrate along the borders of the zones. This compromises the accuracy and precision of the assay. Another drawback is the short shelf life of reagents kept in paper-based microfluidic devices. Enzymes are among the many substances that quickly deteriorate when stored on paper substrates. However, by incorporating various nanomaterials into μPADs, detection limits have been greatly increased by up to 1,000 times, thereby resolving the issues of low sensitivity and selectivity (Das et al. 2022).

6.6 Conclusion

The important role of harnessing tiny devices in water observation is highlighted by the remarkable quality and annual growth in the number of publications. Despite significant efforts to include microfluidics in environmental monitoring, conventional techniques such as mass spectroscopy and liquid chromatography continue to be the major tools for water analysis. The miniaturization of analytical devices, on-site analysis, multiplexing, easy device manipulation, and cost-effectiveness in fabrication are key features that will propel advancing microfluidics to the forefront of studies on water monitoring. However, utilizing microfluidic devices necessitates a thorough analysis of the use of functional equipment and supplies to address miniaturization challenges related to valves, sensors, LOD improvement, and multiple-analyte detection. Microfluidic devices reduce dependency on large external components by utilizing novel actuators and sensors that improve automation through the use of functional materials. This development will make it possible to monitor and control water sources continuously, quickly, affordably, and in real time. This will allow for more prompt and effective action against contaminated water, even in an emergency.

References

Agustini D, Caetano FR, Quero RF, da Silva JAF, Bergamini MF, Marcolino-Junior LH, de Jesus DP (2021) Microfluidic devices based on textile threads for analytical applications: state of the art and prospects. Anal Methods 13(41):4830–4857

Akhtarian S, Doostmohammadi A, Archonta DE, Kraft G, Brar SK, Rezai P (2023) Microfluidic sensor based on cell-imprinted polymer-coated microwires for conductometric detection of bacteria in water. Biosensors 13(10):943

Aladese AD, Jeong HH (2021) Recent developments in 3D printing of droplet-based microfluidics. BioChip J 15(4):313–333. https://doi.org/10.1007/s13206-021-00032

Allioux FM, Kapruwan P, Milne N, Kong L, Fattaccioli J, Chen Y, Dumée LFv (2018) Electro-capture of heavy metal ions with carbon cloth integrated microfluidic devices. Sep Purif Technol 194: 26–32. https://doi.org/10.1016/j.seppur.2017.10.064

Almeida MIG, Jayawardane BM, Kolev SD, McKelvie ID (2018) Developments of microfluidic paper-based analytical devices (μPADs) for water analysis: a review. Talanta 177:176–190. https://doi.org/10.1016/j.talanta.2017.08.072

Al-Osman M, Yang F, Massey IY (2019) Exposure routes and health effects of heavy metals on children. Biometals 32:563–573

Alsaeed B, Mansour FR (2020) Distance-based paper microfluidics; principle, technical aspects and applications. Microchem J 155:104664. https://doi.org/10.1016/j.microc.2020.104664

Andriukaitis D, Vargalis R, Šerpytis L, Drevinskas T, Kornyšova O, Stankevičius M et al (2022) Fabrication of microfluidic Tesla valve employing femtosecond bursts. Micromachines 13(8):1180

Aryal P, Hefner C, Martinez B, Henry CS (2024a) Microfluidics in environmental analysis: advancements, challenges, and future prospects for rapid and efficient monitoring. Lab Chip 15:452

Aryal P, Hefner C, Martinez B, Henry CS (2024b) Microfluidics in environmental analysis: advancements, challenges, and future prospects for rapid and efficient monitoring. Lab Chip 39:871. https://doi.org/10.1039/d3lc00871a

Bamshad A, Cho HJ (2021) Laserjet printed micro/nano sensors and microfluidic systems: a simple and facile digital platform for inexpensive, flexible, and low-volume devices. Adv Mater Technol 6(12):2100401. https://doi.org/10.1002/admt.202100401

Baranwal J, Barse B, Gatto G, Broncova G, Kumar A (2022) Electrochemical sensors and their applications: a review. Chemosensors 10(9):363. https://doi.org/10.3390/chemosensors10090363

Barocio ME, Hidalgo-Vázquez E, Kim Y, Rodas-Zuluaga LI, Chen WN, Barceló D et al (2021) Portable microfluidic devices for in-field detection of pharmaceutical residues in water: recent outcomes and current technological situation—a short review. Case Stud Chem Environ Eng 3:100069. https://doi.org/10.1016/j.cscee.2020.100069

Behroodi E, Latifi H, Bagheri Z, Ermis E, Roshani S, Salehi-Moghaddam M (2020) A combined 3D printing/CNC micro-milling method to fabricate a large-scale microfluidic device with the small size 3D architectures: an application for tumor spheroid production. Sci Rep 10(1):22171. https://doi.org/10.1038/s41598-020-79015-5

Bhattacharya S, Gupta AB, Gupta A, Pandey A (eds) (2018) Water remediation. Springer, Singapore

Buchroithner B, Mayr S, Hauser F, Priglinger E, Stangl H, Santa-Maria AR et al (2021) Dual channel microfluidics for mimicking the blood–brain barrier. ACS Nano 15(2):2984–2993. https://doi.org/10.1021/acsnano.0c09263

Caetano FR, Carneiro EA, Agustini D, Figueiredo-Filho LCS, Banks CE, Bergamini MF, Marcolino-Junior LH (2018) Combination of electrochemical biosensor and textile threads: a microfluidic device for phenol determination in tap water. Biosens Bioelectr 99:382–388

Carrell C, Kava A, Nguyen M, Menger R, Munshi Z, Call Z et al (2019) Beyond the lateral flow assay: a review of paper-based microfluidics. Microelectron Eng 206:45–54. https://doi.org/10.1016/j.mee.2018.12.002

Catalan-Carrio R, Saez J, Cuadrado LÁF, Arana G, Basabe-Desmonts L, Benito-Lopez F (2022) Ionogel-based hybrid polymer-paper handheld platform for nitrite and nitrate determination in water samples. Anal Chim Acta 1205:339753

Charbaji A, Heidari-Bafroui H, Anagnostopoulos C, Faghri M (2021) A new paper-based microfluidic device for improved detection of nitrate in water. Sensors 21(1):1–15. https://doi.org/10.3390/s21010102

Charles PT, Wadhwa V, Kouyate A, Mesa-Donado KJ, Adams AA, Deschamps JR, Kusterbeck AW (2018) A high aspect ratio bifurcated 128-microchannel microfluidic device for environmental monitoring of explosives. Sensors 18(5):1568. https://doi.org/10.3390/s18051568

Chen QY, DesMarais T, Costa M (2019a) Metals and mechanisms of carcinogenesis. Annu Rev Pharmacol Toxicol 59(1):537–554. https://doi.org/10.1146/annurev-pharmtox-010818-021031

Chen X, Li T, Gao QI (2019b) A novel method for rapid fabrication of PMMA microfluidic chip by laser cutting and sealing integration. Surf Rev Lett 26(08):1950042. https://doi.org/10.1142/s0218625x19500422

Das S, Gagandeep R, Bhatia R (2022) Based microfluidic devices: fabrication, detection, and significant applications in various fields. Rev Anal Chem 41(1):112–136. https://doi.org/10.1515/revac-20220037

Dharmasiri U, Witek MA, Adams AA, Osiri JK, Hupert ML, Bianchi TS et al (2010) Enrichment and detection of *Escherichia coli* O157: H7 from water samples using an antibody modified microfluidic chip. Anal Chem 82(7):2844–2849. https://doi.org/10.1021/ac100323k

Ding R, Cheong YH, Zhao K, Lisak G (2021) Acidified paper substrates for microfluidic solution sampling integrated with potentiometric sensors for determination of heavy metals. Sens Actuat B Chem 347:130567. https://doi.org/10.1016/j.snb.2021.130567

Elgohary A, Block E, Squier J, Koneshloo M, Shaha RK, Frick C et al (2020) Fabrication of sealed sapphire microfluidic devices using femtosecond laser micromachining. Appl Opt 59(30):9285–9291. https://doi.org/10.1364/ao.400184

Fan Y (2018) Low-cost microfluidics: materials and methods. Micro Nano Lett 13(10):1367–1372. https://doi.org/10.1049/mnl.2018.5169

Farajikhah S, Cabot JM, Innis PC, Paull B, Wallace G (2019) Life-saving threads: advances in textile-based analytical devices. ACS Combin Sci 21(4):229–240

Fernández-Ramos MD, Ogunneye AL, Babarinde NAA, Erenas MM, Capitán-Vallvey LF (2020) Bioactive microfluidic paper device for pesticide determination in waters. Talanta 218:121108

Fu LM, Wang YN (2018) Detection methods and applications of microfluidic paper-based analytical devices. TrAC Trends Anal Chem 107:196–211

Gao B, Yang Y, Liao J, He B, Liu H (2019) Bioinspired multistructured paper microfluidics for POCT. Lab Chip 19(21):3602–3608. https://doi.org/10.1039/c9lc00907h

Gilmore J, Islam M, Martinez-Duarte R (2016) Challenges in the use of compact disc-based centrifugal microfluidics for healthcare diagnostics at the extreme point of care. Micromachines 7(4):52. https://doi.org/10.3390/mi7040052

Gonçalves IM, Madureira M, Miranda I, Schütte H, Moita A, Minas G et al (2021) Separation microfluidic device fabricated by micromilling techniques. Eng Proceed 4(1):37

Guckenberger DJ, De Groot TE, Wan AM, Beebe DJ, Young EW (2015) Micromilling: a method for ultra-rapid prototyping of plastic microfluidic devices. Lab Chip 15(11):2364–2378. https://doi.org/10.1039/c5lc00234f

Gupta PK, Raghunath SS, Prasanna DV, Venkat P, Shree V, Chithananthan C et al (2019) An update on overview of cellulose, its structure and applications. Cellulose 201(9):84727. https://doi.org/10.5772/intechopen.84727

Guselnikova O, Postnikov P, Elashnikov R, Miliutina E, Svorcik V, Lyutakov O (2019) Metal-organic framework (MOF-5) coated SERS active gold gratings: a platform for the selective detection of organic contaminants in soil. Anal Chim Acta 1068:70–79. https://doi.org/10.1016/j.aca.2019.03.058

Hossain MM, Rahman T (2018) Low cost micro milling machine for prototyping plastic microfluidic devices. In: Proceedings. MDPI, p 707

Hwang H, Kim Y, Cho J, Lee JY, Choi MS, Cho YK (2013) Lab-on-a-disc for simultaneous determination of nutrients in water. Anal Chem 85(5):2954–2960. https://doi.org/10.1021/ac3036734

Ibrahim S, El-Liethy MA, Abia ALK, Abdel-Gabbar M, Al Zanaty AM, Kamel MM (2020) Design of a bioaugmented multistage biofilter for accelerated municipal wastewater treatment and deactivation of pathogenic microorganisms. Sci Total Environ 703:134786. https://doi.org/10.1016/j.scitotenv.2019.134786

Islam M, Natu R, Martinez-Duarte R (2015) A study on the limits and advantages of using a desktop cutter plotter to fabricate microfluidic networks. Microfluid Nanofluid 19:973–985

Islam MM, Loewen A, Allen PB (2018) Simple, low-cost fabrication of acrylic based droplet microfluidics and its use to generate DNA-coated particles. Sci Rep 8(1):8763. https://doi.org/10.1038/s41598-018-27037-5

Jarujamrus P, Meelapsom R, Pencharee S, Obma A, Amatatongchai M, Ditcharoen N et al (2018) Use of a smartphone as a colorimetric analyzer in paper-based devices for sensitive and selective determination of mercury in water samples. Anal Sci 34(1):75–81. https://doi.org/10.2116/analsci.34.75

Javidanbardan A, Azevedo AM, Chu V, Conde JP (2021) A systematic approach for developing 3D high-quality PDMS microfluidic chips based on micromilling technology. Micromachines 13(1):6. https://doi.org/10.3390/mi13010006

Jaywant SA, Arif KM (2019) A comprehensive review of microfluidic water quality monitoring sensors. Sensors 19(21):4781

Jemmeli D, Marcoccio E, Moscone D, Dridi C, Arduini F (2020) Highly sensitive paper-based electrochemical sensor for reagent free detection of bisphenol A. Talanta 216:120924

Jiménez-Díaz E, Cano-Jorge M, Zamarrón-Hernández D, Cabriales L, Páez-Larios F, Cruz-Ramírez A et al (2019) Micro–macro: selective integration of microfeatures inside low-cost macromolds for PDMS microfluidics fabrication. Micromachines 10(9):576. https://doi.org/10.3390/mi10090576

Jin L, Hao Z, Zheng Q, Chen H, Zhu L, Wang C et al (2020) A facile microfluidic paper-based analytical device for acetylcholinesterase inhibition assay utilizing organic solvent extraction in rapid detection of pesticide residues in food. Anal Chim Acta 1100:215–224

Kaneta T, Alahmad W, Varanusupakul P (2019) Microfluidic paper-based analytical devices with instrument-free detection and miniaturized portable detectors. Appl Spectrosc Rev 54(2):117–141. https://doi.org/10.1080/05704928.2018.1457045

Kamnoet P, Aeungmaitrepirom W, Menger RF & Henry CS (2021) Highly selective simultaneous Cu (II), Co (II), Ni (II), Hg (II), Mn (II) determination in water samples on microfluidic paper-based analytical devices. Analyst 146(7):2229. https://doi.org/10.1039/D0AN02200D

Khoshbin Z, Housaindokht MR & Verdian A (2020) A low-cost paper-based aptasensor for simultaneous trace-level monitoring of mercury (II) and silver (I) ions. Anal Biochem 597: 113689. https://doi.org/10.1016/j.ab.2020.113689

Khorsandi D, Nodehi M, Waqar T, Shabani M, Kamare B, Zare EN et al (2021) Manufacturing of microfluidic sensors utilizing 3d printing technologies: a production system. J Nanomater 2021(1):5537074. https://doi.org/10.1155/2021/5537074

Kinuthia GK, Ngure V, Beti D, Lugalia R, Wangila A, Kamau L (2020) Levels of heavy metals in wastewater and soil samples from open drainage channels in Nairobi, Kenya: community health implication. Sci Rep 10:1–13. https://doi.org/10.1038/s41598-020-65359-5

Kosoff D, Yu J, Suresh V, Beebe DJ, Lang JM (2018) Surface topography and hydrophilicity regulate macrophage phenotype in milled microfluidic systems. Lab Chip 18(19):3011–3017. https://doi.org/10.1039/c8lc00431e

Ku X, Zhang Z, Liu X, Chen L, Li G (2018) Low-cost rapid prototyping of glass microfluidic devices using a micromilling technique. Microfluid Nanofluid 22:1–8. https://doi.org/10.1007/s10404-018-2104-y

Kudr J, Zitka O, Klimanek M, Vrba R, Adam V (2017) Microfluidic electrochemical devices for pollution analysis: a review. Sens Actuat B Chem 246:578–590

Li W, Zhang X, Li T, Ji Y, Li R (2021) Molecularly imprinted polymer-enhanced biomimetic paper-based analytical devices: a review. Anal Chim Acta 1148:238196. https://doi.org/10.1016/j.aca.2020.12.071

Li H, Zhao B, Wang D, Zhang K, Tan X, Zhang Q (2023) Effect of multiple spatial scale characterization of land use on water quality. Environ Sci Pollut Res 30(3):7106–7120

Lim H, Jafry AT, Lee J (2019) Fabrication, flow control, and applications of microfluidic paper-based analytical devices. Molecules 24(16):2869. https://doi.org/10.3390/molecules24162869

Lin H, Zhao Y, Lin S, Wang B, Yeung C, Cheng X et al (2019) A rapid and low-cost fabrication and integration scheme to render 3D microfluidic architectures for wearable biofluid sampling, manipulation, and sensing. Lab Chip 19(17):2844–2853. https://doi.org/10.1039/c9lc00418a

Liu C, Gomez FA, Miao Y, Cui P, Lee W (2019) A colorimetric assay system for dopamine using microfluidic paper-based analytical devices. Talanta 194:171–176

Loo JF, Ho AH, Turner AP, Mak WC (2019) Integrated printed microfluidic biosensors. Trends Biotechnol 37(10):1104–1120

Luy E, Smith J, Grundke I, Sonnichsen C, Furlong A, Sieben V (2022) Two chemistries on a single lab-on-chip: Nitrate and orthophosphate sensing underwater with inlaid microfluidics. Front Sens 3:1080020

Ma J, Yan S, Miao C, Li L, Shi W, Liu X et al (2019) Paper microfluidics for cell analysis. Adv Healthc Mater 8(1):1801084. https://doi.org/10.1002/adhm.201801084

Madureira M, Faustino V, Schütte H, Pinho D, Minas G, Gassmann S, Lima R (2019) Red blood cells separation in a curved T-shaped microchannel fabricated by a micromilling technique. In: VipIMAGE 2019: proceedings of the VII ECCOMAS thematic conference on computational vision and medical image processing, October 16–18, 2019, Porto, Portugal. Springer, Berlin, pp 585–593

Maguire I, Okennedy R, Ducrée J, Regan FJAM (2018) A review of centrifugal microfluidics in environmental monitoring. Anal Methods 10(13):1497–1515

Mahmud MA, Blondeel EJ, Kaddoura M, MacDonald BD (2018) Features in microfluidic paper-based devices made by laser cutting: how small can they be? Micromachines 9(5):220. https://doi.org/10.3390/mi9050220

Mako TL, Levenson AM, Levine M (2020) Ultrasensitive detection of nitrite through implementation of N-(1-Naphthyl) ethylenediamine-grafted cellulose into a paper-based device. ACS Sens 5(4):1207–1215. https://doi.org/10.1021/acssensors.0c00291

Mesquita P, Gong L, Lin Y (2022) Low-cost microfluidics: towards affordable environmental monitoring and assessment. Front Lab Chip Technol 1:1074009

Morbioli GG, Mazzu-Nascimento T, Stockton AM, Carrilho E (2017) Technical aspects and challenges of colorimetric detection with microfluidic paper-based analytical devices (μPADs): a review. Anal Chim Acta 970:1–22

Mukherjee S, Bhattacharyya S, Ghosh K, Pal S, Halder A, Naseri M et al (2021) Sensory development for heavy metal detection: a review on translation from conventional analysis to field-portable sensor. Trends Food Sci Technol 109:674–689

Nguyen TQ, Mah J, Park WT, Lee S (2019) Rapid and versatile micromold fabrication using micromilling and nanopolishing for microfluidic devices. In: Fluids engineering division summer meeting. American Society of Mechanical Engineers, p V004T06A011

Nishat S, Jafry AT, Martinez AW, Awan FR (2021) Based microfluidics: simplified fabrication and assay methods. Sens Actuat B Chem 336:129681. https://doi.org/10.1016/j.snb.2021.129681

Nogueira SA, Lemes AD, Chagas AC, Vieira ML, Talhavini M, Morais PA, Coltro WK (2019) Redox titration on foldable paper-based analytical devices for the visual determination of alcohol content in whiskey samples. Talanta 194:363–369. https://doi.org/10.1016/j.talanta.2018.10.036

Nxumalo NL, Madikizela LM, Kruger HG, Onwubu SC, Mdluli PS (2020) Development of a paper-based microfluidic device for the quantification of ammonia in industrial wastewater. Water SA 46(3):506–513

Oliveira ACM, Araújo DAG, Pradela-Filho LA, Takeuchi RM, Trindade MAG, Dos Santos AL (2022) Threads in tubing: an innovative approach towards improved electrochemical thread-based microfluidic devices. Lab Chip 22(16):3045–3054. https://doi.org/10.1039/d2lc00387b

Olmos CM, Vaca A, Rosero G, Peñaherrera A, Perez C, de Sá Carneiro I et al (2019) Epoxy resin mold and PDMS microfluidic devices through photopolymer flexographic printing plate. Sens Actuat B Chem 288:742–748. https://doi.org/10.1016/j.snb.2019.03.062

Owens CE, Hart AJ (2018) High-precision modular microfluidics by micromilling of interlocking injection-molded blocks. Lab Chip 18(6):890–901

Pan Y, Hui Q, Yang Z (2020) Low-cost and rapid paper based microfluidic device for wastewater surveillance at low-resource settings

Parisi J, Dong Q, Lei Y (2015) In situ microfluidic fabrication of SERS nanostructures for highly sensitive fingerprint microfluidic-SERS sensing. RSC Adv 5(19):14081–14089. https://doi.org/10.1039/c4ra15174g

Persson H, Park S, Mohan M, Cheung KK, Simmons CA, Young EW (2022) Rapid assembly of PMMA microfluidic devices with PETE membranes for studying the endothelium. Sens Actuat B Chem 356:131342

Pollard M, Hunsicker E, Platt M (2020) A tunable three-dimensional printed microfluidic resistive pulse sensor for the characterization of algae and microplastics. ACS Sens 5:2578–2586. https://doi.org/10.1016/j.sna.2022.113926

Prince R, Monsur I, Ankur S (2019) Microfluidic devices for the detection of contamination in water samples: a review

Quinn CW, Cate DM, Miller-Lionberg DD, Reilly T, Volckens J, Henry CS (2018) Solid-phase extraction coupled to a paper-based technique for trace copper detection in drinking water. Environ Sci Technol 52(6):3567–3573. https://doi.org/10.1021/acs.est.7b05436

Rahim MS, Ehsan AA (2021) Micro milling process for the rapid prototyping of microfluidic devices. Adv Microfluid Nanofluids 72:91560. https://doi.org/10.5772/intechopen.91560

Rai PK, Islam M, Gupta A (2022) Microfluidic devices for the detection of contamination in water samples: a review. Sens Actuat A Phys 347:113926

San Park T, Yoon JY (2014) Smartphone detection of Escherichia coli from field water samples on paper microfluidics. IEEE Sen J 15(3): 1902–1907

Sankar K, Lenisha D, Janaki G, Juliana J, Kumar RS, Selvi MC, Srinivasan G (2020) Digital image-based quantification of chlorpyrifos in water samples using a lipase embedded paper based device. Talanta 208:120408

Santana HS, Silva JL Jr, Aghel B, Ortega-Casanova J (2020) Review on microfluidic device applications for fluids separation and water treatment processes. SN Appl Sci 2(3):395. https://doi.org/10.1007/s42452-020-2176-7

Santangelo MF, Shtepliuk I, Filippini D, Puglisi D, Vagin M, Yakimova R, Eriksson J (2019) Epitaxial graphene sensors combined with 3D-printed microfluidic chip for heavy metals detection. Sensors 19(10):2393. https://doi.org/10.3390/s19102393

Saptaji K, Triawan F, Sai TK, Gebremariam A (2021) Deburring method of aluminum mould produced by milling process for microfluidic device fabrication. Ind J Sci Technol 6(1):123–140

Scala-Benuzzi ML, Raba J, Soler-Illia GJ, Schneider RJ, Messina GA (2018) Novel electrochemical paper-based immunocapture assay for the quantitative determination of ethinylestradiol in water samples. Anal Chem 90(6):4104–4111. https://doi.org/10.1021/acs.analchem.8b00028

Sekar A, Yadav R, Basavaraj N (2021) Fluorescence quenching mechanism and the application of green carbon nanodots in the detection of heavy metal ions: a review. New J Chem 45(5):2326–2360

Sharifi H, Tashkhourian J, Hemmateenejad B (2020) A 3D origami paper-based analytical device combined with PVC membrane for colorimetric assay of heavy metal ions: application to determination of Cu(II) in water samples. Anal Chim Acta 1126:114–123

Sharma N, Nigam A, Lobanov D, Gupta A, Novikov A, Kumar M (2021) Mercury(II) ion detection using AgNWs-MoS 2 nanocomposite on GaN HEMT for IoT-enabled smart water quality analysis. IEEE Internet Things J 9(16):14317–14324. https://doi.org/10.1109/JIOT.2021.3071382

Shimazu R, Tomimuro K, Ni Y, Malegori C, Hamedpour V, Hiruta Y et al (2022) Microfluidic thread-based analytical devices for point-of-care detection of therapeutic antibody in blood. Sens Actuat B Chem 352:131002. https://doi.org/10.1016/j.snb.2021.131002

Shu J, Tang D (2017) Current advances in quantum-dots-based photoelectrochemical immunoassays. Chem Asian J 12(21):2780–2789

Silva CG, Pereira MF, Órfão JJ, Faria JL, Soares OS (2018) Catalytic and photocatalytic nitrate reduction over Pd–Cu loaded over hybrid materials of multi-walled carbon nanotubes and TiO_2. Front Chem 6:632

Soum V, Park S, Brilian AI, Kwon OS, Shin K (2019) Programmable paper-based microfluidic devices for biomarker detections. Micromachines 10(8):516. https://doi.org/10.3390/mi10080516

Suqi W, Rong CLJ, Abdulbari HA, Mahmood WK (2024) Microfluidic approaches in water quality monitoring: an insight and a comprehensive review. ChemBioEng Rev 11(2):215–230

Tan W, Powles E, Zhang L, Shen W (2021) Go with the capillary flow simple thread-based microfluidics. Sens Actuat B Chem 334:129670

Waghwani B, Balpande S, Kalambe J (2019) Development of microfluidic paper based analytical device for detection of phosphate in water. Int J Innov Technol Explor Eng 8(6):592–595

Wang G, Chu LT, Hartanto H, Utomo WB, Pravasta RA, Chen TH (2019) Microfluidic particle dam for visual and quantitative detection of lead ions. ACS Sens 5(1):19–23. https://doi.org/10.1021/acssensors.9b01945

Wang G, Li J, Wu S, Jiang T, Chen TH (2022a) A fully integrated, ready-to-use distance-based chemosensor for visual quantification of multiple heavy metal ions. Anal Chem 94(46):15925–15929

Wang L, Li B, Wang J, Qi J, Li J, Ma J, Chen L (2022b) A rotary multi-positioned cloth/paper hybrid microfluidic device for simultaneous fluorescence sensing of mercury and lead ions by using ion imprinted technologies. J Hazard Mater 428:128165. https://doi.org/10.1016/j.jhazmat.2021.128165

Weng X, Kang Y, Guo Q, Peng B, Jiang H (2019) Recent advances in thread-based microfluidics for diagnostic applications. Biosens Bioelectr 132:171–185. https://doi.org/10.1016/j.bios.2019.03.009

Xia J, Khaliliazar S, Hamedi MM, Sonkusale S (2021) Thread-based wearable devices. MRS Bull 46(6):502–511. https://doi.org/10.1557/s43577-021-00116-1

Xiao G, He J, Chen X, Qiao Y, Wang F, Xia Q et al (2019) A wearable, cotton thread/paper-based microfluidic device coupled with smartphone for sweat glucose sensing. Cellulose 26:4553–4562. https://doi.org/10.1007/s10570-019-02396-y

Yan X, Li H, Su X (2018) Review of optical sensors for pesticides. TrAC Trends Anal Chem 103:1–20

Yuan Y, Jia H, Xu D, Wang J (2023) Novel method in emerging environmental contaminants detection: fiber optic sensors based on microfluidic chips. Sci Total Environ 857:159563

Zhang Z, Ma X, Jia M, Li B, Rong J, Yang X (2019) Deposition of CdTe quantum dot on microfluidic paper chips for rapid fluorescence detection of pesticide 2, 4-D. Analyst 144(4):1282–1291. https://doi.org/10.1039/c8an02051e

Zhang Y, Li J, Jiao S, Li Y, Zhou Y, Zhang X et al (2024) Microfluidic sensors for the detection of emerging contaminants in water: a review. Sci Total Environ 34:172734

Zhao WW, Xu JJ, Chen HY (2015) Photoelectrochemical bioanalysis: the state of the art. Chem Soc Rev 44(3):729–741

Zheng L, Liu Y, Zhang C (2021) A sample-to-answer, wearable cloth-based electrochemical sensor (WCECS) for point-of-care detection of glucose in sweat. Sens Actuat B Chem 343:130131. https://doi.org/10.1016/j.snb.2021.130131

Part II
Approaches for Wastewater Management

Chapter 7
Potential of Artificial Intelligence and Machine Learning in Wastewater Management

N. Nirmala, J. Merlin, J. Arun, J. Emmanuel, and S. Sanjay Kumar

Abstract Water is a vital aspect in contemporary civilization, and it is impacted in two ways these days. One is caused by water pollution, also referred to as wastewater, while the other is a result of overpopulation, which is referred to as water shortage. Therefore, it is beneficial to monitor the aforementioned issues and find remedies. Recycling water is the only option to prevent water scarcity; however, wastewater comes in a variety of forms and mixes. It is simpler to identify the underlying causes and thereafter address them with the aid of artificial intelligence (AI) and machine learning (ML). This chapter discusses the advantages of applying Artificial Intelligence (AI) and Machine Learning (ML) in wastewater management and treatment, along with the outcomes achieved by countries that have implemented these technologies. Using ML and AI in wastewater treatment provides high accuracy rate, reduces process costs and duration, and produces results that can be monitored globally, allowing us to resolve this problem.

Keywords Artificial intelligence · Internet of Things (IoT) · Machine learning · Predictive analysis · Wastewater treatment

N. Nirmala (✉) · J. Merlin · J. Arun · J. Emmanuel
Centre for Waste Management, Sathyabama Institute of Science and Technology, Chennai, India
e-mail: nirmala.amutha1189@gmail.com

N. Nirmala
Centre of Excellence for Energy Research, Sathyabama Institute of Science and Technology, Chennai, India

S. Sanjay Kumar
Indian Institute of Technology Madras, Chennai, India

A. M. Yatoo et al. (eds.), *Wastewater Monitoring and Management by Advanced Approaches*, Advances in Wastewater Research,
https://doi.org/10.1007/978-981-95-2601-7_7

7.1 Introduction

Water is an essential resource for production, and a company's profitability and operational efficiency may be significantly impacted by its availability, reliability, and quality (Li et al. 2019). Factory wastewater that leaks into waterways has the potential to contaminate the environment and harm both human health and the ecosystem in the area. Since the world's population is growing at the same time as it is becoming more developed, more needs to be done and more resources should be available for everyone to use. The two main sources of pollution are air and water, but the latter can be divided into two categories: water scarcity and polluted water, both of which are caused by the discharges from factories and other industries, in which people are severely impacted. Therefore, the trend should be reversed in order to develop more plans that reduce the time and cost of the process so that everyone can use the final product.

Wastewater management plays a crucial role in safeguarding public health, promoting environmental sustainability, and maintaining urban infrastructure, yet it faces a number of significant challenges. One of the most pressing issues is the rising volume of wastewater due to population growth, industrial development, and urban expansion. As cities grow, the need for efficient systems to treat and manage wastewater becomes more urgent. In addition, many existing sewer systems, built decades ago, are outdated and require expensive upgrades to meet the demands of modern urban populations. Another challenge lies in the complex composition of wastewater, which can contain a broad spectrum of pollutants, including organic matter, chemicals, pathogens, and heavy metals (Lowe et al. 2022). This diversity complicates the treatment process, as different types of contaminants necessitate specialized methods of removal. Many treatment plants are also ill-equipped to address newer pollutants, such as pharmaceuticals, personal care products, and microplastics, which are increasingly present in wastewater (Chowdhury and Karanfil 2024).

Resource limitations, such as insufficient funding for infrastructure maintenance and upgrades, also hinder progress in wastewater management. In low-income regions, access to adequate sanitation and wastewater treatment remains a significant challenge, leading to direct contamination of water sources and increasing public health risks (Zhu et al. 2022). Climate change further intensifies these issues by altering the precipitation patterns, increasing the frequency of extreme weather events, and heightening the risk of flooding, which can overwhelm existing wastewater systems. Water scarcity is another concern, especially in regions where freshwater resources are limited. Wastewater treatment and recycling offer potential solutions to this problem, but the necessary technologies and infrastructure for safe and effective reuse are often lacking. Public awareness about the importance of proper wastewater management is often insufficient, which impedes efforts to reduce pollution and improve systems. Inconsistent or poorly enforced regulatory frameworks contribute to inadequate wastewater treatment and insufficient protection of water resources (Chowdhury and Karanfil 2024). Tackling these challenges requires a multifaceted approach that includes technological innovation, stronger policy enforcement, and enhanced cooperation among governments, industries, and local communities.

Technology plays a crucial role in modernizing wastewater treatment by enhancing efficiency, reducing costs, and promoting more sustainable practices to meet growing demands on water infrastructure (Behera et al. 2024). Advances in chemical and biological treatment techniques, such as using enzymes, probiotics, and advanced oxidation processes, have significantly improved the ability to break down complex pollutants—like pharmaceuticals, personal care products, and microplastics—that were previously difficult to treat. These innovations help transform harmful pollutants into nontoxic substances, resulting in safer water being discharged into natural ecosystems. The integration of digital technologies, including automation, sensors, and data analytics, is also revolutionizing wastewater treatment. Real-time monitoring allows for quick detection of changes in water quality, enabling faster responses to contamination or operational issues. Predictive analytics and machine learning (ML) models help operators anticipate maintenance needs, optimize energy use, and lower operational costs. These tools also improve the management of essential resources—like water, chemicals, and energy—ensuring that treatment processes are both cost-efficient and environmentally sound (Kumari et al. 2024).

Artificial intelligence (AI) and machine learning (ML) are increasingly being incorporated into environmental management to tackle a wide range of challenges, including climate change, pollution control, resource conservation, and biodiversity preservation. AI enables machines to replicate human intelligence, while ML, a subset of AI, allows systems to learn from data and improve over time without explicit programming. These technologies are especially valuable in environmental management, where they can process large volumes of data, detect patterns, and make predictions that would be difficult for humans to achieve (Sharafi et al. 2024). For instance, AI algorithms can analyze real-time data from sensors and satellite imagery to detect pollution, monitor deforestation, or track changes in ecosystems. ML models can forecast trends, such as the effects of climate change on specific areas, or predict weather patterns that may impact natural resources.

AI and ML are increasingly recognized for their ability to revolutionize wastewater management by boosting efficiency, cutting costs, and supporting sustainability. These technologies have the potential to enhance various stages of wastewater treatment, including monitoring, data analysis, predictive maintenance, and resource recovery. A major area where AI and ML are making a difference is in real-time monitoring and contaminant detection. By leveraging sensors and advanced analytics, AI can process vast amounts of data from treatment plants, identifying pollutants or irregularities in water quality that may not be detected through traditional methods (Lowe et al. 2022). This allows operators to respond more quickly, improving treatment efficiency and ensuring that the effluent complies with regulatory standards. AI algorithms can also optimize treatment processes, such as chemical dosing, energy use, and aeration, in systems like activated sludge and membrane bioreactors (Mahardhika and Putriani 2023). ML models can analyze historical data and real-time inputs to automatically adjust treatment parameters, ensuring that operations are running at peak efficiency while reducing energy and chemical consumption. This article highlights the integration of AI and ML into wastewater management, offering significant opportunities to optimize treatment processes, reduce costs, recover valuable resources, and improve system resilience. As these technologies evolve, they hold great potential to create more sustainable, efficient, and cost-effective wastewater management solutions, helping to address the challenges posed by urbanization and environmental sustainability.

7.2 Wastewater Management

Wastewater treatment is a vital process that ensures water is purified by removing contaminants before being released back into the environment or reused. The primary objectives are to safeguard public health, protect the environment, and enable water recycling for various uses. This process typically consists of three main stages: collection, treatment, and disposal. The collection phase involves gathering wastewater from residential, industrial, and commercial sources through a network of pipes, known as sewers. The wastewater is then transported to treatment plants, where it undergoes further processing, such as primary, secondary, and tertiary treatment (Pérez-Beltrán et al. 2024).

Primary treatment is the first step, where large solids and debris are physically removed, usually through processes like screening and sedimentation. In this stage, wastewater flows into large tanks where heavier solids settle at the bottom, forming sludge, while lighter materials like oils and grease rise to the surface and are skimmed off. Secondary treatment focuses on removing dissolved and suspended organic matter using biological processes. Bacteria and microorganisms are employed to break down these organic pollutants. One common method is the activated sludge process, where aeration tanks supply oxygen to promote the growth of microorganisms that consume organic matter in the water. Tertiary treatment is used to further purify the water by removing any remaining contaminants, such as nitrogen, phosphorus, or pathogens. This stage may involve chemical treatments, filtration, or disinfection methods like chlorination or ultraviolet light (Alaswad and Almatrafi 2024).

After treatment, the wastewater is ready for disposal or reuse. If it is being returned to the environment, the treated water is typically discharged into rivers, lakes, or oceans. The sludge generated during the treatment process, especially in the primary and secondary stages, must also be managed. It can be further processed, for example through anaerobic digestion to reduce its volume, stabilize it, and produce biogas (Yuvaraju and Deena 2024). The treated sludge is then either disposed of safely or used as fertilizer, depending on its composition and safety standards. Wastewater treatment is a multistep procedure designed to ensure that wastewater is effectively collected, treated, and safely disposed of or reused. Each stage plays a crucial role in removing contaminants, protecting public health, and minimizing environmental harm. Ongoing advancements in treatment technologies continue to enhance the

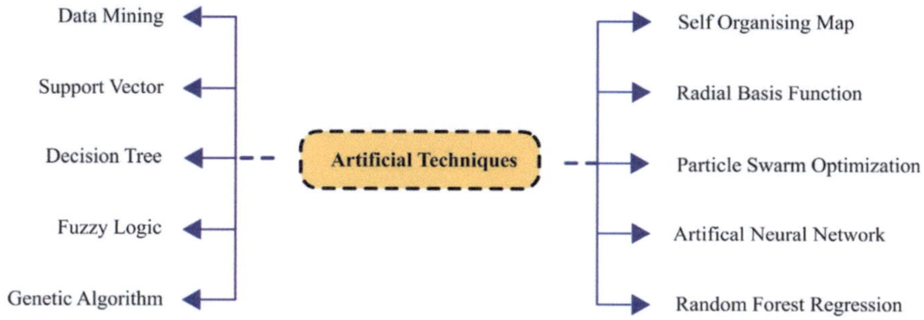

Fig. 7.1 Classifications of AL modeling in wastewater treatment

efficiency and sustainability of wastewater management (Li et al. 2021). Figure 7.1 shows different types of AI modeling used in wastewater treatment plants.

Wastewater management faces several significant challenges that hinder effective treatment and disposal while ensuring the protection of public health and the environment. One of the most urgent issues is pollution control. Wastewater contains a diverse mix of contaminants, such as organic matter, chemicals, heavy metals, pharmaceuticals, and pathogens (Kavianpour et al. 2024). Removing these contaminants effectively requires advanced treatment technologies, which can be costly and resource-intensive. Furthermore, emerging pollutants like microplastics and personal care products are not always adequately addressed by conventional treatment methods, presenting risks to both human health and aquatic ecosystems. Energy efficiency is another major challenge. Wastewater treatment, particularly biological methods like activated sludge, consumes large amounts of energy, especially for aeration. As energy prices rise, these treatments become more expensive for wastewater treatment facilities to maintain. Many plants also operate with outdated infrastructure, which further increases energy demand (Maurya et al. 2024b). Real-time monitoring of wastewater quality presents another critical challenge. Traditional monitoring, which relies on periodic sampling and laboratory analysis, can be slow and may not provide the real-time data needed to detect contaminants or system malfunctions. Delays in identifying problems can lead to significant environmental or health risks. Real-time monitoring, powered by sensors and advanced data analytics, offers more timely detection, but implementing these technologies can be expensive, and integrating them into existing systems can be difficult, especially in regions with limited resources or outdated infrastructure. Figure 7.2 shows the possible application and classifications of ML algorithms. Table 7.1 shows the applications of AI and ML in wastewater treatment and management.

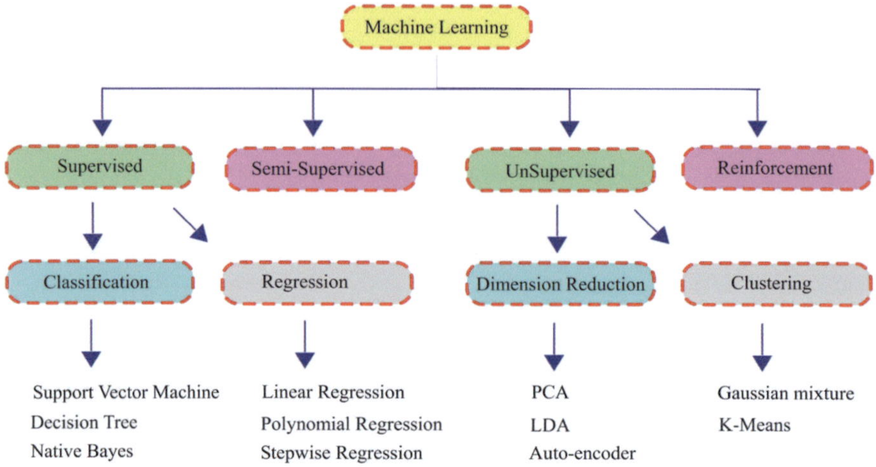

Fig. 7.2 Classifications of ML algorithm

7.3 Applications of Artificial Intelligence and Machine Learning

7.3.1 Real-Time Monitoring and Predictive Analysis

AI and ML can be applied to predict effluent water quality based on input factors such as influent flow, temperature, pH, and pollutant levels. These predictions help guide the treatment process, ensuring that the effluent complies with regulatory standards while minimizing the need for excessive chemicals or operational costs (Alam et al. 2022). Machine learning techniques like decision trees, neural networks, and ensemble methods can forecast future water quality based on historical data, enabling plant operators to adjust treatment processes proactively instead of reacting to issues after they arise (Ding et al. 2024). AI can forecast sludge production rates by analyzing factors like influent characteristics, seasonal trends, and operational conditions. Table 7.2 gives the applications of AI algorithms, along with their merits and demerits, and these algorithms help optimize dewatering processes and lower disposal costs (Durgun 2024).

7.3.2 Process Optimization

AI can also optimize maintenance intervals by identifying the most cost-effective times for repairs or replacements, helping to avoid unnecessary servicing and extend the lifespan of equipment. AI-driven control systems can enhance the efficiency of

Table 7.1 Applications of AI and ML in wastewater treatment

S. No.	Treatment process	AI and ML model	Objective of prediction	Model performance (R^2)	References
1	Anaerobic toxic biological treatment	SDAE	TN	0.955	Zhao et al. (2020)
2	Activated sludge process	MLP-BP-ANN	COD, SS, and MLSS	0.85, 0.68, and 0.88	Baskar et al. (2024b)
3	Sequence batch reactor	ANN	NH_4 COD	0.958	
4	Anaerobic and aerobic-activated sludge system	MLP-BP-ANN	BOD	> 0.998	Nagpal et al. (2024)
5	Adsorption	ANN	Zn^{2+}, Cu^{2+}, Pb^{2+}, and Cd^{2+}	0.991	Baskar et al. (2024b)
6	Cadmium, lead, malachite green	Machine learning process, ANN	Adsorption	1.000, 0.964	El Alaoui El Fels et al. (2023)
7	Sequential batch reactor	MLP-BP-ANN	COD, BOD, and TSS	0.90317	
8	Aeration, nitrification and denitrification	ANN	COD		Zhu et al. (2022)
9	Sulfidogenic CSTR	MLP-BP-ANN	Effluent sulfate, COD, acetate, and Zn concentrations	0.998,	Baskar et al. (2024b)
10	Adsorption	BTR, ANN	Methylene blue	0.993, 0.948	Malviya and Jaopal (2021)
11	Upflow anaerobic filter reactor	MLP-ANN, RBF-ANN, GRANN	Effluent COD	0.983, 0.995, and 0.951	Zhao et al. (2020)
12	Aeration diffusion	ANN	BOD	0.976	Altowayti et al. (2022)
13	Biological	NNE	BOD	0.999	Viet et al. (2022)
14	Membrane	RBF, ANN	Cu^{2+}	0.990, 0.966	Baskar et al. (2024b)
15	Photodegradation	ANN	Naphthalene	0.990	Ibrahim et al. (2024)
16	Phytoremediation	ANN	As^{5+}	1.278	Zhao et al. (2020)
17	Moving-bed sequential batch biofilm reactor	MLP-BP-ANN	Oxidized nitrogen concentration	0.992, 0.596	Zhang et al. (2023)

(continued)

Table 7.1 (continued)

S. No.	Treatment process	AI and ML model	Objective of prediction	Model performance (R^2)	References
18	Extraction	ANN, BPPSO	Mn^{2+}	0.996, 0.998	Ismail et al. (2021)
19	Airlift multilevel circulation MBR	MLP-BP-ANN	pH, MLSS, and COD	1.0	Zhao et al. (2020)
20	Adsorption	ANN	Malachite green, bisphenol A, methylene blue, Zn^{2+}, Cu^{2+}, Pb^{2+}, and Cd^{2+}	0.992	Nagpal et al. (2024)

SDAE stack denoising autoencoder,*MLP multilayer perception*, *BP* backpropagation, *ANN* artificial neural work, *BTR* boost regression tree, *RBF* radial basis function, *BPPSO* backpropagation particle swarm optimization, *CSTR* continuous stirrer tank reactor, *MLSS* mixed liquor suspended solids

treatment processes such as aeration, filtration, chemical dosing, and sludge management (Viet and Jang 2023). For instance, machine learning models can adjust aerator speeds or chemical dosages in real time based on sensor data, ensuring optimal treatment while minimizing energy and chemical consumption (Ibnouf et al. 2025). Since wastewater treatment plants are highly energy-intensive, AI and ML can help reduce energy usage by optimizing processes like aeration, which consumes a large portion of energy, pump operations, and flow management (Ejaz et al. 2024). Figure 7.3 shows the classification system and prediction flowchart of ML.

7.4 Automation in Decision-Making

7.4.1 AI-Based Decision Support Systems to Automate Control and Management Decisions

Data integration: Sensors embedded in the wastewater treatment plant can generate massive volumes of data, including parameters like pH levels, chemical oxygen demand (COD), biological oxygen demand (BOD), turbidity, temperature, and more. AI and ML algorithms can process this data in real time to monitor the system's performance and detect anomalies.

Anomaly detection: Machine learning models, particularly unsupervised learning approaches like clustering and outlier detection, can automatically identify deviations in system behavior, such as equipment failure or unusual pollution levels, before they become critical (Wang et al. 2023).

Table 7.2 Applications of different AI algorithms

AI algorithms	Applications and categories	Pros	Cons	Reference
Artificial neural network (ANN)	Classification and regression	Demonstrates complicated relationships, deals with chaotic data, and adapts well to fresh information	Training is computationally costly and potentially prone to overfitting	Maurya et al. (2024a)
Fuzzy neural network (FNN)	Regression, categorization, and prediction	Easy to understand and apply; ideal for challenging nonlinear issues	High computational cost and complex model architecture	El Alaoui El Fels et al. (2023)
Radial basis function neural network (RBFNN)	Categorization and regression; both wavelet and Gaussian basis functions are suitable; worldwide approximation	High ability to tolerate loudness; high-quality rapid training capability in generalization	Needs a substantial quantity of data to train; requires numerous hidden neurons	Girones et al. (2010)
Particle swarm optimization (PSO)	Clustering, regression, and classification	Reliable universality, excellent computational effectiveness, and usability and simplicity	Defective for certain problems and sensitive to initial conditions	Ismail et al. (2021)
Multilayer perceptron neural network (MLPNN)	Classification, regression, and instruction under supervision; backpropagation algorithm, frequently used to train models	Incredibly accurate and dependable predictions despite fluctuations	Trial and error are a reliable method of determining the number of hidden neurons	Nguyen et al. (2022)
Decision tree (DT)	Regression, classification, and optimization	Easy to understand, interpret, and classify; no need for prior processing	Not appropriate for training with uneven datasets	Goap et al. (2018)
Generalized regression neural network (GRNN)	Classification, regression, and neural nonparametric supervised learning; iterative optimization, not required with this single-pass learning approach	Quick training because all that needs to be remembered is the training set; ease of implementation, computational efficiency during prediction, and straightforward architecture	Inefficient for big datasets because it needs more memory to store all the data; memory-intensive because it stores all the training data for prediction	Nguyen et al. (2022)

(continued)

Table 7.2 (continued)

AI algorithms	Applications and categories	Pros	Cons	Reference
Principal component analysis (PCA)	Clustering	Cuts down on complexity and is easy to apply	Loss of some important data; sensitive to data noise	Ibrahim et al. (2024)
Genetic algorithm (GA)	Clustering, sorting, and prediction; mimics the mechanisms of natural selection and genetics involved in biological evolution; estimation at worldwide level, a heuristic-based algorithm	Greater ability to optimize globally; provides practical solutions and gives researchers several ways to enhance their findings; compatible with multiple models and techniques	Took several attempts and errors to reach the best new generation; computationally heavy	Adeleke et al. (2023)
Convolution neural network (CNN)	Segmentation, regression, and classification	Ideal for extracting important features from pictures and modeling snapshots	Analytically expensive and challenging to understand	Maurya et al. (2024a)
Support vector machine (SVM)	Pattern recognition, sorting, and regression organized risk reduction; nonlinear computing challenges with support vectors	Bare minimum of information required the capacity to search internationally; greater resilience to noise	High memory and computation requirements during batch training	Viet et al. (2022)
Random forest (RF)	Classification, regression, and prediction	It is straightforward to use, appropriate for high-dimensional datasets, and easy to comprehend	Expensive to calculate; extensive decision trees are needed to guarantee accuracy and resilience	Maurya et al. (2024a)
Artificial neural network coupled with genetic algorithm (GA-ANN)	Classification and regression. Neural network sensitivities and weights are optimized by GA. The genetic algorithms can utilize the projected output value of the neural network as an indicator of fitness function	Exceeds the local minimum; quick convergence; outstanding precision	Heaviness related to calculation; not able to count the total number of neurons hidden	Girones et al. (2010)

(continued)

Table 7.2 (continued)

AI algorithms	Applications and categories	Pros	Cons	Reference
Deep neutral network (DNN)	Regression, prediction, and classification	Prompt and precise prediction suitable for challenging nonlinear issues	There is no need to process it beforehand, and understanding, interpreting, and categorizing it are easy	El Alaoui El Fels et al. (2023)
Adaptive network-based fuzzy inference system (ANFIS)	Classification and regression; has an outcome and a prediction; combining the gradient descent and least square approaches for parameter training	Reliability of nonlinear function approximations picks up the ropes quickly, attaining the required result quickly	The most effective structures are found via trial and error	Zhu et al. (2022)

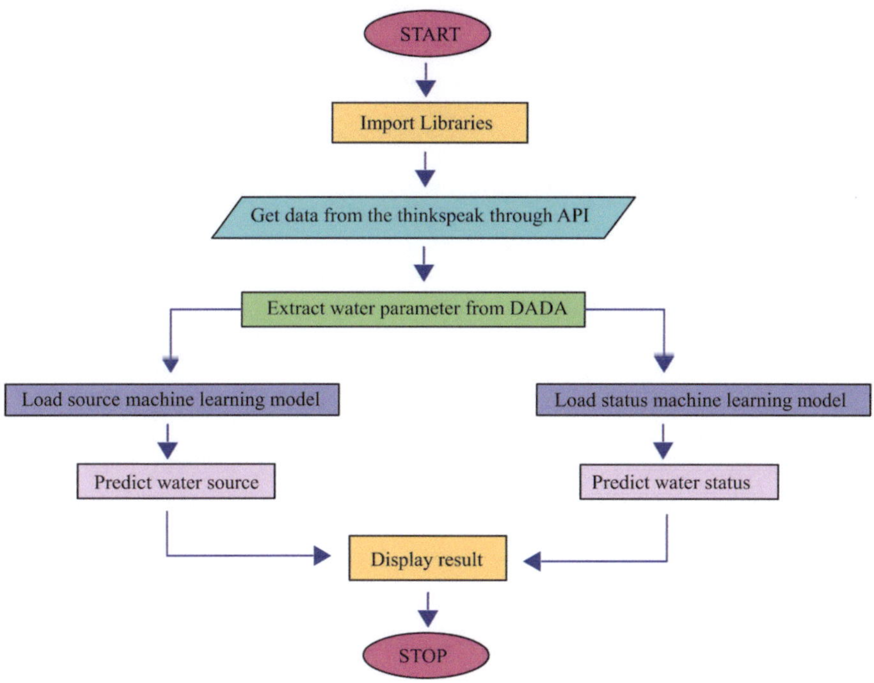

Fig. 7.3 Flowchart for machine learning prediction and classification system

7.5 Data Analytics for Forecasting

7.5.1 Using Historical Data to Predict Future Wastewater Flow and Pollutant Load

AI helps manage the effects of industrial growth and changing economic activities. As industries evolve, the nature and concentration of pollutants can vary. Machine learning algorithms can track and analyze real-time data from various industrial sectors, identifying shifts in pollution levels and adjusting treatment processes as needed. For instance, if an industrial sector expands or changes its output, AI systems can predict how these changes will affect effluent characteristics and modify treatment parameters to efficiently handle the new pollutant load. By continually learning from historical data and adapting to emerging trends, AI enables wastewater treatment plants to effectively manage these complex and ever-changing variables. Incorporating factors like seasonal fluctuations, population growth, and industrial changes into predictive models enhances decision-making and operational efficiency (Bagheri et al. 2023). This not only helps optimize treatment processes but also supports long-term planning, ensuring that wastewater management systems are resilient to future challenges and continue to meet both regulatory and sustainability objectives.

7.5.2 AI for Handling Complex Variables

AI plays a vital role in managing the complex factors that influence wastewater treatment, such as seasonal changes, population growth, and other dynamic elements. These variables can significantly impact wastewater flow, pollutant concentrations, and treatment requirements, making it difficult to maintain efficient operations without real-time predictions and adjustments. Machine learning algorithms, which are capable of processing large volumes of data and recognizing patterns, offer an effective way to handle these challenges, ensuring more accurate forecasts and improved plant performance (Singh et al. 2023). For example, seasonal variations, such as changes in rainfall or temperature, can cause fluctuations in wastewater flow and quality. These changes can result in higher wastewater volumes or shifts in pollutant concentrations, influenced by factors like increased runoff or industrial activities. AI models can be trained to detect these seasonal trends and adjust treatment processes accordingly. By analyzing historical data, such as weather patterns, water quality metrics, and past seasonal trends, AI can predict how these factors will influence future wastewater conditions. This enables treatment plants to prepare for these changes in advance, adjusting chemical dosing, aeration, and other treatment parameters to maintain optimal efficiency.

7.6 Anomaly Detection

AI-driven anomaly detection systems have become essential in enhancing safety, efficiency, and sustainability in waste management, especially when it comes to detecting leaks, identifying contaminants, and automating corrective measures. In waste management systems, leaks in landfills, sewage networks, or industrial waste storage can lead to significant environmental and health risks. AI aids in detecting these leaks by analyzing data collected from various sensors and monitoring systems. AI can work with a network of Internet of Things (IoT) sensors that monitor parameters such as pressure, temperature, and humidity. Table 7.3 shows the different types of IoT-based studies in monitoring the quality of water. By quickly identifying contaminants and leaks, AI helps prevent harmful substances from entering the environment, improving sustainability in waste management (Yogi and Chakravarthy 2024). AI can assist in adhering to environmental regulations by continuously monitoring compliance, automatically reporting incidents, and ensuring that operations remain within legal limits.

7.7 Benefits of AI/ML in Wastewater Management

The use of predictive models, real-time data feedback, and advanced control systems enables significant improvements in wastewater treatment and manufacturing. The integration of these technologies represents an exciting avenue for innovation, with the potential to deliver lasting environmental and economic benefits.

Reinforcement learning (RL): **RL** can be used for real-time adaptive chemical dosing by treating the problem as a control system where an agent continuously adjusts dosing rates based on system feedback (e.g., water quality metrics such as turbidity, pH, or contaminants). Over time, the RL agent learns the most efficient dosing strategy to minimize cost or achieve desired treatment levels while avoiding overdosing.

Optimization algorithms: Genetic algorithms or particle swarm optimization can be employed to solve the chemical dosing problem, ensuring the right balance between cost, efficiency, and environmental compliance (Zhao et al. 2020).

7.8 Challenges and Limitations

Data availability and quality pose substantial challenges when integrating AI and ML into wastewater management, directly impacting the accuracy, reliability, and overall success of AI- based solutions. Real-world processes have a wide range of operating conditions, mostly due to inherent constraints and distinct process dynamics. Due to

Table 7.3 Different IoT-based studies for water quality monitoring

Author	Studies	Parameter	Findings	Limitations
Prasad et al. (2015)	A sophisticated system that tracks water quality using remote sensing and Internet of Things technologies	Temperature, ORP, electrical conductivity, pH	Tested in four water sources every 12 h; uses GSM technology to notify consumers of any anomalies	An effective system requires more water parameters, and no water treatment mechanism was included. There was no use of prediction models
Adeleke et al. (2023)	Internet of Things infrastructure for river water quality monitoring in Prayagraj, India. Data collection is done using a Libelium smart water kit	CO_2, pH, turbidity	Waterways are ideal for fishing and agricultural irrigation	Restricted to particular geographical areas and sensor kinds
Kumar et al. (2023)	Potential sensitivity to parameter selection not explored	Temperature, turbidity, pH, TDS	The combination of polynomial regression and gradient boosting both worked well for WQI. MLP attained a high classification accuracy in WQC	The potential impact of parameter selection sensitivity was not examined
Lalithadevi et al. (2019)	A system that uses the machine learning algorithm, known as the linear regression model to forecast water quality and offer monitoring of the clouds	Salinity, pH, TDS	Analyzes water variables, transmits data to the cloud, and employs a wireless sensor network with both on-site and distant modules. The linear regression model technique is then used to forecast future water levels based on utilization rate	An effective system requires more water parameters, and no water treatment mechanism was included

(continued)

Table 7.3 (continued)

Author	Studies	Parameter	Findings	Limitations
Rahu et al. (2024)	A model for predicting agricultural water quality by optimizing logistic regression using the momentum algorithm; increased efficacy and accuracy of predictions	CO_2, pH, turbidity, temperature	Prediction accuracy has improved by an average of 1.11 percentage points when compared to current algorithms	The potential sensitivity to the parameters of the Momentum algorithm was not addressed
Kumar et al. (2023)	An Internet of Things system for well water quality monitoring	TDS, temperature	Measures the properties of water using two sensors that are interfaced with a microcontroller; in the event of an abnormal condition, consumers receive an SMS	An effective system requires more water parameters, and no water treatment mechanism was included. There was no use of models of prediction
Ramadhan et al. (2020)	Infrastructure for analyzing water quality that is supported by IoT and ML. An integrated Internet of Things (IoT) system using several sensors is used at Pakistan's Rohri Canal to analyze the water quality	CO_2, pH, turbidity, temperature	In classification, random forest performed better than MLP in regression; superior outcomes over earlier studies	Beyond the Rohri Canal, applicability is not addressed

(continued)

Table 7.3 (continued)

Author	Studies	Parameter	Findings	Limitations
Alobaidy et al. (2022)	To link water parameters obtained from sensors with the ThingSpeak program for real-time water quality monitoring, the study used a microcontroller	pH, turbidity, water level detection, temperature, humidity	The study used Wi-Fi to update the ThingSpeak server every 20 s while measuring water parameters in a storage tank at 10-s intervals	To ensure that water quality is monitored effectively and efficiently, prediction techniques like machine learning algorithms should be used in conjunction with other water metrics including electrical conductivity, nitrates, and dissolved oxygen
Duangwongsa et al. (2021)	A combined IoT and ML-based platform for water quality monitoring and evaluation	pH, turbidity	While Naive Bayes performed exceptionally well in tasks involving classification, polynomial regression showed remarkable success with regression challenges	Only a few water quality criteria were used

WQI water quality index, *WQC* water quality class, *IoT* internet of things, *ML* machine learning, *MLP* multilayer perceptron, *TDS* total dissolved solids

variations in the information utilized for training, the model trained using historical data produces erroneous results (Baskar et al. 2024a). A key issue is data scarcity. AI and ML models require large volumes of high-quality data to function effectively, but many wastewater treatment plants lack comprehensive, consistent data collection systems (Oruganti et al. 2023). The data is spilled across different departments or stored in various systems, making it difficult to integrate for analysis. As a result, AI models may not have enough data or may lack the necessary variety to make accurate predictions or optimize treatment processes.

Along with scarcity, data quality is another major concern. The data used to train AI models must be precise, complete, and representative of real-world conditions in wastewater treatment. However, in practice, sensor data can be flawed due to issues like calibration errors, sensor malfunctions, or environmental factors like temperature changes, humidity, or chemical interference. Such factors can lead to inaccurate readings, resulting in poor-quality data that can misguide AI systems (Nasir and Li 2024).

Additionally, gaps in data or incomplete records can hinder AI's ability to generate reliable predictions or make informed decisions. Another challenge is data consistency. Wastewater treatment processes are highly dynamic and vary due to factors such as weather, industrial discharge, seasonal changes, or sudden shifts in wastewater inflow. This variability can make the data inconsistent or difficult to interpret. Since AI and ML models need consistent data to detect patterns and make accurate predictions, this unpredictability in wastewater characteristics can complicate the process of model development and deployment.

Data privacy and security are also critical concerns, particularly in facilities where data may be sensitive or proprietary. Many wastewater treatment plants are part of larger municipal or industrial operations, meaning that the data collected could contain confidential information related to operational procedures, financial details, or compliance matters. Ensuring data protection and adhering to privacy regulations is vital for the successful application of AI and ML technologies.

7.9 Conclusion and Future Prospects

Wastewater treatment plants are typically energy-intensive, especially due to aeration and pumping operations. AI can optimize these processes, leading to reduced electricity consumption and lower operational costs. By minimizing chemical use, energy consumption, and environmental impact, AI can make wastewater treatment more sustainable. Additionally, AI can enhance the utilization of by-products like biogas from anaerobic digestion, turning them into valuable resources for the plant. Finally, socioeconomic implications of AI adoption in wastewater management should be carefully considered. The introduction of AI could lead to automation of certain manual tasks, potentially disrupting the workforce. This raises concerns about job displacement and the need to reskill workers affected by these changes. Ethical considerations regarding job creation, worker development, and equitable distribution of AI's benefits must be factored into decision-making, especially for communities that may be more vulnerable to these shifts. While AI has the potential to transform wastewater management, its application must be carefully managed to address ethical and regulatory challenges. By prioritizing data privacy, transparency, fairness, regulatory adherence, and socioeconomic equity, AI can be harnessed responsibly to maximize benefits while minimizing risks and ensuring the sustainable management of public resources.

The potential advancements in AI and machine learning technologies for wastewater management are vast and could bring transformative improvements to both operational efficiency and environmental sustainability of treatment plants. As AI and ML continue to evolve, they will likely play a larger role in optimizing various aspects of wastewater treatment, from real-time monitoring to predictive maintenance. Future developments in AI could lead to even more sophisticated predictive models, allowing plants to forecast treatment needs with even greater accuracy based on a wider range of variables, such as weather patterns, seasonal fluctuations, and emerging contaminants. The combination of AI and ML with sensors and Internet of Things (IoT) devices is reshaping the way we monitor water quality in real time, making it more efficient, accurate, and proactive in managing water resources. By utilizing real-time data from IoT sensors, AI-driven systems can identify anomalies, forecast potential problems, automate corrective actions, and enhance the overall management of water resources. As technology evolves and data collection grows more widespread, the role of AI and ML in water quality monitoring will expand, leading to improved environmental protection, more effective resource management, and better public health outcomes (Baskar et al. 2024a). In the future, wastewater will arise from diverse sources, including domestic, industrial, and other sectors. As simultaneous treatment of all wastewater may not be feasible and certain types may be difficult to manage manually,this could pose considerable challenges in wastewater management. Therefore, AI and ML should handle separating the wastes according to their harmful index with the assistance and provision of necessary data with minimal expense and time, endeavor to steer clear of harmful and cancer-causing substances, and primarily refrain from disposing of any plant waste—though if it is released, it should be harmless.

References

Adeleke IA, Nwulu NI, Ogbolumani OA (2023) A hybrid machine learning and embedded IoT-based water quality monitoring system. Int Things 22:100774

Alam G, Ihsanullah I, Naushad M, Sillanpää M (2022) Applications of artificial intelligence in water treatment for optimization and automation of adsorption processes: recent advances and prospects. Chem Eng J 427:130011

Alaswad SO, Almatrafi E (2024) Artificial intelligence applications in forward osmosis for water treatment: recent developments and research directions. Desalination Water Treat 319:100551. https://doi.org/10.1016/j.dwt.2024.100551

Alobaidy HAH, Nordin R, Singh MJ et al (2022) Low-altitude-platform-based airborne IoT network (LAP-AIN) for water quality monitoring in harsh tropical environment. IEEE Int Things J 9:20034–20054

Altowayti WAH, Shahir S, Othman N et al (2022) The role of conventional methods and artificial intelligence in the wastewater treatment: a comprehensive review. Processes 10:1832

Bagheri M, Farshforoush N, Bagheri K, Shemirani AI (2023) Applications of artificial intelligence technologies in water environments: from basic techniques to novel tiny machine learning systems. Process Saf Environ Protect 180:10–22. https://doi.org/10.1016/j.psep.2023.09.072

Baskar G, Nashath Omer S, Saravanan P et al (2024a) Status and future trends in wastewater management strategies using artificial intelligence and machine learning techniques. Chemosphere 362. https://doi.org/10.1016/j.chemosphere.2024.142477

Baskar G, Omer SN, Saravanan P et al (2024b) Status and future trends in wastewater management strategies using artificial intelligence and machine learning techniques. Chemosphere 142477

Behera SK, Karthika S, Mahanty B et al (2024) Application of artificial intelligence tools in wastewater and waste gas treatment systems: recent advances and prospects. J Environ Manage 370:122386. https://doi.org/10.1016/j.jenvman.2024.122386

Chowdhury S, Karanfil T (2024) Applications of artificial intelligence (AI) in drinking water treatment processes: possibilities. Chemosphere 356:141958. https://doi.org/10.1016/j.chemosphere.2024.141958

Ding Y, Sun Q, Lin Y et al (2024) Application of artificial intelligence in (waste)water disinfection: Emphasizing the regulation of disinfection by-products formation and residues prediction. Water Res 253:121267. https://doi.org/10.1016/j.watres.2024.121267

Duangwongsa J, Ungsethaphand T, Akaboot P et al (2021) Real-time water quality monitoring and notification system for aquaculture. In: 2021 joint international conference on digital arts, media and technology with ECTI Northern section conference on electrical, electronics, computer and telecommunication engineering. IEEE, pp 9–13

Durgun Y (2024) Real-time water quality monitoring using AI-enabled sensors: detection of contaminants and UV disinfection analysis in smart urban water systems. J King Saud Univ Sci 36:103409. https://doi.org/10.1016/j.jksus.2024.103409

Ejaz U, Khan SM, Jehangir S et al (2024) Monitoring the Industrial waste polluted stream—integrated analytics and machine learning for water quality index assessment. J Clean Prod 450:141877. https://doi.org/10.1016/j.jclepro.2024.141877

El Alaoui El Fels A, Mandi L, Kammoun A, et al (2023) Artificial intelligence and wastewater treatment: a global scientific perspective through text mining. Water (Basel) 15:3487

Girones R, Ferrús MA, Alonso JL et al (2010) Molecular detection of pathogens in water–the pros and cons of molecular techniques. Water Res 44:4325–4339

Goap A, Sharma D, Shukla AK, Krishna CR (2018) An IoT based smart irrigation management system using Machine learning and open source technologies. Comput Electron Agric 155:41–49

Ibnouf M, Jaber H, Abukhalifeh H et al (2025) A comprehensive review of AI algorithms for performance prediction, optimization, and process control in desalination systems. Desalination Water Treat 321:100892. https://doi.org/10.1016/j.dwt.2024.100892

Ibrahim M, Haider A, Lim JW et al (2024) Artificial neural network modeling for the prediction, estimation, and treatment of diverse wastewaters: a comprehensive review and future perspective. Chemosphere 142860

Ismail W, Niknejad N, Bahari M et al (2021) Water treatment and artificial intelligence techniques: a systematic literature review research. Environ Sci Pollut Res 1–19

Kavianpour B, Piadeh F, Gheibi M et al (2024) Applications of artificial intelligence for chemical analysis and monitoring of pharmaceutical and personal care products in water and wastewater: a review. Chemosphere 368:143692. https://doi.org/10.1016/j.chemosphere.2024.143692

Kumar M, Singh T, Maurya MK et al (2023) Quality assessment and monitoring of river water using IoT infrastructure. IEEE Int Things J 10:10280–10290

Kumari S, Chowdhry J, Kumar M, Garg MC (2024) Machine learning (ML): An emerging tool to access the production and application of biochar in the treatment of contaminated water and wastewater. Groundw Sustain Dev 26:101243. https://doi.org/10.1016/j.gsd.2024.101243

Lalithadevi B, Yadav A, Pandey A, Adhikari M (2019) IoT based wsn ground water monitoring system with cloud-based monitoring as a service (maas) and prediction using machine learning. Int J Innov Tech Explor Eng 9:816–821

Li J, See KF, Chi J (2019) Water resources and water pollution emissions in China's industrial sector: a green-biased technological progress analysis. J Clean Prod 229:1412–1426. https://doi.org/10.1016/j.jclepro.2019.03.216

Li L, Rong S, Wang R, Yu S (2021) Recent advances in artificial intelligence and machine learning for nonlinear relationship analysis and process control in drinking water treatment: a review. Chem Eng J 405:126673

Lowe M, Qin R, Mao X (2022) A review on machine learning, artificial intelligence, and smart technology in water treatment and monitoring. Water (Basel) 14:1384

Mahardhika SP, Putriani O (2023) Deployment and use of Artificial Intelligence (AI) in water resources and water management. In: IOP conference series: earth and environmental science. IOP Publishing, p 012056

Malviya A, Jaspal D (2021) Artificial intelligence as an upcoming technology in wastewater treatment: a comprehensive review. Environ Technol Rev 10:177–187

Maurya BM, Yadav N, Amudha T et al (2024a) Artificial intelligence and machine learning algorithms in the detection of heavy metals in water and wastewater: methodological and ethical challenges. Chemosphere 353:141474

Maurya BM, Yadav N, T A et al (2024b) Artificial intelligence and machine learning algorithms in the detection of heavy metals in water and wastewater: methodological and ethical challenges. Chemosphere 353:141474. https://doi.org/10.1016/j.chemosphere.2024.141474

Nagpal M, Siddique MA, Sharma K et al (2024) Optimizing wastewater treatment through artificial intelligence: recent advances and future prospects. Water Sci Technol 90:731–757

Nasir FB, Li J (2024) Understanding machine learning predictions of wastewater treatment plant sludge with explainable artificial intelligence. Water Environment Research 96:e11136

Nguyen XC, Nguyen TTH, Tran QB et al (2022) Artificial intelligence for wastewater treatment. In: Current developments in biotechnology and bioengineering. Elsevier, pp 587–608

Oruganti RK, Biji AP, Lanuyanger T et al (2023) Artificial intelligence and machine learning tools for high-performance microalgal wastewater treatment and algal biorefinery: a critical review. Sci Total Environ 876:162797

Pérez-Beltrán CH, Robles AD, Rodriguez NA et al (2024) Artificial intelligence and water quality: from drinking water to wastewater. TrAC Trends Anal Chem 172:117597. https://doi.org/10.1016/j.trac.2024.117597

Prasad AN, Mamun K Al, Islam FR, Haqva H (2015) Smart water quality monitoring system. In: 2015 2nd Asia-pacific world congress on computer science and engineering (APWC on CSE). IEEE, pp 1–6

Rahu MA, Shaikh MM, Karim S et al (2024) Water quality monitoring and assessment for efficient water resource management through internet of things and machine learning approaches for agricultural irrigation. Water Resour Manag 1–42

Ramadhan AJ, Ali AM, Kareem HK (2020) Smart water-quality monitoring system based on enabled real-time internet of things. J Eng Sci Technol 15:3514–3527

Sharafi M, Rezaverdinejad V, Behmanesh J, Samadianfard S (2024) Development of long short-term memory along with differential optimization and neural networks for coagulant dosage prediction in water treatment plant. J Water Process Eng 65:105784

Singh S, Mishra S, Singh T, Thakur S (2023) A study on machine learning-based water quality assessment and wastewater treatment. In: Artificial intelligence applications in water treatment and water resource management. IGI Global, pp 146–162

Viet ND, Jang A (2023) Machine learning-based real-time prediction of micropollutant behaviour in forward osmosis membrane (waste)water treatment. J Clean Prod 389:136023. https://doi.org/10.1016/j.jclepro.2023.136023

Viet ND, Jang D, Yoon Y, Jang A (2022) Enhancement of membrane system performance using artificial intelligence technologies for sustainable water and wastewater treatment: a critical review. Crit Rev Environ Sci Technol 52:3689–3719

Wang Y, Cheng Y, Liu H et al (2023) A review on applications of artificial intelligence in wastewater treatment. Sustainability 15:13557

Yogi MK, Chakravarthy ASN (2024) Novel security mechanism for ai enabled wastewater treatment systems. In: The AI cleanse: transforming wastewater treatment through artificial intelligence: harnessing data-driven solutions. Springer, pp 283–312

Yuvaraju M, Deena D (2024) Artificial intelligence-based control for membrane bioreactor in sewage treatment. Appl Nanosci 14:943–953

Zhang S, Jin Y, Chen W et al (2023) Artificial intelligence in wastewater treatment: a data-driven analysis of status and trends. Chemosphere 336:139163

Zhao L, Dai T, Qiao Z et al (2020) Application of artificial intelligence to wastewater treatment: a bibliometric analysis and systematic review of technology, economy, management, and wastewater reuse. Process Saf Environ Prot 133:169–182

Zhu M, Wang J, Yang X et al (2022) A review of the application of machine learning in water quality evaluation. Eco-Environ Health 1:107–116

Chapter 8
Nanotechnology Applications in Wastewater Treatment: A Paradigm Shift

Sippy Singh, Shivangi Pathak, Pooja Yadav, Atifa Haseeb Ansari, and Durgesh Singh

Abstract Pure water is the major leading issue nowadays and is the concerned status for wastewater treatment via a variety of techniques. Various toxic substances including heavy metals, hydrocarbons, pesticides and chemicals have a negative impact either directly or indirectly on both biotic and abiotic components. Sometimes it may also show carcinogenic effects on human beings; thus, their removal must be the interest of researchers. To minimize the risk of harmful effects of such hazardous substances, nanomaterials could be one of the options for wastewater management. Nanoparticles can be categorized into different groups: metal oxide, silver-, iron-, cellulose- and carbon-based nanoparticles. Silver-based nanoparticles have antimicrobial activity and are widely used to overcome problems related to the organic material. Metallic oxide has promising potential to remove heavy metals including copper, nickel, chromium and zinc easily. Nanoadsorbents such as hematite, magnetite and maghemite have antimicrobial activity; hence, they are widely used for the decontamination of water. Nanotechnology has been emerging as a powerful weapon in the advancement of industries and agricultural fields. Carbon nanotubes (CNTs) are more effective than activated carbon as they interact with pollutants. Membranes are typically utilized to reduce impurities from water such as bacteria, viruses and other microbes. This chapter highlights the potential of nanomaterials for wastewater treatment. Nanotechnology could be considered an eco-friendly approach for wastewater management. Thus, nanomaterial-based research is ongoing to minimize the risk of pollutants in an eco-friendly manner.

Keywords Nanomaterials · Carcinogenic · Nanotubes · Wastewater · Metallic · Contaminants

---

S. Singh · S. Pathak · P. Yadav · A. H. Ansari · D. Singh (✉)
Department of Zoology, S.S. Khanna Girls' Degree College, (A Constituent College of University of Allahabad, Prayagraj), Prayagraj, U.P, India
e-mail: durgeshsingh.ssk@gmail.com

A. M. Yatoo et al. (eds.), *Wastewater Monitoring and Management by Advanced Approaches*, Advances in Wastewater Research,
https://doi.org/10.1007/978-981-95-2601-7_8

8.1 Introduction

Recently, preserving freshwater is one of the big tasks for scientific technology (Moustafa et al. 2023). The ongoing deterioration of high-quality water resources is a serious worldwide issue brought on by urbanization, fast industrial development, rapid population increase and agricultural activities (Rathi and Kumar 2021). The most urgent environmental problem of the modern era is the lack of drinking water (Singh et al. 2022). A degree of water stress affects about 25% of big cities, and about 2.2 billion people globally lack access to the essential liquid (Salehi 2022). Treatments should be conducted in accordance with the kind of wastewater and the level of contamination (Baskar et al. 2022). Numerous contaminants including heavy metals, pesticides, antibiotics and hydrocarbons are found in wastewater, and these contaminants have shown negative impacts on both human health and ecosystems. Many technologies, including reverse osmosis, adsorption and photodegradation, have been developed to overcome this situation. However, they have a few disadvantages such as high installation and running costs in the effective removal of hazardous residues (Sikiru et al. 2022). Nanotechnology is one of the most effective methods for wastewater treatment (Marimuthu et al. 2023). It has better quality and salient features such as high surface-to-volume ratios, superior adsorption capabilities and increased reactivity (Nasrollahzadeh et al. 2021); high mobility and low consumption of energy (Sheoran et al. 2022) make it a more powerful catalyst for wastewater treatment (Zahid and Abd-Elsalam 2021). Nanoadsorbents, nanocatalysts and nanomembranes could be better options for minimal contamination (Boulkhessaim et al. 2022). There are variety of nanomaterials including metal oxides, carbon- and cellulose-based nanomaterials. It has been observed that wastewater treatment can be done through three successive stages: primary (physicochemical), secondary (biological) and tertiary (advanced physicochemical treatment) (Urgina and Milojkovic 2024). In this chapter, more focus has been given to the potential of nanotechnology in treating wastewater.

8.2 Applications of Nanomaterials

Composition of wastewater varies greatly on the basis of origination, and it has been observed that the main wastewater contaminants might exist in the source of wastewater and even in radically different amounts: non-biodegradable substances (such as insecticides and detergents), metallic ions and microbes are found in industrial as well as domestic sources (Bartolomeu et al. 2018). The toxicity level of wastewater generally depends upon the chemical composition, although it is known that even a trace amount of chemicals can show dangerous effect on the ecosystem and human health. Conventional water decontamination techniques mainly involve oxidation or reduction (chemical methods) and physicochemical processes using various separation techniques such as adsorption and ion exchange. Although these technologies are good, there are still several serious issues that need to be resolved. There is a requirement for more advanced technology for energy consumption because traditional methods of water decontamination are not selective enough against certain contaminants (Lu et al. 2017). It indicates inefficiency and vulnerability to interference during operation, as they are unable to eliminate one molecule over another (Malato et al. 2009). To sum up, some of the shortcomings of decontamination technologies demonstrate the necessity of developing new materials and procedures for treatment approaches. The advancement in the nanotechnology applications may lead to the improvement or replacement of conventional methods in the field of water treatment (Bishoge et al. 2018). It has been reported that nanomaterials have shown significant role in water purification (Table 8.1) (Khan and Malik 2019). Materials less than 100 nm, at least having one dimension, are often referred to as nanomaterials because of their reduced size and their volume-to-specific surface area (VSSA) ratio; these materials have special utility. Materials with a VSSA of at least 60 m^2/cm^3 have been suggested to be considered nanoparticles (NPs) (Kreyling et al. 2010). With a large specific surface area and footprint, nanomaterials provide a great potential for liquid–solid interactions and possess the required qualities and properties for effective water treatment (Fig. 8.1). Various nanoparticles with different physical, chemical and biological properties can be found in organic, inorganic, carbon and composite materials. The most studied nanomaterials are metallic oxides, carbon nanotubes (CNTs), magnetic nanoparticles, metal–organic frameworks (MOFs), layered double hydroxides (LDHs), graphene and polymer-based materials. These exhibit the potential for nanotechnology to produce more feasible and innovative methods in the treatment of wastewater (Gudainiyan et al. 2024).

Table 8.1 Nanoparticles, their properties and uses

S. No.	Nanoparticles	Properties	Uses
1	Silver nanoparticles (AgNPs)	Photocatalytic activity and antibacterial activity	Organic pollutants and dyes
2	Iron oxides	Adsorption and catalytic efficiency	Toxic metal ions
3	Carbon nanotubes	High porosity, high affinity and active sites	Pollutants and microorganisms
4	Membrane	High energy and selectivity	Impurities and pathogens
5	Cellulose	High energy consumption and adsorption	Heavy metals
6	Graphene	Specific surface area and mobility	Cations and anions

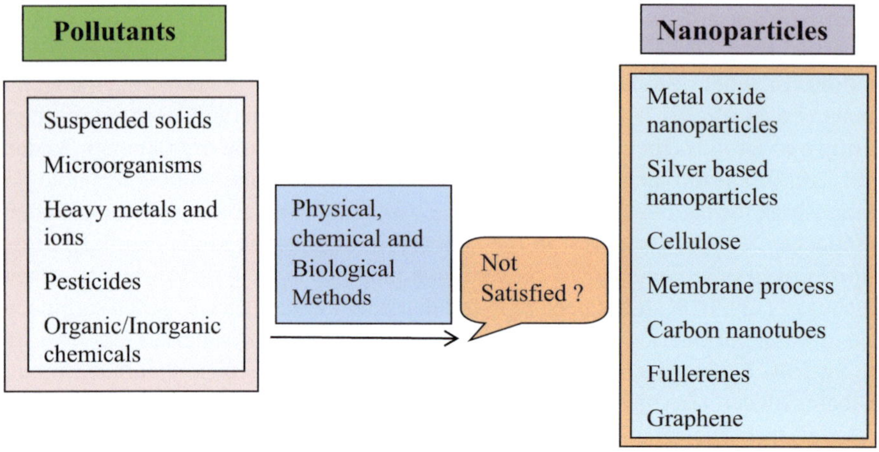

Fig. 8.1 Nanoparticles as detoxifiers for wastewater pollutants

8.3 Metal Oxide Nanoparticles

Metal oxide nanoparticles (MNPs) have shown great prospects as a sustainable, affordable and ecologically acceptable approach for wastewater treatment. According to Corsi et al. (2018), these nanoparticles are compressible without reducing surface area and, due to high adsorption capacity, simple separation and regeneration, are economically and technologically efficient (Abdelbasir and Shalan 2019). Additionally, few of them have shown superparamagnetic properties which lead to better adsorption than activated carbon. Rate-limiting intraparticle diffusion is known to initiate the effective adsorption of heavy metals on the exterior of MNPs along their walls. Because of their multiple surface reaction sites and high specific surface area, these nanoparticles have a high adsorption capacity (Singh et al. 2019). As the size reduces, the rate of adsorption capacity can increase up to 100 times and represent the nanoscale effect (Nassar 2012). MNPs are used to remove heavy metals, including chromium, nickel, cadmium, copper, mercury and arsenic. A significant degree of antibacterial potential has been demonstrated by metallic nanoparticles, including magnesium and other metal oxides, mainly TiO_2 and MgO, due to different physical and chemical properties and small size (Atacan et al. 2022). For antimicrobial action, those materials are used which have the capability to inhibit growth without damaging other tissues. Nanomaterials have antimicrobial properties, mainly used to produce reactive oxygen species (ROS).

8.3.1 Silver Nanoparticles (AgNPs)

The main advantage of AgNPs is the removal of organic pollutants and dyes from wastewater. The formation of AgNPs relies on reactants, repeating frequency and cost-effectiveness. AgNPs have been utilized to degrade and reduce pollutants by the process of catalytic reduction. Additionally, AgNPs are employed in the catalytic degradation as an induced light degradation technique. Various studies suggest AgNPs obtained from biological or natural sources are used to remove dyes from wastewater. AgNPs obtained from *Citrus paradisi* decompose and minimize the hazardous waste from textiles and factories. Consequently, the silver-based nanocomposite can be considered as powerful photocatalyst for organic dyes present in wastewater and industrial effluents. Direct Blue 14 (DB-14) obtained from green-synthesized nanomaterial (*Brassica nigra*) shows efficient photocatalytic and antibacterial activity, and one of the major benefits of these particles is that it can be reused without significantly losing its photocatalytic performance due to their extreme toxicity against bacteria, fungi and viruses. AgNPs have potent antimicrobial properties against variety of microbes and are thus used as disinfectants. AgNPs have been explained as an adhesive present in bacterial cell walls and provide more permeability. They have the ability to destroy cell membranes and are considered the main cause of cellular death (Janovák et al. 2014). AgNPs can interact with DNA and ultimately damage the membrane, as they contain phosphorus and sulphur components. Breakdown of silver nanoparticles produces silver ions which attach to the thiol groups of many essential enzymes in order to deactivate them and interfere with regular cellular processes (Ferreira et al. 2015). Much of the research content is based upon the nanoforms of AgNPs or the released silver ions (Nowack et al. 2011). One such study used silver nanoparticle-impregnated paper which has antibacterial properties against *Escherichia coli* and *Enterococcus faecalis* (Dankovich and Grey 2011). Many effects have been observed such as decrease in intracellular ATP levels, degradation of cell membranes, inhibition of respiration and metabolism and inhibition of cell wall production and cell division (Khan et al. 2020).

8.3.2 Iron-Based Nanoparticles (IONPs)

IONPs have demonstrated an extensive spectrum across multiple disciplines; for instance, they are used as adsorbents and catalysts in the treatment of wastewater. Furthermore, these materials are indispensable for the production of coatings, gas sensors, ion exchangers, magnetic recording devices, magnetic resonance imaging (MRI), separation and pharmaceutical applications. Currently, the sustainable remediation methodology for environmental remediation and waste reclamation is the focus of ongoing research (Guerra et al. 2018). Due to their unique physicochemical characteristics, IONPs are distinct from other nanoparticles. Therefore, research has been done to create superficial production techniques for these particles and to turn them into nanoadsorbents and biocompatible materials. Three groups of contaminants are commonly identified in water and wastewater: inorganic, hazardous organic compounds and pathogenic microorganisms. The wastewater produced from many sources is highly contaminated; for instance, the most common environmental pollutants are non-biodegradable dyes, leather and tile industry effluents that include heavy metals. Both humans and aquatic ecosystems may suffer greatly due to their toxic or carcinogenic effects; thus, effective treatment of these pollutants is required. Currently, industrial water pollution control and wastewater treatment are on a path of profound understanding and considerable scientific vision. Nowadays, nanomaterials are widely used to reduce ecological disturbances. Typically, oxides of iron (magnetite, hematite and maghemite) and hydroxides (goethite [a-FeOOH], akaganeite [b-FeOOH] and lepidocrocite [c-FeOOH]) are evaluated for their use in wastewater treatment to reduce the heavy metal ions. Nanoscale zerovalent iron (nZVI) is useful in the remediation of toxic groundwater pollutants (Shojaei et al. 2017). There are now 16 recognized forms of iron oxides like $Fe(OH)_3$, $Fe(OH)_2$, Fe_3O_4 and FeO (Bahadar et al. 2011). The most well-known oxide forms of iron are magnetite (Fe_3O_4), magnethemite (c-Fe_2O_3) and hematite (aFe_2O_3). Numerous products have been identified currently by studies on ZVI oxidation, siderite ($FeCO_3$), iron sulphide, magnetite (Fe_3O_4), lepidocrocite (c-FeOOH), green rusts (a-FeOOH) and akaganeite (b-FeOOH). Apart from the previously mentioned oxyhydroxide forms, other forms of iron oxides include feroxyhyte (d-FeOOH), ferrihydrite (Fe_5HO_8 ~ $4H_2O$ or approximately $5Fe_2O_3$- ~ $9H_2O$), better represented as FeOOH ~ $0.4H_2O$, and high-pressure FeOOH. Researchers have been using iron oxide nanoadsorbents such as hematite (a-Fe_2O_3), magnetite (Fe_3O_4) and maghemite (c-Fe_2O_3) extensively to remove a variety of pollutants from industrial or environmental effluents including arsenic(V), chromium(VI), copper(II), lead(II) and mercury(II).

8.3.2.1 Zerovalent Iron Nanoparticle (ZVI)

Zerovalent iron (ZVI) nanoparticles are among the most often used metallic reducing agents when it comes to removing harmful contaminants from wastewater. Research is being conducted to determine the effectiveness of nZVI in meeting waste-treatment standards and enriching multiple metal ions from wastewater under laboratory conditions. In a study, the quantitative elimination of Cu(II) and Ni(II) following half an hour of treatment with a set quantity of ZVI in acidic form was examined using batch-based experiment techniques; pH had a crucial role in the adsorption, allowing the nanoparticles to efficiently remove the metals (Sen and Shah 2014). With a contact duration of 30 min, the maximum removal of approximately 90% for copper and chromium was achieved in a pH range of around 2–2.5. The ZVI nanoparticle has efficiency to quickly and simultaneously remove metallic ions. About 99.5% of the hazardous ions including copper and arsenic can be removed through nanoparticles; thus, it can be considered a better treatment than other techniques (Li et al 2017). Nitrate in water has also been eliminated with the use of ZVI. This process involves two treatment units that were set up in succession; each unit contained a clarifier, a recirculation pump and a reactor for nanoparticles. Following the addition and mixing of ZVI with the wastage in the reactor, the clarifier was allowed to settle down and the circulation pump was used to reverse the process.

8.3.2.2 Iron Hydroxide Nanoparticles

$Fe(OH)_3$ has been proven as a detoxifier in wastewater treatment due to its stability, high adsorbent capability and catalytic efficiency. Three-dimensional flower-like $Fe(OH)_3$ nanomaterial was created using an environmentally friendly and inexpensive injection technique to remove organic colours and harmful metal ions (Zhao et al. 2018). As an adsorbent, $Fe(OH)_3$ has substantially better capabilities to selectively remove neutral dyes from wastewater than conventional a-FeOOH and a-Fe_2O_3 nanoparticles, suggesting a tremendous ability for usage in wastewater treatment. When compared to ferrous salts, such particles have insufficient bioavailability which promotes their behaviour in wastewater treatment.

8.3.2.3 Iron Oxyhydroxide

Iron oxyhydroxides are the most often employed nanoparticles for wastewater treatment. It has been suggested that iron oxyhydroxide nanoparticles could act as adsorbents of fluoride (F ~) ions from water and have photoreduction efficiency for Cr(VI) for the transformation into less hazardous Cr(III).

8.3.2.4 Magnetite Nanoparticles

Magnetite (Fe_3O_4) is synthesized in a laboratory by using solutions having ferric and ferrous ions. Such molecules have magnetism and are widely used as a magnetic substance to produce magnetic adsorbents which can be scaled up or down by adjusting the synthesis conditions (Taylor et al. 2007). Fe_3O_4 nanoparticles are highly promising materials utilized for separation of heavy metals and dyes in water treatment. As a result of their composite materials or surface functions, these nanomaterials can be employed as coagulating agents, photocatalysts, and adsorbents for water remediation. It is known that magnetite nanoparticles containing Al-Abbas' (a fruit extract) as a reductant and stabilizer were prepared in an eco-friendly manner. It has been observed that 12 h of exposure to magnetite can reduce bacterial load, while 24 h of treatment can kill bacteria in the water.

8.4 Carbon-Based Nanomaterials

After oxygen, carbon is the second most prevalent element in the human body and the sixth on the list of most abundant elements. Coal reserves are the main source of carbon. Nowadays, carbon-based nanomaterials are widely used as detoxifiers for industrial effluents. Their versatile nature, stability, porosity and surface area make them more proficient in decontamination of wastewater treatment (Apul et al. 2013). Adsorption is the process used to minimize the hazardous toxicity of chemicals. The process of adsorption can be categorized into two groups: physisorption and chemisorption. Adsorption and absorption are collectively called sorption. According to Huang and Chen (2014), physisorption occurs when the adsorbate and adsorbent are both attached through van der Waals interactions, while in case of chemisorption, molecules are bonded through chemical bonds (Krishnamurthy and Agarwal 2013). The adsorption efficiency is determined by the quantity of impurities removed and is correlated with the adsorbent's adsorption capacity. Carbon-based nanomaterials are efficient adsorbents because of their high specific surface area and effective active sites (Ibrahim et al. 2016). The main determining factors of an adsorbent's adsorption capability are surface area, active sites and affinity for the pollutants.

8.4.1 Carbon Nanotubes (CNTs)

CNTs have a more effective adsorption capability than activated carbon and are especially widely used in organic substances. The outer surface area of CNTs is responsible for adsorption. The hydrophobic nature of the graphite surface lowers the effective surface area of CNTs, causing them to form loose aggregates in solution. On the structural basis, CNTs may be categorized into two types: single-walled carbon nanotubes (SWNTs), which consist of a single layer of graphene sheets, and multi-walled carbon nanotubes (MWNTs), which consist of multiple layers of graphene sheets. Moreover, Ni^{2+} adsorption from aqueous solutions was carried out employing oxidized MWCNTs and SWCNTs with a NaClO solution (Lu and Liu 2006), incorporating HNO_3-oxidized MWCNTs. Numerous techniques including chemical vapour deposition may be used to make these composites on a wide scale, which lowers production costs. CNT-based composites have the ability to adsorb dyes from aqueous environments. Wastewater remediation based on CNTs has remarkable properties for multiple water pollutants (Upadhyayula et al. 2009). A variety of CNT-based nanoadsorbents including CNT-cellulose, CNT-graphene, CNT-Fe_3O_4, CNT-chitosan and CNT-activated carbon fibre have been used for dye removal (Alwash et al. 2018). Additionally, organic molecules with carboxylic, hydroxyl and amino groups attach themselves to the surface of CNTs to create H-bonds that provide electrons. Electrochemical potential, ion exchange and adsorption processes are all governed by surface properties. Van der Walls force of attraction leads to the formation of pores through clustering of nanotubes, and these pores function as mesopores by providing bigger surface area to the structure, which helps to concentrate microorganisms (viruses, bacteria and parasites) with a higher affinity. Some antibiotics can be adsorbed more easily by electrostatic forces at the right pH. Due to link formation and electrostatic attraction, oxidized CNTs efficiently adsorb metal ions, including cadmium (AlSaadi et al. 2016), lead (Kabbashi et al. 2009), zinc (Zhao et al. 2015), arsenic (Naghizadeh et al. 2012), cobalt and copper (Ouni et al. 2019). Redox reactions take place between various adsorbates and on the surface of the adsorbent, and thus surface oxidation increases the adsorption capability. The remarkable mechanical, optical and electrical characteristics of graphitic carbon-based nanocomposites, especially CNTs, have potential for hybrid nanocatalysts, enhancing their efficiency in wastewater treatment. Although some semiconductors like TiO_2 and ZnO have large band gaps, slow ultraviolet photoresponse and low quantum efficiency, they are attractive adsorbents for applications pertaining to environmental protection.

8.5 Nanoporous-Activated Carbon

Recently, there has been advancement in the manufacture of adsorbent-dependent biomaterials, especially from agricultural wastes (Liu et al. 2018a, b). This mainly involves sawdust, crab shells, bagasse, leftover olive pits and activated carbon fabrics (ACFs), which are utilized to form carbon-associated adsorbents (Thines et al. 2017). When it comes to eliminating pollutants from wastewater, these are widely used because of their strong adsorption capacity. Nonetheless, there is still a need to increase their adsorption efficiency. Nanoporous forms of activated carbon are much more efficient than any other ACF forms. In terms of functional aspects, the activated form of carbon may be changed into nanoporous form. It has been observed that ACFs had a greater adsorption efficiency than granular activated carbon (GAC), based on the Freundlich adsorption isotherm. In order to verify and analyse GAC's efficiency as an adsorbent, Valderrama et al. (2008) studied its application in the aqueous solution for the adsorption of naphthalene, fluorene and fluoranthene. Teimouri et al. (2019) used a low-cost microwave approach to convert agricultural wastage, such as dry fruit shells, into activated carbon in an environmentally acceptable manner.

8.6 Graphene Oxide (GO)

According to Shan et al. (2017), GO is a two-dimensional (2D) layer of sp^2-hybridized carbon atoms that is one atom thick. The atoms form a hexagonal crystal lattice resembling honeycomb, using sigma (σ) and pi (π) bonding. GO is formed when graphite sheets (3D) are divided into graphene: a layer that can be rolled into a carbon nanotube (1D) or modified into a fullerene (0D). GO and its derivatives have developed into innovative and effective wastewater treatment materials. For decontamination of wastewater, graphene-based nanomaterials are used because of their specific properties such as surface area, functional group and carrier charge mobility (Ali et al. 2019). Wastewater laden with pharmaceuticals can be treated efficiently by the use of graphene-based materials, as adsorption efficiency may be changed by experimental alteration. Graphene is widely used in remediation of toxic heavy metals and dyes (Zhao et al. 2011), and the adsorption capacity can be affected by various parameters including dosage, pH, temperature, exposure time, acidity, ligands and ions. According to Sabzevari et al. (2018), strong adsorption ability for MB (402.6 mg g^{-1}) could be produced, in comparison to others from the original Hummers' method, by sonication. Iron oxide and silver nanoparticles were grown on the surface of GO to create a recyclable nanocomposite. This nanocomposite has higher antibacterial efficacy than silver nanoparticles as it works against both Gram-positive and Gram-negative bacteria. Additionally, nanocomposite can be easily recycled by the process of magnetic separation using magnetic iron oxide nanoparticles (NPs) which allow it for proper functioning. Furthermore, the produced quantum dots function as sensors to expedite the analytical application of 4-NP detection with the limiting values in deionized water and wastewater being 0.7 and 3.5 nm, respectively. A paper sensor based on sulphur-doped graphene quantum dots was able to rapidly detect 4-NP in wastewater in less than a minute. Numerous metals including lead (Pb), cobalt (Co), gold (Au), nickel (Ni), copper (Cu) and others have been used to create a variety of (nano)catalysts by physicochemical and/or biological techniques. For the purpose of elimination, stable lanthanum-substituted manganese ferrites (LMFs) have been synthesized and applied to graphene oxide (GO). These nanohybrids have the potential to reduce the harmful impact of cations, anions and other organic compounds (Elanchezhiyan et al. 2020).

8.7 Fullerenes

Fullerenes represent a significant advancement in the synthesis of carbon-based particles. Fullerenes appear as tubes, ellipsoids and spheres and belong to a group of carbon allotropes made up of even number of carbon atoms referred to as convex closed polyhedra (Thilgen & Diederich 2006). Due to high specific surface area and low aggregation propensity (Lucena et al. 2011) of fullerene, these are widely used as adsorbent detoxifiers of industrial wastes, especially heavy metals and organic compounds in polluted waters. It has been reported that adsorbent C60 fullerene has the capacity to adsorb naphthalene (Cheng et al. 2004). Recent findings demonstrate that naphthalene is more efficient in adsorption capability in comparison to ACFs. The hydrophobic surface of C60 fullerene makes it suitable for better adsorption efficiency of organic molecules. Researchers are working to increase the adsorption affinity of toxic materials by adding trace amount of fullerene.

8.8 Nanomaterials of Cellulose

Cellulose is the primary component of plants and one of the most plentiful compounds on Earth (Olivera et al. 2016). It is the main constituent of long-chain polysaccharides. Cellobiose (core unit of cellulose) is a dimer formed of β-D-glucopyranose units connected by β-1,4-glycosidic linkages having chemical formula $C_6H_{10}O_5$. Many lignocellulosic materials including wood (including hardwoods and softwoods), annual plants (flax) and grasses (bamboo), as well as additional sources like animal or bacterial matter, can be used to separate cellulose (Choudhury et al. 2020). A range of processing techniques has been performed to produce cellulose nanoparticles (CNs) with diverse morphology and physical characteristics. CNs have been classified into two major categories: cellulose nanocrystals (CNCs) and cellulose nanofibrils (CNFs) by a number of Canadian organizations. During the mechanical extraction of CNCs, processes including high-pressure homogenization, ball milling, grinding and microfluidization subject cellulose to shear stresses. These processes involve high energy consumption and may alter cellulose morphology, which reduces the purity of CNCs and affects crystallization. Chemical hydrolysis is the most affordable and effective method, which dissolves the amorphous fraction while leaving the crystalline portions of cellulose intact. Chemical hydrolysis also shows few limitations such as the need to neutralize environmentally hazardous waste solutions and its thermal instability. Biological hydrolysis involves the polymerization and crystallization of glucose during the fermentation of some non-pathogenic, Gram-negative bacteria such *Acetobacter* species. This method uses very little energy and generates an effective biocompatible material, which minimizes the risk of toxicity and satisfies the needs of ecological and sustainable development. There are some variable factors that might affect this biological process such as fermentation duration, inoculum concentration, culture media and strain. Qualities such as high surface area and intrinsic inertness make CNs suitable for adsorption. CNs have a high surface-to-volume ratio and intrinsic environmental inertness making them a suitable substitute as adsorbents. In this instance, CNCs have been utilized to adsorb Cd(II), Pb(II) and Ni(II) from rice straw fibres. The synthesis of CNCs involves the elimination of amorphous domains from the cellulose chains which contribute to increase the adsorption efficiency by increasing the number of primary hydroxyl groups. Additionally, during the manufacture of CNCs, acid hydrolysis of cellulose fibres adds negatively charged sulphate ester groups to the surface of CNCs, acting as binding sites for heavy metallic ions. Adding succinic acid to CNCs greatly boosts binding affinity (Yu et al. 2013). Conversion of carboxylic acid groups to sodium carboxylate enhances the rate of binding efficiency for metallic ion extractions. Mahfoudhi and Boufi (2020) used an aerogel modified with CNF-$Al(OH)_3$ to adsorb fluoride in an aqueous solution. The cellulose nanofiber gel was freeze-dried, then impregnated with an $AlCl_3$ solution and rinsed with water to create the adsorbent, and the range of the adsorption capacity was 20–35 mg/g. Five adsorption–desorption cycles of the CNF-$Al(OH)_3$ adsorbent were done without a discernible decrease in adsorption capability. Multi-N-methylol groups were added to the surface of the CNF-based textiles by modification using reactive polyhedral oligomeric silsesquioxane (R-POSS). Reactive dye adsorption capability was greater in the R-POSS-modified CNF hybrid fabric than in the new CNF-derived textiles.

8.9 Membrane Process

Membrane process falls under the category of physical process that involves the movement of particles on the basis of their concentration and sizes, especially metals, bacteria or viruses (Kunduru et al. 2017). This process is generally used to reduce water impurities. Membrane permeability and selectivity are the two essential components that affect the function of a membrane. Higher permeability membranes reduce the membrane area which is a necessary requirement for wastewater treatment. Selectivity and higher permeability reduce energy consumption which is required for membrane process. Biofouling is one of the primary challenges to membrane integrity. Therefore, the development of specialized anti-biofouling is necessary for membrane technology, which is used in wastewater treatment. More effort is required to minimize long-term biofouling, which is caused by continuous bioadsorption of bacteria, growth and hydrophilic membranes. The goal of current research is to integrate biocidal nanoparticles through surface function and mixing to create antibacterial membranes (Mansouri et al. 2016). These substances provide extra desirable characteristics to the membrane such as hydrophilicity and smooth surface, which increase the membrane's resistance to fouling. Generally, nanoparticles function against microbes by a number of methods including directly damaging microbial membranes (chitosan NPs) or by releasing toxic metal ions (Ag^+) and forming reactive oxygen species (ROS) (photocatalytic TiO_2). Bionanomaterials have great potential to eliminate microorganisms from contaminated water.

8.9.1 Nanobioremediation (NBR)

Nanobioremediation (NBR) involves the removal of environmental contaminants (organic and inorganic pollutants, heavy metals) from the contaminated locations by using nanomaterials obtained from plants or microbes (fungi and bacteria) (Koul and Taak 2018). Environmentally sustainable clean-up is gaining a lot of attention as it is a flexible method for detoxification of contaminated water. According to recent developments, bioremediation provides an affordable and eco-friendly beneficial method to reduce the level of pollutants from the ecosystem. There are mainly three primary methods that are used in bioremediation, namely enzymatic remediation, plant-based methods and microbial utilization (Jorgensen 2007). Nanomaterials have evolved into a viable substitute and play a vital role in current treatment due to their specific properties such as high efficiency, low cost and environmentally friendly nature. As per recent findings, iron-based nanoparticles were the first nanoparticles to be applied to environmental clean-up. There are a number of intriguing iron-based technologies available for remediation of contaminated soil/groundwater. Zinc nanoparticles (ZnNPs) have a remarkable ability to break down organic dyes and phenolic compounds. Noble metal nanoparticles, especially gold (Au), silver (Ag) and copper (Cu), are also efficient in the degradation of organic dyes and organic pigments. In comparison to other noble metals, copper is much more effective. Moreover, phytoremediation can be coupled with nanoparticles in an enzyme-based bioremediation. The combination of biotechnology and nanotechnology involves the enzymatic reaction which catalyses the rapid breakdown of complex organic chemicals into simpler ones with the help of microorganisms and plants. Extensive research has been done to date in attempt to build cost-effective and efficient methods to address environmental challenges. Remedial technologies have been continuously expanded, developed and evolved in terms of remediation of contaminated groundwater, wastewater, leachate and soil using the in situ and ex situ methods. Due to unique characteristics of nanomaterials, they are very applicable in in situ methods. Their tiny size makes them suitable for wider distribution and allows them to easily enter into extremely small underground crevices and stay suspended in groundwater. In the case of in situ nanoremediation, there is no movement of soil from one location to another.

8.9.2 Green Nanoparticles

Green nanotechnology has great importance in the field of manufacturing industries and factories. However, green-synthesized metal/metallic oxide nanoparticles have been widely employed to treat a variety of wastewater types. In a study endeavour, it has been reported that *Piliostigma thonningii* leaf extract is used in the formation of silver nanoparticles, which are very efficient in detoxification of metals. Recent findings demonstrated that leaf extract of *Ocimum tenuiflorum,* commonly known as holy basil (Tailor et al. 2020), have the ability to produce silver nanocomposites to overcome the problems of textile dyes. These are widely used in the treatment of wastewater in residential and commercial environments. Tangerine peel extract is widely used in the preparation of green synthetic-based iron oxide, which is effective in the removal of cadmium from wastewater. Current research reports that this peel extract acts as a stabilizing agent in an eco-friendly manner. Few points must be considered during synthesis, especially the regulation of size and shape of nanoparticles. Photochemicals present in leaf extracts are used for the stabilization of nanoparticles. It has been observed that these photochemicals can be used in the removal of chemical oxygen demand (COD) and chromium (Cr) simultaneously. Rapid green chemistry techniques utilize leaf extract of *Lagerstroemia speciosa* to synthesize gold-based nanoparticles (AuNPs). Solid photocatalytic activity is observed by the reduction of dyes such as methylene blue, methyl orange, bromophenol blue, bromocresol green and 4-nitrophenol in the presence of sunlight during the green approach for the development of AuNPs. As previously mentioned, research has been conducted on the synthesis and separation of metallic NPs, but relatively there is less information about the nanostructure extraction from plants. These nanomaterials can be used to develop biocompatible and bioactive nanodevices for novel potential uses. Nanoparticles and nanostructures with biocompatible properties derived from plants act as sustainable and biorenewable sources.

8.10 Conclusion

In this era, wastewater is one of the concerned issues as it contains hazardous wastes from household garbage and industrial effluents including toxic metals, chemicals, fertilizers and a variety of pathogens. To overcome this situation and minimize their toxicity, different efforts are ongoing. Although many biotechnological methods are good enough to remove the pollutants, nanotechnology is evolving as a potent activator for wastewater treatment. The advancement in the field of nanotechnology is increasing day by day in agricultural and industrial aspects. Metal oxide nanoparticles are mainly used to remove heavy metals such as zinc, molybdenum, chromium, cobalt, iron and other harmful metals. The main application of AgNPs is to remove organic pollutants, while iron-based nanoparticles such as magnetite and maghemite are commonly used in detoxification of metals. Carbon-based nanoparticles are utilized in adsorption of both organic and inorganic materials. Graphene is responsible for the remediation of radionuclides, while fullerene is an excellent absorbent and is used in the extraction of heavy metals from industrial effluents. Membrane process minimizes impurities including microbes from water due to their selective nature of permeability. Green nanoparticles have proven to be the best methodology, with least side effects and improved performance. Current research focuses on eco-friendly techniques to ensure no negative effects on the biotic and abiotic components of water.

Acknowledgements The authors wish to thank the Principal of S. S. Khanna Girls' Degree College (a constituent college of the University of Allahabad, Prayagraj) and the DST-CURIE Grant of the Department of Science and Technology, New Delhi, GOI (DST/CURIE-PG/2022/10 G) for providing all the necessary facilities in the college.

References

Abdelbasir SM, Shalan AE (2019) An overview of nanomaterials for industrial wastewater treatment. Korean J Chem Eng 36(8):1209–1225

Ali I, Mbianda XY, Burakov A, Galunin E, Burakova I, Mkrtchyan E, Tkachev A, Grachev V (2019) Graphene based adsorbents for remediation of noxious pollutants from wastewater. Environ Int 127:160–180

AlSaadi MA, Al Mamun A, Alam MZ, Amosa MK, Atieh MA (2016) Removal of cadmium from water by CNT-PAC composite: effect of functionalization. NANO 11:1650011

Alwash A, Adil H, Hussain Z, Yousif E (2018) Potential of carbon nanotubes in enhance of photocatalyst activity. Arch Nano Op Acc J 1:65–70

Apul OG, Wang Q, Zhou Y, Karanfil T (2013) Adsorption of aromatic organic contaminants by graphene nanosheets: comparison with carbon nanotubes and activated carbon. Water Res 47(4):1648–1654

Atacan K, Güy N, Özacar M (2022) Recent advances in photocatalytic coatings for antimicrobial surfaces. Curr Opin Chem Eng 36:100777

Bahadar S, Jamal A, Faisal M (2011) Iron oxide nanoparticles. Nanomater

Bartolomeu M, Neves S, Faustino F, Almeida A (2018) Wastewater chemical contaminants: remediation by advanced oxidation processes. Photochem Photobiol Sci 17(11):1573–1598

Baskar AV, Bolan N, Hoang SA, Sooriyakumar P, Kumar M, Singh L, Siddique H (2022) Recovery, regeneration and sustainable management of spent adsorbents from wastewater treatment streams: a review. Sci Total Environ 822:153555

Bishoge OK, Zhang L, Suntu SL, Jin H, Zewde A, Qi Z (2018) Remediation of water and wastewater by using engineered nanomaterials: a review. J Environ Sci Health Part A 53(6):537–554

Boulkhessaim S, Gacem A, Khan H, Amari A, Yadav K, Harharah N, Jeon H (2022) Emerging trends in the remediation of persistent organic pollutants using nanomaterials and related processes: a review. Nanomaterials 12(13):2148

Cheng X, Kan T, Tomson B (2004) Naphthalene adsorption and desorption from aqueous C60 fullerene. J Chem Eng Data 49(3):675–683

Choudhury R, Sahoo K, Gohil M (2020) Potential of bioinspired cellulose nanomaterials and nanocomposite membranes thereof for water treatment and fuel cell applications. Cellulose 27:6719–6746

Corsi I, Nielsen M, Sethi R, Punta C, Torre C, Libralato G, Lofrano L, Sabatini L, Aiello M, Fiordi L, Cinuzzi A, Caneschi A, Pellegrini D, Buttino I (2018) Ecofriendly nanotechnologies and nanomaterials for environmental applications: key issue and consensus recommendations for sustainable and ecosafe nanoremediation. Ecotoxicol Environ Saf 154:237–244

Dankovich TA, Gray DG (2011) Bactericidal paper impregnated with silver nanoparticles for point-of-use water treatment. Environ Sci Technol 45(5):1992–1998

Elanchezhiyan SS, Prabhu SM, Kim Y, Park CM (2020) Lanthanum-substituted bimetallic magnetic materials assembled carboxylate-rich graphene oxide nanohybrids as highly efficient adsorbent for perfluorooctanoic acid adsorption from aqueous solutions. Appl Surf Sci 509:144716

Ferreira AM, Roque ÉB, Fonseca VD, Borges CP (2015) High flux microfiltration membranes with silver nanoparticles for water disinfection. Desalin Water Treat 56(13):3590–3598

Gudainiyan J, Kumar R, Singh D, Sing DP, Srivastava A, Srivastava AP, Khan AK (2024). Nanotechnology-enabled Solutions for water purification and Environmental Sustainability. E3S Web Conferen 511:01011

Guerra FD, Attia MF, Whitehead DC, Alexis F (2018) Nanotechnology for environmental remediation: materials and applications Molecules

Huang Y, Chen X (2014) Carbon nanomaterial-based composites in wastewater purification. Nano Life 4(03):1441006

Ibrahim RK, Hayyan M, AlSaadi MA, Hayyan A, Ibrahim S (2016) Environmental application of nanotechnology: air, soil, and water. Environ Sci Pollut Res 23(14):13754–13788

Janovák L, Tallósy SP, Sztakó M, Deak A, Bitó T, Buzas N, Bartfai G, Dékany I (2014) Synthesis of pH-sensitive copolymer thin solid films embedded with silver nanoparticles for controlled release and their fungicide properties. J Drug Deliv Sci Technol 24(6):628–636

Jørgensen KS (2007) In Situ bioremediation. Adv Appl Microbiol 61:285–305

Kabbashi NA, Atieh MA, Al-Mamun A, Mirghami ME, Alam M, Yahya N (2009) Kinetic adsorption of application of carbon nanotubes for Pb(II) removal from aqueous solution. J Environ Sci 21:539–544

Khan T, Malik A (2019) Engineered nanomaterials for water decontamination and purification: from lab to products. J Hazard Mater 363:295–308

Khan T, Ahmad F, Shahadat M, Rehman U, Khan M (2020) Metal and metal oxide nanoparticles for water decontamination and purification. Environ Nanotechnol Water Purific 1:151–186
Koul B, Taak (2018) Biotechnological strategies for effective remediation of polluted soils, XVII 240
Kreyling G, Behnke M, Chaudhry Q (2010) A complementary definition of nanomaterial. Nano Today 5(3):165–168
Krishnamurthy G, Agarwal S (2013) Optimization of reaction conditions for high yield synthesis of carbon nanotube bundles by low temperature solvothermal process and study of their H_2 storage capacity. Bull Korean Chem Soc 34(10):3046–3054
Kunduru R, Nazarkovsky M, Farah S, Pawar P, Basu A, Domb J (2017) Nanotechnology for water purification: applications of nanotechnology methods in wastewater treatment. Water Purific 1:33–74
Li S, Wang W, Liang F, Zhang X (2017) Heavy metal removal using nanoscale zero-valent iron (nZVI): theory and application. J Hazard Mater 322:163–171
Liu A, Liu J, Pan B, Zhang X (2018a) Formation of lepidocrocite (c-FeOOH) from oxidation of nanoscale zero-valent iron (nZVI) in oxygenated water. RSC Adv 4:57377–57382
Liu X, He C, Yu X, Bai Y, Ye L, Wang B, Zhang L (2018b) Net-like porous activated carbon materials from shrimp shell by solution processed carbonization and H_3PO_4 activation for methylene blue adsorption. Powd Technol 326:181–189
Lucena R, Simonet M, Cárdenas S, Valcárcel M (2011) Potential of nanoparticles in sample preparation. J Chromatogr A 1218(4):620–637
Lu C, Liu C (2006) Removal of nickel (II) from aqueous solution by carbon nanotubes. J Chem Tech Biotech: Int Res Process Environ Clean Technol 81:1932–1940
Lu H, Wang J, Wang T, Wang N, Bao Y, Hao H (2017) Crystallization techniques in wastewater treatment: an overview of applications. Chemosphere 173:474–484
Mahfoudhi N, Boufi S (2020) Porous material from cellulose nanofibrils coated with aluminum hydroxide as an effective adsorbent for fluoride. J Environ Chem Eng 8(3):103779
Malato S, Ibáñez P, Maldonado I, Blanco J, Gernjak W (2009) Decontamination and disinfection of water by solar photocatalysis: recent overview and trends. Catal Today 147(1):1–59
Mansouri J, Charlton T, Chen V, Weiss T (2016) Biofouling performance of silver-based PES ultrafiltration membranes. Desalin Water Treat 57(58):28100–28114
Marimuthu T, Chee Y, Sulaiman N (2023) A review on the use of cellulose nanomaterials for wastewater remediation of heavy metal ions. Int J Environ Sci Technol 20(3):3421–3436
Moustafa GS, Hassan MH, Jianhua W, Zawrah MF, Abouelmagd A (2023). Nanotechnology for water treatment: is it the best solution now? J Earth Sci 34(5):1616–1620
Naghizadeh A, Yari AR, Tashauoei HR, Mahdavi M, Derakhshani E, Rahimi R, Bahmani P, Daraei H, Ghahremani E (2012) Carbon nanotubes technology for removal of arsenic from water. Archiv Hygiene Sci 1:6–11
Nasrollahzadeh M, Sajjadi M, Iravani S, Varma S (2021) Carbon-based sustainable nanomaterials for water treatment: state-of-art and future perspectives. Chemosphere 263:128005
Nassar N (2012) Iron oxide nanoadsorbents for removal of various pollutants from wastewater: an overview. Appl Adsorb Water Pollut Control 1:81–118
Nowack B, Krug F, Height M (2011) 120 years of nanosilver history: implications for policy makers. Environ Sci Technol 45:1177–1183
Olivera S, Muralidhara B, Venkatesh K, Guna K, Gopalakrishna K, Kumar Y (2016) Potential applications of cellulose and chitosan nanoparticles/composites in wastewater treatment: a review. Carbohyd Polym 153:600–618
Ouni L, Ramazani A, Fardood ST (2019) An overview of carbon nanotubes role in heavy metals removal from wastewater. Front Chem Sci Eng 1–22
Rathi BS, Kumar PS (2021) Application of adsorption process for effective removal of emerging contaminants from water and wastewater. Environ Pollut 280:116995
Sabzevari M, Cree E, Wilson D (2018) Graphene oxide-chitosan composite material for treatment of a model dye effluent. ACS Omega 3:13045–13054

Salehi M (2022) Global water shortage and potable water safety; today's concern and tomorrow's crisis. Environ Int 158:106936

Sen S, Shah P (2014) Application of nanoscale zero-valent iron for wastewater treatment. Int Conf Multidiscip Res Pract I

Shan J, Zhao Y, Tang H, Cui Y (2017) A mini-review of carbonaceous nanomaterials for removal of contaminants from wastewater. IOP Conferen Series: Earth Environ Sci 68:1–7

Sheoran K, Kaur H, Siwal S, Saini K, Vo N, Thakur K (2022) Recent advances of carbon-based nanomaterials (CBNMs) for wastewater treatment: synthesis and application. Chemosphere 299:134364

Shojaei S, Khammarnia S, Shojaei S, Sasani M (2017) Removal of reactive Red 198 by nanoparticle zero valent iron in the presence of hydrogen peroxide. J Water Environ Nanotechnol 2:129–135

Sikiru S, Ayodele A, Sanusi K, Adebukola Y, Soleimani H, Yekeen N, Haslija A (2022) A comprehensive review on nanotechnology application in wastewater treatment a case study of metal-based using green synthesis. J Environ Chem Eng 10:108065

Singh K, Singh A, Rai S (2022) A study on nanomaterials for water purification. Mater Today: Proceed 51:1157–1163

Singh S, Kumar V, Romero R, Sharma K, Singh J (2019) Applications of nanoparticles in wastewater treatment. Springer, London, UK

Tailor G et al (2020) Green synthesis of silver nanoparticles using Ocimum canum and their antibacterial activity. Biochem Biophys Rep 24:100848

Taylor P, Majewski P, Thierry B (2007) Critical reviews in solid state and materials sciences functionalized magnetite nanoparticles—synthesis, properties, and bio-applications. Small 8436:37–41

Teimouri Z, Salem A, Salem S (2019) Clean and new strategy for catalytic conversion of agriculture waste shells to activated carbon via microwave-assisted impregnation: applied and eco-friendly aspect for decoloration of industrial corn syrup and process identifications. J Environ Chem Eng 7(3):103161

Thilgen C, Diederich F (2006) Structural aspects of fullerene chemistry a journey through fullerene chirality. Chem Rev 106(12):5049–5135

Thines K, Mubarak M, Nizamuddin S, Sahu N, Abdullah C, Ganesan P (2017) Application potential of carbon nanomaterials in water and wastewater treatment: a review. J Taiwan Inst Chem Eng 72:116–133

Ugrina M, Milojkovic J (2024) Advances in wastewater treatment 2024. Energies 17(6):1400

Upadhyayula K, Deng S, Mitchell C, Smith B (2009) Application of carbon nanotube technology for removal of contaminants in drinking water: a review. Sci Total Environ 408(1):1–13

Valderrama C, Gamisans X, Heras l, Farran X, Cortina A (2008) Sorption kinetics of polycyclic aromatic hydrocarbons removal using granular activated carbon: intraparticle diffusion coefficients. J Hazard Mater 157(2–3):386–396

Yu X, Tong S, Ge M, Wu L, Zuo J, Cao C, Song W (2013) Adsorption of heavy metal ions from aqueous solution by carboxylated cellulose nanocrystals. J Environ Sci 25(5):933–943

Zahid M, Abd-Elsalam A (2021) Applications of nanomaterials in water remediation: a note from the editors. Elsevier, Amsterdam, the Netherlands

Zhao H, Jiao P, Yu G, Xi Y, Jiang Y, Chen Q (2015) Removal of Cu(II) from aqueous solutions by tartaric acid modified multi-walled carbon nanotubes. Colloid Surf Physicochem Eng Aspect 476:35–41

Zhao G, Li J, Ren X, Chen C, Wang X (2011) Few-layered graphene oxide nanosheets as superior sorbents for heavy metal ion pollution management. Environ Sci Technol 45:10454–10462

Zhao X, Su Y, Li S, Bi Y, Han X (2018) A green method to synthesize flowerlike $Fe(OH)_3$ microspheres for enhanced adsorption performance toward organic and heavy metal pollutants. J Environ Sci (China) 73:47–57

Chapter 9
Microbial Fuel Cells: Harnessing Bacteria's Power for Wastewater Treatment

Ganesh Swain and Ravi Kumar

Abstract The steep rise in energy demand worldwide has induced several countries to use renewable and clean energy sources. In addition to solar, wind, and hydrogen energies, microbial fuel cell (MFC) technology has become a rising alternative source. MFC, a bioelectrochemical system, has a unique property of converting carbon-rich wastewater into bioelectricity using microorganisms. The membrane used in MFC acts as a barrier between the two electrode (i.e., anode and cathode) compartments while allowing the exchange of radicals across the chambers. Additionally, minimal sludge generation compared to conventional activated sludge processes is a major advantage. This chapter provides a brief overview of MFC technologies and their various components for the simultaneous mineralization of pollutants and bioelectricity generation. The working principle and the various designs of MFCs are briefly described. The electrodes are considered to be a vital part as they provide surface area for biofilm growth and chemical reactions (oxidation and reduction). Hence, the ideal characteristics of electrodes and conventional electrodes that were used are described briefly. Recent modifications and their performances are presented. Moreover, the role of the proton exchange membrane in MFC and its recent modifications are addressed. Further, the major bottlenecks, which restrict the real-field application of MFC for energy production during wastewater treatment, are included in this chapter.

Keywords Microbial fuel cell · Electrode modification · Proton exchange membrane · Design aspects · Bioelectricity · Electrode properties

---

G. Swain (✉)
Department of Chemical and Biochemical Engineering, IIT, Patna, Bihar, India
e-mail: ganeshswain24@iitp.ac.in

R. Kumar
Department of Chemical Engineering, Indian Institute of Technology (ISM), Dhanbad, Jharkhand, India
e-mail: 21dr0144@ce.iitism.ac.in

A. M. Yatoo et al. (eds.), *Wastewater Monitoring and Management by Advanced Approaches*, Advances in Wastewater Research,
https://doi.org/10.1007/978-981-95-2601-7_9

9.1 Introduction

The steep increase in population and urbanization has led to a high demand for industrialization. In recent decades, various industrial sectors have been rapidly growing to meet the requirements of the rising population. Thus, energy and water are two major utilities used in every manufacturing industry. Moreover, energy becomes a limiting factor as more than 80% of total energy is generated from nonrenewable sources (Opeyemi 2021; Shah et al. 2024). Further, the increasing human activity has disturbed the ecosystem causing air, water, and soil pollution. The carbon footprint and limitations in extraction of fossil fuel due to stringent regulations to control climate change are major shortcomings of nonrenewable energy. Alternative sources for energy production, such as renewable energy, are gaining more attraction to meet the current demand. Hence, various renewable sources, including solar, wind, nuclear, hydrogen, and bioenergy, have been examined and utilized in large-scale applications (Matsena et al. 2021; Awad et al. 2024). It is estimated that the contribution of renewable energy will rise from 20 to 75% by 2050 (Vidhyeswari et al. 2022). Moreover, energy production through waste is an effective and novel approach, which gives a value-added product along with waste management. In this direction, wastewater is a suitable source for energy generation. Various technologies such as physiochemical (adsorption, membrane separation, coagulation, chlorination, etc.), biological (bioremediation, phytoremediation, mycoremediation, etc.), and advanced oxidation processes have been employed for the treatment purpose (Swain et al. 2022; Ramalingam et al. 2022). However, each treatment technology has at least one limitation, such as, high energy consumption, operating cost, chemical cost, and secondary sludge generation (Agrahari and Kumar 2024; Apollon et al. 2022; Swain et al. 2024).

In the meantime, microbial fuel cells (MFCs), a bioelectrochemical technology, are preferred as a suitable method for the degradation of pollutants along with electricity generation. The wastewater containing various organic matter acts as a substrate for electroactive microorganisms, which produce electricity by consuming these pollutants under anaerobic conditions. The MFC behaves as a battery-type device in which the microorganisms form a thin layer on the surface of the anode and decompose the carbon source into simpler substances such as CO_2 and H_2O (Yaqoob et al. 2022; Vidhyeswari et al. 2022). In addition, the generated protons and electrons are allowed to move via the proton exchange membrane (PEM) and external circuit, respectively, to the cathode chamber, where they reduce to water by reacting with oxygen. The MFC technology has several advantages, including less sludge production (compared to conventional activated sludge system), generation of value-added products (such as electricity, hydrogen, and biofuel), no need for aeration (in case of single-chamber MFC [SCMFC]), and applicability over a wide range of operating conditions (pH, temperature, and microorganisms) (Gupta et al. 2021; Opeyemi 2021; Hassan et al., 2024).

Despite huge efforts, the MFC has several limitations in terms of power output per anode surface area and cost of power production, which hinder its large-scale implementation. However, the physical configuration and properties of the electrodes of MFC play a major role in its modeling and scale-up. This chapter provides a brief idea about the necessity of modifications of MFC in terms of its exterior and interior design.

Moreover, low efficiency, poor biocompatibility, limited durability, low conductivity, and high cost of electrodes and membranes are the bottlenecks of MFC (Wang et al. 2024). These factors should be addressed for possible improvements and research in MFC. Electrodes and membranes are two major components of an MFC whose properties and configuration control the power generation in MFC. In this chapter, a detailed report is provided about the conventional electrodes and membranes along with their recent developments and modifications.

The objectives of this chapter are to demonstrate (i) the basic mechanisms and working principles of MFC, (ii) the design aspects of MFC, and (iii) the recent developments and modifications of electrodes and proton exchange membranes. In addition, the key challenges and bottlenecks of MFC toward its practical applicability are discussed. Moreover, the major research gaps that restrict the commercialization of MFCs, including the up-to-date research works, are summarized in this chapter.

9.2 Overview of MFC

9.2.1 Working Principle of MFC

MFC is a bioelectrochemical device that is generally associated with two compartments, namely anode and cathode, which are separated by a proton exchange membrane (PEM) (Fig. 9.1) (Dwivedi et al. 2022). The anode chamber should be free from oxygen (that means anaerobic condition), and the electron-generating bacteria are permitted to develop colonies on the anode surface. The microorganisms utilize the carbon source and decompose it to generate protons and electrons. The protons transfer from the anode chamber to the cathode through the PEM, while the electrons are transferred via an external circuit. The electrons and protons react with oxygen in the cathode chamber to convert into water (Apollon et al. 2022).

The electrode reactions for substrate glucose are illustrated below.

$$\text{Anode reaction}: \ C_6H_{12}O_6 + 6H_2O \rightarrow 6CO_2 + 24H^+ + 24e^- \quad (9.1)$$

$$\text{Cathode reaction}: \ 6O_2 + 24H^+ + 24e^- \rightarrow 12H_2O \quad (9.2)$$

$$\text{Overall reaction}: \ C_6H_{12}O_6 + 6H_2O \rightarrow 6CO_2 + 24H^+ + 24e^- \quad (9.3)$$

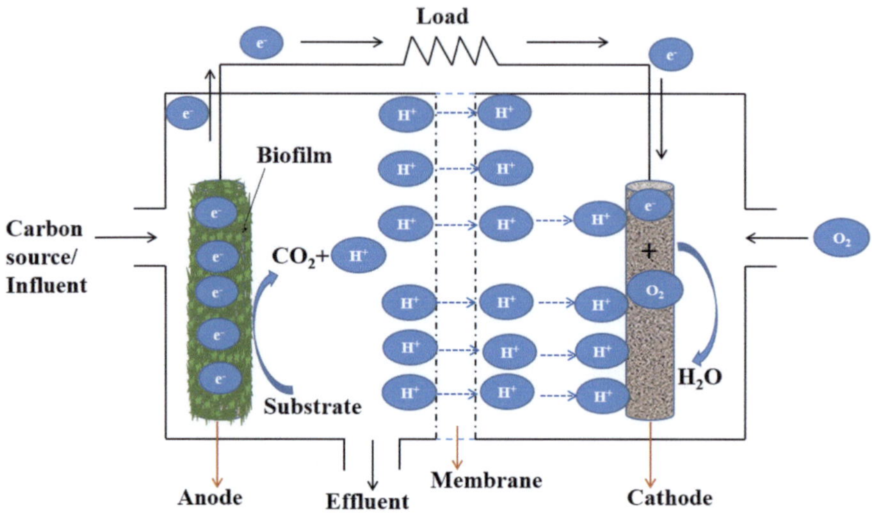

Fig. 9.1 Schematic presentation of a microbial fuel cell showing its various components

9.2.2 Extracellular Electron Transfer Mechanism

The generated electrons are transferred from the bacterial cell to the anode surface via two mechanisms, i.e., direct electron transfer (DET) and mediated electron transfer (MET) (Fig. 9.2) (Nawaz et al. 2020). In the DET mechanism, the extracellular electrons are transferred directly to the anode surface. The number of transferred electrons depends on the quantity of biomass attached to the electrode surface. Some bacteria, namely *Geobacter* sp., *Shewanella* sp., *Rhodoferax* sp., *Pseudomonas* sp., etc., use their locomotive organ (known as flagella or nanowires) to transfer electrons (Wang et al. 2023). In addition, C-type cytochromes, which are present on the external surface of the cells, are responsible for the DET. The major bottleneck of this type of mechanism is that the bacterial cells need to adhere to the anode surface. However, a recent study revealed that some specific bacterial cells can transfer electrons over a distance of a few millimeters.

However, most bacterial species cannot transfer electrons through the DET mechanism. In that case, a redox mediator (either exogenous or endogenous) helps to transport the electrons from the cell membrane to the anode surface (Fig. 9.2). This mechanism is known as mediated electron transfer (MET) (Arkatkar et al. 2019). Mediators that are secreted from the metabolic activity of microorganisms are called endogenous mediators. In addition, there are some artificial mediators used for this purpose, which are known as exogenous mediators. These mediators get reduced by accepting an electron from the bacterial site, moving to the cathode surface, and finally getting oxidized to transfer the electrons to the anode. It is reported that some metabolites, such as flavins, riboflavins, pyocyanins, and phenazines, can act

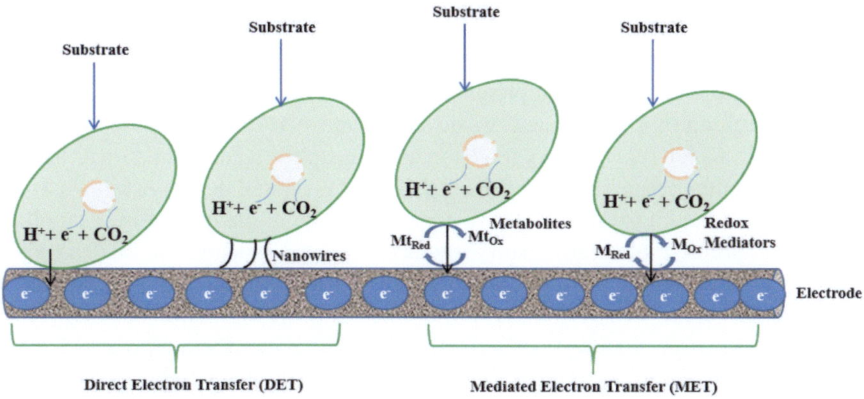

Fig. 9.2 Extracellular electron transport from bacterial cell to the electrode surface via direct and mediated electron transfer mechanisms

as MET (Prathiba et al. 2022). There are some artificial mediators, including methylene blue, neutral red, thionine, 2-hydroxy-1,4-napthoquinone, EDTA, etc., utilized for the above purpose. However, toxicity and adverse impact of artificial mediators on bacterial growth and the environment restrict their use in MFC.

9.3 Various Designs of MFC

9.3.1 Dual-Chamber MFC

Dual-chamber MFC generally consists of two compartments, namely anode and cathode chambers, which are separated by either a salt bridge or a porous membrane (Gupta et al. 2021). The proton exchange membrane allows the transfer of proton ions from the anode to the cathode chamber. In the meantime, the membrane acts as a barrier for water, oxygen, and other ions to transfer from one chamber to another. The electrodes, made up of mostly carbon or any other metal, have conducting properties. This kind of MFC is mostly used on a laboratory scale for preliminary studies. Moreover, the examination of the developed electrodes or proton exchange membrane for real application can be analyzed in this setup. It has several advantages, including a modest structure, ease of fabrication, and simple operation (Hidayat et al. 2022). However, it still has some disadvantages, such as large space requirements, high energy consumption, and membrane costs. Thus, alternative designs have been developed to mitigate the demerits of dual-chamber MFC, which are discussed in the following sections.

9.3.2 Single-Chamber (Air-Breathing) MFC

The single-chamber MFC comprises a single compartment containing anode, cathode, membrane (optional), and wastewater. One side of the cathode is exposed to ambient air, which receives oxygen from the atmosphere. The reduction of oxygen takes place on the porous surface of the cathode. This type of design is known as air-breathing MFC, and it is more effective compared to dual-chamber MFC as it eliminates the requirement of aeration and cathode chamber (Flimban et al. 2019). Hence, the addition of catholyte is not required in this case. Therefore, the major advantages of single-chamber MFC include low fixed and operational costs (Saravanan et al. 2022). However, the single-chamber MFC (without membrane) possesses some limitations, such as diffusion of oxygen through the cathode, which results in short circuits and microbiological contamination. In this direction, several researchers developed and used membrane electrode assembly (MEA) in single-chamber MFC to counterattack the above problem. In this type of design, one surface of the cathode is attached to the membrane and the other side is exposed to the outside air. Hence, the membrane restricts the diffusion of air to the anode chamber, and at the same time, it allows the proton to reach the cathode surface. For large applications, the long space between the anode and cathode increases the internal resistance of MFC. Therefore, the single-chamber MFC is a suitable design for low-volume wastewater treatment.

9.3.3 Rectangular MFC

Rectangular designs are primarily focused on constructing dual-chamber MFC. Both chambers are comprised of rectangular boxes, which are made up of acrylic sheets or plexiglass. This type of module is more favorable for the continuous operation of wastewater. However, large space consumption is the major shortcoming of this type of design. In addition, the membrane and aeration costs add up to the fixed and operational costs, respectively.

There are some limitations in this type of MFC including the biofilm attachment to the anode surface, which can be sorted out by providing an agitator to facilitate the better attachment of the biolayer on the electrode surface. The membrane used in MFC also contributes to higher internal resistance resulting in the decline of MFC performance (Nawaz et al. 2022).

9.3.4 Tubular MFC

The tubular design has a cylindrical configuration that consists of an anode chamber (bottom) and a cathode chamber (top). Generally, there is a provision for wastewater recirculation that enhances diffusivity and antifouling. Initially, this type of configuration is used for wastewater treatment with little power generation. However, the stacked or series arrangement of tubular MFCs enhances the power density. Nguyen and Min (2020) developed an air-breathing tubular MFC of 280 mL that offered a power density of 1560 mW/m^3 and a maximum chemical oxygen demand (COD) removal of 93.5% using sewage wastewater. A series connection of 32 tubular MFCs having a total working volume of 527 mL showed an effective power density of 18 mW/m^2 (Cristiani et al. 2019).

9.3.5 Flat-Plate MFC

In this type of configuration, the cathode is attached to the membrane, and the combination is known as electrode–membrane assembly. The operational and monitoring easiness makes flat-plate design more convenient. The distance between the anode and cathode is minimal in this case, which ultimately reduces the internal resistance for electron transfer. The major advantages of flat-plate MFC include compact size and higher power density compared to other configurations of MFC. The flat-plate MFC has attained a maximum power density of 236 mW/m^2 using municipal wastewater.

9.3.6 Stacked MFC

The arrangement of multiple single MFCs in either series or parallel is called stacked MFC, which is feasible for large-scale applications. Additionally, this type of configuration adds up the current density as well as pollutant removal. It was reported that a stacked MFC consisting of 40 individual MFC units with a total volume of 4 L could achieve a maximum power and current density of 2500 mW/m^2 and 500 mA/m^2, respectively. Similarly, a power and current density of 5.8 mW/m^2 and 24 mA/m^2, respectively, were obtained in a parallel design. The study revealed that higher power and current density resulted in a series connection due to the higher anodic potential. However, voltage loss was observed in stacked MFC, which may be due to the ambient oxygen diffusion to the anode chamber. Another researcher observed that the series connection of MFCs did not guarantee a higher power density due to the increase of internal resistance and decline of substrate concentration, although the series design offered a higher open-circuit voltage (OCV) (Chen et al. 2019). Yuvraj and Aranganathan (2018) designed connected dual-chamber MFCs in three ways, including series, parallel, and series–parallel. The maximum power densities of 2418, 1546, and 813 mW/m^2 were evaluated for stacked MFC connected in series–parallel, parallel, and series, respectively. In addition, they observed that the stacked MFC showed a greater power density compared to a single MFC (288 mW/m^2). The series–parallel combination showed better COD removal (81%) and power density compared to the other configuration due to the higher redox potential. Moreover, the parallel design may be prone to short circuits, which is unavoidable in series connections. The voltage reversal in stacked MFC deteriorates the biofilm layer, which further reduces the biochemical conversion rate.

9.4 Characteristics of Electrode Materials

The electrode materials play a significant role in bioelectricity production in MFC. Therefore, the proper selection of electrodes is necessary for optimum output. In this direction, the anode and cathode should share some basic properties, which are discussed below.

9.4.1 *Conductivity*

The electrical conductivity is one of the crucial criteria when choosing the electrode. The anode should have high electrical conducting ability so that the electrons generated from the biochemical process can easily be transferred to the cathode via an external circuit. Hence, the high-conducting material can minimize the bulk liquid resistance and enable the rapid transfer of electrons.

9.4.2 *Surface Area*

The morphological structure of the anode is very crucial for bacterial growth. A large surface area is desirable for higher biocell attachment, which ultimately generates electrons. Although, microorganisms can grow in both suspended liquid as well as on the anode surface, the biomass attached to the conducting material contributes a major portion to the bioelectricity generation. The presence of pores on the electrode surface is an advantage for higher biomass growth. Hence, a porous and conducting material is essential for the bioelectricity generation in MFC (Myung et al. 2019).

9.4.3 *Biocompatibility*

The electrodes used in MFC should have high bioaffinity. The biochemical activity in MFC depends upon the extent of incompatibleness of the electrode. High growth of microorganisms was observed with a biocompatible material; otherwise, a decline in bio growth was found in the case of a nonbiocompatible material. It was studied that microorganisms have a positive affinity toward carbon-based electrodes. However, some electrode materials made up of silver, gold, and copper have a negative impact on the growth of microorganisms. Moreover, carbon nanotubes and stainless steel have low biocompatibility, which is further enhanced by some modifications.

9.4.4 Physical and Chemical Stability

The longevity of the electrodes is another bottleneck for the practical application of MFC on a larger scale. The mechanical strength and surface roughness play a major role in an ideal electrode. In addition, highly porous electrode provides better microbial growth; however, it offers fouling, which inhibits the performance of the MFC (Ibrahim et al. 2020). The material should be chemically inert to acid and base. Generally, corrosion and swelling in anode materials are seen due to low mechanical strength and thermal stability. Hence, the conducting electrode should be chemically inert and durable.

9.4.5 Cost-Effectiveness

The electrode constitutes a major portion of the total cost of an MFC. Hence, a cost-effective and abundant electrode material is highly preferable. For instance, platinum (Pt), titanium, gold, etc., are very expensive, which hinders their use in MFCs (Yaqoob et al. 2021). Mostly, carbon-based materials like graphite rods, carbon cloth, carbon felt, graphite foam, etc., are widely used in MFC. However, the search for alternative low-cost electrodes is essential for the commercialization of MFC.

9.5 Types of Anodes and Their Modification

The morphological structure and physical and chemical properties of anode materials are the crucial factors for bioelectricity generation. Moreover, the bioaffinity or adhesion of biofilm on the anode surface controls the easiness of the transfer of the extracellular electrons from the bacterial site to the conducting surface. Besides, fragility and durability are considered key parameters when choosing suitable anodes for MFC. In this direction, various materials, including carbon, metal, and composite-based materials, have been studied previously (Dessie et al. 2020). A comparative analysis of their structure, performance, and bioaffinity is provided in Table 9.1. Recent modifications of primitive anode materials, such as thermal, chemical, and ammonia treatments, are briefly described in the next section.

Table 9.1 Use of various types of electrode materials in MFC for bioelectricity generation

Anode	Cathode	Surface area of anode (cm^2)	Power density (mW/m^2)	References
Polyaniline–SA–carbon brush hydrogel	Graphite rod	8.0	515	Wang et al. (2020b)
Activated carbon-coated carbon cloth	Activated carbon-coated carbon cloth	14	1287.7 mW/m^3	Matsena et al. (2021)
Graphene oxide–ZnO	Graphite rod	76	1.21	Yaqoob et al. (2022)
Graphene oxide–TiO_2	Graphite rod	76	0.784	
Mango wood-based activated carbon–carbon felt	Pt/C air cathode	75.3	589.8	Li et al. (2020)
Coffee waste-derived activated carbon–carbon cloth	Carbon cloth	1.0	3927	Hung et al. (2019)
Graphite felt	Carbon brush	19.8	253	Ye et al. (2020)
Polypyrrole–SS	Pt/C air cathode	4.0	1190	Pu et al. (2018)
Titanium rod	Carbon cloth	-	6119	Firdous et al. (2018)
Polypyrrole–CMC–TiO_2–carbon brush	Carbon rod	8.0	14.11 mW/m^3	Wang et al. (2020a)
MnO_2-coated carbon felt	Carbon brush	17	1534	Dessie et al. (2020)

SA Sodium alginate, *SS* Stainless steel, *CMC* Carboxymethyl cellulose

9.5.1 Carbon-Based Anodes

In the last decade, the use of carbon-based anodes has become more popular due to their abundance, high surface area, excellent electrical conductivity, bioaffinity, and thermal properties. In this regard, various materials, i.e., carbon cloth, carbon paper, carbon rod, carbon brush, carbon felt, reticulated vitreous carbon, and graphene, have been utilized in MFC. However, low porosity and mechanical strength (brittleness) are the major shortcomings of carbon-based materials.

The carbon brush is a preferred anode due to its unique surface morphology and excellent conductive properties. The carbon brush (resistivity of 15.5 μΩm) looks like a twisted wire brush in which the carbon wires are bound together with a curled titanium wire. The entangled structure of the carbon brush facilitates a large surface area for biofilm growth, whereas the conducting and antimicrobial properties of titanium help in the transfer of an electron from the anode to the external circuit. Hence, the carbon brush shows superior performance compared to other carbon-based anodes. A comparative study between carbon brush and carbon cloth was investigated by Sayed et al. (2020). They reported that the carbon brush showed 72% higher power density (maximum) compared to that of carbon cloth (projected surface area of 7.0 cm^2). The loading of conducting fabrics in carbon brush plays an important role. However, recently, the modification of carbon brush with sodium alginate and polyaniline showed a higher power density (1.4 times) and coulombic efficiency (1.14 times) compared to the former one (Wang et al. 2020a). Similarly, Wang et al. (2020b) investigated the electrochemical properties of anode comprised of carbon brush, polypyrrole, titanium nitride, and carboxymethyl cellulose hydrogel. They observed a higher biomass growth on the modified anode, which resulted in higher power density.

Carbon paper, carbon rods, graphite plates, carbon cloth, and granular activated carbon are some of the simple structures that are generally used as anode materials in MFCs. Among these, carbon cloth is generally preferred in MFC system as it provides a large surface area along with a porous structure for biofilm growth. Additionally, its high electrical conductivity, better mechanical strength, and flexibility make it more suitable than carbon paper. Both carbon paper and carbon cloth have common demerits such as high cost and fouling properties. Similarly, carbon felt and carbon veil are two interesting conducting materials, which are used in MFCs. The carbon felt looks like a fabric blanket whereas the appearance of a carbon veil is a thin ply of continuous fibers that are twisted around to form a tubular shape. Both have high porosity and electrical conductivity. However, the high cost is a major concern for both materials. Moreover, granular activated carbon is a low-cost and biocompatible conducting material. However, the microporous nature of the activated carbon and the interparticle void hinder the electron transfer performance. Matsena et al. (2021) developed a composite anode by attaching activated carbon (0.6–1.1 mm diameter) deposited on carbon cloth. They found that the anode having activated carbon of diameter 0.6 mm showed higher power density. However, the above anode material offered less biomass attachment due to high porosity. Therefore, the size of activated carbon should be optimized while using it as the anode in MFC.

9.5.2 Metal-Based Anodes

Apart from carbon-based materials, metals including gold, nickel, aluminum, stainless steel, iron, titanium, etc., are utilized in MFCs. Despite good conductivity, poor bacterial adhesion is the major limitation of the above materials. Moreover, corrosion affinity, long-term usability, and biocompatibility are the key concerns for the utilization of metal-based materials as anode in MFCs. Recently, modification of these materials via chemical oxidation, ammonia treatment, and surface modification was studied to mitigate the above problems. For instance, the electrochemical deposition of polypyrene on stainless steel mesh enhanced the power density by 29 times higher compared to the unmodified stainless-steel anodes (Pu et al. 2018). In addition, improvements in corrosion resistance and capacitance were found in the above case. In another study, the biocompatibility and conductivity of titanium, stainless steel, and copper sheets were significantly enhanced by the deposition of carbon nanoparticles made up of candle soot (Pu et al. 2022). They reported that the maximum power densities were increased by 18, 29, and 986 times for stainless steel, copper, and titanium sheets, respectively. The long-term stability of titanium electrodes is a great concern despite a remarkable voltage of 5.839 V obtained from an MFC using vegetable oil wastewater (Firdous et al. 2018). Moreover, the long-term usability of the metal-based anode was investigated by developing a hydrogel bioanode composed of titanium dioxide–alginate–polyaniline–graphite. It was found that the composite anode showed a greater power density and stability during the continuous operation of MFC.

9.5.3 Biowaste-Based Anode

The high manufacturing cost of carbon-based and metal-based anodes has promoted the exploration of alternative, less expensive conducting materials. Moreover, the endorsement of green energy in developed countries induces the use of recyclable and renewable sources for anode development. In this direction, biomass waste-derived materials have gained superior attraction in the last few years. Additionally, the researchers have extensively investigated the electrochemical and mechanical properties of the biowaste due to its carbon-rich composition, abundance, low cost, and eco-friendly characteristics. Various biowaste materials including coconut shells, waste papers, bamboo, corn straw, banana peel, palm kernel shells, sacred fig leaves, almond shells, cashew nut shells, wooden waste, etc., have been studied to fabricate anode for fuel cells. The employment of biomass in the direction of anode development mainly depends upon local availability. Moreover, several developed countries are trying to find alternative (i.e., renewable) energy sources for sustainability and economic purposes. For instance, Malaysia produces approximately 20 million metric tons of palm oil every year. Palm oil waste materials are rich in cellulose, hemicellulose, and lignin, which can be utilized to prepare conducting and biocompatible electrodes. The graphene oxide derived from palm oil waste was investigated as an electrode in MFC (Yaqoob et al. 2022). The modified electrode resulted in 0.148 mW/m^2 of power density and 39.47 mA/m^2 of current density. Moreover, the morphological analysis of the graphene electrode showed excellent bioaffinity toward microorganisms. However, the modification of graphene electrodes with metal oxide showed significant improvement in the electrochemical performance of the MFC system.

Biochar and activated carbon have gained significant attention in the fabrication of electrodes due to their low preparation cost, abundance of raw materials, and remarkable biocompatibility. The porous nature, high carbon content, and biocompatibility have been a great advantage in the fabrication of electrodes for use in MFC. Idris et al. (2023) discussed the preparation of biochar from various biowastes and their application for anode fabrication. Various biowastes including leaves, fruits, microalgae, etc., have been utilized for anode preparation purposes. Hung et al. (2019) prepared an anode based on the activated carbon prepared from coffee waste and used in MFC. They found that the prepared activated carbon-based anode (3927 mW/m^2)

showed four times higher power density compared to the commercial activated carbon (975 mW/m^2). Similarly, other researchers have investigated the efficacy of activated carbon or biochar prepared from sewage sludge, wheat straw, fruit peels, corncob, shells, etc. (Li et al. 2020; Moradian et al. 2022).

9.6 Development of Cathode Materials

Bioelectricity generation in MFC also depends upon the nature of the cathode materials, as they play a key role in oxygen reduction reaction (ORR). Some major classifications of cathode materials are air-breathing cathodes, aqueous air cathodes, and biocathodes. Although the materials that have been used as an anode can be utilized as a cathode in an MFC, some crucial parameters must be considered while choosing the cathode material. For example, catalyst coating on the cathode material influences the performance of MFC, although it is not always recommended due to its high cost.

Various materials including carbon cloth, graphite plate, graphite rod, carbon paper, graphite felt, etc., have been used as air-breathing and aqueous air cathodes in MFCs without any additional catalysts. One side (facing the air) of the air-breathing cathode is coated with binder and catalyst, while the other side (facing the electrolyte) remains without catalysts. Similarly, the aqueous air cathode contains a conducting surface coated with binder and catalyst. Generally, polytetrafluoroethylene (PTFE), polyvinyl alcohol (PVA), polyvinylidene fluoride (PVDF), and Nafion are used as binders in MFC. In the meantime, carbon black, activated carbon, biochar powder, etc., are utilized as catalysts. For instance, Simeon et al. (2022) studied the effect of four binders (PTFE, PVA, PVDF, and epoxy) on the stainless-steel cathode performance using carbon black as the catalyst. They reported that the PTFE binder showed a higher initial start-up voltage compared to the other binders. However, during long-term operation, PVDF showed 1.33, 3.19, and 6.36 times higher current density than epoxy, PTFE, and PVA, respectively. In an air-breathing cathode, the catalyst layer acts as a semipermeable membrane for the diffusion of oxygen to the cathode side and also acts as a sealant for water. The internal resistance in SCMFC can be decreased with 20% (w/w) of PTFE binding (Song et al. 2019). Various researchers have reported the use of Pt-coated air cathode for better performance.

However, the Pt catalyst-coated cathode bears more than 45% of the fixed cost which is mostly unacceptable for larger applications. The MFC associated with a non-catalyst-coated carbon brush resulted in a maximum power density of 253.84 mW/m^2 due to the high surface area of the carbon electrode (Ye et al. 2020). Activated carbon cathode with PTFE coating showed better performance compared to the activated carbon (without catalyst) electrode. Therefore, modification of cathode materials is required to increase the ORR kinetics. In this direction, the chemical treatment method may be useful, as the conversion of the metal valence state accelerates the electron transmission and improves the cathode performance. Moreover, the cathode materials have been treated with phosphoric acid (H_3PO_4), nitric acid

(HNO_3), potassium hydroxide (KOH), and hydrogen peroxide (H_2O_2) to increase the electrochemical stability.

9.7 Development of Proton Exchange Membranes

The proton exchange membrane plays a vital role in MFC as it separates the anolyte from the catholyte and restricts the diffusion of oxygen. The characteristics including proton conductivity, ion exchange capacity, water uptake, swelling ratio, oxygen, and substrate diffusivity are the major engineering parameters of a proton exchange membrane. Previously, agar–agar salt bridge, filtration membranes, Nafion 117, and Nafion 112 were used in MFCs for bioelectricity generation and wastewater treatment. Low proton conductivity, high biofouling, and poor durability are the limitations of agar–agar salt bridges and filtration membranes. Nafion membranes are mostly preferred due to their excellent proton conductivity, less substrate diffusivity, and better performance. Nafion is mostly composed of a polytetrafluoroethylene (PTFE) backbone containing ether and sulfonic groups. However, the high cost of Nafion membrane induces the exploration of alternative proton exchange membranes for use in MFC. In this direction, sulfonated polymeric membranes, polymer-based membranes, sulfonated poly(ether ether ketone) (SPEEK)-based composite membranes, and ceramic-based membranes have been fabricated and used in biofuel systems (Ekambaram et al. 2021). The properties including water uptake capacity, ion exchange capacity, and swelling ratio of different membranes are summarized in Table 9.2.

9.7.1 Nafion Membrane

Nafion membranes are classified as perfluorinated sulfonic acid membranes comprised of a PTFE backbone. They have various advantages including chemical stability, excellent proton conductivity, good water uptake, and ion exchange capacity. Despite good performance, high cost and nonbiocompatibility hinder their long-term use. Hence, the addition of various polymers and inorganic components may help overcome these issues.

Further, Fan et al. (2020) modified the Nafion membrane using polyvinylidene fluoride (PVDF) and acetone. The blending of PVDF and acetone significantly improved the water absorption capacity, which ultimately increased the ion conductivity of the membrane. Moreover, the modified membrane resulted in a maximum voltage and COD removal of 96.3% and 75.1%, respectively, compared to the raw Nafion membrane. In addition, the modification of the Nafion membrane with polyvinyl alcohol and borosilicate was carried out to fabricate a low-cost membrane with performance equal to that of the Nafion membrane (Das et al. 2021). The study

Table 9.2 Analysis of various physical properties of the prepared membranes

Types of membrane	IEC (mmol/g) or proton conductivity (mS/cm)	WUC (%)	SR (%)	References
SPEEK	1.63 mS/cm	15.85	–	Venkatesan and Dharmalingam (2015)
Chondroitin sulfate	1.0 mS/cm	160	–	Erenler et al. (2024)
PVA/Nafion/ borosilicate	70 mS/cm	70	105	Tiwari et al. (2016)
Graphite oxide/ PVA/styrene	0.33 mmol/g	50	10.5	Rudra et al. (2017)
PVA/sulfosuccinic acid	1.34 mmol/g	70.8	–	Hou et al. (2014)
PVA/TiO_2/Si	0.81 mmol/g	60.9	–	Bhowmick et al. (2020)
SPVA/SCS	0.209 mmol/g	268.13	20	Atkar et al. (2024)
ACCS/clay	2.22 mmol/g	19.4	1.9	Neethu et al. (2019)
Ceramic blended with rice husk	0.091 mmol/g	29.28	–	Raychaudhuri and Behera (2020)
Nafion 117	0.92 mmol/g	17.1	9.3	Kumar et al. (2016)

SPEEK Sulfonated poly(ether ether ketone), *IEC* Ion exchange capacity, *WUC* Water uptake capacity, *SR* Swelling ratio, *SPVA* Sulfonated polyvinyl alcohol, *SCS* Sulfonated chitosan, *ACCS* Activated carbon derived from coconut shell

revealed that the biocompatibility and hydrophilicity of the PVA polymer matrix showed good performance during operation.

9.7.2 *Sulfonated Polymeric Membrane*

Due to the high cost of Nafion membranes, non-fluorinated membranes have become quite popular among researchers. It was found that sulfonation of the membrane adds up sulfonate groups into the polymer matrix by electrophilic substitution mechanisms. Moreover, the low cost and easy process of sulfonation make it more popular. Further, the sulfonate group increases the hydrophilicity characteristic of the membrane, which ultimately enhances the proton conductivity. In this direction, sulfonated poly(ether sulfone) (SPES) and sulfonated poly(ether ether ketone) (SPEEK) membranes have been extensively studied in the MFC system. The previous research work showed that SPEEK- and SPES-based membranes outperformed Nafion membranes in terms of electrical performance and chemical stability. However, the performance of the SPEEK membrane mainly depends upon the degree of sulfonation. Earlier, it was found that COD removal, power density, and coulombic efficiency significantly improved to a certain degree of sulfonation. With further rise in the degree of sulfonation, the deterioration of the membrane performance resulted in the release of unstable cations into the system (Li et al. 2024). Hence, to overcome the above issue, several attempts were carried out by doping the membrane with other inorganic components.

The SPEEK–metal oxide-based composite membranes have been fabricated and employed in MFCs. In this direction, various metal oxides such as TiO_2, zeolite, SiO_2, sulfated zirconia titania, etc., have been utilized for the composite membrane preparation (Vidhyeswari et al. 2021; Ekambaram et al. 2021). The fabricated membranes show greater power density compared to raw SPEEK and Nafion membranes due to a further increase in sulfonated functional groups. In the meantime, the high doping of metal oxides blocks the pores of the membrane, which decreases the membrane characteristics.

9.7.3 Polymer-Based Membrane

The search for alternative low-cost membranes has made the researchers explore various synthetic and natural polymers for membrane preparation. The thin-film formation ability and compatibility are the major advantages of polymeric membranes. In this direction, polyvinyl alcohol, sodium alginate, chitosan, cellulose, etc., have been investigated. Despite their low cost and eco-friendly nature, low proton conductivity remains a major issue. Hence, the fabrication of membranes by either blending two polymers or with other components is mostly preferred nowadays. Previously, Musa et al. (2022) prepared polymeric membranes of chitosan–alginate and chitosan–PVA, crosslinked with plasticizers, for wastewater treatment in MFCs. They reported that the former increased the power density and voltage compared to the chitosan–PVA membrane. In addition, the dual-chamber MFC associated with the chitosan–alginate membrane offered a maximum voltage and power density of 355 mV and 365 mW/m^3, respectively. Rosli et al. (2020) reviewed membranes made of chitosan combined with other chemicals and their characteristics for applications in various energy devices. They reported that the composite membrane of chitosan-silica oxide-polyaniline membranes showed a maximum water uptake capacity of 72.87% and proton conductivity of 8.39×10^{-3} S/cm. Moreover, sulfonation of various polymers enhanced the proton conductivity.

9.7.4 Ceramic-Based Membranes

Researchers have found that ceramic materials also possess crucial properties such as proton exchange capacity, water uptake capacity, and chemical and thermal stability. However, the abovementioned properties solely depend upon the type, chemical composition, porosity, and thickness of the soil. Hence, these factors must be considered before the fabrication of a cost-effective ceramic-based membrane. The presence of silica oxide (SiO_2) in the ceramic material is mainly responsible for proton conduction. Besides, the crystallinity of the clay soil also influences the performance of the membrane. Previously, Cheraghipoor et al. (2022) explored the performance of a ceramic membrane composed of zirconium orthosilicate ($ZrSiO_4$). The maximum current density (440 mA/m^2) and coulombic efficiency (69%) were obtained at a 10% loading of $ZrSiO_4$. Additionally, they reported that the present ceramic-based membrane could produce 10 times higher power density than native soil-based membrane. Similarly, other ceramic materials, i.e., montmorillonite, kaolinite, vermiculite, illite, mullite, etc., have been employed for the fabrication of membranes for use in MFC (James 2022; Rao et al. 2023; Jadhav et al. 2022).

9.8 Conclusions and Future Perspectives

The microbial fuel cell (MFC) technology has become an emerging research topic because of its environmentally friendly nature and energy production ability. The continuous harvesting of bioelectricity along with wastewater treatment from MFC requires a cost-effective and efficient system. The major hindrance includes the initial energy requirement to start the system. Recent studies have focused on reducing or eliminating the need for external power during MFC operation, which is highly beneficial for sustaining large-scale MFC processes over extended periods. There is growing interest in applying MFC technology to wastewater treatment due to its potential for energy recovery and reduction of excess sludge from waste. The COD removal efficiency and electricity production are significantly influenced by factors like wastewater concentration and composition. However, MFCs using real or complex wastewater as a substrate often face the issue of low coulombic efficiency. This may be due to the diversion of some electrons toward suspended microorganisms (for example, through fermentation and methanogenesis).

Comparatively, the traditional active sludge treatment system requires 30 times less capital cost than MFC due to the high fixed cost (i.e., cost of electrodes and membranes), complex design, fuel crossover, membrane fouling, and high internal resistance. Hence, lowering the operational cost of MFC must be employed for real-life applications and commercialization. The usual approach for scaling up an MFC system involves stacking individual MFC units and ensuring a continuous supply of wastewater. Among several research works, a few laboratory scale designs such as tubular or flat designs with continuous flow have been successfully adopted for scale-up operation. To ensure steady and robust performance by MFC, real-time and online monitoring of process parameters is essential. The key challenges encompass activation losses, mass transfer losses, and ohmic losses. Additionally, there are concerns regarding substrate crossover, associated costs, and technical barriers related to robustness and stability. In conclusion, the future of advanced integrated MFC systems holds great promise as reliable sources of energy creation and wastewater treatment, with the added benefit of recovering and reusing valuable nutrients from the treated wastewater. Ongoing research and development in these areas are vital to realizing the full potential of MFCs in sustainable energy generation and wastewater treatment applications.

References

Agrahari S, Kumar S (2024) Emerging and futuristic phyto-technologies for sustainable wastewater treatment with resource recovery and economical aspects. J Water Process Eng 105753

Apollon W, Rusyn I, González-Gamboa N, Kuleshova T, Luna-Maldonado AI, Vidales-Contreras JA, Kamaraj SK (2022) Improvement of zero waste sustainable recovery using microbial energy generation systems: a comprehensive review. Sci Total Environ 153055

Arkatkar A, Mungray AK, Sharma P (2019) Effect of treatment on electron transfer mechanism in microbial fuel cell. Energy Sour Part A: Recov Utiliz Environ Effects 1–16

Atkar A, Sridhar S, Deshmukh S, Dinker A, Kishor K, Bajad G (2024) Synthesis and characterization of sulfonated chitosan (SCS)/sulfonated polyvinyl alcohol (SPVA) blend membrane for microbial fuel cell application. Mater Sci Eng B 116942

Awad, M, Said, A, Saad, MH, Farouk, A, Mahmoud, MM, Alshammari, MS, Alghaythi, ML, Aleem, SHEA, Abdelaziz, AY, Omar AI (2024) A review of water electrolysis for green hydrogen generation considering PV/wind/hybrid/hydropower/geothermal/tidal and wave/biogas energy systems, economic analysis, and its application. Alexandr Eng J 213–239

Bhowmick GD, Dhar D, Ghangrekar MM, Banerjee R (2020) TiO_2-Si-or $SrTiO_3$-Si-impregnated PVA–based low-cost proton exchange membranes for application in microbial fuel cell. Ionics 6195–6205

Chen S, Patil SA, Brown RK, Schröder U (2019) Strategies for optimizing the power output of microbial fuel cells: transitioning from fundamental studies to practical implementation. Appl Energy 15–28

Cheraghipoor M, Mohebbi-Kalhori D, Noroozifar M, Maghsoodlou MT (2022) Enhancing the efficiency of ceramic native soil membrane using Zircon in a continuous microbial fuel cell for wastewater treatment and sustainable energy. J Environ Chem Eng 108255

Cristiani P, Gajda I, Greenman J, Pizza F, Bonelli P, Ieropoulos I (2019) Long term feasibility study of in-field floating microbial fuel cells for monitoring anoxic wastewater and energy harvesting. Front Energy Res 119

Das B, Gaur SS, Katha AR, Wang CT, Katiyar V (2021) Crosslinked poly (vinyl alcohol) membrane as separator for domestic wastewater fed dual chambered microbial fuel cells. Int J Hydro Energ 7073–7086

Dessie Y, Tadesse S, Eswaramoorthy R (2020) Review on manganese oxide based biocatalyst in microbial fuel cell: nanocomposite approach. Mater Sci Energy Technol 136–149

Dwivedi KA, Huang SJ, Wang CT (2022) Integration of various technology-based approaches for enhancing the performance of microbial fuel cell technology: a review. Chemosphere 132248

Ekambaram R, Paraman V, Raja L, Suresh MK, Dharmalingam S (2021) Design and development of electrospun SPEEK incorporated with aminated zirconia and curcumin nanofibers for periodontal regeneration. J Mech Behav Biomed Mater 104796

Erenler SA, Unver T, Ozaslan BF, Koytepe S, Sezer S (2024) Development of microbial chondroitin sulfate-based proton exchange membranes for microbial fuel cells. Fuel 130976

Fan L, Shi J, Xi Y (2020) PVDF-modified Nafion membrane for improved performance of MFC. Membranes 185

Firdous S, Jin W, Shahid N, Bhatti ZA, Iqbal A, Abbasi U, Mahmood Q, Ali A (2018) The performance of microbial fuel cells treating vegetable oil industrial wastewater. Environ Technol Innov 143–151

Flimban SG, Ismail IM, Kim T, Oh SE (2019) Overview of recent advancements in the microbial fuel cell from fundamentals to applications: design, major elements, and scalability. Energies 3390

Gupta S, Srivastava P, Patil SA, Yadav AK (2021) A comprehensive review on emerging constructed wetland coupled microbial fuel cell technology: potential applications and challenges. Bioresource Technol 124376

Hassan, M, Kanwal, S, Singh, RS, SA, MA, Anwar, M, Zhao C (2024) Current challenges and future perspectives associated with configuration of microbial fuel cell for simultaneous energy generation and wastewater treatment. Int J Hydro Energy 323–350

Hidayat ARP, Widyanto AR, Asranudin A, Ediati R, Sulistiono DO, Putro HS, Sugiarso D, Prasetyoko D, Purnomo AS, Bahruji H, Ali BTI, Caralin IS (2022) Recent development of double chamber microbial fuel cell for hexavalent chromium waste removal. J Environ Chem Eng 107505

Hou Y, Li K, Luo H, Liu G, Zhang R, Qin B, Chen S (2014) Using crosslinked polyvinyl alcohol polymer membrane as a separator in the microbial fuel cell. Front Environ Sci Eng 137–143

Hung YH, Liu TY, Chen HY (2019) Renewable coffee waste-derived porous carbons as anode materials for high-performance sustainable microbial fuel cells. ACS Sustain Chem Eng 16991–16999

Ibrahim RSB, Zainon Noor Z, Baharuddin NH, Ahmad Mutamim NS, Yuniarto A (2020). Microbial fuel cell membrane bioreactor in wastewater treatment, electricity generation and fouling mitigation. Chem Eng Technol 1908–1921

Idris MO, Guerrero-Barajas C, Kim HC, Yaqoob AA, Ibrahim MNM (2023) Scalability of biomass-derived graphene derivative materials as viable anode electrode for a commercialized microbial fuel cell: a systematic review. Chinese J Chem Eng 277–292

Jadhav DA, Park SG, Eisa T, Mungray AK, Madenli EC, Olabi AG, Abdelkareem MA, Chae KJ (2022) Current outlook towards feasibility and sustainability of ceramic membranes for practical scalable applications of microbial fuel cells. Renew Sustain Energy Rev 112769

James A (2022) Ceramic-microbial fuel cell (C-MFC) for waste water treatment: a mini review. Environ Res 112963

Kumar V, Kumar P, Nandy A, Kundu PP (2016) Fabrication of laminated and coated Nafion 117 membranes for reduced mass transfer in microbial fuel cells. RSC Adv 21526–21534

Li J, Cui N, Liu D, Zhao Z, Yang F, Zhong J, Pang J (2024) SPEEK-co-PEK-x proton exchange membranes with controllable sulfonation degree for proton exchange membrane fuel cells. Int J Hydro Energy 606–617

Li M, Li YW, Cai QY, Zhou SQ, Mo CH (2020). Spraying carbon powder derived from mango wood biomass as high-performance anode in bio-electrochemical system. Bioresource Technol 122623

Matsena MT, Mabuse M, Tichapondwa SM, Chirwa EM (2021) Improved performance and cost efficiency by surface area optimization of granular activated carbon in air-cathode microbial fuel cell. Chemosphere 130941

Moradian JM, Wang S, Ali A, Liu J, Mi J, Wang H (2022). Biomass-derived carbon anode for high-performance microbial fuel cells. Catalysts 894

Musa MT, Shaari N, Kamarudin SK, Wong WY (2022) Recent biopolymers used for membrane fuel cells: Characterization analysis perspectives. Int J Energy Res 16178–16207

Myung Y, Jung S, Tung TT, Tripathi KM, Kim T (2019) Graphene-based aerogels derived from biomass for energy storage and environmental remediation. ACS Sustain Chem Eng 3772–3782

Nawaz A, Hafeez A, Abbas SZ, Haq IU, Mukhtar H, Rafatullah M (2020) A state of the art review on electron transfer mechanisms, characteristics, applications and recent advancements in microbial fuel cells technology. Green Chemistr Lett Rev 365–381

Nawaz A, Haq IU, Qaisar K, Gunes B, Raja SI, Mohyuddin K, Amin H (2022) Microbial fuel cells: insight into simultaneous wastewater treatment and bioelectricity generation. Process Saf Environ Protect 357–373

Neethu B, Bhowmick GD, Ghangrekar MM (2019) A novel proton exchange membrane developed from clay and activated carbon derived from coconut shell for application in microbial fuel cell. Biochem Eng J 170–177

Nguyen HT, Min B (2020) Using multiple carbon brush cathodes in a novel tubular photosynthetic microbial fuel cell for enhancing bioenergy generation and advanced wastewater treatment. Bioresour Technol 123928

Opeyemi BM (2021) Path to sustainable energy consumption: the possibility of substituting renewable energy for non-renewable energy. Energy 120519

Prathiba S, Kumar PS, Vo DVN (2022) Recent advancements in microbial fuel cells: a review on its electron transfer mechanisms, microbial community, types of substrates and design for bio-electrochemical treatment. Chemosphere 131856

Pu KB, Gao JY, Cai WF, Chen QY, Guo K, Huang Y, Gao SH, Wang YH (2022) A new modification method of metal substrates via candle soot to prepare effective anodes in air-cathode microbial fuel cells. J Chem Technol Biotechnol 189–198

Pu KB, Ma Q, Cai WF, Chen QY, Wang YH, Li FJ (2018). Polypyrrole modified stainless steel as high performance anode of microbial fuel cell. Biochem Eng J 255–261

Ramalingam G, Pachaiappan R, Kumar PS, Dharani S, Rajendran S, Vo DVN, Hoang TK (2022) Hybrid metal organic frameworks as an exotic material for the photocatalytic degradation of pollutants present in wastewater: a review. Chemosphere 132448

Rao A, Kaushik A, Kuppurangan G, Selvaraj G (2023) Characterization and application of novel fly ash blended ceramic membrane in MFC for low-cost and sustainable wastewater treatment and power generation. Environ Sci Pollut Res 45872–45887

Raychaudhuri A, Behera M (2020) Ceramic membrane modified with rice husk ash for application in microbial fuel cells. Electrochimica Acta 137261

Rosli NAH, Loh KS, Wong WY, Yunus RM, Lee TK, Ahmad A, Chong ST (2020) Review of chitosan-based polymers as proton exchange membranes and roles of chitosan-supported ionic liquids. Int J Mol Sci 632

Rudra R, Kumar V, Pramanik N, Kundu PP (2017) Graphite oxide incorporated crosslinked polyvinyl alcohol and sulfonated styrene nanocomposite membrane as separating barrier in single chambered microbial fuel cell. J Pow Sourc 285–293

Saravanan A, Kumar PS, Srinivasan S, Jeevanantham S, Kamalesh R, Karishma S (2022) Sustainable strategy on microbial fuel cell to treat the wastewater for the production of green energy. Chemosphere 133295

Sayed ET, Alawadhi H, Elsaid K, Olabi AG, Almakrani MA, Tamim STB, Alafranji GHM, Abdelkareem MA (2020) A carbon-cloth anode electroplated with iron nanostructure for microbial fuel cell operated with real wastewater. Sustainability 6538

Shah, WUH, Hao, G, Yan, H, Zhu, N, Yasmeen, R, Dincă G (2024) Role of renewable, non-renewable energy consumption and carbon emission in energy efficiency and productivity change: evidence from G20 economies. Geosci Front 101631

Simeon IM, Herkendell K, Pant D, Freitag R (2022) Electrochemical evaluation of different polymer binders for the production of carbon-modified stainless-steel electrodes for sustainable power generation using a soil microbial fuel cell. Chem Eng J Adv 100246

Song BY, He Y, He YL, Huang D, Zhang YW (2019) Experimental study on anode components optimization for direct glucose fuel cells. Energy 15–22

Swain G, Maurya KL, Sonwani RK, Singh RS, Jaiswal RP, Rai BN (2022) Effect of mixing intensity on biodegradation of phenol in a moving bed biofilm reactor: process optimization and external mass transfer study. Bioresour Technol 126921

Swain G, Dash T, Srivastava Y, Kumar R, Dutta S (2024) Development and application of modified biocarrier immobilized with Bacillus sp. for azo dye degradation: batch and continuous study. Bioremed J 1–13

Tiwari BR, Noori MT, Ghangrekar MM (2016) A novel low cost polyvinyl alcohol-Nafion-borosilicate membrane separator for microbial fuel cell. Mater Chemistr Phys 86–93

Venkatesan, PN, Dharmalingam S (2015) Effect of cation transport of SPEEK–Rutile TiO_2 electrolyte on microbial fuel cell performance. J Membr Sci 518–527

Vidhyeswari D, Surendhar A, Bhuvaneshwari S (2021) Evaluation of power generation and treatment efficiency of dairy wastewater in microbial fuel cell using TiO2–SPEEK as proton exchange membrane. Water Sci Technol 3388–3402

Vidhyeswari D, Surendhar A, Bhuvaneshwari SS (2022) General aspects and novel pems in microbial fuel cell technology: a review. Chemosphere 136454

Wang C, Xing Y, Zhang K, Zheng H, Zhang YN, Zhu X, Yuan X, Qu J (2023) Evaluation of photocathode coupling-mediated hydroxychloroquine degradation in a single-chamber microbial fuel cell based on electron transfer mechanism and power generation. J Power Sour 232625

Wang Y, Wen Q, Chen Y, Li W (2020a) Conductive polypyrrole-carboxymethyl cellulose-titanium nitride/carbon brush hydrogels as bioanodes for enhanced energy output in microbial fuel cells. Energy 117942

Wang Y, Wen Q, Chen Y, Zheng H, Wang S (2020b) Enhanced performance of microbial fuel cell with polyaniline/sodium alginate/carbon brush hydrogel bioanode and removal of COD. Energy 117780

Wang, Y, Wang, Z, Kong, X, Song, Y, Tian, Y, Lin J (2024) Capacitive and biocompatibility composite anode material for enhanced renewable energy conversion for MFCs. Fuel 132736

Yaqoob AA, Ibrahim MNM, Guerrero-Barajas C (2021) Modern trend of anodes in microbial fuel cells (MFCs): an overview. Environ Technol Innov 101579

Yaqoob AA, Ibrahim MNM, Yaakop AS, Rafatullah M (2022) Utilization of biomass-derived electrodes: a journey toward the high performance of microbial fuel cells. Appl Water Sci 99

Ye Y, Ngo HH, Guo W, Chang SW, Nguyen DD, Zhang X, Zhang S, Luo G, Liu Y (2020) Impacts of hydraulic retention time on a continuous flow mode dual-chamber microbial fuel cell for recovering nutrients from municipal wastewater. Sci Total Environ 139220

Yuvraj C, Aranganathan V (2018) Configuration analysis of stacked microbial fuel cell in power enhancement and its application in wastewater treatment. Arabian J Sci Eng 101–108

Chapter 10
Electrocoagulation: A Sustainable Approach to Wastewater Treatment

Suvendu Manna

Abstract Electrocoagulation (EC) is an advanced water treatment technology that uses electrochemical reactions to remove a wide range of contaminants present in water, such as heavy metals, organic matter, suspended solids, and bacteria. EC discharges an electrical current through water, which develops coagulant ions from the electrodes, and these electrodes can be made up of aluminum or iron. EC produces coagulant ions, which neutralize contaminants' charges to form flocs. This technology provides a cost-effective and environmentally friendly alternative to traditional water treatment methods by minimizing the use of chemical additives, reducing sludge generation, and improving contaminant removal efficiencies. EC has been used in a variety of industries, including textiles, food processing, metal production, agriculture, and municipal water treatment. Its flexibility and scalability make it suitable for both small-scale decentralized systems and large-scale industrial applications. While it has many benefits, some limitations exist, such as electrode fouling, high energy consumption, and challenges in scaling up. However, recent advancements, such as new-generation electrode materials, hybrid treatment technology, and coupling with renewable energy sources, show promise in overcoming these limitations and further optimizing the sustainability of EC. This chapter offers an in-depth review of the working principles, applications, and future directions of electrocoagulation as a green water treatment technology, emphasizing its potential to address global water quality issues.

Keywords Electrocoagulation · Water treatment · Wastewater treatment · Sustainable technology · Heavy metal removal · Electrochemical processes

---

S. Manna (✉)
Department of Microbiology, Health Sceinces Cluster, SOHST, UPES University, Dehradun, Uttarakhand, India
e-mail: smanna@ddn.upes.ac.in

A. M. Yatoo et al. (eds.), *Wastewater Monitoring and Management by Advanced Approaches*, Advances in Wastewater Research,
https://doi.org/10.1007/978-981-95-2601-7_10

10.1 Introduction

Water is a basic commodity for life, industry, agriculture, and the environment (Jain et al. 2021). As the world's population increases and industrialization gathers momentum, however, the need for safe and clean water has increased further. At the same time, the quality of available water resources has been declining progressively due to pollution, global warming, and bad management practices (Jain et al. 2021). Domestic, industrial, and agricultural contaminants often enter natural water bodies, posing threats to aquatic ecosystems, human health, and water security in general (Kesari et al. 2021). Water pollution is among the world's most significant environmental issues today. Heavy metals, toxic chemicals, organic waste, leachate, and microorganisms often contaminate drinking water sources, leading to health hazards and decreasing the supply of clean water for drinking (Yatoo et al. 2024; Singh et al. 2022). According to the World Health Organization (WHO), waterborne diseases continue to be a leading cause of death worldwide, especially in developing nations. Besides pollution, water shortage is another challenge. The United Nations approximates that two-thirds of the world's population will be under water-stressed conditions by 2025 (WWAP 2012). Climate change, groundwater over-extraction, and poor water management increase the pressure on freshwater. Therefore, efficient water treatment technologies that are able to supply clean water from contaminated sources are key to a sustainable future.

Conventional water treatment technologies, including chemical coagulation, filtration, and biological treatment, have been commonly employed for eliminating contaminants from water (Ahmed et al. 2021; Leiknes 2009). Although these technologies have been successful in numerous applications, they tend to be associated with high energy demands, the production of excessive sludge, and the utilization of harmful chemicals (Rosenwinkel et al. 2001). Additionally, such traditional systems may be costly to run and maintain, especially in remote or resource-constrained settings. There is a need for increasingly sustainable water treatment technologies that not only remove pollutants efficiently but are also economical, energy-saving, and environmentally friendly. The most desirable treatment process should use fewer chemicals, produce less waste, and be easily scalable to serve different communities, industries, and cities. Among the new water treatment technologies, electrocoagulation (EC) is one of the most promising options (Shahedi et al 2020; Akter et al. 2022; Tahreen et al 2020; Boinpally et al 2023; Das et al. 2022; Yan et al. 2024). EC is an electrochemical process in which electric current is used to destabilize water impurities so that they come together in the form of large particles that can be easily removed by sedimentation or filtration (Tegladza et al. 2021). This process uses electrodes, usually made of metals such as aluminum or iron, to discharge coagulants in place without the need for additional chemical additives. The aspect that makes electrocoagulation particularly attractive is its sustainability. Electrocoagulation has low energy requirements, reduces the need for external chemicals, and generates less waste than conventional water treatment technologies (Shah et al. 2024; Phu et al. 2025). It is effective against a wide range of contaminants such as heavy metals, organic contaminants, and microorganisms, and can be used for various water treatment purposes (Zhang et al. 2024; Wang et al. 2025; Boulanouar et al. 2025).

In this chapter, we will explore the principles of electrocoagulation, its effectiveness in removing various water pollutants, its sustainability benefits, and its growing role in modern water treatment systems. We will also examine the technological advancements in EC systems, the challenges associated with their implementation, and the future prospects of this promising technology.

10.2 Understanding Electrocoagulation

Electrocoagulation (EC) is a water treatment technology that uses electricity to cause the coagulation of impurities in water so that they can be removed by physical agglomeration. The technology relies on electrochemical reactions to destabilize suspended solids, heavy metals, and organic contaminants so that they aggregate into larger clumps or "flocs" that can be easily removed by sedimentation or filtration (Tegladza et al. 2021). EC is a robust, effective, and ecologically safe solution for conventional water treatment processes (Das et al. 2023). The electrocoagulation process basically revolves around electrochemistry and coagulation chemistry principles. The process steps primarily include the following reactions:

A. **Electrochemical reactions**:

- When an electric current is conducted through water, electrochemical reactions take place at the cathode and anode surfaces of the electrodes. At the anode, metal ions are released into the water. These metal ions usually originate from electrodes that are made of aluminum (Al) or iron (Fe), which dissolve in the water and serve as coagulants (Fig. 10.1).
- At the cathode, hydroxide ions (OH^-) are produced, increasing the pH of the water, which helps in precipitating other contaminants like heavy metals (Fig. 10.2).

B. **Coagulation and flocculation**:

- The metal ions from the anode (e.g., Al^{3+} or Fe^{2+}) combine with impurities in water, which neutralizes their charge and enables them to come together in masses, referred to as flocs (Fig. 10.1).
- These flocs are dense enough to settle to the bottom of the water tank, where they can be removed easily by sedimentation or filtration. In a few instances, more flocculants can be added to enhance the aggregation process, but most EC systems need no chemical additions.

C. **Electrode reactions**:

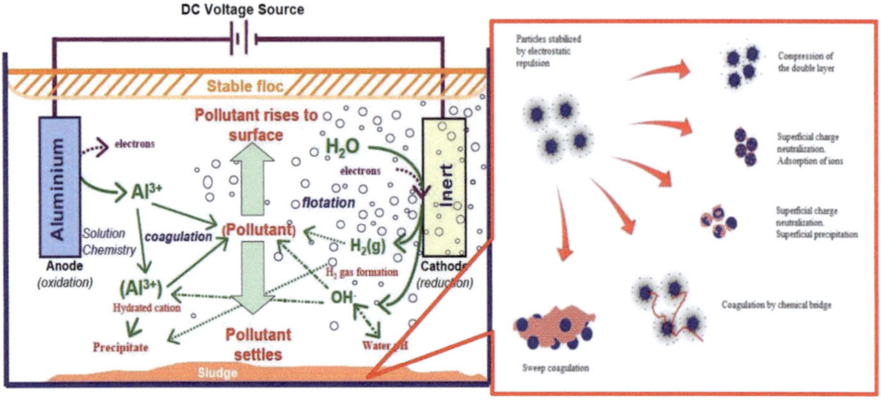

Fig. 10.1 Overall EC method used for wastewater treatment. 2020 Adapted from Shahedi et al. (), with permission, license number 5973440916571

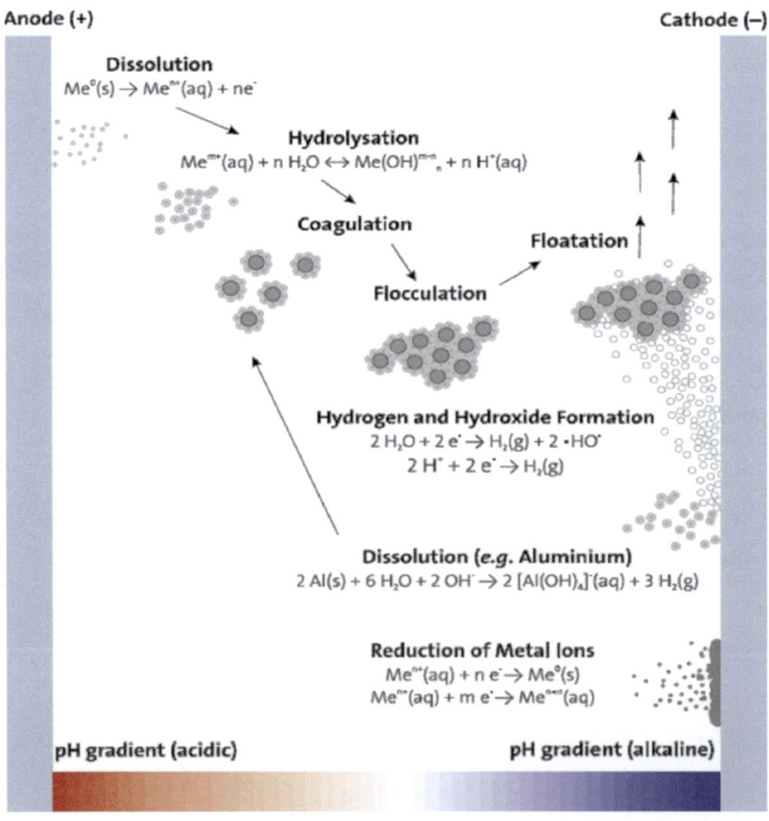

Fig. 10.2 Impact of pH on the EC. 2020 Adapted from Shahedi et al. (), with permission, license number 5973440916571

a. During EC, both anodic and cathodic reactions contribute significantly. Anode reaction tends to be the oxidation of the electrode material, which liberates metal ions to form coagulating agents. The cathode reaction tends to be the reduction of water, with the release of hydrogen gas and hydroxide ions. The latter enhances the alkalinity of water, which assists in precipitating contaminants (see Eqs. 10.1–10.9).

Anode:

$$Fe(s) \rightarrow Fe^{+2}(Aq) + 2e^- \tag{10.1}$$

$$4Fe \rightarrow 4Fe^{+2}(Aq) + 8e^- \tag{10.2}$$

$$Al(s) \rightarrow Al^{+3}(Aq) + 3e^- \tag{10.3}$$

Cathode:

$$2H_2O(1) + 2e^- \rightarrow H_2(g) + 2OH^-(Aq) \tag{10.4}$$

$$8H^+(s) + 8e^- \rightarrow 4H_2(g) \tag{10.5}$$

Overall:

$$Fe(s) + 2H_2O \rightarrow Fe(OH)_2(s) \downarrow +H_2(g) \tag{10.6}$$

$$3H_2O + 3e^- \rightarrow 1.5H_2(g) + 3OH^-(Aq) \tag{10.7}$$

$$3H_2O + Al(s) \rightarrow 1.5H_2(g) + Al(OH)_3(s) \downarrow \tag{10.8}$$

$$4Fe(s) + 10H_2O + 10O_2 \rightarrow 4Fe(OH)_3(s) + 4H_2(g) \tag{10.9}$$

10.3 Types of Electrodes

The selection of electrode material has a major impact on the performance and cost-effectiveness of EC systems. Electrodes play a significant role in different technologies for the treatment of water, such as electrocoagulation, electrolysis, and electrochemical oxidation processes. The major role of electrodes in these systems is the support of redox reactions that eliminate contaminants from water. Various kinds of electrodes are utilized depending on parameters such as conductivity, corrosion resistance, cost, and the type of contaminants being treated. For this purpose, the primary classes of electrodes are anodes, cathodes, and sacrificial electrodes, for which materials are selected based on their resistance to harsh environmental conditions and the efficient treatment.

10.3.1 Anode Electrodes

Anode electrodes are used for oxidation reactions in electrochemical water treatment. The choice of material for anodes is important as they must be corrosion-resistant and, at the same time, have a high electrical conductivity. Titanium is often used as a material for anode electrodes in water treatment. Due to its high corrosion resistance, particularly in salty and acidic conditions, it is suitable for long-term use. Platinized titanium electrodes are also used due to their high electrochemical performance, including stability and conductivity. Although these electrodes are more expensive, they have a long service life and provide effective treatment by stimulating oxidation reactions. Graphite electrodes are another widely accepted option for anodes and have the advantage of being inexpensive and good conductors. However, graphite tends to corrode over time and release carbon particles into the water, which can cause poor water quality. Although graphite is less resistant, it is still widely used in low-cost and small electrocoagulation processes (Sillanpää and Shestakova 2017). Mixed metal oxide (MMO) electrodes are a type of anode electrode used in water treatment. MMO electrodes are made by coating titanium with a metal oxide mixture such as ruthenium or iridium. The electrodes are more resistant to corrosion and tend to be applied in processes that demand higher voltages and currents. MMO electrodes are preferred due to their effectiveness in oxidative water treatment processes like organic compound removal and disinfection. The most critical benefit of MMO electrodes is stable performance over very long durations of time, such that they fit continuous industrial purposes (Ganzoury et al. 2022).

10.3.2 Cathode Electrodes

Cathodes, in turn, execute the reduction reaction. The most frequently used material for cathodes is stainless steel due to its mechanical strength and resistance to corrosion. Stainless steel cathodes are extensively utilized as they are cheaper compared to noble metals, which makes them a suitable choice for large-scale water treatment facilities. Besides stainless steel, graphite can also be utilized for cathodes, particularly in small systems. Copper is another candidate material, with good conductivity and corrosion resistance, but potentially less frequently used than stainless steel in water treatment plants because of the risk of copper ion leaching (Dube et al. 2025).

10.3.3 Sacrificial Electrodes

Sacrificial electrodes are electrodes that corrode over time, releasing ions such as aluminum or iron, which help remove impurities such as suspended solids and heavy metals from the water. They are most commonly used in electrocoagulation systems, where the release of metal ions is responsible for the formation of flocs that trap the impurities and aid their removal. Iron electrodes are particularly efficient in removing a wide range of contaminants, such as arsenic, chromium, and organic substances. Aluminum electrodes, on the other hand, are used in electrocoagulation to discharge aluminum ions that aid in the clumping of impurities and facilitate their subsequent removal by filtration or settling. A major advantage of sacrificial electrodes is that they are inexpensive, although they do need to be replaced from time to time as they wear out. They are particularly effective in treating wastewater with high concentrations of impurities (Sillanpää and Shestakova 2017).

10.4 Influencing Parameters in EC Processes

Electrocoagulation (EC) has found wide application due to its ease of use, high efficiency, and ability to treat a broad group of contaminants ranging from heavy metals and organic contaminants to suspended solids (Dina et al. 2017). Although very useful, the efficiency of electrocoagulation depends on several parameters that need to be optimized for maximum performance. These parameters are electrode material and geometry, current and voltage, pH, treatment time, flow rate, contaminant concentration, and temperature. These factors are discussed in detail in the following sections.

10.4.1 Electrode Material and Configuration

The choice of material used as an electrode is perhaps the most important element in electrocoagulation. Aluminum (Al) and iron (Fe) are most commonly used as electrode materials because they dissolve easily in water and produce coagulants that can be used to remove impurities (Kabdaşlı and Tünay 2023). When a direct current is applied across the electrodes, metal ions are released into the solution, where they react with the impurities and aggregate to form flocs that can be easily removed from the water. The electrode material not only influences the type of coagulant formed but also the dissolution rate of the electrodes and the efficiency of the electrocoagulation process (Das et al. 2024). For example, aluminum electrodes tend to form aluminum hydroxide ($Al(OH)_3$) and iron electrodes tend to form iron hydroxide ($Fe(OH)_3$). The dissolution rate of the electrodes must be regulated, as too rapid dissolution can lead to excessive sludge production and insufficient dissolution can result in insufficient coagulation (Kobya et al. 2003). The electrode arrangement, including electrode spacing and geometry, also affects the electrocoagulation process. Shorter electrode spacing increases the electric field between the electrodes, which improves ion mobility and enables faster coagulation (Ebba et al. 2022). The electrode surface area is also important. A larger surface area provides more active sites for electrochemical reactions, which can make the treatment more effective overall.

10.4.2 Current and Voltage

Voltage and current are of central importance for the process of electrocoagulation, as they influence the speed of electrolysis and the coagulant produced. Higher current densities tend to increase the production rate of the coagulant, which results in impurities being removed at a higher rate (Solak et al. 2009). On the other hand, higher current strength tends to lead to side reactions such as the formation of chlorine gas (from chloride) and higher energy consumption (Das et al. 2022). Therefore, the current must be optimized to avoid unwanted reactions. The voltage, which determines the electrical potential between the electrodes, also influences the energy efficiency of the process. Although higher voltages can produce more coagulants, they can also cause higher energy consumption and higher temperatures, which can have a detrimental effect on the treatment process (Holt et al. 2005). The best operating voltage depends on many variables, such as the type of impurities and water quality, and is usually chosen to minimize effective treatment while maintaining energy efficiency. Another important parameter in electrocoagulation is the current density, a measure of the current applied per unit area of the electrodes. The coagulation rate can be increased by high current densities, but this can lead to high energy consumption and thermal effects that can damage the equipment. To achieve effective coagulation without high energy costs, a suitable current density range must be set (Devlin et al. 2019).

10.4.3 pH of the Water

The pH value of the water has a major influence on the process of electrocoagulation, as it changes the solubility and speciation of the coagulants produced during electrolysis. Both iron hydroxide and aluminum hydroxide work best as coagulants under slightly acidic to neutral pH conditions (Jing et al. 2021). Ideally, the pH is between 5 and 8, a wide range suitable for most electrocoagulation procedures. At high pH values, metal hydroxides precipitate too quickly and can be ineffective in removing impurities (Jing et al. 2021). At low pH values, aluminum and iron become more soluble, leading to the formation of soluble metal ions that are less efficient for coagulation (Syam Babu et al. 2020). In addition, pH also influences the development of gas bubbles during electrolysis, which can affect flocculation and the removal of contaminants. In certain situations, external pH adjustment may be required to ensure optimization of the electrocoagulation process.

10.4.4 Treatment Time

The treatment time or the time of electrocoagulation is another important parameter that determines the efficiency of pollutant removal. The longer the treatment time, the more opportunities coagulants and contaminants have to interact with each other and form larger flocs that are less difficult to remove (Boulanouar et al. 2025). However, a longer treatment time can lead to high energy consumption and the generation of additional sludge, which can make disposal after treatment difficult (Shrestha et al. 2021). The optimal treatment time depends on variables such as contaminant concentration, electrode material, and flow rate. For example, highly contaminated water requires a longer treatment time to achieve satisfactory removal, while low levels of contamination can be optimally treated within a limited time. A balance between energy efficiency and optimum treatment time is crucial for effective treatment.

10.4.5 Flow Rate and Hydraulic Conditions

The flow rate of the water through the electrocoagulation reactor is an important parameter, as it determines the residence time of the water in the reactor and thus the contact time between the coagulants and the impurities. Increased flow rates reduce the residence time, which can lead to insufficient coagulation and lower removal efficiency (Bagastyo 2021). On the other hand, a lower flow rate increases residence time, which can improve coagulation, but can also lead to excessive energy consumption. Hydraulic conditions, including the type of flow (e.g., laminar or turbulent), also influence the coagulation process. A turbulent flow promotes the mixing of coagulants and impurities, which increases the efficiency of the coagulation process. However, excessive turbulence can destabilize the formed flocs, which in turn impairs their settling and removal ability (Hossain et al. 2013). Therefore, to maximize the efficiency of electrocoagulation, the flow rate and hydraulic conditions must be optimized.

10.4.6 Contaminant Concentration and Nature

The level and type of impurities in the water have a major influence on the electrocoagulation process. Increased contaminant concentrations usually require higher current densities and longer treatment times to achieve efficient removal (Bagastyo 2021). Different pollutants can also react differently to electrocoagulation. For example, heavy metals such as copper, lead, and cadmium are usually efficiently removed with iron-based or aluminum-based coagulants, while organic pollutants may require alternative treatment methods (Hossain et al. 2013). The properties of pollutants, such as their charge, size, and solubility, also influence the coagulation process. Charged particles are more likely to coagulate in the presence of coagulants, whereas neutral or uncharged particles may require additional measures to enhance coagulation. Understanding the specific properties of the impurities is important for optimizing electrocoagulation conditions.

10.4.7 Temperature of the Water

The temperature of the water influences the process of electrocoagulation through the speed of the electrochemical reaction and the dissolution of the electrode material. Higher temperatures tend to increase the rate of dissolution of the electrode and the production of the coagulant, making the process more efficient overall (Hossain et al. 2013). But extreme temperatures can lead to unsatisfactory side reactions, including the release of chlorine gas from chloride ions, which can be harmful to the environment and human health (Naje et al. 2017). The best water temperature for electrocoagulation is usually between 20 and 40 °C. At low temperatures, the coagulation rate can slow down, requiring longer treatment times or higher current densities. Conversely, very high temperatures can increase energy consumption and lead to the formation of unwanted by-products (Khaled et al. 2019). Therefore, it is important to maintain a suitable temperature range to ensure proper treatment and reduce operating costs.

10.4.8 Ionic Strength of the Water

The electrical conductivity of water depends on its ionic strength, which also affects the efficiency of the electrocoagulation process. Water with low ionic strength (e.g., deionized or distilled water) has a low conductivity, and a higher voltage or current is required to achieve the same degree of coagulation (Hossain et al. 2013). The addition of salts such as sodium chloride ($NaCl$) or potassium chloride (KCl) increases the ionic strength and enhances the conductivity of the water, making the process of electrocoagulation more effective. However, exclusively high ionic strength can lead to the formation of undesirable by-products, such as chlorine gas, so careful control is necessary.

10.5 Electrocoagulation for Different Water Contaminants

Electrocoagulation is extremely versatile and can be used to treat a wide range of pollutants commonly found in domestic and industrial wastewater. This technology has been shown to effectively remove suspended solids, heavy metals, oils, organic chemicals and microorganisms. Below is an overview of how EC treats various water contaminants.

10.5.1 Removal of Heavy Metals

One of the most remarkable applications of electrocoagulation is the extraction of heavy metals from polluted water. Heavy metals such as lead (Pb^{2+}), mercury (Hg^{2+}), arsenic (As^{3+}), cadmium (Cd^{2+}), and chromium (Cr^{3+}) are dangerous for humans and the environment. They are normally found in industrial wastewater, mining wastewater, and polluted drinking water sources. The conventional techniques for removing heavy metals, including chemical precipitation and adsorption, are often challenged with high costs, inefficiency, and the production of secondary waste. However, EC has some advantages, such as in situ generation of coagulants, efficiency in removal, and reduced need for chemicals (Hossain et al. 2013).

10.5.1.1 Mechanisms of Electrocoagulation for Heavy Metal Removal

In the EC method, an electric current is passed through electrodes, usually made of aluminum (Al) or iron (Fe), which are inserted into the polluted wastewater. The current triggers electrochemical reactions between the anode and cathode. At the anode, the metal electrode is oxidized, releasing metal cations (e.g., Al^{3+} or Fe^{2+}) into the solution. These metal cations are then hydrolyzed to produce metal hydroxides (e.g., $Al(OH)_3$ or $Fe(OH)_3$), which have coagulating properties. These coagulants neutralize the charge of the heavy metal ions and enable their aggregation into larger flocs, which in turn can be removed by sedimentation or flotation (Hossain et al. 2013). Apart from the electrochemical reactions, the EC process generates hydrogen gas at the cathode, which can facilitate the flotation of the resulting flocs, further aiding the removal of the heavy metals. The flocs developed during the process encapsulate heavy metal ions, resulting in effective removal of pollutants from wastewater (Naje et al. 2017). Hydroxide formation and precipitation of metal ions in flocs significantly reduce the concentration of toxic heavy metals in the treated water.

10.5.1.2 Removal Efficiency for Various Heavy Metals

Electrocoagulation has proven to be very efficient in removing various heavy metals from wastewater. The efficiency of removal depends on several factors, including the metal type, initial metal concentration, electrode material, applied voltage, and treatment time (Naje et al. 2017). Research has shown that aluminum and iron electrodes are very efficient at removing metals such as lead, chromium, copper, and cadmium (Karchiyappan et al. 2024). For example, aluminum electrodes are often used for the removal of heavy metals because they have a higher production rate of coagulants, which increases the effectiveness of the EC process (Hossain et al. 2013). In a study by Naje et al. (2017), for example, it was shown that EC is very effective in removing lead (Pb) from simulated wastewater and has a removal efficiency of more than 90% under ideal conditions. It was also found to be highly effective for cadmium (Cd), copper (Cu), and chromium (Cr).

10.5.1.3 Factors Affecting the Removal of Heavy Metals

Some parameters determine the efficiency of electrocoagulation for heavy metal removal. Wastewater pH is among the key factors. EC treatment is best conducted in a narrow pH range since metal hydroxide solubility varies with pH. A higher pH usually increases the formation of metal hydroxides, which assists in heavy metal coagulation and removal (Karchiyappan et al. 2024). But the pH has to be closely regulated because extreme pH values cause metal hydroxides to dissolve, thereby lowering removal efficiency (Hossain et al. 2013). Another essential parameter is the voltage or current density applied. Increased voltage typically leads to more coagulant production and higher rates of heavy metal removal, but can also contribute to greater energy consumption and more electrode wear at high voltage (Naje et al. 2017). The material and the nature of the electrodes also contribute to a large extent. Iron and aluminum electrodes are widely used because they have high reactivity and the ability to produce coagulants. But iron electrode formation of a passive oxide layer can make the electrodes less efficient in the long run, and maintenance will cost more (Hossain et al. 2013).

10.6 Removal of Dyes

The electrocoagulation process is highly effective for treating industrial effluents, especially from the textile, paper, and dye industries, where synthetic dyes are among the most persistent pollutants (Jaiswar and Saroha 2025; Mehellou et al. 2025). It offers several advantages over conventional treatment methods like chemical coagulation, including simplicity, low operating costs, and the ability to treat a wide range of contaminants. In electrocoagulation, an electric current is passed between two metal electrodes—usually aluminum or iron—immersed in wastewater. These electrodes oxidize and release metal cations into the solution. The cations react with dye molecules, causing them to destabilize and agglomerate into larger particles, which can then be removed by settling or flotation. The process also forms hydroxide flocs that adsorb dye molecules, further aiding in their removal. Dye removal via electrocoagulation is influenced by several factors. The dye's properties—such as molecular size, charge, and hydrophobicity—affect removal efficiency. Key operational parameters include pH, current density, electrode material, and electrolysis time. Acidic or neutral pH levels generally enhance the release of metal cations, which are essential for coagulation. Current density controls the rate of ion production and thus the effectiveness of the process. Electrode material also plays a crucial role, as different metals release different cations with varying coagulating properties. For example, iron electrodes release Fe^{2+} and Fe^{3+} ions, while aluminum electrodes release Al^{3+} ions, all of which help destabilize dye particles.

Several studies have shown that electrocoagulation can effectively remove various dyes from wastewater (Cruz et al. 2024; Jegathambal et al. 2024; Jasim et al. 2024). For instance, Khandegar and Saroha (2013) investigated the removal of reactive dyes from textile effluent using aluminum electrodes. They found that up to 98% of the dye was removed, with optimal results at a current density of 15 mA/cm^2 and a treatment time of 60 min. Similarly, Hossain et al. (2013) studied the removal of synthetic dyes like methylene blue and rhodamine B. Their results showed that electrocoagulation not only removed dyes effectively but also reduced chemical oxygen demand (COD) and total suspended solids (TSS). One of electrocoagulation's key advantages is its ability to treat high-strength effluents with minimal chemical input. Unlike traditional coagulation methods, which require large amounts of added chemicals and generate significant sludge, electrocoagulation only consumes the metal electrodes, making it a greener alternative. Its simplicity also makes it suitable for both small-scale and large-scale applications. Electrocoagulation is also effective in treating complex wastewater with high levels of salts or organic matter. Naje et al. (2017) demonstrated its success in removing up to 95% of dyes from textile effluents containing a mix of dyes, salts, and organic contaminants, highlighting its versatility and effectiveness. Despite its advantages, electrocoagulation faces challenges such as energy consumption and electrode wear, which can impact cost-effectiveness. Recent research has focused on optimizing these factors—such as using pulsed or intermittent currents and experimenting with different electrode geometries—to improve the process. Further study in these areas could help make electrocoagulation more viable for large-scale industrial use.

10.7 Sustainability of Electrocoagulation

Electrocoagulation is a highly sustainable water treatment technology due to its low environmental impact, energy efficiency, and minimal chemical use. This section discusses the environmental, economic, and social sustainability aspects of electrocoagulation (EC). EC offers several environmental advantages over traditional water treatment methods. Unlike conventional coagulation, which often relies on large amounts of chemicals like aluminum sulfate or ferric chloride—resulting in significant hazardous sludge—EC generates coagulants on-site. This reduces the need for chemical additives and lowers sludge production and disposal issues. Additionally, EC is highly effective at removing harmful contaminants such as heavy metals, organic pollutants, and suspended solids, which can otherwise pose long-term risks to ecosystems. The process produces minimal waste, and the residual sludge it does create is typically easier to manage and dispose of compared to conventional treatment sludge. From an economic standpoint, electrocoagulation is an economical water treatment technology. The cost of operating EC systems is low because the consumption of chemicals is minimized and the technology is simple. In addition, EC systems are simple to scale to meet various applications, ranging from small-scale rural water treatment facilities to large-scale industrial purposes. Although the initial capital expense of an EC installation may differ based on the installation size and complexity, the long-term chemical cost savings and decreased sludge management make EC a sound economic decision in most instances. Also, the lower energy consumption of EC systems over other high-end treatment technologies (e.g., reverse osmosis and advanced filtration) results in lower operational expenses.

Social sustainability refers to a technology's ability to meet community needs while improving quality of life and reducing social inequalities. Electrocoagulation (EC) is particularly well-suited for rural or remote areas where access to clean water is limited and traditional treatment methods may not be feasible due to infrastructure constraints. EC systems are relatively simple to operate and require minimal technical expertise, making them ideal for decentralized water treatment. In many cases, they can be managed by local communities, empowering them to take control of their own water treatment needs and increasing their self-reliance.

10.8 Recent Advancements in Electrocoagulation Technology

10.8.1 Improved Electrode Materials

Another of the key elements affecting the performance of EC systems is the electrode. The most widely used electrodes in the past have been aluminum and iron, but their performance can be reduced with time because they become fouled, corroded, or scaled. Alternative electrode materials like carbon-based electrodes, composite electrodes, and conductive ceramics are under development to improve the service life and performance of EC systems.

- **Carbon-based electrodes**: These products are corrosion-resistant and fouling-resistant, ensuring longer electrode life and fewer maintenance requirements. They also have increased current efficiency, which can result in more efficient removal of contaminants (Ogedey and Oguz 2024).
- **Composite electrodes**: By combining metals like aluminum with nonmetallic materials (e.g., carbon or graphite), composite electrodes can exhibit improved conductivity, corrosion resistance, and higher mechanical strength, making them ideal for longer-term use in EC systems (Lu et al. 2024).

10.8.2 Electrode Configurations

Research to optimize the electrode design and configuration has improved energy efficiency and treatment performance. Advanced EC systems incorporate different electrode geometries, including parallel plate, cylindrical, and mesh structures, to enhance the distribution of the electric field and increase coagulation rates. New configurations provide more consistent current distribution, which enhances overall treatment effectiveness. Multilayer electrode structures enhance the contact area with water, facilitating better release of ions and enhancing coagulation efficiency (El-Naggar et al. 2025; Gök and Gülyaşar 2025).

10.8.3 Integration with Renewable Energy Sources

The merging of EC systems with renewable power sources, including solar, wind, and hydropower, represents a significant advancement that increases the technology's sustainability. Solar-powered EC systems are especially suitable for use in far-off or off-grid areas, minimizing reliance on conventional power grids and bringing water treatment to remote communities. Wind power has also been considered for application in EC systems, particularly where there are abundant wind resources. Small EC units can be energized by solar panels, keeping the process cost-effective and reliable for rural areas and remote zones, where power access might not be readily available. Where winds are consistent in the region, wind turbines generate power to provide electricity for running EC systems, which is an off-the-grid and sustainable system for water filtration (Sanjay and Savithiri 2025; Mohammed and Abdulhameed 2025; Biswas et al. 2025; Biswas Çetinkaya 2025).

10.8.4 Automated Control Systems and Smart Monitoring

Automation and remote monitoring technologies have changed the way EC systems operate and are maintained. Intelligent monitoring systems and smart sensors can monitor parameters like pH, electrical conductivity, current density, and water flow rates in real time. Operators can modify settings automatically to improve performance and provide consistent quality treatment. Internet of Things (IoT)-integrated sensors can transmit information to cloud-based systems, enabling remote monitoring and control of EC units. This technology increases the responsiveness and efficiency of the treatment process. Machine learning algorithms can forecast system behavior, optimize energy usage, and identify faults or inefficiencies, making EC systems more reliable and easier to operate (Öztel et al. 2025; Mallick and Poddar 2025).

10.9 Challenges and Future Prospects

Despite its many advantages, electrocoagulation faces several challenges that need to be addressed to make the technology more widely adopted. Below, we discuss the primary challenges and explore potential solutions.

10.9.1 Challenges in Electrocoagulation

- **Electrode fouling and corrosion**: With time, electrodes in EC systems can become contaminated or corroded, lowering their efficiency. Replacing and maintaining them regularly is expensive and time-consuming. Work on corrosion-resistant materials and methods of regenerating electrodes continues, though.
- **Energy consumption**: Although EC systems are energy-efficient, large-scale use can still involve a lot of energy input. Maximizing the electrical current and incorporating renewable energy sources (e.g., solar or wind) could reduce this problem.
- **Scaling up for large-scale applications**: Whereas EC works very well in small-scale to medium-scale operations, expanding the technology for application at large municipal wastewater treatment plants or industrial processes poses challenges. There is a need for more studies on enhancing electrode design, managing flow rates, and energy efficiency for large-scale EC systems.
- **Sludge management**: While EC produces less sludge than traditional approaches, treating the residual sludge may also prove to be a problem in high-contaminant-load applications. Having effective sludge recycling or disposal technologies could alleviate the situation.

10.9.2 Future Prospects of Electrocoagulation

- **Hybrid systems**: Hybridization of electrocoagulation with other treatment technologies, including activated carbon filtration, reverse osmosis, or advanced oxidation processes, may provide an improved overall performance of water treatment systems, especially for multiplex wastewater streams.
- **Nanotechnology and advanced materials**: The inclusion of nanomaterials, such as nanostructured electrodes or nanocatalysts, can enhance the functioning of EC systems by improving coagulation efficiency, increasing surface area, and reducing energy requirements.
- **Decentralized and off-grid applications**: With the increasing need for decentralized water treatment solutions, EC presents a sustainable option for rural, off-grid, and remote communities. Future research could be on low-cost, solar-powered EC units that produce safe drinking water for underserved communities.

10.10 Conclusion

Electrocoagulation is a creative, sustainable, and efficient process for water treatment that has the potential to overhaul the method with which we respond to water contamination and scarcity. With its capabilities to remove all manner of impurities, ranging from heavy metals to organic substances, suspended matter, and microbes, EC has emerged as an adaptable and environment-friendly technology to be applied across industrial and municipal sectors. By minimizing the use of chemical coagulants, reducing waste generation, and involving relatively low energy input, electrocoagulation is well in line with the principles of sustainability. It is especially useful in remote and off-grid locations, where access to traditional water treatment technologies is limited. Though problems like fouling of electrodes, energy needs, and scale-up for wide applications are present, research efforts and technological breakthroughs are in progress to eliminate these problems and make EC an ever more real and desirable candidate for water treatment. With increased water pollution and scarcity problems being faced by the world, electrocoagulation is one potential solution that could ensure safe, clean, and sustainable water in the future.

References

Ahmed SF, Mofijur M, Nuzhat S, Chowdhury AT, Rafa N, Uddin MA, Show PL et al (2021) Recent developments in physical, biological, chemical, and hybrid treatment techniques for removing emerging contaminants from wastewater. J Hazard Mater 416:125912

Akter S, Suhan MBK, Islam MS (2022) Recent advances and perspective of electrocoagulation in the treatment of wastewater: a review. Environ Nanotechnol, Monit Manag 17:100643

Bagastyo AY (2021) Electrocoagulation for drinking water treatment: a review. IOP Conf Ser: Earth Environ Sci 623(1):012016

Biswas B, Joshy A, Goel S (2025) Solar-powered electrocoagulation for the removal of atrazine with and without microplastics. Environ Sci: Water Res Technol 11(4):942–958

Biswas Çetinkaya AY (2025) Integration of electrocoagulation and solar energy for sustainable wastewater treatment: a thermodynamic and life cycle assessment study. Environ Monit Assess 197(2):224

Boinpally S, Kolla A, Kainthola J, Kodali R, Vemuri J (2023) A state-of-the-art review of the electrocoagulation technology for wastewater treatment. Water Cycle 4:26–36

Boulanouar L, Louhichi B, Hamdi W, Jellali S, L'taief B, Hamdi N, Houas A et al (2025) Parametric study of cadmium and nickel removal from synthetic and actual industrial wastewater industry by electrocoagulation using solar energy. J Water Process Eng 71:107261

Cruz KD, Camacho SLA, Rodenas HMP (2024) Electrocoagulation treatment for Remazol Brilliant Blue R (Reactive Blue 19) removal of textile dye simulated wastewater using iron-aluminum electrodes. IOP Conf Ser: Earth Environ Sci 1372(1):012026

Das PP, Sharma M, Purkait MK (2022b) Recent progress on electrocoagulation process for wastewater treatment: a review. Sep Purif Technol 292:121058. https://doi.org/10.1016/j.seppur.2022.121058

Das PP, Sontakke AD, Purkait MK (2023) Electrocoagulation process for wastewater treatment: applications, challenges, and prospects. In: Development in wastewater treatment research and processes, pp 23–48

Das PP, Baite TN, Purkait MK (2024) Industrial wastewater treatment by electrocoagulation process, In: Shah MP, Shah N (eds) Development in wastewater treatment research and processes, pp 55–73. https://doi.org/10.1016/B978-0-443-13884-3.00019-6

Devlin TR, Kowalski MS, Pagaduan E, Zhang X, Wei V, Oleszkiewicz JA (2019) Electrocoagulation of wastewater using aluminum, iron, and magnesium electrodes. J Hazard Mater 368:862–868. https://doi.org/10.1016/j.jhazmat.2018.10.017

Dina T, Moussa MH, El-Naas MN, Mohammed J-M (2017) A comprehensive review of electrocoagulation for water treatment: potentials and challenges. J Environ Manage 186(2017):24–41. https://doi.org/10.1016/j.jenvman.2016.10.032

Dube A, Malode SJ, Alshehri MA, Shetti NP (2025) Electrochemical water treatment: review of different approaches. J Environ Manage 373:123911. https://doi.org/10.1016/j.jenvman.2024.123911

Ebba M, Asaithambi P, Alemayehu E (2022) Development of electrocoagulation process for wastewater treatment: optimization by response surface methodology. Heliyon 8(5):e09383. https://doi.org/10.1016/j.heliyon.2022.e09383

El-Naggar MA, Attia SA, Zatout AA, El-Ashtoukhy EZ, Abdel-Aziz MH, Sedahmed GH, Fathalla AS (2025) Removal of sulfides via electrocoagulation using a porous anode with special reference to petroleum wastewater treatment. Int J Environ Res 19(4):113

Ganzoury MA, Ghasemian S, Zhang N, Yagar M, de Lannoy C-F (2022) Mixed metal oxide anodes used for the electrochemical degradation of a real mixed industrial wastewater. Chemosphere 286:131600. https://doi.org/10.1016/j.chemosphere.2021.131600

Gök Z, Gülyaşar HC (2025) Preliminary laboratory assessment of the removal of heavy metals from metal plating wastewater by electrocoagulation. Water Air Soil Pollut 236(6):1–15

Holt PK, Barton GW, Mitchell CA (2005) The future for electrocoagulation as a localised water treatment technology. Chemosphere 59(3):355–367. https://doi.org/10.1016/j.chemosphere.2004.10.023

Hossain M, Mahmud I, Parvez S, Cho HM (2013) Impact of current density, operating time and pH of textile wastewater treatment by electrocoagulation process. Environ Eng Res 18(3):157–161

Jain K, Patel AS, Pardhi VP, Flora SJS (2021) Nanotechnology in wastewater management: a new paradigm towards wastewater treatment. Molecules 26(6):1797

Jaiswar VK, Saroha AK (2025) Effect of different electrode configuration modes on the performance of electrocoagulation. J Indian Chem Soc 102(2):101569

Jasim RA, Salman RH, Zabar MK (2024) Removal of artificial anionic dye by electrocoagulation and electro-oxidation combined system using aluminum and nano (Cu-Mn-Ni) composite electrodes. Iraqi J Chem Petrol Eng 25(4):35–47

Jegathambal P, Mayilswamy C, Parameswari K (2024) Reuse and recovery of water from industrial textile dyeing effluent using high-performance electrodes continuous flow electrocoagulation reactor. Nat Environ Pollut Technol 23(3)

Jing G, Ren S, Pooley S, Sun W, Kowalczuk PB, Gao Z (2021) Electrocoagulation for industrial wastewater treatment: an updated review. Environ Sci: Water Res Technol 7(7):1177–1196

Kabdaşlı I, Tünay O (2023) Hexavalent chromium removal from water and wastewaters by electrochemical processes: review. Molecules 28(5):2411. https://doi.org/10.3390/molecules28052411

Karchiyappan T, Bilińska L, Karri RR, Gmurek M (2024) Electrocoagulation as a promising treatment for textile industry wastewater: effect of process parameters and reuse approach. In: Dehghani MH, Karri RR, Tyagi I (eds) Sustainable remediation technologies for emerging pollutants in aqueous environment, pp 179–201. https://doi.org/10.1016/B978-0-443-18618-9.00001-2

Kesari KK, Soni R, Jamal QMS, Tripathi P, Lal JA, Jha NK, Ruokolainen J et al (2021) Wastewater treatment and reuse: a review of its applications and health implications. Water Air Soil Pollut 232:1–28

Khaled B, Wided B, Béchir H, Elimame E, Mouna L, Zied T (2019) Investigation of electrocoagulation reactor design parameters effect on the removal of cadmium from synthetic and phosphate industrial wastewater. Arab J Chem 12(8):1848–1859. https://doi.org/10.1016/j.arabjc.2014.12.012

Khandegar V, Saroha AK (2013) Electrocoagulation for the treatment of textile industry effluent—a review. J Environ Manage 128:949–963. https://doi.org/10.1016/j.jenvman.2013.06.043

Kobya M, Can OT, Bayramoglu M (2003) Treatment of textile wastewaters by electrocoagulation using iron and aluminum electrodes. J Hazard Mater 100(1–3):163–178. https://doi.org/10.1016/S0304-3894(03)00102-X

Leiknes T (2009) The effect of coupling coagulation and flocculation with membrane filtration in water treatment: a review. J Environ Sci 21(1):8–12

Lu H, Wang H, Gao C, Liu L, Yu T, Li Y, Lin S (2024) Enhanced lanthanum and cerium removal of self-powered electro-membrane bioreactor (SPEMBR) using high-active bi-functional MN-CNTs@ Fe/Mo electrode. Chem Eng J 490:151877

Mallick R, Poddar S (2025) Future directions in water quality management: integrating advanced technologies and sustainable practices. In: Computational automation for water security. Elsevier, pp 215–227

Mehellou A, Aitbara A, Djellabi R, Delimi R, Innocent C (2025) Unravelling the potential of combined electrodialysis/electrocoagulation for boosted dye removal and membrane anti-fouling activity. Environ Sci Pollut Res 32(13):8098–8116

Mohammed ON, Abdulhameed HM (2025) Solar powered electrocoagulation system for groundwater remediation at Al-Raaed Station for irrigation purposes: batch mode 20(3):633–645. http://iieta.org/journals/ijdne

Naje AS, Chelliapan S, Zakaria Z, Ajeel MA, Alaba PA (2017) A review of electrocoagulation technology for the treatment of textile wastewater. Rev Chem Eng 33(3):263–292

Ogedey A, Oguz E (2024) Leachate treatment via electrocoagulation–coal-based powdered activated carbon process: efficiencies, mechanisms, kinetics, and costs. Water Environ Res 96(6):e11060

Öztel MD, Alver A, Akbal F, Altaş L, Kuleyin A (2025) Predictive insights into arsenic remediation: Advancing electro and chemical coagulation through machine learning models. Journal of Water Process Engineering 72:107498

Phu TKC, Nguyen PL, Phung TVB (2025) Recent progress in highly effective electrocoagulation-coupled systems for advanced wastewater treatment. iScience

Rosenwinkel K, Austermann-Hann U, Meyer H (2001) Industrial wastewater sources and treatment strategies, In: Rehm H-J, Reed G (eds) Industrial wastewater sources and treatment strategies, biotechnology set, vol 2, pp 191–215. https://doi.org/10.1002/9783527620999.ch10l

Sanjay D, Savithiri V (2025) Comparative study of novel aluminium mesh electrode and solid iron aluminium for turbidity removal from domestic wastewater by solar powered electrocoagulation. AIP Conf Proc 3270(1):020196

Shah AA, Walia S, Kazemian H (2024) Advancements in combined electrocoagulation processes for sustainable wastewater treatment: a comprehensive review of mechanisms, performance, and emerging applications. Water Res 252:121248

Shahedi A, Darban AK, Taghipour F, Jamshidi-Zanjani A (2020) A review on industrial wastewater treatment via electrocoagulation processes. Curr Opin Electrochem 22(2020):154–169. https://doi.org/10.1016/j.coelec.2020.05.009

Shrestha R, Ban S, Devkota S, Sharma S, Joshi R, Tiwari AP, Joshi MK et al (2021) Technological trends in heavy metals removal from industrial wastewater: a review. J Environ Chem Eng 9(4):105688

Sillanpää M, Shestakova M (2017) Electrochemical water treatment methods fundamentals, methods and full scale applications, 1st edn. Butterworth-Heinemann, Elsevier

Singh A, Sharma A, Verma RK, Chopade RL, Pandit PP, Nagar V, Sankhla MS et al (2022) Heavy metal contamination of water and their toxic effect on living organisms. In: The toxicity of environmental pollutants. IntechOpen

Solak M, Kılıç M, Hüseyin Y, Şencan A (2009) Removal of suspended solids and turbidity from marble processing wastewaters by electrocoagulation: comparison of electrode materials and electrode connection systems. J Hazard Mater 172(1):345–352. https://doi.org/10.1016/j.jhazmat.2009.07.018

Syam Babu D, Anantha Singh TS, Nidheesh PV, Suresh Kumar M (2020) Industrial wastewater treatment by electrocoagulation process. Sep Sci Technol 55(17):3195–3227

Tahreen A, Jami MS, Ali F (2020) Role of electrocoagulation in wastewater treatment: a developmental review. J Water Process Eng 37:101440

Tegladza ID, Xu Q, Xu K, Lv G, Lu J (2021) Electrocoagulation processes: a general review about role of electro-generated flocs in pollutant removal. Process Saf Environ Prot 146:169–189

Wang RS, Shan LL, Zhu ZB, Liu ZQ, Liao ZM, Cui YH (2025) Current status of electrode corrosion passivation and its mitigation strategies in electrocoagulation. Chem Eng Process-Process Intensification, 110192

WWAP (World Water Assessment Programme) (2012) The United Nations world water development report 4: managing water under uncertainty and risk. UNESCO, Paris

Yan F, An L, Xu X, Du W, Dai R (2024) A review of antibiotics in surface water and their removal by advanced electrocoagulation technologies. Sci Total Environ 906:167737

Yatoo AM, Hamid B, Sheikh TA, Ali S, Bhat SA, Ramola S, Kumar S et al (2024) Global perspective of municipal solid waste and landfill leachate: generation, composition, eco-toxicity, and sustainable management strategies. Environ Sci Pollut Res 31(16):23363–23392

Zhang Y, Tang X, Zhang J, Zhang Y, Yu R, Wang W, Yu J et al (2024) Valuable components recovery from wastewater and brine using electrocoagulation-based coupled process: a systematic review. Desalination, 117732

Chapter 11
Introduction to Wastewater Management: For Wastewater Reuse and Resource Recovery

Shabnam Parveen and Zeba Khan

Abstract Water reuse has become a dire necessity, not a mere option. Wastewater serves as a substantial and stable source of alternative water supplies. The depletion and variability of water resources, coupled with uneven global distribution, population growth, water pollution, and water stress in numerous nations, have spurred efforts to explore unconventional water sources for sustainable development. The utilization of wastewater as a dependable and copious water source in sectors like agriculture and industry has gained significant attention. To optimize water resource management and secure future water needs, measures such as demand management, public awareness campaigns, supply management, consumption prioritization, recycled water use, and increased water productivity are pivotal. The recycling and reuse of wastewater emerges as a crucial strategy for reducing the consumption of surface and groundwater resources. Harnessing treated wastewater offers multiple benefits, including enhanced water availability, sustainable resource utilization, reduced energy demand, diminished nutrient loads, and increased crop production. Additionally, wastewater reuse fosters environmental preservation, promotes employment opportunities, and supports local economies.

Keywords Environmental pollution · Sustainable agriculture · Wastewater reuse · Resource recovery

S. Parveen (✉)
Department of Botany, S.S. Khanna Girls Degree College, Allahabad, India
e-mail: p_shabnam786@rediffmail.com

Z. Khan
Department of Family and Community Science, University of Allahabad, Prayagraj, India
e-mail: zeba.khan09@gmail.com

A. M. Yatoo et al. (eds.), *Wastewater Monitoring and Management by Advanced Approaches*, Advances in Wastewater Research,
https://doi.org/10.1007/978-981-95-2601-7_11

11.1 Introduction

Global water security is facing a looming crisis due to a confluence of factors: freshwater scarcity, rising consumption, and climate change (Jain et al. 2021). This escalating demand for a vital resource, coupled with global water stress, underscores the critical requirement for common implementation of water reuse and recycling (Bera et al. 2022). As urban populations swell, judicious use, recycling, and renewal of resources become paramount. Wastewater reuse emerges as a crucial solution in addressing potential shortages of conventional water sources (Kesari et al. 2021). Scientific forecasts predict a doubling of global water consumption by 2050. Recognizing wastewater as a key element in the water cycle, people must implement proper management practices throughout the whole water management cycle, treatment, encompassing freshwater extraction, usage, distribution, collection, posttreatment and reuse, and finally come back to the environment (UN Water 2017). According to Jodar-Abellan et al. (2019), this underscores the rationale behind wastewater treatment for resource revival as a viable way out to mitigate water and drought shortages, particularly in water-restricted regions.

Wastewater, commonly known as sewage, is a by-product of diverse uses of water (Mishra et al. 2023). From household activities like showering, dishwashing, laundry, and toilet flushing, to industrial processes, product manufacturing, and cleaning operations, water is used extensively. Once utilized, this water becomes wastewater and enters the sewage system, eventually flowing to a wastewater treatment plant. These treatment plants, often exceeding public expectations, are intricate facilities designed to produce high-quality treated water (Djandja et al. 2021).

Treated wastewater and household water contain a variety of minerals, including potassium, nitrogen, phosphorus, and sulfur. However, the primary sources of nitrogen and phosphorus found in wastewater can be voluntarily absorbed by plants, for irrigation, which are commonly utilized (Drechsel et al. 2010; Duncan 2009; Poustie et al. 2020; Sengupta et al. 2015). In domestic wastewater, the abundant minerals present help to boost crop yields, reduce the need for fertilizers, enhance soil richness, and potentially lower production costs for farmers (Chen et al. 2013; Jeong et al. 2016). Over 10% of the global population relies on agricultural products that are grown using wastewater irrigation (WHO 2006). Worldwide, agriculture accounts for 92% of total water consumption (Clemmens et al. 2008; Hoekstra and Mekonnen 2012; Tanji and Kielen 2002). Discharging wastewater into natural water bodies can significantly harm aquatic ecosystems. Untreated wastewater often contains elevated levels of nutrients like nitrogen and phosphorus, which can trigger eutrophication, resulting in oxygen depletion and posing threats to aquatic life (Bhat and Qayoom 2021). Furthermore, toxic substances found in wastewater can accumulate in the food chain, adversely affecting the health and diversity of aquatic organisms. Various parasites present in wastewater can also lead to serious health concerns. Effective wastewater treatment is essential to reduce these ecological risks, thereby maintaining the balance and integrity of ecosystems (Nasir et al. 2024).

The reuse of treated wastewater has seen remarkable growth, with annual increases of approximately 10–29% in regions such as the USA, Europe, China, and in Australia up to 41% (Aziz and Farissi 2014). China leads among Asian countries in wastewater reuse, with an estimated area of 1.3 million hectares, surpassing nations like India, Vietnam, and Pakistan (Zhang and Shen 2017). Currently, it is estimated that only 37.6% of urban wastewater in India is being treated (Singh et al. 2019). Israel is the largest consumer of treated wastewater for agricultural irrigation, utilizing 90% of its reclaimed water (Angelakis and Snyder 2015). In the past, the introduction of various traditional wastewater treatment methods, including thermal treatment, ion exchange, adsorption, electrochemical degradation, coagulation, and chemical precipitation (Saravanan et al. 2021). However, these methods often fall short in effectively purifying water due to several limitations, such as high energy consumption, the generation of unwanted by-products, negative side effects, and cost concerns. The development of innovative wastewater treatment approaches, consisting of advanced oxidation processes and nanofibrous membrane filtration, promotes sustainable environments and the production of biofertilizers. Notably, activated sludge generated during wastewater treatment can be utilized to create biofertilizers (Rani et al. 2024). Wastewater treatment plays a crucial role in safeguarding public health, preserving ecosystems, ensuring the sustainability of water resources, reducing pollution, and promoting sustainable development (Connor 2015). Moreover, it is vital for achieving various Sustainable Development Goals. Primarily, it significantly enhances the availability of safe water resources by ensuring access to clean water and sanitation (Obaideen et al. 2022). Additionally, effective treatment techniques directly contribute to addressing health and well-being by preventing the spread of waterborne diseases. The promotion of industry, innovation, and infrastructure aligns with the integration of sustainable and innovative technologies in wastewater treatment (Awan et al. 2022). Furthermore, it encourages responsible consumption and production by promoting water efficiency and reducing environmental impacts. The protection of aquatic and terrestrial ecosystems from pollution is closely linked to the preservation of life below water and life on land. Lastly, wastewater treatment also aids in climate action during the implementation of energy-efficient processes and greenhouse gas emissions reduction (Delanka-Pedige et al. 2021).

11.1.1 Why Treat Wastewater?

The main aim of the treatment of wastewater is to eliminate pollutants, filter out large particles, eradicate toxic substances, and destroy potential pathogens, allowing the resulting clean water, referred to as effluent, to be safely released back into the environment for various uses (Alagha et al. 2020; Villarín and Merel 2020). Societal water usage leads to water contamination with pollutants. Untreated pollutants threaten water ecosystems if left unaddressed. Organic matter depletion in water bodies causes oxygen depletion, resulting in fish mortality and unpleasant odors. Effective wastewater treatment eliminates waterborne diseases. Pollutants pose toxic threats to both aquatic life and humans. The demand for freshwater is rapidly increasing while supplies are dwindling, making wastewater reclamation and reuse essential in today's context. Al-Juaidi et al. (2014) consider that the increasing human population will place greater demands on natural resources, particularly clean water for residential and industrial purposes. In the absence of a sustainable water supply, the world may encounter severe water shortages and illnesses linked to contaminated water. Treated wastewater is typically utilized for nondrinkable applications such as agricultural irrigation, land irrigation, replenishing groundwater, irrigating golf courses, washing vehicles, flushing toilets, firefighting, and various building construction activities. Additionally, it may be used for cooling in thermal power plants (Katsoyiannis et al. 2017; Mohsen 2004; Smith 1995; Yang et al. 2017). The reuse of wastewater for crop irrigation has raised several health concerns (Ungureanu et al. 2020). Using industrial wastewater for irrigation, whether directly or mixed with domestic water, poses a greater risk (Chen et al. 2013). The risk factors increase due to contamination from heavy metals and pathogens, as heavy metals are nonbiodegradable and possess a long biological half-life (Chaoua et al. 2019; WHO 2006). These wastewater sources can contain various toxic elements, including copper (Cu), chromium (Cr), manganese (Mn), iron (Fe), lead (Pb), zinc (Zn), and nickel (Ni) (Mahfooz et al. 2020). These heavy metals tend to accumulate in the topsoil (up to a depth of 20 cm) and are taken up by plant roots, eventually entering the bodies of humans and animals through the consumption of leafy vegetables and inhalation of contaminated soil (Mahmood et al. 2014). Consequently, conducting health risk assessments for wastewater irrigation is essential, particularly for adults (Mehmood et al. 2019; Njuguna et al. 2019; Xiao et al. 2017). To mitigate these risks, it is crucial to implement advanced wastewater treatment methods before discharging wastewater into rivers, agricultural lands, and soils. As the demand for water rises and drought conditions worsen, treating wastewater emerges as one of the most effective solutions for improving water sustainability.

11.2 Wastewater Management

The process of eliminating impurities from wastewater or sewage before it enters aquifers or natural bodies of water, ensuring its purity for environmental conservation. As pure water doesn't exist naturally, the distinction between clean and polluted water lies in the presence and concentration of impurities relative to its intended use. It is considered polluted water when mixed with excessive impurities that render it unsuitable for specific purposes like swimming, drinking, and fishing. While usual factors can impact water quality, pollution commonly originates from human activities that contaminate the water (Kesari et al. 2021). Water pollution primarily comes from the release of polluted wastewater into surface or groundwater, emphasizing the importance of wastewater treatment in pollution control. The wastewater progresses through a collection system and enters the treatment plant, where it undergoes initial

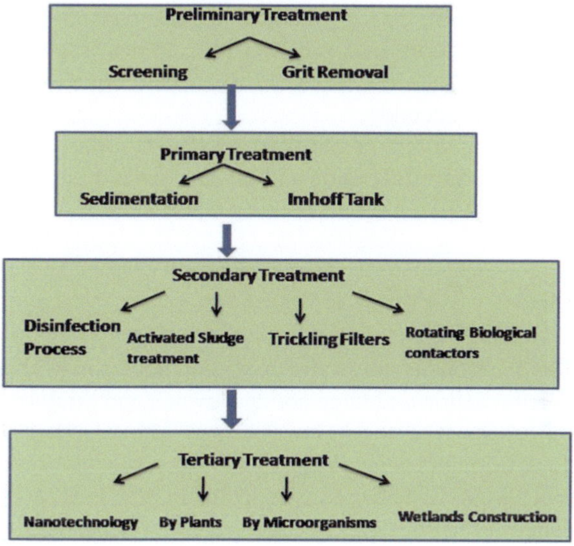

Fig. 11.1 Process of wastewater treatment

treatment, followed by primary, secondary, and potentially complex or tertiary treatment stages. Solid waste extracted from the stream of wastewater requires additional treatment for proper disposal.

Waste treatment involves the following steps (Fig. 11.1):

Primary treatment: The exclusion of insoluble particulate materials involves processes like screening, settling, alum addition, coagulation, and other physical procedures.

Secondary treatment: Animate elimination of liquified organic matter, activated sludge, trickling filters, lagoons, protracted ventilation systems, and anaerobic digesters.

Tertiary treatment: This type of treatment is biological. Biochemical removal of inorganic nutrients, trace chemicals, and viruses.

11.2.1 Initial Treatment

The purpose of initial treatment: Before wastewater can be effectively treated, it's crucial to remove large and heavy debris that could damage equipment or disrupt subsequent treatment processes. This action comprises screening and grit elimination stages.

Screening: It removes large floating debris such as rags, paper, and plastics, ensuring these materials don't clog equipment or interfere with later stages. The removed material is known as screening.

Grit removal: The grit tackles densely packed inorganic particles such as gravel, sand, and other bulky debris (e.g., bone fragments and corn kernels). Grit is commonly classified as fine sand (diameter of 0.2 mm) and has a settling velocity of 20 mm per second.

Preliminary treatment acts as a vital first line of defense in wastewater treatment, removing harmful debris and preparing the wastewater for the subsequent stage.

11.2.2 Primary Treatment

The first step in wastewater treatment, known as primary treatment, aims to eliminate floating debris and large solids (Awaleh and Soubaneh 2014). Key processes in this stage consist of the removal of grit and screening. During screening, wastewater flows through barriers designed to catch big items such as rags and plastics (Markin et al. 2020). Grit chambers or sedimentation tanks are used to settle heavier inorganic materials, like sand and gravel, which could potentially harm equipment downstream (Jasim 2020). Primary treatment uses physical settling to eliminate solid impurities. In the primary settling tank (clarifier), heavy solids sink to the bottom due to the reduced velocity of wastewater. Grease, which is a buoyant material, accumulates at the surface of the tank. Settling tanks incorporate mechanisms for removing both settled and floating solids. Circular and rectangular primary clarifiers perform equally well when optimized and maintained. Not every wastewater treatment facility uses primary treatment. Primary treatment produces primary sludge, which is transferred to the treatment process for eventual disposal. Following the removal of settleable and floatable pollutants, wastewater still contains solids, both dissolved and suspended. Dissolved solids are minute particles (similar to sugar dissolved in water) that are invisible to the naked eye. Suspended solids behave like opposing ends of a magnet, repelling each other and observable to the human eye. Suspended and dissolved solids are removed in the subsequent treatment stage (Kesari et al. 2021). The primary treatment involves the following steps (Fig. 11.2):

11.2.3 Secondary Treatment

This treatment utilizes biological processes to stabilize dissolved solids in wastewater.

Microbes consume organic solids and convert them into cellular mass, which can be subsequently removed in wastewater. Oxygen is required for the above biological processes to function effectively, necessitating its provision. Secondary treatment involves the use of settling tanks, known as secondary clarifiers, to remove the biological mass formed during biological treatment. Activated sludge is a prevalent secondary treatment method involving mixing wastewater with microorganisms returned from secondary clarifiers. Activated sludge treatment entails a continuous

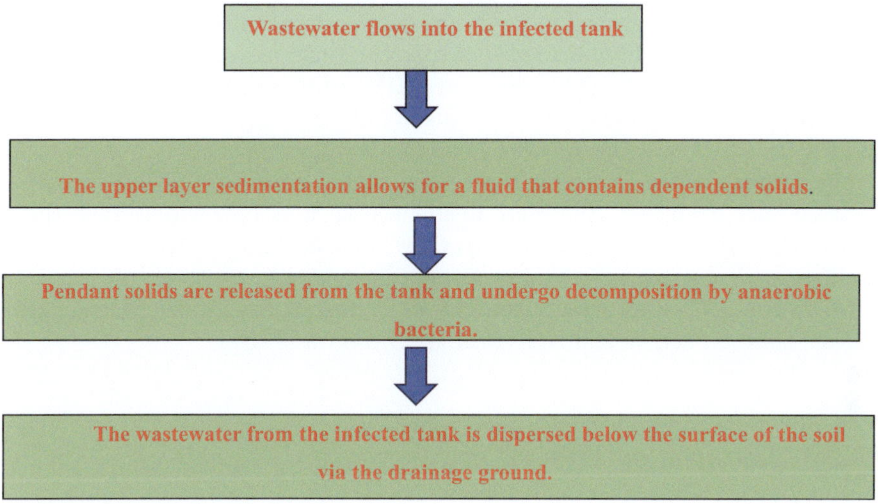

Fig. 11.2 Steps of primary treatment

recirculation of microorganisms from secondary clarifiers, referred to as activated return sludge. In the aeration tank, oxygen (O_2) is introduced through diffusers and blowers for mechanical mixing. A popular variation of the activated sludge treatment process is the sequential batch reactor (SBR), which uses the aeration tank for settling by suspending aeration and allowing solids to separate. During the settling phase in SBRs, the wastewater flow is diverted to another SBR tank to allow continuous treatment.

SBR processes offer compensation such as nutrient (phosphorus and nitrogen) removal capabilities and a compact footprint (Kesari et al. 2021). Activated sludge, a by-product of secondary wastewater treatment, is transformed into a valuable agricultural resource through a meticulously planned process for producing bio-fertilizer (Engida et al. 2020). Following secondary treatment, dewatering is performed to reduce the moisture content of the activated sludge. Next, stabilization techniques, such as aerobic or anaerobic digestion, are used to eliminate pathogens and decrease the organic matter present (Seruga et al. 2020). Once stabilized, the sludge can undergo composting to enhance its nutrient profile and ensure the removal of any potential contaminants. The formulation of the bio-fertilizer can be optimized by incorporating additional organic or mineral additives (Chojnacka et al. 2021). Rigorous quality control measures are implemented to meet legal standards, ensuring the efficacy and safety of the final product. After the bio-fertilizers are produced, they are tested on various plants to evaluate their growth rate and yield.

Techniques such as sand filters, multimedia filters, and membrane filtration (including microfiltration and ultrafiltration) are utilized to eliminate residual suspended solids and fine particles (Bairagi and Ali 2020). To eradicate pathogens and undesirable nutrients and to guarantee the microbial safety of treated wastewater, chemical disinfection methods (such as chlorination and ozonation) or physical disinfection methods (like ultraviolet [UV] irradiation) are used (Shi, Chen et al. 2021). Various wastewater treatments in full-scale reactor techniques are used besides the activated sludge process. Ponds (maturation, facultative, anaerobic, and aerobic) offer alternative methods. Trickling filters provide a different approach to wastewater treatment. Anaerobic treatments, such as upflow anaerobic sludge blanket (UASB) reactors, remove impurities without oxygen. Artificial wetlands mimic natural ecosystems for wastewater purification.

Microbial fuel cells generate electricity while treating wastewater. Methanogenic reactors harness bacteria to produce methane from wastewater.

11.2.4 Disinfection Process

In many plants, after secondary treatment, wastewater undergoes a procedure known as disinfection, which inactivates organisms of causal disease. This is sometimes confused with sterilization, which completely eliminates all organisms. Two methods involve chlorination or ultraviolet radiation.

11.2.5 Tertiary Treatment

Tertiary or advanced treatment processes are used to eliminate extra solids, nutrients, or biochemical oxygen demand (BOD). Some treatment plants may be mandated to extract nutrients (nitrogen and phosphorus) due to the adverse effects on the receiving water bodies, such as ammonia's toxic impact on fish. Advanced treatment offers a significantly higher level of processing compared to secondary treatment. The removal of nitrogen relies on biological processes, while the extraction of phosphorus necessitates the use of chemical additives (Kesari et al. 2021). Tertiary treatment primarily focuses on purifying water to render it suitable for drinking through various methods.

Tertiary wastewater treatment involves the following process: A comprehensive approach tertiary wastewater treatment goes beyond primary and secondary treatment by using advanced techniques to achieve a high level of purity in the effluent. These techniques include: Physicochemical and Biological Processes, Absorptive Process: Activated carbon is used to remove impurities, Efficient Oxidation: Ozone is utilized for enhanced oxidation.

Foam Separation: Impurities are removed through foam separation, and Demineralization: Reverse osmosis or distillation is used to remove minerals.

While some tertiary treatment systems utilize standard equipment, others may incorporate more specialized and expensive technologies like electro dialysis units or ion exchange columns. The primary benefit of tertiary treatment in the final effluent is the significant reduction of phosphate, nitrite, and nitrate. This minimized nutrient load prevents extensive microbial growth in receiving waters. Despite being the most comprehensive method for treating sewage, tertiary treatment hasn't gained widespread adoption due to its relatively high costs associated with the complete nutrient removal.

11.2.5.1 Tertiary Treatment of Wastewater Through Nanotechnology

Nanofillers: A promising solution for tertiary wastewater treatment: Nanotechnology's emergence presents a compelling opportunity to utilize nanofillers as a feasible method for tertiary wastewater treatment. Their exceptionally small pore size (1–5 nm) enables the effective removal of diverse contaminants, including pharmaceutical active compounds (PhACs), pathogenic microorganisms, inorganic and organic pollutants, and heavy metals (Mohammad et al. 2015; Vergili 2013). Nanofillers' efficacy has gained widespread recognition in various industries, including textile, pharmaceutical, dairy, and wastewater treatment. According to Abdel-Fatah (2018), their application ranges from pulp bleaching to microbial elimination and heavy metal removal. The effectiveness of carbon nanotube-based water filters in eliminating Poliovirus Sabin-1 and bacterial pathogens (*Escherichia coli* and *Staphylococcus aureus*) from wastewater was demonstrated by Srivastava et al. (2004). These highly efficient filters are reusable, highlighting the potential for nanofillers in addressing the challenges of wastewater treatment.

Nanomaterials are used effectively for purifying wastewater from microbial contaminants. Wastewater which are contaminated with various microorganisms, including *E. coli* and *Salmonella.* Carbon nanotubes (CNTs) are widely used (Akasaka and Watari 2009). Silver nanoparticles have demonstrated remarkable effectiveness against microorganisms in wastewater, leading to their extensive use for the removal of microbes (Inoue et al. 2002). Carbon nanotubes possess magnetic properties and exhibit a high affinity for bacterial cells (Pan and Xing 2008). Studies have established the application of carbon nanotubes for eliminating *E. coli* from wastewater (Melanta 2008). For removing total coliform bacteria from municipal wastewater, zinc oxide (ZnO) nanoparticles have been proposed as a potential antibacterial agent (Mostafaii et al. 2017). It is necessary to recognize the challenges and drawbacks of nanotechnology. While nanotechnology offers promising applications in wastewater treatment, various nanoengineered techniques are still in research, awaiting further development (Gehrke et al. 2015). Despite these challenges, nanomaterials and nanotechnology have excellent properties for contaminant exclusion and water cleansing. Nanotechnology has the potential to become a crucial solution for wastewater treatment, enabling its reuse for drinking purposes (Zekić et al. 2018).

11.2.5.2 Treatment of Wastewater Through Plants

In the wastewater treatment, some plants have the potential to utilize contaminants and pollutants as nutrients for their growth (Zimmels et al. 2004). Plant species incorporating can be commercial, energy-efficient, and straightforward to operate. Additionally, it allows for on-site wastewater treatment (Vogelmann et al. 2016). Phytoremediation capabilities of these five plants (*Ipomoea aquatic, Centella asiatica, Eichhornia crassipes, Salvinia molesta*, and *Pistia stratiotes*) were assessed by Nizam et al. (2020). Their analysis revealed a significant reduction in the levels of three pollutants: ammoniacal nitrogen, suspended solids, and phosphate. Five plants exhibited competent elimination rates, ranging from 63.9–98% for ammoniacal nitrogen (NH_3-N), total suspended solids (TSS), and phosphate. The impact on treatment of household wastewater through three Appalachian plant species: gray club-rush (*Scirpus validus* L.), common rush (*Juncus effusus* L.), and (*Typha latifolia* L.) broadleaf cattail, studied by Coleman et al. (2001). Significant reductions 50–60% nitrogen, ammonia, and phosphate levels, 70%, suspended solids (TSS), biochemical oxygen demand (BOD), and substantial reduction in fecal coliform populations revealed by researchs. Greater removal efficiencies using three ornamental plant species: *Hedychium coronarium, Canna indica,* and *Cyperus papyrus* were observed by Zamora et al. (2019). They achieved 20–60% higher removal rates for nitrogen as ammonium (N-NH_4), chemical oxygen demand (COD), phosphate (P-PO_4), nitrate (N-NO_3), and total suspended solids (TSS). These studies highlight the potential of many plant species for effective treatment of wastewater, with both common and ornamental plants demonstrating significant contaminant removal capabilities.

11.2.5.3 Utilization of Microorganisms in Wastewater Treatment

A varied array of bacteria, such as *Pseudomonas putida, Pseudomonas fluorescens,* and species of *Bacillus,* possess the ability to participate in biological wastewater treatment systems (Table 11.1). These bacteria aggregate as biofilms, flocs, or granules to facilitate wastewater treatment. The recognition of bacterial exopolysaccharides (EPS) as effective adsorbents has led to potential applications in heavy metal removal (Gupta and Diwan 2017). Commercially available EPS examples include alginate (produced by *Azotobacter vinelandii* and *Pseudomonas aeruginosa*), hyaluronan (by attenuated strains of *P. aeruginosa, Pasteurella multocida* and *Streptococci*), gellan (by *Sphingomonas paucimobilis*), galactopol (by *Pseudomonas oleovorans*) and xanthan (by *Xanthomonas campestris*) (Freitas et al. 2011a, b). Freitas et al. (2009) achievement of significant reductions in synthetic wastewater: 70% for Sludge Hammer, 54% for *B. subtilis*, 52% for *Brevibacillus laterosporus*, and 42% for *P. aeruginosa* Hesnawi et al. (2014) through biodegradation of municipal wastewater using both local and commercial bacteria (Sludge Hammer). Bioaugmentation of wastewater treatment reactors with selected or mixed bacterial strains can significantly enhance treatment efficiency.

Table 11.1 Microorganisms involved in metal removal from wastewater

S. No.	Microorganisms	Metals removed
1	*Alcaligenes faecalis*	Cadmium, zinc
2	*Aspergillus niger*	Copper, cadmium, zinc
3	Aspergillus oryzae	Cadmium
4	*Bacillus* sp.	Heavy metal
5	*Chlorella vulgaris*	Gold, copper, mercury, zinc
6	*Clostridium*	Heavy metal
7	*Corynebacterium*	Mercury, lead
8	*Mycobacterium*	Heavy metal
9	*Neurospora*	Heavy metal
10	*Penicillium spinulosum*	Copper, cadmium, zinc
11	*Pseudomonas fluorescens*	Heavy metal
12	*Pseudomonas putida*	Heavy metal
13	*Rhizopus arrhizus*	Uranium
14	*Trichoderma viride*	Copper
15	*Scenedesmus* sp.	Heavy metal
16	*Serratia*	Heavy metal
17	*Staphylococcus aureus*	Cadmium, mercury
18	*Stigeoclonium tenue*	Heavy metal
19	*Thiobacillus ferrooxidans*	Cadmium, mercury
20	*Yersinia enterocolitica*	Cadmium, mercury

In recent years, microalgae have garnered research interest as a promising alternative for the treatment of wastewater. Microalgae are photosynthetic microorganisms, unicellular or multicellular, that inhabit water bodies, including both freshwater and saltwater environments, as well as moist soil. For growth and metabolism, algae utilize excessive nutrients such as nitrogen, phosphorus, and carbon through their anaerobic systems. The ability of algae to consume nutrients can mitigate eutrophication, reducing nutrient overload in the water bodies. Algae produce oxygen during the nutrient digestion process that supports the growth of aerobic heterotrophic bacteria, which can further degrade inorganic and organic contaminants. A study by Kim et al. (2014) demonstrated that when algae were used in a consortium of municipal wastewater, significant reductions in 93% nitrogen, COD (86%), and 83% phosphorus were achieved. Microalga *Scenedesmus* sp. has the capabilities of heavy metal removal in contaminated river water from the River Melaka in Malaysia, reported by Nmaya et al. (2017). Their findings indicated efficient removal of Zn (97–99%) within a 3–7 day experimental period.

11.2.5.4 Wetland Construction and Natural Treatment Systems

Constructed wetland systems are purifying water successfully and improving biodiversity and ecological balance. For the treatment of wastewater, constructed wetlands are planned (Gopal 1999). Hybrid systems, subsurface flow wetlands (SSF), and free water surface wetlands (FWS) are types of natural treatment systems and constructed wetlands. To develop pollutant removal efficiency, hybrid systems merge activated sludge or filtration-type treatment technologies and constructed wetlands (Liu, Zhao et al. 2015). Subsurface flow wetlands treatment includes physical, chemical, and biological processes (Vymazal 2011) and free water surface wetlands promoting microbial activity, which helps in removing contamination (Zhang, Zheng et al. 2010).

11.2.5.5 Treatment of Wastewater Through Membrane Bioreactors (MBRs)

The membrane filtration technique includes reverse osmosis (RO), nanofibrous membrane filtration, microfiltration (MF), and ultrafiltration (UF). To remove dissolved organic compounds, salts, and solids, reverse osmosis utilizes highly selective membranes for the reuse of water. For the refinement of water in wastewater treatment, nanofibrous membrane filtration techniques use nanofibrous (less than 100 nm) membranes, which are helpful in eliminating organic compounds, divalent ions, and specific small molecules (Adusei-Gyamfi et al. 2019). To remove bacteria, protozoa, and suspended solids, microfiltration uses large pore size membranes (Branch et al. 2021).

11.2.5.6 Treatment of Wastewater Through Microbial Fuel Cells (MFCs)

Microbial fuel cells (MFCs) serve as a renewable energy source, functioning by converting chemical energy into electricity through bacterial metabolic processes. These systems consist of an anode and a cathode (Srivastava et al. 2022). A salt bridge, also known as a proton exchange membrane, separates the cathode chamber from the anode chamber. MFCs are primarily focused on treating domestic, industrial, agricultural, and municipal wastewater while simultaneously generating electricity and hydrogen (Saravanan et al. 2022). Researchers and scientists are exploring the development of various types of microbial fuel cells, including single-chamber microbial fuel cells (SCMFCs) and double-chamber microbial fuel cells (DCMFCs).

11.3 Treated Wastewater Applications

11.3.1 Crop Irrigation Scope

In several major sectors, including public parks, agriculture, landscaping, golf course irrigation, cooling water for power plants and refineries, industrial processing dust control, toilet flushing, concrete mixing, construction activities, and the creation of synthetic lakes, numerous studies have examined the utilization of recycled and treated wastewater. While secondary treatment typically produces wastewater suitable for reuse due to heavy metal levels comparable to natural sources (Ayers and Westcot 1985), research has demonstrated that soil fertility enhances and alters its chemical characteristics when irrigation is done with treated wastewater (Mohammad and Mazahreh 2003). The planned model incorporating ultraviolet, ultrasound (US), nanoparticle, and ozone treatment builds upon previous laboratory-scale testing using an ultrasonic bath for wastewater treatment from various sources (Kesari and Behari 2008; Kesari et al. 2011a, b; Kumar et al. 2010). The treated water's suitability for crop irrigation is confirmed by recent studies (Bhatnagar et al. 2016). The model incorporates individual and combined treatments (nanoparticle and US) aiming to achieve chief water excellence values for irrigation and drinking purposes (Kesari et al. 2011a, b).

11.3.2 Economical and Energy Management

Wastewater treatment plants are crucial for ensuring wastewater sanitation and safeguarding communal health. Household wastewater is no longer viewed solely as waste; it is now recognized as a valuable resource for energy recovery and plant fertilization nutrients (e.g., phosphorus and nitrogen). Utilizing domestic wastewater for crop irrigation and industrial consumption has gained widespread acceptance as a means to address water scarcity. Anaerobic conversion of wastewater's organic matter has emerged as a promising source of energy through methane gas production. Traditional wastewater treatment technologies, such as anaerobic sludge digestion, are energy-intensive and only partially recover wastewater's energy potential. Municipal wastewater treatment plants aim to maximize energy and nutrient recovery, particularly phosphorus and nitrogen, in the next generation. In addition, these plants are pivotal in removing and recovering emerging pollutants and valuable products, including heavy metals, radioactive elements, fertilizers, hormones, and pharmaceuticals.

Enhancing wastewater treatment plants to recover and reuse valuable compounds remains a key area for improvement. Numerous methods in wastewater are under progress to transform organic matter into bioenergy, such as biodiesel, biohydrogen, microbial fuel cells, and bioethanol. All methods hold promise for generating electricity from wastewater; they require further development and refinement. Wastewater offers significant potential for energy generation, producing ten times more energy than conventional chemical, thermal, and hydraulic methods. For stabilizing sludge from wastewater treatment plants, vermicomposting presents a viable solution. The significant environmental and economic benefits of vermicomposting in converting solid waste materials into valuable organic fertilizers have been highlighted by Kesari and Jamal (2017). One of the key issues is the processing and management of dried municipal sewage sludge (Contreras-Ramos et al. 2005) for vermicomposting, which could potentially serve as a method for producing organic fertilizers in future studies. Vermicomposting of household solid waste, agricultural residues, or by-products from the pulp and sugarcane industries demonstrates significant promise as a fertilizer for enhancing crop yields (Bhatnagar et al. 2016; Kesari and Jamal 2017).

11.3.3 Environmental Sustainability

Wastewater treatment significantly contributes to sustainable resource management by minimizing waste. Gernaey et al. (2004) emphasize that waste poses a substantial threat to sustainability as it leads to the depletion of natural resources. Through the treatment of wastewater, potential waste transforms into a valuable resource applicable for both domestic and industrial purposes. Additionally, this process safeguards natural water resources from the risk of depletion. According to Kamali et al. (2019), sustainability aims to guarantee the continuous availability of natural resources to satisfy the needs of both present and future generations. One of the most pressing global challenges is the potential exhaustion of natural resources, driven by increasing populations and increasing demands (Herrero and Stuckey 2015; Łuczkiewicz et al. 2010; Panepinto et al. 2016; Stottmeister et al. 2003). For example, many forests that once flourished with greenery have been converted into agricultural land, industrial sites, and urban developments to accommodate human habitation. These once-lush areas have been replaced by concrete structures to fulfil the requirements of society (Chaúque and Rott 2021). Trees and forests play a crucial role in this ecosystem. It is essential to recognize that the selection of wastewater treatment methods may differ based on the specific characteristics and context of the wastewater involved. Choosing the most appropriate treatment strategy depends on technical, economic, social, and environmental factors, all of which should be evaluated in each unique case.

11.3.4 Innovative Technologies in Wastewater Treatment

Innovative technologies are revealing the newest advancements and potential future developments in wastewater treatment. This discussion focuses on emerging technologies, processes, research trends, and future outlooks that could revolutionize the sector. Sensor technologies, including remote sensing and real-time monitoring devices, along with advanced analytical tools, are enhancing wastewater treatment by delivering precise and timely data regarding process parameters, water quality, and system performance (Chang et al. 2015). These sensor technologies play a crucial role in optimizing treatment processes, promoting automation, and enabling proactive decision-making for process control, monitoring, and maintenance.

Furthermore, artificial intelligence (AI) and machine learning (ML) techniques are increasingly being utilized in wastewater treatment to boost process optimization, improve system performance, and enhance predictive modeling (Bagherzadeh et al. 2021). The potential of AI and ML algorithms lies in their ability to analyze data, recognize patterns, and optimize processes, paving the way for more efficient and adaptive wastewater treatment systems. Additionally, the emergence of novel materials and technologies presents new opportunities for improving treatment efficiency and addressing emerging challenges (Moore and Shi 2014). This emphasizes advancements in various areas, such as membrane materials, catalysts, nanotechnology, and adsorbents, all of which have the potential to enhance pollutant removal, increase energy efficiency, and boost resource recovery in wastewater treatment processes.

11.4 Conclusion

Water scarcity is a growing concern due to population growth. Developing countries, particularly India and other Asian nations, face a significant challenge in wastewater generation (municipal, industrial, and agricultural). Wastewater can potentially be repurposed for rice irrigation if proper treatment and distribution strategies are implemented. Wastewater treatment poses a significant hurdle for many countries due to the increasing presence of harmful pollutants. This chapter investigated current and future research on wastewater treatment. Conventional treatment of wastewater involves primary, secondary, and tertiary procedures, while advanced techniques offer potential for improved water quality. The future model can both effectively treat wastewater and utilize solid waste as fertilizer. Future advancements aim to develop efficient wastewater treatment methods for potable water use. Adhering to WHO standards and guidelines for wastewater treatment is strongly advised. Computational tools such as artificial intelligence, fluid dynamics, and Geographic Information System (GIS) may contribute to future wastewater treatment advancements. Wastewater treatment continues to encounter various challenges, including the necessity for improved treatment efficiency, the emergence of new contaminants, and the need to close the gap between research and practical application. Future research efforts should concentrate on creating technologies to tackle these emerging contaminants, encourage resource recovery, optimize treatment methods, and improve adaptability to climate change. The looming freshwater crisis is a major concern in the twenty-first century. Each year, the world generates approximately 330 cubic kilometers of municipal wastewater alone. This staggering figure underscores the critical need for treated wastewater reuse to address water scarcity. Utilizing treated industrial, municipal, or even seawater for irrigation holds enormous potential for meeting future water demands. To mitigate risks and maximize the benefits of wastewater reuse, the implementation of standardized and advanced treatment methods is essential. Wastewater treatment is a dynamic area that necessitates ongoing research, innovation, and collaboration between academic institutions and industries to develop sustainable and effective solutions. By adopting new technologies, addressing emerging challenges, and establishing sound policies, wastewater treatment can effectively safeguard public health, enhance resource production, protect the environment, and support the sustainable management of water resources. Establishing robust policies and practices that incentivize safe water reuse will unlock its full potential in agriculture, benefiting both farmers and the environment.

References

Abdel-Fatah MA (2018) Nanofiltration systems and applications in wastewater treatment: review article. Ain Shams Eng J 9:3077–3092

Adusei-Gyamfi J, Ouddane B, Rietveld L, Cornard JP, Criquet J (2019) Natural organic matter-cations complexation and its impact on water treatment: a critical review. Water Res 160:130–147

Akasaka T, Watari F (2009) Capture of bacteria by flexiblecarbon nanotubes. Acta Biomater 5:607–612

Alagha O, Allazem A, Bukhari AA, Anil I, Mu'azu ND (2020) Suitability of SBR for wastewater treatment and reuse: pilot-scale reactor operated in different anoxic conditions. Int J Environ Res Public Health 17:1617

Al-Juaidi AE, Kaluarachchi JJ, Mousa AI (2014) Hydrologic-economic model for sustainable water resources management in a coastal aquifer. J Hydrol Eng 19

Angelakis A, Snyder S (2015) Wastewater treatment and reuse: past, present, and future. Water 7:87–95

Awaleh MO, Soubaneh YD (2014) Waste water treatment in chemical industries: the concept and current technologies. Hydrol Current Res 5(1):1–12

Awan U, Sroufe R, Bozan K (2022) Designing value chains for Industry 4.0 and a circular economy: a review of the literature. Sustainability 14:7084

Ayers RS, Westcot DW (1985) Water quality for agriculture. Food and Agriculture Organization of the United Nations

Aziz F, Farissi M (2014) Reuse of treated wastewater in agriculture: solving water deficit problems in arid areas. Ann West Univ Timişoara Ser Biol 17:95–110

Bagherzadeh F, Mehrani MJ, Basirifard M, Roostaei J (2021) Comparative study on total nitrogen prediction in wastewater treatment plant and effect of various feature selection methods on machine learning algorithms performance. J Water Process Eng 41:102033

Bairagi S, Ali SW (2020) Conventional and advanced technologies for wastewater treatment. Environ Nanotechnol Water Purif, pp 33–56

Bera SP, Godhaniya M, Kothari C (2022) Emerging and advanced membrane technology for wastewater treatment: a review. J Basic Microbiol 62(3–4):245–259

Bhat SU, Qayoom U (2021) Implications of sewage discharge on freshwater ecosystems

Bhatnagar A, Kesari KK, Shurpali N (2016) Multidisciplinary approaches to handling wastes in sugar industries. Water Air Soil Pollut 11:1–30

Branch A, Trinh T, Ta TM, Leslie G, Le-Clach P (2021) Log removal values in membrane bioreactors: correlation of surrogate monitoring and operational parameters. J Water Process Eng 41:102032

Chang N-B, Imen S, Vannah B (2015) Remote sensing for monitoring surface water quality status and ecosystem state in relation to the nutrient cycle: a 40-year perspective. Crit Rev Environ Sci Technol 45(2):101–166

Chaoua S, Boussaa S, Gharmali AE, Boumezzough A (2019) Impact of irrigation with wastewater on accumulation of heavy metals in soil and crops in the region of Marrakech in Morocco. J Saudi Soc Agric Sci 18:429–436

Chaúque BJ, Rott MB (2021) Solar disinfection (SODIS) technologies as alternative for large-scale public drinking water supply: advances and challenges. Chemosphere 281:130754

Chen W, Lu S, Pan N, Jiao W (2013) Impacts of long-term reclaimed water irrigation on soil salinity accumulation in urban green land in Beijing. Water Resour Res 49:7401–7410

Chojnacka K, Krowiak AK, Moustakas KD, Skrzypczak K, Mikula D, Loizidou M (2021) A transition from conventional irrigation to fertigation with reclaimed wastewater: prospects and challenges. Renew Sustain Energy Rev 130:109959

Clemmens AJ, Allen RG, Burt CM (2008) Technical concepts related to conservation of irrigation and rainwater in agricultural systems. Water Resour Res 44:1–16

Coleman J, Hench K, Garbutt K, Sexstone A, Bissonnette B, Skousen J (2001) Treatment of domestic wastewater by three plant species in constructed wetlands. Water Air Soil Pollut 3:283–295

Connor R (2015) The United Nations world water development report 2015: water for a sustainable world. UNESCO Publishing

Contreras-Ramos SM, Escamilla-Silva EM, Dendooven L (2005) Vermicomposting of biosolids with cow manure and wheat straw. Biol Fertil Soils 41:190–198

Delanka-Pedige HMK, Munasinghe-Arachchige SP, Abeysiriwardana Arachchige ISA, Nirmalakhandan N (2021) Evaluating wastewater treatment infrastructure systems based on UN Sustainable Development Goals and targets. J Clean Prod 298:126795

Djandja OS, Yin LX, Wang ZC, Duan PG (2021) From wastewater treatment to resources recovery through hydrothermal treatments of municipal sewage sludge: a critical review. Process Saf Environ Prot 151:101–127

Drechsel P, Scott A, Sally R, Redwood M, Bachir A (2010) Wastewater irrigation and health: assessing and mitigating risk in low-income countries. International Water Management Institute (ed). Earthscan, London, UK

Duncan M (2009) Waste stabilization ponds: past, present and future. Desalin Water Treat 4:85–88

Engida TM, Mekonnen A, Wu JM, Dong XU, Wu ZB (2020) Review paper on beverage agro-industrial wastewater treatment plant bio-sludge for fertilizer potential in Ethiopa. Appl Ecol Environ Res 18(1):33–57

Freitas F, Alves VD, Reis MA (2011a) Advances in bacterial exopolysaccharides: from production to biotechnological applications. Trends Biotechnol 29:388–398

Freitas F, Alves VD, Torres CA, Cruz M, Sousa I, Melo MJ (2011b) Fucose-containing exopolysaccharide produced by the newly isolated *Enterobacter* strain A47DSM 23139. Carbohyd Polym 83:159–165

Freitas F, Alves VD, Pais J, Costa N, Oliveira C, Mafra L (2009) Characterization of an extracellular polysaccharide produced by a *Pseudomonas* strain grown on glycerol. Bioresour Technol 100:859–865

Gehrke I, Geiser A, Somborn-Schulz A (2015) Innovationsin nanotechnology for water treatment. Nanotechnol Sci Appl 8:1–17

Gernaey KV, Van Loosdrecht MC, Henze M, Lind M, Jørgensen SB (2004) Activated sludge wastewater treatment plant modeling and simulation: state of the art. Environ Model Softw 19:763–783

Gopal B (1999) Natural and constructed wetlands for wastewater treatment: potentials and problems. Water Sci Technol 40(3):27–35

Gupta P, Diwan B (2017) Bacterial Exopolysaccharide mediated heavy metal removal: a review on biosynthesis, mechanism and remediation strategies. Biotechnol Rep 13:58–71

Herrero M, Stuckey DC (2015) Bioaugmentation and its application in wastewater treatment: a review. Chemosphere 140:119–128

Hesnawi R, Dahmani K, Al-Swayah A, Mohamed S, Mohammed SA (2014) Biodegradation of municipal wastewater with local and commercial bacteria. Procedia Eng 70:810–814

Hoekstra AY, Mekonnen MM (2012) The water footprint of humanity. Proc Natl Acad Sci 109:3232–3237

Inoue Y, Hoshino M, Takahashi H, Noguchi T, Murata T, Kanzaki Y (2002) Bactericidal activity of Ag-zeolite mediated by reactive oxygen species under aerated conditions. J Inorg Biochem 92:37–42

Jain K, Patel AS, Pardhi VP, Flora SJS (2021) Nanotechnology in wastewater management: a new paradigm towards wastewater treatment. Molecules 26(6):1797

Jasim NA (2020) The design for wastewater treatment plant (WWTP) with GPS X modelling. Cogent Eng 7(1):1723782

Jeong H, Kim H, Jang J (2016) Irrigation water quality standards for indirect wastewater reuse in agriculture: a contribution toward sustainable wastewater reuse in South Korea. Water 8:169

Jodar-Abellan A, López-Ortiz MI, Melgarejo-Moreno J (2019) Wastewater treatment and water reuse in Spain. Current situation and perspectives. Water 11:1551

Kamali M, Persson KM, Costa ME, Capela I (2019) Sustainability criteria for assessing nanotechnology applicability in industrial wastewater treatment: current status and future outlook. Environ Int 125:261–276

Katsoyiannis IA, Gkotsis P, Castellana M, Cartechini F, Zouboulis AI (2017) Production of demineralized water for use in thermal power stations by advanced treatment of secondary wastewater effluent. J Environ Manag 1(190):132–139

Kesari KK, Behari J (2008) Ultrasonic impact on bacterial population in sewage sample. Int J Environ Waste Manag 2:233–244

Kesari KK, Kumar S, Verma HN, Behari J (2011a) Influence of ultrasonic treatment in sewage sludge. Hydrol: Curr Res 2:115

Kesari KK, Verma HN, Behari J (2011b) Physical methods in wastewater treatment. Int J Environ Technol Manag 14:43–66

Kesari KK, Soni R, Jamal QMZ, Pooja T, Lal JA, Jha NK, Siddiqui MH, Kumar P, Tripathi V, Ruokolainen J (2021) Wastewater treatment and reuse: a review of its applications and health implications. Water Air Soil Pollut 232:208

Kesari KK, Jamal QMS (2017) Review processing, properties and applications of agricultural solid waste: effect of an open burning in environmental toxicology. In: Kesari K (ed) Perspectives in environmental toxicology. Environmental science and engineering. Springer, Cham, pp 161–181. Chapter 8

Kim BH, Kang Z, Ramanan R, Choi JE, Cho DH, Oh HM et al (2014) Nutrient removal and biofuel production in high rate algal pond using real municipal wastewater. J Microbiol Biotechnol 24:1123–1132

Kumar M, Kesari KK, Behari J (2010) Low frequency ultrasonic irradiation of activated sludge. Int J Environ Pollut 43:52–65

Łuczkiewicz A, Jankowska K, Fudala-Książek S, Olańczuk-Neyman K (2010) Antimicrobial resistance of fecal indicators in municipal wastewater treatment plant. Water Res 44:5089–5097

Markin A, Lepikhin AV, Ulianov SV (2020) Sewage treatment. In: Innovatika-2020 Proceedings XVI international school-conference of students, postgraduates, and young scientists, Tomsk, Russia, April 2020, pp 23–25

Mahfooz Y, Yasar A, Guijian L, Islam QU, Akhtar ABT, Rasheed R, Irshad S, Naeem U (2020) Critical risk analysis of metals toxicity in wastewater irrigated soil and crops: a study of a semi-arid developing region. Sci Rep 10:12845

Mahmood A, Malik RN (2014) Human health risk assessment of heavy metals via consumption of contaminated vegetables collected from different irrigation sources in Lahore, Pakistan. Arab J Chem 7(1):91–99

Mehmood A, Mirza AS, Choudhary MA, Kim KH, Raza W, Raza N, Lee SS, Zhang M, Lee JH, Sarfraz M (2019) Spatial distribution of heavy metals in crops in a wastewater irrigated zone and health risk assessment. Environ Res 168:382–388

Melanta S (2008) Aquatic bacteria removal using carbon nanotubes. Biological and Agricultural Engineering Undergraduate Thesis. University of Arkansas

Mishra S, Kumar R, Kumar M (2023) Use of treated sewage or wastewater as an irrigation water for agricultural purposes-environmental, health, and economic impacts. Total Environ Res Themes 6:100051

Mohammad AW, Teow YH, Ang WL, Chung YT, Oatley-Radcliffe DL, Hila N (2015) Nanofiltration membranes review: recent advances and future prospects. Desalination 356:226–254

Mohammad M, Mazahreh N (2003) Changes in soil fertility parameters in response to irrigation of forage crops with secondary treated wastewater. Commun Soil Sci Plant Anal 34:181–1294

Mohsen MS (2004) Treatment and reuse of industrial effluents: case study of a thermal power plant. Desalination 167:75–86

Moore AL, Shi L (2014) Emerging challenges and materials for thermal management of electronics. Mater Today 17(4):163–174

Mostafaii G, Chimehi E, Gilasi HR, Iranshahi L (2017) Investigation of zinc oxide nanoparticles effects on removal of total coliform bacteria in activated sludge process effluent of municipal wastewater. J Environ Sci Technol 1:49–55

Nasir O, Ashraf S, Rizwan MA, Javed A, Amjad K, Bibi R, Nazeer MN (2024) Conventional and advanced wastewater treatment techniques; development of sustainable environment and production of biofertilizers. Int J Res Publ Rev 5(1):5491–5499

Nizam NUM, Hanafiah MM, Noor IM, Karim HIA (2020) Efficiency of five selected aquatic plants in phytoremediation of aquaculture wastewater. Appl Sci 10:2712

Nmaya MM, Agam MA, Matias-Peralta HM, Yabagi JA, Kimpa MI (2017) Freshwater green microalga for bioremediation of river Melaka heavy metals contamination

Njuguna SM, Makokha VA, Yan X, Gituru RW, Wang Q, Wang J (2019) Health risk assessment by consumption of vegetables irrigated with reclaimed wastewater: a case study in Thika (Kenya). J Environ Manag 231:576–581

Obaideen K, Shehata N, Taha Sayed Taha E, Ali Abdelkareem Ali M, Mohamed S, Olabi AGM (2022) The role of wastewater treatment in achieving sustainable development goals (SDGs) and sustainability guideline. Energy Nexus 7:100112. https://doi.org/10.1016/j.nexus.2022.10011

Pan B, Xing BS (2008) Adsorption mechanisms of organic chemicals on carbon nanotubes. Environ Sci Technol 42:9005–9013

Panepinto D, Fiore S, Zappone M, Genon G, Meucci L (2016) Evaluation of the energy efficiency of a large wastewater treatment plant in Italy. Appl Energy 161:404–411

Poustie A, Yang Y, Verburg P, Pagilla K, Hanigan D (2020) Reclaimed wastewater as a viable water source for agricultural irrigation: a review of food crop growth inhibition and promotion in the context of environmental change. Sci Total Environ 739:139756

Rani M, Keshu, Shanker U (2024) Green construction of biochar@$NiFe_2O_4$ nanocomposite for highly efficient photocatalytic remediation of pesticides from agriculture wastewater. Chemosphere 352

Saravanan A, Kumar PS, Jeevanantham S, Karishma S, Tajsabreen B, Yaashikaa PR, Reshm B (2021) Effective water/wastewater treatment methodologies for toxic pollutants removal: processes and applications towards sustainable development. Chemosphere 280:130595

Saravanan A, Deivayanai VC, Kumar PS, Rangasamy G, Hemavathy RV, Harshana T, Gayathri N, Alagumalai K (2022) A detailed review on advanced oxidation process in treatment of wastewater: mechanism, challenges and future outlook. Chemosphere 308(3):136524

Sengupta S, Nawaz T, Beaudry J (2015) Nitrogen and phosphorus recovery from wastewater. Curr Pollut Rep 1:155–166

Seruga P, Krzywonos M, Wilk M (2020) Treatment of by-products generated from anaerobic digestion of municipal solid waste. Waste Biomass Valorization 11(9):4933–4940

Singh A, Sawant M, Kamble SJ, Herlekar M, Starkl M, Aymerich E (2019) Performance evaluation of a decentralized wastewater treatment system in India. Environ Sci Pollut 26:21172–21188

Smith RG (1995) Water reclamation and reuse. Water Environ Res 67:488–495

Srivastava A, Srivastava ON, Talapatra S, Vajtai R, Ajayan PM (2004) Carbon nanotube filters. Nat Mater 3:610–614

Srivastava A, Singh R, Rajput VD, Tatiana Minkina T, Agarwal S, Garg MC (2022) A systematic approach towards optimization of brackish groundwater treatment using nanofiltration (NF) and reverse osmosis (RO) hybrid membrane filtration system. Chemosphere 303(3):135230

Stottmeister U, Wießner A, Kuschk P, Kappelmeyer U, Kästner M, Bederski O, Müller RA, Moormann H (2003) Effects of plants and microorganisms in constructed wetlands for wastewater treatment. Biotechnol Adv 22:93–117

Tanji KK, Kielen NC (2002) Agricultural drainage water management in arid and semi-arid areas. In: FAO Irrigation and Drainage Paper. Food and Agriculture Organization, p 61

UN World Water Development Report (2017) Wastewater: the untapped resource UNESCO's world water assessment programme (WWAP)

Ungureanu N, Vlăduţ V, Voicu G (2020) Water scarcity and wastewater reuse in crop irrigation. Sustainability 12(21): 9055

Vergili I (2013) Application of nanofiltration for the removal of carbamazepine, diclofenac and ibuprofen from drinking water sources. J Environ Manage 127:177–187

Villarín MC, Merel S (2020) Paradigm shifts and current challenges in wastewater management. J Hazard Mater 390:122–139

Vogelmann ES, Awe GO, Prevedello J (2016) Selection of plant species used in wastewater treatment. In: Wastewater treatment and reuse for metropolitan regions and small cities in developing countries, pp 1–10

Vymazal J (2011) Constructed wetlands for wastewater treatment: five decades of experience. Environ Sci Technol 45(1):61–69

WHO-World Health Organization (2006) Guidelines for the safe use of wastewater, excreta and greywater. In: Wastewater use in agriculture, 2nd ed. WHO-World Health Organization

Xiao R, Wang S, Li R, Wang JJ, Zhang Z (2017) Ecotoxicology and environmental safety soil heavy metalcontamination and health risks associated with artisanal gold mining in Tongguan, Shaanxi, China. Ecotoxicol Environ Saf 141:17–24

Yang J, Jia R, Gao Y, Wan W, Cao P (2017) Thereliability evaluation of reclaimed water reused in power plant project. IOP conference series: earth and environmental science, vol 100. pp 012189

Zamora S, Marín-Muñíz JL, Nakase-Rodríguez C, Fernández-Lambert G, Sandoval L (2019) Wastewater treatment by constructed wetland eco-technology: influence of mineral and plastic materials as filter media and tropical ornamental plants. Water 11:2344

Zekić E, Vuković Z, Halkijević I (2018) Application of nanotechnology in wastewater treatment. GRAĐEVINAR 70:315–323

Zhang Y, Shen Y (2017) Wastewater irrigation: past, present and future. Wastewater treatment: aims and challenges. Water 6(3):e1234

Zhang B, Zheng JS, Sharp RG (2010) Phytoremediation in engineered wetlands: mechanisms and applications. Procedia Environ Sci 2:1315–1325
Zimmels Y, Kirzhner F, Roitman S (2004) Use of naturally growing aquatic plants for wastewater purification. Water Environ Res 76:220

Chapter 12
Advanced Biological Nutrient Removal Techniques for Enhanced Water Quality

Aqsa Arooj, BiLong Chen, Shumaila Batool, M. D. Sohrab Ali, Jia-Cheng E. Yang, and Murad Muhammad

Abstract Nutrients, including nitrogen, phosphorus, hydrogen, oxygen, and carbon, are essential chemicals required by living organisms for growth, development, and the execution of many catalytic, functional, structural, and biochemical processes crucial for survival. However, discharging nutrient-laden waste streams into aquatic environments results in eutrophication. Hence, removing nutrients from wastewater is essential to adhere to the stringent nutrient discharge regulations. Equally important is the extraction of nutrients from waste streams to achieve a circular economy and prevent the exhaustion of limited resources. Advanced techniques for biological nutrient removal (BNR) are a key innovation in wastewater treatment to improve water quality by efficiently eliminating nitrogen and phosphorus compounds. This chapter thoroughly evaluates current knowledge regarding advancements in biological BNR technologies, exploring various traditional and innovative treatment methods widely used for nutrient removal from household wastewater. Moreover, strategies to address critical challenges such as operational stability, cost-effectiveness, and scalability are explored. In addition, the chapter delves into the environmental and regulatory

---

A. Arooj (✉) · B. Chen · M. D. Sohrab Ali · J.-C. E. Yang
Key Laboratory of Urban Pollutant Conversion, Institute of Urban Environment, Chinese Academy of Sciences, Fujian, China
e-mail: arooj@iue.ac.cn

A. Arooj
College of Earth and Environmental Sciences, University of Punjab, Quaid-E-Azam Campus, Lahore, Pakistan

S. Batool
Department of Zoology, Wildlife and Fisheries, University of Agriculture, Faisalabad, Pakistan

M. Muhammad
State Key Laboratory of Desert and Oasis Ecology, Key Laboratory of Ecological Safety and Sustainable Development in Arid Lands, Xinjiang Institute of Ecology and Geography, Chinese Academy of Sciences, Beijing, China

Xinjiang Key Laboratory of Biodiversity Conservation and Application in Arid Lands, Xinjiang Institute of Ecology and Geography, Chinese Academy of Sciences, Beijing, China

A. Arooj · B. Chen · M. D. Sohrab Ali · J.-C. E. Yang · M. Muhammad
University of Chinese Academy of Sciences, Beijing, China

A. M. Yatoo et al. (eds.), *Wastewater Monitoring and Management by Advanced Approaches*, Advances in Wastewater Research,
https://doi.org/10.1007/978-981-95-2601-7_12

consequences of utilizing advanced BNR techniques, highlighting their importance in sustainable water management and safeguarding aquatic ecosystems. The chapter underscores the significance of improvements in monitoring and control systems for maximizing BNR performance. Overall, this comprehensive manuscript underscores the critical role of advanced BNR techniques in enhancing water quality and promoting environmental sustainability.

Keywords Nutrient removal · Biological technology · Enhanced water quality · Sustainability · Wastewater treatment

12.1 Introduction

Organisms require nutrients for growth, development, and the execution of different functional, structural, catalytic, and biochemical processes vital for life. The fundamental elements of living organisms comprise hydrogen, oxygen, nitrogen, phosphorus, and carbon (Rout et al. 2021). Phosphorus and nitrogen are crucial for evolution and ecology, while carbon is present in every living thing and is involved in almost every chemical and metabolic process. All land-dwelling creatures, from microbes to plants and animals, rely on nitrogen and phosphorus as macronutrients for development and reproduction (Weissbrodt et al. 2020). Most water sources possess exceedingly low nutrient concentrations, and even minimal fluctuations in nutrient input can significantly alter the structure and activities of various ecosystem organisms (Poikane et al. 2019). Therefore, constant additions of phosphorus (> 0.5–1 mg/L) and/or nitrogen compounds (NO_3^--N > 10 mg/L) to water channels lead to higher eutrophication rates (N'Goran et al. 2019). This significant ecological and environmental hazard harms human health and threatens aquatic ecosystem biodiversity. Table 12.1 summarizes some of the detrimental consequences of nutrients on both the environment and human health. Algal blooms are the easiest-to-understand expression of this phenomenon. Algal blooms, the only drivers of toxins that poison fish and infect people and other animals, are triggered by a fast build-up in the algal population (Brown et al. 2020). Increased algal biomass means drinking water needs larger chlorine dosages, raising disinfection by-products (Park et al. 2019). The degradation and decay of algal biomass cause an oxygen shortage in aquatic environments, which lowers the biodiversity of the populations of microorganisms and macroorganisms. Moreover, the presence of nutrients in wastewater promotes the growth of *Pfiesteria* and speeds up the synthesis of microsystem, which in humans results in hepatocellular carcinoma (Verma et al. 2020; Park et al. 2019). Similarly, excess NO_3^--N in drinking water can induce severe poisoning in animals, harm the thyroid gland by preventing iodine absorption, and result in miscarriages in expectant mothers. It can also cause blue baby syndrome in neonates (Awuchi et al. 2020; Coffman et al. 2021). Nitrosoamides and nitrosoamines are carcinogenic chemicals that can be created when NO_2^- and amines interact chemically or enzymatically

(Badid and Hadji 2023). Ammonia nitrogen can have a toxic effect on the reproduction and mortality of fish, amphibians, and unionid mussels. Its concentration ranges from 0.1–10 mg/L (Gibson 2019).

Increased ammonia levels in inland waters have been linked to the significant decline in the unionidae mussel population. The World Health Organization (WHO) and the U.S. Environmental Protection Agency (USEPA) have set maximum discharge thresholds for certain nutrient species, acknowledging the possible environmental risks (Boyd and Boyd 2020b). These limits are as follows: < 10 mg/L for

Table 12.1 Environmental and health effects of various nutrients

Nutrients	Health and environmental effects	References
Ammonia	• It is very toxic to aquatic organisms like fish, amphibians, invertebrates, etc. • Responsible for acidifying ground and water bodies, thus harming plant and animal life • Causes lung and chronic respiratory diseases, etc.	Castell et al. (2020)
Nitrogen	Nitrogen determines the productivity of aquatic ecosystems. So, increased nitrogen concentrations result in eutrophication	Boyd and Boyd (2020a)
Nitrous oxide	It is a potential greenhouse gas whose emission contributes to global climate change	Jones et al. (2023)
Nitric oxide	Associated with a variety of respiratory diseases and also adversely affects vegetation	Lundberg and Weitzberg (2022)
Nitrite	Carcinogenic nitrosoamines and nitrosoamides are formed due to enzymatic or chemical reactions with amines	Nikolova and Belichovska (2024)
Nitrate	Excess nitrate causes methemoglobinemia in infants, acute poisoning in livestock, miscarriages among pregnant women, etc.	McInnes et al. (2023)
Phosphorus	Phosphorus is the limiting nutrient, and its rising quantities in surface waters enhance the proliferation of phosphate-dependent species, such as algae and duckweed, resulting in eutrophication • Excess total phosphorus boosts algal blooms	Bhambri and Karn (2020)
Organic phosphate	• Many organic phosphates are essential for life, whereas some are very toxic • Exposure to organic phosphates represses the antimicrobial defenses of sea urchins	Abbas et al. (2020)
Polyphosphate	• Present water treatment processes and in detergents as water softeners cause water pollution	Toledo et al. (2021)
Orthophosphate	• High concentrations of phosphate cause kidney damage, osteoporosis, etc. • Reduced phosphate levels may lead to diminished appetite, anxiety, weariness, brittle bones, irregular respiration, irritability, numbness, and weakness	Malik et al. (2023)

nitrate, < 0.1 mg/L for nitrite, < 5 mg/L for ammonia, 0.5–1.0 mg/L for phosphate, etc. (EPA 2009; WHO 2011).

Domestic wastewater, the most prevalent form, plays a substantial role in global biogeochemical cycles by serving as a fertilizer supply. The principal source of nonpoint nutrient contamination in most of the developing countries, constituting over 50% of the total nonpoint nutrient strain in aquatic ecosystems, is the direct release of domestic wastewater, whether untreated or inadequately treated (Mushtaq et al. 2020). According to Lee et al. (2021), household wastewater provided roughly 7.7 Tg of nitrogen and 1.0 Tg of phosphorus to the world's oceans yearly by the end of the twentieth century. Residential wastewater accounts for 6% and 50% of the flux of total nitrogen (TN) and total phosphorus (TP), respectively. Consequently, it is essential to mitigate the environmental degradation caused by nutrient runoff into aquatic systems, particularly surface water (Kamilya et al. 2022). This seems a significant and challenging undertaking for academics working in the field today. Even though academics have recently focused a lot of attention on nutrient elimination to control eutrophication, water quality problems still rank among the worst worldwide (Akinnawo 2023; Zhang et al. 2020b; Shortle et al. 2020).

Research indicates that for wastewater treatment to be sustainable, nutrients must be recovered from waste streams, facilitating a circular bioeconomy and averting the depletion of scarce resources. The fundamental principles of the circular bioeconomy are reduction, reuse, and recycling. Extracting recyclable nutrients from diverse wastewater sources may significantly contribute to meeting the escalating demands resulting from population growth (Kundu et al. 2022). Thus, from the circular bioeconomy's perspective, nutrients enriched in wastewater can be recovered and used again. For instance, the leading cause of the quickly depleting phosphate rock, the only global phosphorus reserve, is the growing phosphorus demand brought on by population growth (De Boer et al. 2019). However, extracting phosphorus from household sewage might meet about 15–20% of the world's total phosphorus needs. As a result, the recycling or recovery of phosphorus from the anthropogenic phosphorus cycle is receiving more attention (Robles et al. 2020). This chapter aims to thoroughly examine biological nutrient removal (BNR) technologies, emphasizing their importance in improving water quality and advancing sustainable development. It aims to delineate the goals of increasing resource recovery, cutting operating costs, and optimizing the removal of nutrients. The chapter also discusses the innovations and difficulties needed to bring these technologies to a broader audience, highlighting their contribution to the advancement of environmental sustainability and the circular economy.

Most wastewater treatment plants (WWTPs) use preliminary treatment to physically eliminate floatable solids by skimming and to separate settleable organic and inorganic materials via sedimentation. The secondary treatment phase guarantees the biological elimination of remaining organic substances and minimally absorbed nutrients (Wang et al. 2021b).

Consequently, following standard secondary treatment, a considerable quantity of nutrients and certain suspended and dissolved substances may remain in the effluent of WWTPs. Components of wastewater that are not amenable to secondary treatment, including nutrients, refractory organics, and heavy metals, necessitate an advanced or tertiary treatment system (Zhou et al. 2022). As previously mentioned, the primary point source of nutrient pollution from these treatment plants is the nutrient-loaded secondary effluents from the conventional WWTPs, since they lack tertiary treatment facilities (Robles et al. 2020). The first attempt at developing techniques for nutrient removal was made with traditional aerobic technologies based on the activated sludge process (ASP) for managing domestic wastewater in anticipation of potential nutrient removal (Ashraf et al. 2021).

Nonetheless, substantial financial obstacles impede the extensive use of these technologies, especially in poor countries, owing to the elevated capital and operational expenses associated with their implementation. Likewise, physical methods are prohibitively expensive as they necessitate membranes, which are inefficient due to their removal of around 10% of total nutrients and the problems arising from membrane fouling (Bagheri et al. 2019).

Furthermore, before the discharge of sewage into surface waterways, chemical dosing is frequently used to reduce nutrient concentration in the effluent effectively. Regrettably, the chemicals are costly, their handling and storage pose significant risks, and the method generates increased sludge, elevating operational and maintenance costs (Saravanan et al. 2021). Because of the previously mentioned reasons, even the traditional chemical and physical treatment methods cannot satisfy the stringent nutrient release standards. Biological nutrient removal has developed as a result.

12.2 Basic Principles of Biological Nutrient Removal

The activated sludge process (ASP) is modified by biological nutrient removal (BNR) methods, which eliminate chemical oxygen demand (COD), TN, and TP by the metabolic processes such as nitrification, denitrification, as well as biological phosphorus removal. These microorganisms can flourish in hypoxic, anaerobic, and aerobic environments (Parde et al. 2024). Oxygen is the electron acceptor in the aerobic zone, whereas nitrogen oxides are the electron acceptor in the anoxic zone. The anaerobic zone is characterized by organic matter, which is an electron acceptor and a donor. These roles are utilized to establish the redox conditions (Ehrlich 2021). Nitrogen removal benefits from aerobic-anoxic conditions, but phosphorus removal requires alternating anaerobic–aerobic conditions (He et al. 2022).

12.2.1 Biological Nitrogen Removal

The method of biological nitrogen removal typically involves a two-step procedure, commencing with nitrification and concluding with denitrification. In the nitrification process, ammonium is initially oxidized to nitrite (termed nitritation) by ammonia-oxidizing bacteria (AOB), subsequently followed by the oxidation of nitrite to nitrate (referred to as nitration) by nitrite-oxidizing bacteria (NOB) (Hernandez-Martinez et al. 2021). The energy sources used by the AOBs and NOBs are ammonia/nitrite, carbon dioxide (CO_2), and oxygen (O_2) as an electron acceptor (Shabbirahmed et al. 2022). The most prominent AOB is *Nitrosomonas*, although other AOBs that can carry out the nitridation process include *Nitrosospira*, *Nitrosovibrio*, *Nitrosococcus*, etc. (Anjali and Sabumon 2022). Similarly, *Nitrobacter* is the most prominent NOB; however, the nitration process is also known to involve *Nitrosospira*, *Nitrococcus*, *Nitrocystis*, and *Nitrospina* (Pattnaik et al. 2023). In the beta subdivision of proteobacteria, AOBs are connected, whereas NOBs are associated with the alpha subdivision. The stoichiometry of the nitrification process is as follows (Rout et al. 2021):

$$2NH_4^+ + 3O_2AOB^- \rightarrow 2NO_2^- + 4H^+ + 2H_2O + \text{New cells} \quad (12.1)$$

$$2NO_2^- + 2O_2NOB^- \rightarrow 2NO_3^- + \text{New cells} \quad (12.2)$$

The overall reaction is:

$$NH_4^+ 1.86O_2 + 0.098CO_2 \rightarrow 0.98NO_3^- + 1.98H^+ + 0.094H_2O + 1.92H_2CO_3 + 0.0196C_5H_7O_2N(\text{new cells}) \quad (12.3)$$

In the nitrification process, 0.08 g of inorganic carbon is necessary to produce 0.16 g of new cells, whereas 4.57 g of O_2 and 7.14 g of $CaCO_3$ (alkalinity) are consumed for each gram of NH_4 + oxidized. Similarly, facultative heterotrophic bacteria complete denitrification by a series of sequential reduction processes (Conca 2021). The enzymatic actions of nitrate, nitrite, nitric oxide, and nitrous oxide reductases convert nitrate to nitrite (denitratation) and then nitrite to di-nitrogen gas (denitritation), involving intermediates such as nitric oxide, nitrous oxide, and nitrogen gas. During denitrification, the bacteria use organic materials as a carbon and energy source, with nitrate or nitrite acting as electron acceptors. Nitrate is either converted into ammonium for biosynthesis in the assimilatory phase of denitrification or used as an electron acceptor for respiration in the dissimilatory process. The most common genera of heterotrophic denitrifying bacteria are *Pseudomonas*, *Bacillus*, *Neisseria*, *Chromobacterium*, *Alcaligenes*, and *Paracoccus* (Ni et al. 2016). The governing equations of the heterotrophic biological denitrification process are as follows (Song et al. 2021):

$$NO_3^- \text{Nitrate reductase} \rightarrow NO_2^- \text{Nitrite reductase} \rightarrow NO(g) \text{Nitric oxide reductase} \rightarrow N_2O(g) \text{Nitrous oxide reductase} \rightarrow N_2(g) \quad (12.4)$$

$$NO_3^- + \text{Organic Carbon Source Heterotrophic Microbes} \rightarrow N_2 + CO_2 + H_2O + OH^- + \text{Cell} \quad (12.5)$$

The specific type of electron donor determines the stoichiometry of the denitrification process. The overall reaction is as follows when organic carbon is used as an electron donor (Aponte-Morales et al. 2013):

$$\begin{aligned} 0.02C_{10}H_{19}O_3N + 0.13429NO_3^- + 0.13429H^+ \rightarrow 0.01429C_5H_7O_2N \\ + 0.06N_2 + 0.10857CO_2 + 0.15714H_2O + 0.02HCO_3^- \end{aligned} \tag{12.6}$$

In the denitrification process, 50% of the alkalinity diminished during nitrification is restored, yielding 3.57 g of $CaCO_3$ for each gram of NO_3^- reduced (Gyawali 2021). Denitrification must follow nitrification to eliminate total nitrogen (TN); Fig. 12.1a depicts the main technique of biological nitrogen removal.

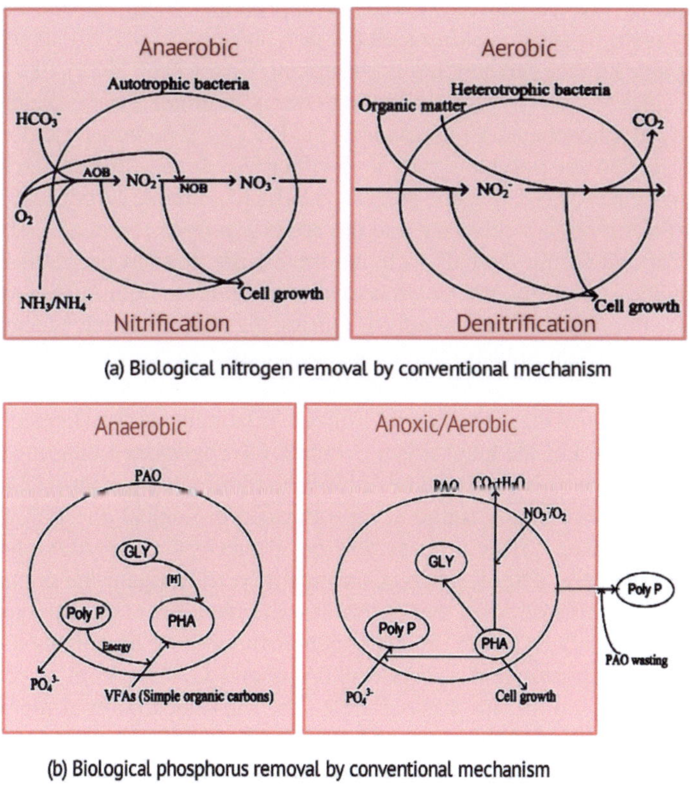

Fig. 12.1 **a** Biological nitrogen removal by conventional mechanism, **b** biological phosphorus removal by conventional mechanism

12.2.2 Biological Phosphorus Removal

Eliminating biological phosphorus from wastewater entails the integration of cellular biomass with organically bound phosphorus, polyphosphate, and orthophosphate. The biomass is eventually discarded by desludging it from the treatment unit (Landa-Acuña et al. 2021). This process, facilitated by phosphate-accumulating organisms (PAOs), is referred to as enhanced biological phosphorus removal (EBPR) or luxuriant phosphorus uptake (Akram et al. 2022). *Accumulibacter* and *Acinetobacter* sp. are two instances of the beta-2 subclass of proteobacteria, which comprises the principal microbial populations involved in the EBPR process. For EBPR, mixtures of anaerobic and anoxic/aerobic conditions are needed (Anjali and Sabumon 2024). The PAOs in the anaerobic phase transform volatile fatty acids (VFAs) within the cells into polyhydroxyalkanoates (PHA). The principal energy source required for this activity is the hydrolysis of intracellular polyphosphate (poly-P), which liberates orthophosphate into the surrounding environment (Zhang et al. 2023). PHA is synthesized from intracellular glycogen, utilizing energy liberated during the degradation of reserved polyphosphates. Consequently, the PAO's PHA concentrations increase, while its glycogen and polyphosphate levels decrease. In the ensuing aerobic/anoxic environment, the PAOs oxidize the preserved PHA to generate carbon and energy for cellular proliferation, concurrently making some glycogen (Ni et al. 2021). Poly-P bonds are formed within the cell, utilizing the energy that has been released. After its solubilization, the orthophosphate is extracted from the liquid phase and assimilated into the microbial cell as the poly-P. During the aerobic/anoxic phase, the PAOs frequently retain more poly-P than what was released in the anaerobic phase. The proportion of biomass discarded with the sludge is utilized for the disposal of the accumulated poly-P following its extraction from the treatment system (Iannacone 2021). Figure 12.1b illustrates the mechanism in detail. Glycogen-accumulating organisms (GAOs), characterized by their limited ability to eliminate phosphorus, are another kind of organism commonly found in the enhanced biological phosphorus removal (EBPR) process. The EBPR process will be unsuccessful if GAOs predominate, as GAOs periodically assault PAOs. Consequently, to mitigate the contamination, biological phosphorus removal requires thorough process control of polyphosphate-accumulating organisms (PAOs) by glycogen-accumulating organisms (GAOs) and to regulate the competitiveness between PAOs and GAOs (Ni et al. 2022). The progress of biological therapy has paralleled the implementation of more stringent regulations on nutrition waste. In recent decades, numerous reactor configurations have been devised to create optimal conditions for microorganisms to metabolize and remove nitrogen, phosphorus, or both concurrently.

12.3 Advanced BNR Techniques

12.3.1 Membrane Bioreactors (MBRs)

Biological, inorganic, and organic contaminants can all be effectively removed from wastewater streams using membrane bioreactors (MBRs). As the name implies, the technology combines membrane filtering with a bioreactor (Rahman et al. 2023). To improve mixing and reduce fouling, the system uses submerged membranes in bioreactors with coarse bubble aeration. The method varies from a conventional wastewater treatment plant in that membrane filtration is used to achieve separation rather than sedimentation in a secondary clarification tank (Du et al. 2020).

Wastewater-dissolved solids can be separated via membrane filtration, an improved type. The pore size is determined for each pressure-driven membrane process, ranging from microfiltration (MF) to reverse osmosis (RO) (El Batouti et al. 2021). Particulate matter or suspended solids with sizes between 0.1 and 10 microns (μm) can be removed with microfiltration (MF). Reverse osmosis (RO) membranes can separate materials less than 0.001 μm. At the same time, Ultrafiltration (UF) can recover macromolecules in the range of 0.01 to 0.1 μm, and Nanofiltration (NF) can remove particulate from 0.001 to 0.01 μm (Khatkar et al. 2023).

Metallic membranes also exist, although most comprise various plastic and ceramic materials. The most often utilized materials include polymers like polyethylsulphone (PES), polyacrylonitrile (PAN), polyethylene (PE), polypropylene (PP), polyvinylidene difluoride (PVDF), celluloses, polyamides, polysulphone, and charged polysulphone (Paul et al. 2023). These polymeric materials are appealing because of their physical and chemical durability. They possess hydrophobic properties, and it is widely recognized that the majority of membrane-foulant interactions are hydrophobic, resulting in a higher propensity for fouling in hydrophobic membranes than hydrophilic ones (Warsinger et al. 2018). Consequently, all commercially available membranes undergo modification through grafting, chemical oxidation, organic chemical reactions, or plasma treatment to attain a more hydrophilic surface. The alteration procedure and the membrane module manufacturing technique often differentiate one membrane from another (Shukla et al. 2022).

Membrane bioreactor (MBR) systems offer numerous advantages compared to conventional biological systems, including enhanced automation, reduced spatial requirements, and superior effluent quality. Specifically, MBRs have reduced hydraulic retention times due to their operation at elevated volumetric loading rates. Compared to a conventional system, reduced space is required due to the brief retention periods (Arroyo and Molinos-Senante 2018). Despite being superfluous, more conventional solids residence durations (SRTs) have been utilized. Historically, MBRs have operated with extended SRTs, hence diminishing sludge production. The MBR effluent exhibits biochemical oxygen demand (BOD), total suspended solids (TSS), low concentrations of phosphorus, and microbes (Qrenawi and Rabah 2023).

This facilitates high-level disinfection. Effluents can be readily marketed for alternative use, such as irrigation, or discharged into surface water bodies. The primary drawback of MBR systems is that they often incur higher operational and capital costs for equivalent throughput than conventional systems. The costs involved include cleaning, fouling management, and ultimately membrane replacement (Das et al. 2022). Air scouring is required to regulate the proliferation of microorganisms on the membranes, which also escalates energy costs. Moreover, the sludge produced by this system may settle gradually, requiring chemicals to generate appropriate biosolids for proper disposal (Hussain et al. 2021).

12.3.2 Anammox (Anaerobic Ammonium Oxidation)

An innovative, autotrophic, and economical replacement for conventional biological nitrogen elimination is anaerobic ammonium oxidation or anammox. The earliest prediction of the bacteria's existence was made using thermodynamic simulations in the 1970s (Dezotti et al. 2018). Anammox bacteria convert ammonium to nitrogen gas in anoxic environments by using nitrite as an electron acceptor. They develop through carbon dioxide fixation (Ponce-Jahen et al. 2024).

Under anaerobic conditions, NH_4^+-N and NO_2^--N are immediately converted into N_2 by anammox bacteria, such as Candidatus *Kuenenia stuttgartiensis* and *Brodia anammoxidans* (GONSALVES 2020). In the preprocess, anammox only needs to oxidize NH_4^+-N to NO_2^--N; oxidizing NO_2^--N to NO_3^--N is not required, lowering the aeration cost. Equation (12.7) provides the anammox reaction equation (Li et al. 2021b). Furthermore, unlike the conventional denitrification process, anammox bacteria do not require an external carbon source, and the organic matter in sewage can be utilized to enhance methanogen recovery rather than oxidizing it to carbon dioxide. Because of its low sludge output, low energy consumption, and low operating costs, anammox technology is regarded as an efficient, environmentally friendly, and economical process (Wu et al. 2022b).

$$NH_4^+ + 1.32NO_2^- + 0.66HCO_3^- + 0.13H^+ \rightarrow 1.02N_2 + 0.26NO_3^- + 2.03H_2O + 0.066CH_2O_{0.5}N_{0.15} \quad (12.7)$$

The traditional biological nitrogen removal from wastewater has undergone significant alterations with the development of the anammox process. The anammox process is a promising and sustainable approach because of its unique qualities, which include minimal biomass yield, needless for aeration, and no need to add foreign carbon sources (Wu et al. 2022a). Although the recently identified anammox process offers new opportunities for nitrogen extraction from wastewater, the main barrier to its application is the anammox microorganisms' slower growth rate (μ max = 0.065d − 1, doubling time (t 1/2 = ln2/μ max) of 11 days), which makes the process difficult for use in wastewater treatment facilities (Srivastava et al. 2024).

Anammox bacteria, on the contrary, have been shown to be quite challenging to grow in pure culture; in fact, *Candidatus Brocadia anammoxidans* has only been seemingly homogenized with the use of Percoll density centrifugation. Researchers concentrate on enriching slowly growing anammox bacteria to satisfy the practical use of the anammox process. Numerous investigations were conducted to enrich anammox organisms using various techniques, including granulation, biofilm formation, and reactor types (Zhang and Okabe 2020; Chen et al. 2019).

The anammox process offers significant advantages over conventional methods for removing nitrogen, which usually involves two energy-intensive steps: denitrification, which uses organic electron donors like methanol to reduce nitrate to nitrogen gas, and nitrification, which requires aeration and the oxidation of ammonia to nitrate. In addition to using a lot of energy, these traditional methods produce a lot of extra sludge and emit many greenhouse gases, such as CO_2, N_2O, and NO, which deplete the ozone layer (Liu et al. 2019). On the other hand, under anaerobic conditions, Anammox bacteria immediately convert ammonium and nitrite into nitrogen gas, negating the need for aeration and additional electron donors. Because only half of the ammonium is converted to nitrite in the partial nitridation/anammox system, less sludge is produced because the bacteria are autotrophic. This further reduces oxygen consumption (Liu et al. 2021).

Additionally, anammox bacteria produce persistent biofilms that maintain high biomass concentrations and effective nitrogen conversion rates, reducing CO_2 emissions and costs by up to 60% (Al-Hazmi et al. 2022). Notwithstanding these advantages, the anammox process is hindered by the bacteria's sluggish growth rate, which doubles every 10 to 2 weeks and causes a delay in reactor startup and recovery. Additionally, the wastewater may need to adjust its pH to maintain an ideal pH of about 8 (Talan et al. 2021). Figure 12.2 presents the process of anammox Bacteria for ammonium-rich wastewater treatment.

12.3.3 Integrated Fixed-Film Activated Sludge (IFAS)

The IFAS process combines suspended activated sludge and fixed-film technologies. Integrating the growth systems that are connected and suspended in IFAS represents a technological advancement over the moving bed biofilm reactor (Kuśnierz et al. 2022). Adding media into the aeration basins allows nitrifying sludge to ages with a much-decreased basin capacity compared to a comparable single-stage activated sludge nitrification process. A surface area that is favorable to microbial proliferation and necessary nitrification is provided by the supplemental medium and mixed liquor suspended solids (MLSS). This leads to MLSS concentrations exceeding 6000 mg/L. However, because the associated biomass stays in the aeration basin, the final clarifiers are not subjected to excessive solid loadings (Ibrahim et al. 2020).

An IFAS system functions analogously to a conventional activated sludge system. Therefore, further action from the operator is unnecessary. As with the old approach, the final clarifiers settle and thicken the MLSS before returning it as return-activated

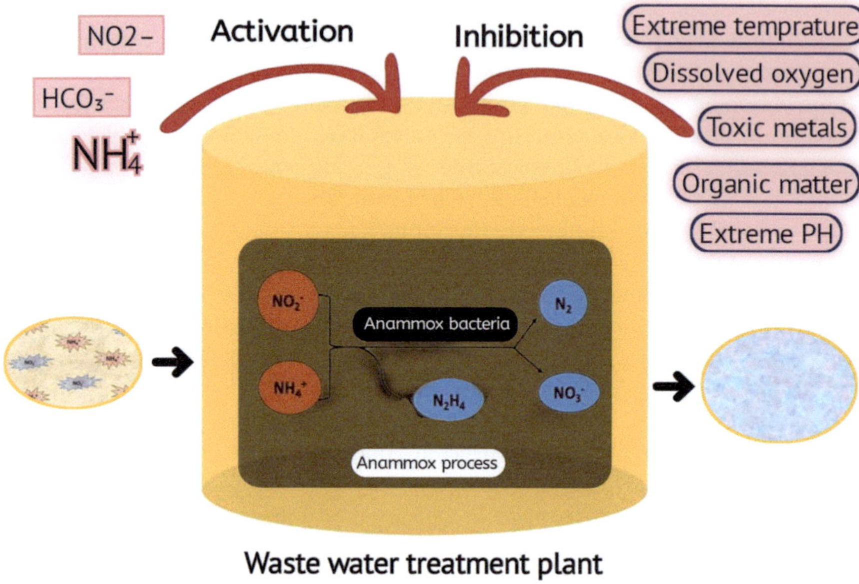

Fig. 12.2 Anammox bacteria in wastewater treatment with high ammonium

sludge (RAS), minus the portion wasted trying to keep the suspended sludge at the right age (Peter 2019). In Fig. 12.3, we can see the primary working mechanism. If related growth factors are included, the actual sludge age can be doubled. Typically, nitrification can be achieved with a suspended sludge with an age of 4–5 days at a temperature of about 10 °C (Karlikanovaite Balıkçı 2019).

During denitrification, heterotrophic denitrifiers must create N_2O as an essential intermediary. No biological nitrogen removal (BNR) process in the IFAS system is complete without it; however, ammonia-oxidizing bacteria (AOB) generate it mainly as a by-product through their denitrification and NH_2OH oxidation pathways, while

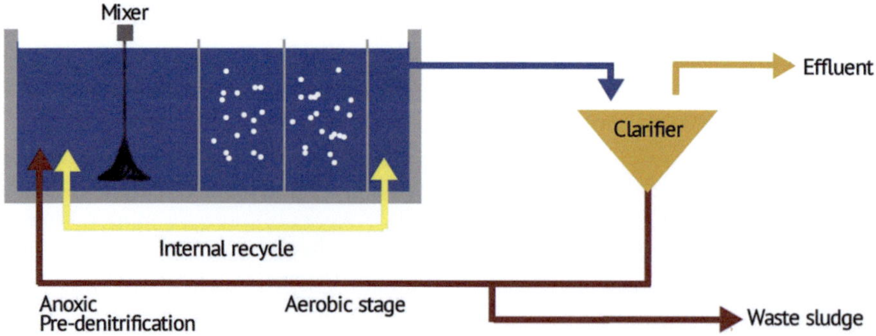

Fig. 12.3 Fixed film activated sludge (IFAS) system

nitrifying (He et al. 2023). Denitrification is the sole biotic sink for N_2O because denitrifiers will inevitably be promoted due to the presence of an oxygen-depleted zone inside a biofilm. One possible mechanism by which biofilms reduce N_2O emissions is this (Jürgens and Taylor 2018). The generation and reduction of N_2O in an IFAS system is anticipated to be significantly more complex than in a suspended sludge system because of the strong relationship between microbial composition, mass transfer, and nutritional competition between biofilm and suspended flocs (Wu et al. 2024). The current consensus holds that, in COD-limited environments like postaerobic tanks, the B phase of the adsorption-biodegradation (A/B) process, or the treatment of specific industrial wastewaters, autotrophic anaerobic bacteria (AOB) are more critical in N_2O production than heterotrophic denitrifiers (Han et al. 2023).

A state-of-the-art approach to wastewater treatment, integrated fixed-film activated sludge (IFAS) is known for its low environmental impact. Anthropogenic composite removal is better, full nitrification is achieved, solids retention duration is longer, and the ecological footprint is lower when using IFAS than the conventional activated sludge approach (Bunce et al. 2018). Numerous pilot and full-scale study outcomes have indicated that IFAS is a desirable alternative. IFAS generally improves sludge settling qualities, increases operating stability, removes ammonia, and combines chemical oxygen demand with > 90% removal rates. Modern IFAS reactors include structures for producing methane, generating energy from algae, or utilizing microbial fuel cells (Waqas et al. 2020).

12.3.4 Bioaugmentation and Microbial Consortia Engineering

Bioremediation breaks down hazardous compounds into their less harmful or nontoxic constituent elements using microbes, and it is regarded as one of the most sustainable remediation techniques (Bala et al. 2022). Although this technology has made progress in recent decades, its advancement is hampered by elements related to microbial consortia, including microbial assembly, functional connections, and the involvement of member species. Its innate issues can be resolved by engineering microbial consortia to reorganize metabolic pathways and social interactions to achieve the intended purpose (Li et al. 2021a). The commercial manufacturing of biomolecules is a common application of microbial consortium engineering. However, there is a need to highlight the engineering of microbial consortia in bioremediation (Gul et al. 2022).

Microbes are ubiquitous in plants, animals, soil, and water. They are extremely important to humanity's survival, well-being, and advancement and are key drivers of the world's biogeochemical cycles (Kaur et al. 2022). Numerous bacteria that may break down organic contaminants and lessen or change heavy metals have been isolated in lab settings. However, when used in the in situ complex contaminated areas, the conversion effectiveness of pollutants by a single species invariably decreases. One species is subjected to stressful circumstances and metabolic hindrances by complex contaminants. In contrast, because different species collaborate to effectively use all types of substrates, microbial consortia typically exhibit resilience and multifunctionality (Paull 2020).

Engineering microbial consortia may achieve long-term stability and optimize interactions between microorganisms and their surroundings. Microbes in microbial communities fight for scarce nutrients and use metabolic products released by other species to attain an edge in fitness. Metabolic interactions greatly impact the use of microbial consortia. Certain species release compounds that can affect how other species interact, promote or inhibit the growth of other species, and even affect how the community functions (Che and Men 2019). A synthetic microbial consortium is characterized as one that is intentionally formed through the co-cultivation of one or more chosen species in a (at least initially) precisely defined medium. In contrast to natural microbial consortia, synthetic microbial consortia may have social networks and metabolic pathways engineered to achieve required functions (Jiang et al. 2023).

Assembling several microbial genera and species is a difficult challenge in engineering microbial consortia. The development of biotechnology facilitates the building of microbial consortia with the intended roles in bioremediation. There are two primary approaches: top-down and bottom-up. Top-down engineering modifies environmental factors to force an already-existing microbial community to carry out the intended tasks. In contrast, bottom-up engineering rearranges the metabolic route and rewires the social networks between bacteria. To engineer microbial consortia for bioremediation, both strategies could be applied (San León and Nogales 2022). For basic and readily degradable pollutants, top-down engineering works well and is free of expense. However, there are benefits to bottom-up engineering for breaking down complicated contaminants. In addition, as engineering advances, ideas like maintaining multifunctionality, dividing up metabolic pathways' labor, and reducing stress must be considered (Lyu et al. 2024).

12.4 Innovative Bioreactor Designs

12.4.1 Sequencing Batch Reactors (SBRs)

An SBR uses a fill-and-draw activated sludge system with irregular aeration and a clarifier to treat suspended-growth biological wastewater in a timed manner. The reactor controls the segregation of solids and liquids in a unit tank that is not steady (Khan et al. 2022). By oxidizing ammonia (NH_3-N) to nitrite (NO_2-N) and subsequently oxidizing NO_2-N to NO_3-N, the typical SBR produces nitrogen gas (N_2) as part of its integrated nitrification–denitrification process. Integrating the anaerobic and aerobic stages achieves simultaneous nitrification, denitrification, and phosphorus removal (SNDPR) (Zaman 2020).

Suspended solids (SS), nutrients, high-strength organic and inorganic contaminants can all be removed from wastewater in a single tank using the highly adaptable sequencing batch reactor (SBR) (Khumalo 2018). Its wide variety of uses involves treating various types of water waste, such as aquaculture, slaughterhouses, biodiesel production, households, and different industrial processes such as petrochemical, pharmaceutical, and textile manufacturing. SBR is also used for biogas production, creating gases such as biohydrogen, nitric oxide, biomethane, and nitrous oxide. SBR efficiently treats biosolids from various sources, such as dairy manure, brewery, pig slurry, and refinery spent caustic in solid waste digestion. The system is essential for granulation, creating valuable granules such as aerobic, nitrifying, Anammox-enriched, and phosphorus-accumulating granules, which improve pollutant breakdown in wastewater treatment. Moreover, SBR shows potential in cost-effective co-digestion of different waste materials like tannery wastewater and solid waste, as well as landfill leachate with domestic or dairy wastewater (Jaramillo et al. 2018). In addition to these applications, SBR is crucial in contemporary waste management and resource recovery through bioremediation of polluted soils, fermentation of food waste and cheese whey, and producing biodiesel, biosurfactants, and biodegradable polymers (Varjani et al. 2020).

The sequencing batch reactor (SBR) process consists of five consecutive stages: fill, react, settle, draw, and idle (Azeez et al. 2023). In the fill phase, wastewater and biomass are placed into the reactor using gravity or pumping methods, utilizing static or mixed filling techniques depending on treatment requirements. This stage usually takes up approximately 25% of the cycle time and is impacted by variables such as reactor volume, effluent flow rate, and desired loading. The react phase comes next, during which the influx of wastewater into the reactor is ceased, and the original reactions that began during the filling stage persist. Aeration and mixing are used to enhance nutrient elimination, with this phase frequently making up more than half of the overall duration of the process. The length of time needed varies depending on the treatment level needed and can be changed based on reactor design or treatment objectives. During the settle phase, solids are allowed to settle at the reactor's top, creating a clear liquid above, lasting 15–20% of the cycle to avoid solids floating due to gas accumulation. The clear supernatant is released during the drawing stage using devices like fixed or floating pipes. Depending on the design requirements, the draw periods might range from 5–30% of the cycle duration. Lastly, when numerous reactors are functioning, the idle stage exists between the draw and fill stages. This allows for the conditioning and mixing of biomass and the removal of surplus sludge once it has settled. The system is made ready for the subsequent cycle at this point (Trần 2023; Askari et al. 2024).

Numerous elements, including influent characteristics, organic loading rate, carbon supply, pH, hydraulic retention time (HRT), temperature, solid retention time (SRT), dissolved oxygen (DO), oxidation reduction potential (ORP), anoxic/oxic ratio, feed pattern, cycle length, settle-ability, may have an impact on the functioning of an SBR system. Rapid-settling bacteria are chosen by short settling times, and sludge with low settleability is successfully eliminated. Since low temperatures negatively affect microbial activity in activated sludge, they are considered a serious problem in systems that remove P and N simultaneously, and are being investigated (Li et al. 2018).

Despite the effects of the previously described factors on the SBR system, it offers the following advantages: (i) It has a compact footprint, works with a single basin and no need for a secondary clarifier, is automatically operated, and offers great process control options. (ii) Easy to set up, inexpensive, highly resilient to different loading shocks, and effective bulking control. (iii) Sturdiness, decreased energy consumption by 20, increased biomass retention, decoupling of SRT from HRT, resistance to toxicity, and concurrent nitrogen, phosphorus, and organics (Jagaba et al. 2021).

12.4.2 Moving Bed Biofilm Reactors (MBBRs)

Moving bed biofilm technology was created in response to early problems with biofilm reactors, such as hydraulic instability and uneven biofilm dispersion. An aeration tank, resembling an active sludge tank, is the centerpiece of the MBBR system. It is equipped with unique plastic carriers that offer a surface on which a biofilm can develop (Wang et al. 2021a). These systems make use of an extensive range of plastic carriers. Each carrier has a different form and surface area, as well as unique benefits and drawbacks. The surface area is crucial to the production of biofilms. A comprehensive interior surface area is essential for contact with water, air, bacteria, and nutrients because free-floating carriers enable biofilms to grow on the surface (Morgan-Sagastume 2018). The aeration system will mix the carriers in the tank, ensuring that the biomass on the carriers and the substrate in the influent wastewater have good contact. High-density polyethylene (HDPE) is currently the most preferred material because of its durability, fluidity, and density (Chen et al. 2022).

The hybrid MBBR systems, in which attached and suspended biomass coexist and support biological functions, have increased biomass concentration in the bioreactors. Anaerobic MBBRs are another type of MBBR, mostly used to treat industrial wastewater (Yang 2021).

Compared to other biofilm systems, moving-bed technology stands out primarily because it attempts to reduce the disadvantages of each system while combining many of the benefits of activated sludge and biofilm systems. Like other submerged-bed biofilm procedures, MBBRs aid in developing an extremely specialized active biofilm that is ideal for the specific circumstances in a reactor (Kulkarni 2024). This extremely active specialized biomass produces a more compact reactor with high volumetric efficiency and improved process stability. Headloss and the operational complex nature of the treatment stage are lowered because the MBBR is a continuous flow-through process, in contrast to the majority of other submerged-bed biofilm methods, which require media backwashing to maintain performance (Gzar et al. 2021).

Moving-bed reactors enable numerous reactors to be designed in a flow-through series arrangement to achieve multiple treatment objectives (i.e., nitrification, pre- and post-nitrification, and BOD removal) (Pickett 2018). These reactors can offer the same flexibility and simplicity as activated sludge procedures. There is no need for intermediary pumping for this to happen. Since most active biomass is continuously held in the reactor, the MBBR's biological performance is independent of the solids separation step, in contrast to suspended growth methods. The treated flow out of the reactor has a solids concentration that is at least an order of magnitude lower. Because of this, MBBRs work with a wide range of separation methods, not just traditional clarifiers (Simm et al. 2021). Due to their adaptability, MBBRs can be used in a wide range of potential reactor geometries. This makes MBBRs ideal for retrofitting installations to existing tanks for modifications at current plants.

12.4.3 Hybrid Systems

The integrated systems for wastewater treatment use several biological and physico-chemical processes to improve nutrient removal, including nitrogen and phosphorus. These systems thus optimize the characteristics of various processes to operate with enhanced efficiency and reliability (Rajesh Banu et al. 2021). Here are examples of some hybrid systems.

12.4.3.1 Integrated Fixed-Film Activated Sludge (IFAS) with Membrane Bioreactors (MBRs)

The hybrid system combines the IFAS-attached growth process with MBR filtration. Biofilm growth is allowed in the stable environment of IFAS, which results in the effective breakdown and removal of organic matter; therefore, nutrients will be removed by the MBR, which separates solids from liquid to provide high-quality effluent (Nowrouzi and Abyar 2021).

This combination improves biomass retention, nutrient removal efficiency and reduces sludge production. The MBR element helps to create a more streamlined system, ideal for space-limited applications (Campo et al. 2020).

12.4.3.2 Partial Nitritation–Anammox (PN/A) Process

This hybrid system combines partial oxidation of ammonia to nitrite, known as partial nitritation, and anammox that converts nitrogen from ammonia and nitrite straight into nitrogen gas. This is a very efficient process and does not require external carbon supplies. The PN/A process yields excellent results when applied to wastewater with comparatively high ammonia levels, for example, of industrial origin. This saves much energy in that full nitrification followed by denitrification is not required, as with the activated sludge process (Trojanowicz et al. 2021).

12.4.3.3 Sequential Batch Reactor (SBR) Combined with Anammox

This system uses the SBR process to treat wastewater in individual batches, creating ideal conditions for the growth of anammox bacteria. However, integrating these two technologies enables better control of the treatment process to meet the variability of wastewater flows. The SBR-anammox system is found to be highly flexible and can achieve enhanced nitrogen concentrations, making it appropriate for facilities with the unsteady inflow of waste (Dai et al. 2021).

12.4.3.4 Anaerobic–Anoxic–Oxic (A2O) with Anammox

A2O is a conventional system for biological nutrient removal that incorporates the anaerobic, anoxic, and oxic processes. This hybrid system can increase nitrogen removal while minimizing the total energy requirement by using the anammox process and a new technique. However, implementing anammox in the A2O process could achieve higher nitrogen removal efficiency and more sustainability since it requires fewer external carbon sources (Rong et al. 2020).

12.4.4 Application of Nanotechnology in BNR

Nanotechnology involves working with objects at a scale of 10^{-9} m, focusing on modifying, designing, and applying nanoscale materials. The field of nanotechnology has experienced rapid expansion and has various environmental applications, including desalination, biosensors, pollution filtering, water purification, and remediation of soil and water using nanoparticles (Tyagi et al. 2018). Compared to more conventional approaches, nanotechnology has recently demonstrated that it is an effective tool for removing nitrates. Because they are minimal and highly reactive, nanomaterials can be utilised as adsorbents and reducing agents due to their enormous surface area (Palani et al. 2021).

Basic chemical reduction processes cannot effectively remove nitrate from aqueous solutions. When strong catalysts are present, effective reduction occurs. Nanomaterials have great reduction potential because of their high catalytic capabilities and limited surface area. Ammonium (NH_4^+), nitrite (NO_2^-), ammonia (NH_3), and nitrogen gas (N_2) are released when nitrate is reduced (Liu et al. 2023). The kind of nanomaterial used determines the different products that are created. Adsorption is a frequent method in the elimination of nitrates. The nitrate molecule is bonded to the surface via chemisorption or physisorption in adsorption. The two primary characteristics of nanomaterials with superior adsorption capabilities are their intrinsic surface area and external functionalization (Taghipour et al. 2019).

A nanomaterial's adsorption capacity is affected by several other properties, such as its adsorption activity, surface area, atomic position on the surface, high surface binding energy, and lack of internal diffusion resistance. One can find the adsorption capacity of a nanomaterial by applying one of many kinetic or isotherm models. A wide range of nanomaterials, such as nanoshells, nanotubes, nanoparticles, nanoclusters, nanofibers, and nanocomposites, have been utilised for nitrate removal. Nanomaterials may have different reduction capacities and adsorption efficiencies (Kharangate-Lad and D'Souza 2021).

The nanoparticles mentioned above have several benefits, including large surface area, simple synthesis, strong catalytic activity, and superior electrical conductivity. The high catalytic activity and huge surface areas of the nanoparticles have led to their usage as reducing agents or as adsorbents. The employed nanoparticles were either immobilized on a support or bare, meaning they were used without any modifications. They were also covered with a coating of carbon-based oxide or a stabilizing capping agent to improve their capacity to eliminate nitrate. These nanomaterials have obtained effective nitrate removal rates at varying pH, temperature, nanomaterial dosage, initial nitrate concentration, and contact time. Ion chromatography, colorimetry, and ultraviolet (UV) spectrophotometry have all been used to measure the amount of nitrate eliminated. Using various nanomaterials, nitrate removal rates up to 100% have been attained (Tyagi et al. 2018). However, these nanomaterials have several drawbacks, including toxicity and particle aggregation. In addition, the nanoparticles might become unstable in different environmental circumstances. It might also be challenging to retrieve, separate, and recycle nanomaterials disseminated in a water supply to remove nitrate. Nitrate removal efficiency may be increased using innovative and environmentally friendly methods like *Halloysite Nanotubes* with their composites (Pandey et al. 2020). Although the restricted separation and recycling from water pose a considerable challenge, immobilizing these nanotubes on a membrane may solve this problem. With the further development of techniques that will not have the drawbacks of nanomaterials, nanotechnology can be the ultimate solution that complements the work required to remove nitrates in the aquatic environment (Somu and Paul 2021). Table 12.2 shows the various applications of nanotechnology in Biological Nanoreactors (BNR), highlighting their descriptions and benefits. Each application demonstrates how nanotechnology enhances efficiency and effectiveness in biological processes.

Table 12.2 Applications of nanotechnology in biological nanoreactors (BNR)

Application	Description	Benefits	References
Drug delivery systems	Nanoparticles are used to enhance the delivery of drugs to specific sites in the body	Improved efficacy and reduced side effects	Yusuf et al. (2023)
Biosensors	Nanosensors detect biological markers for diagnostics and monitoring	High sensitivity and specificity in detection	Perdomo et al. (2021)
Biocatalysis	Nanomaterials serve as catalysts to accelerate biochemical reactions in BNR	Increased reaction rates and efficiency	Rodríguez-de la Garza et al. (2023)
Gene therapy	Nanocarriers facilitate the delivery of genetic material into cells	Targeted and efficient gene delivery	Yu et al. (2021)
Imaging and diagnostics	Nanoparticles enhance imaging techniques for better visualization of biological processes	Improved resolution and contrast in imaging	Han et al. (2019)
Environmental remediation	Nanotechnology is used to remove contaminants from biological systems	Effective cleanup of pollutants and toxins	Guerra et al. (2018)
Tissue engineering	Nanostructured scaffolds support cell growth and tissue regeneration	Enhanced cell adhesion and proliferation	Perumal and Doble (2021)
Antimicrobial agents	Nanoparticles exhibit antimicrobial properties for use in coatings and treatments	Reduced microbial resistance and enhanced efficacy	Staroń and Długosz (2021)

12.5 Challenges and Solutions

One of the challenges in biological nutrient removal (BNR) technologies is process optimization and operational stability. A primary challenge is that wastewater quality is not well-controlled because the concentrations of microorganisms, nitrogen, phosphorus, pH, and temperature vary in the influent, and may cause inhibition and change of microbial populations and their activities, so nutrient removal is unstable. Furthermore, the biological processes—particularly in BNR systems—are intricate, requiring heightened management. This factor can easily pose a challenge since varied microbial populations need to be nurtured under the best conditions, as elaborated in the case of hybrid structures. Other risks include equipment failures, which may involve mechanical parts or monitoring systems, and are likely to affect the stability of operations (Cai et al. 2020). Monitoring and automated control must be used to automatically manipulate variables in real time to accommodate these challenges. This means it is possible to maintain a steady state without locating a specific point of influent quality characteristics. Also, enhancing redundancy and sound engineering of designs can help minimize the effects of equipment failures and continuous operations, along with minimizing downtimes (Zhang et al. 2020a). Another way to improve the general BNR process is by integrating different processes, for instance, linking the anammox process with the traditional nitrification process, as the latter is efficient in specific situations (Li et al. 2023).

Tackling issues like energy usage and resource recovery contributes significantly to the sustainability and environmental advantages of BNR technologies. Certain BNR methods require a lot of energy, especially those that heavily utilize aeration or chemical additions. Moreover, the efficient recovery of resources such as phosphorus from wastewater poses challenges, mainly due to the requirement of specialized processes and technologies. To address these difficulties, implementing energy-saving methods like the anammox process, which functions with minimal oxygen, can significantly reduce energy usage in wastewater treatment. Integrating resource recovery systems that extract valuable nutrients from wastewater, like phosphorus recovery, is in line with circular economy principles and improves the sustainability of BNR systems. Combining BNR systems with renewable energy sources such as solar or biogas can decrease their environmental footprint and enhance overall sustainability (Osman et al. 2024).

12.6 Future Directions in BNR Research

The rapid development and modernization of nutrient removal and recovery technologies have been driven by the need to recycle nitrogen and phosphorus from waste resources in recent years, which is in line with sustainable development and circular economy objectives. The recycling of nutrients not only helps alleviate the negative impacts caused by nutrient-heavy wastewater, like eutrophication, but also presents a sustainable remedy for diminishing non-renewable resources by offering substitute fertilizers for farming (Arumugham et al. 2024).

Conventional energy-intensive methods like chemical precipitation and activated sludge are increasingly substituted with modern BNR technologies such as the anaerobic–anoxic–oxic (A2O) process and sequencing batch reactors (SBR). Despite difficulties in efficiently recovering nutrients from various types of wastewater and environmental conditions, these contemporary approaches are better suited to sustainability objectives.

Many countries banned the direct use of sludge from older technologies as fertilizer due to concerns about toxic by-products. On the other hand, modern approaches to reclaiming nutrients, such as microbial fuel cells and growing microalgae, demonstrate great promise. Nevertheless, these techniques have drawbacks as they may struggle to distinguish between harmful metals and valuable nutrients, which could lead to environmental hazards (Kumar et al. 2021).

New technologies like struvite crystallization, electrodialysis, and bio-electrochemical systems are very effective in recovering nutrients and reducing heavy metal pollution, making them a better choice for sustainable wastewater treatment (Siciliano et al. 2020). Nevertheless, the high costs of operation and establishment are slowing down the implementation of these technologies. Furthermore, the initial stages of developing new nutrient recovery models and technologies require more research and innovation. Adopting these advances will necessitate a holistic strategy that incorporates technology, finances, society, and the environment, thereby strengthening the principles of the circular economy. Moving ahead, it is vital to adopt strategies that support sustainable solutions, guaranteeing that new nutrient recovery technologies meet sustainable development standards.

12.7 Conclusion and Future Perspective

This chapter thoroughly examines biological nutrient removal (BNR) technologies, highlighting their significant influence on wastewater treatment and sustainable development. The conversation delved into various BNR methods, from conventional systems to more sophisticated techniques. This involves the A2O (Anaerobic/Anoxic/Oxic) process, which aims to remove nutrients through different treatment stages; sequencing batch reactors (SBRs), which are flexible and effective in treating varying wastewater flows; moving bed biofilm reactors (MBBRs), which improve biological treatment using biofilm carriers; and integrated fixed-film activated sludge (IFAS) systems, which enhance treatment performance by combining fixed-film and suspended growth processes.

Together, these technologies improve nutrient removal, lower operating expenses, and support retrieving valuable resources like nitrogen and phosphorus. This is in line with the ideals of a circular economy as it encourages the repurposing and recycling of nutrients, leading to reduced waste and resource preservation. Advancements in BNR technologies enhance treated wastewater quality and help protect the environment by decreasing nutrient pollution in water bodies.

The potential of BNR technologies for sustainable water management is considerable in the future. Their extensive deployment offers both chances and difficulties. Ongoing innovation is necessary to tackle problems like effectively eliminating harmful by-products and enhancing resource recovery processes. Furthermore, incorporating sophisticated monitoring and control systems can improve the efficiency and dependability of BNR processes. BNR technologies will be crucial in promoting environmental sustainability and resource efficiency as the need for clean water increases worldwide.

In the end, this chapter emphasizes the crucial role of BNR in improving water quality management and provides a direction toward a more sustainable and environmentally conscious future. We can strike a harmony between technological progress and environmental protection by incorporating these technologies into more extensive circular economy models. This integration will help accomplish sustainability objectives and create opportunities for future water treatment and resource management advancements.

References

Abbas Q, Yousaf B, Ullah H, Ali MU, Ok YS, Rinklebe J (2020) Environmental transformation and nano-toxicity of engineered nano-particles (ENPs) in aquatic and terrestrial organisms. Crit Rev Environ Sci Technol 50:2523–2581

Akinnawo SO (2023) Eutrophication: Causes, consequences, physical, chemical and biological techniques for mitigation strategies. Environ Challenges 12:100733

Akram F, Aqeel A, Ahmed Z, Zafar J, ul Haq I (2022) Role of polyphosphate accumulating organisms in enhanced biological phosphorous removal. In: Microbial consortium and biotransformation for pollution decontamination. Elsevier

Al-Hazmi HE, Hassan GK, Maktabifard M, Grubba D, Majtacz J, Mąkinia J (2022) Integrating conventional nitrogen removal with anammox in wastewater treatment systems: microbial metabolism, sustainability and challenges. Environ Res 215:114432

Anjali G, Sabumon P (2022) Diversity and versatility of ammonia-oxidizing bacteria. In: Development in wastewater treatment research and processes. Elsevier

Anjali G, Sabumon P (2024) Revolutionizing wastewater treatment: cutting-edge technologies for biological nutrient removal. In: Biological and hybrid wastewater treatment technology: recent developments in India. Springer

Aponte-Morales V, Cunningham J, Ergas S (2013) Modification of the zeolite-sequencing batch reactor (Zeo-SBR) for ammonium removal of anaerobically digested swine waste. In: Proceedings of WEF/IWA nutrient removal and recovery, Vancouver, British Columbia, Canada

Arroyo P, Molinos-Senante M (2018) Selecting appropriate wastewater treatment technologies using a choosing-by-advantages approach. Sci Total Environ 625:819–827

Arumugham T, Khudzari J, Abdullah N, Yuzir A, Iwamoto K, Homma K (2024) Research trends and future directions on nitrification and denitrification processes in biological nitrogen removal. J Environ Chem Eng, 111897

Ashraf A, Ramamurthy R, Sayavedra SM, Bhatt P, Gangola S, Noor T, Desmarais M, Rabbani A, Rene ER (2021) Integrating photobioreactor with conventional activated sludge treatment for nitrogen removal from sidestream digestate: current challenges and opportunities. J Environ Chem Eng 9:106171

Askari SS, Giri BS, Basheer F, Izhar T, Ahmad SA, Mumtaz N (2024) Enhancing sequencing batch reactors for efficient wastewater treatment across diverse applications: a comprehensive review. Environ Res, 119656

Awuchi CG, Igwe VS, Amagwula IO (2020) Nutritional diseases and nutrient toxicities: a systematic review of the diets and nutrition for prevention and treatment. Int J Adv Acad Res 6:1–46

Azeez G, Al-Jaberi F, Ahmed S, Hussain A (2023) Sequencing batch reactor (SBR) technology in wastewater treatment: a mini-review. In: AIP conference proceedings, 2023. AIP Publishing

Badid N, Hadji F (2023) Health risks from desalinated seawater used for human consumption: the case of Honaine plant (Northwest Algeria). Electron Physician 15:7917–7925

Bagheri M, Akbari A, Mirbagheri SA (2019) Advanced control of membrane fouling in filtration systems using artificial intelligence and machine learning techniques: a critical review. Process Saf Environ Prot 123:229–252

Bala S, Garg D, Thirumalesh BV, Sharma M, Sridhar K, Inbaraj BS, Tripathi M (2022) Recent strategies for bioremediation of emerging pollutants: a review for a green and sustainable environment. Toxics 10:484

Bhambri A, Karn SK (2020) Biotechnique for nitrogen and phosphorus removal: a possible insight. Chem Ecol 36:785–809

Boyd CE, Boyd CE (2020a) Nitrogen. In: Water quality: an introduction, pp 269–290

Boyd CE, Boyd CE (2020b) Water quality protection. In: Water quality: an introduction, pp 379–409

Brown AR, Lilley M, Shutler J, Lowe C, Artioli Y, Torres R, Berdalet E, Tyler CR (2020) Assessing risks and mitigating impacts of harmful algal blooms on mariculture and marine fisheries. Rev Aquac 12:1663–1688

Bunce JT, Ndam E, Ofiteru ID, Moore A, Graham DW (2018) A review of phosphorus removal technologies and their applicability to small-scale domestic wastewater treatment systems. Front Environ Sci 6:8

Cai J, Liang P, Qiu Y, Cao J, Huang X (2020) A mass balance framework to evaluate the new pathways of biological nutrient removal in wastewater treatment. J Water Process Eng 37:101523

Campo R, Cosenza A, Di Bella G, Di Trapani D, Mannina G (2020) New applications in integrated fixed film activated sludge-membrane bioreactor (IFAS-MBR) systems. In: Current developments in biotechnology and bioengineering. Elsevier

Castell J-F, Faburé J, Pernelet-Joly V, Huber L, Lathière J (2020) The main pollutants and their impacts on agriculture, ecosystems and health. In: Agriculture and air quality: investigating, assessing and managing, pp 31–60

Che S, Men Y (2019) Synthetic microbial consortia for biosynthesis and biodegradation: promises and challenges. J Ind Microbiol Biotechnol 46:1343–1358

Chen R, Ji J, Chen Y, Takemura Y, Liu Y, Kubota K, Ma H, Li Y-Y (2019) Successful operation performance and syntrophic micro-granule in partial nitritation and anammox reactor treating low-strength ammonia wastewater. Water Res 155:288–299

Chen S, Yang D, Wang Q, Huang X, Ren H, Xu K (2022) Study on the advanced nitrogen removal under low temperature by biofilm on weak magnetic carriers. Biores Technol 360:127580

Coffman VR, Jensen AS, Trabjerg BB, Pedersen CB, Hansen B, Sigsgaard T, Olsen J, Schaumburg I, Schullehner J, Pedersen M (2021) Prenatal exposure to nitrate from drinking water and markers of fetal growth restriction: a population-based study of nearly one million Danish-born children. Environ Health Perspect 129:027002

Conca V (2021) Scale-up of innovative and sustainable via-nitrite biological processes for resource recovery in existing wastewater treatment plants

Dai J, Peng Y, Zhang J, Zhang L (2021) Anammox bacteria enrichment in fixed biofilm successfully enhanced nitrogen removal of domestic wastewater in a sequencing biofilm batch reactor (SBBR). J Water Process Eng 42:102154

Das S, O'Connell MG, Xu H, Bernstein R, Kim J-H, Sankhala K, Segal-Peretz T, Shevate R, Zhang W, Zhou X (2022) Assessing advances in anti-fouling membranes to improve process economics and sustainability of water treatment. Acs Est Eng 2:2159–2173

De Boer MA, Wolzak L, Slootweg JC (2019) Phosphorus: reserves, production, and applications. In: Phosphorus recovery and recycling, pp 75–100

Dezotti M, Lippel G, Bassin JP, Bassin JP (2018) New processes for biological nitrogen removal. In: Advanced biological processes for wastewater treatment: emerging, consolidated technologies and introduction to molecular techniques, pp 143–203

Du X, Shi Y, Jegatheesan V, Haq IU (2020) A review on the mechanism, impacts and control methods of membrane fouling in MBR system. Membranes 10:24

Ehrlich HL (2021) Bacterial mineralization of organic carbon under anaerobic conditions. In: Soil biochemistry CRC Press

El Batouti M, Al-Harby NF, Elewa MM (2021) A review on promising membrane technology approaches for heavy metal removal from water and wastewater to solve water crisis. Water 13:3241

Gibson RL (2019) A metabolic approach to assessing interactions between climate change, agrochemicals, and freshwater mussel reproduction. Auburn University

Gonsalves V (2020) An overview of the microbial transformation of nitrogen. In: Global implications of the nitrogen cycle, 1

Guerra FD, Attia MF, Whitehead DC, Alexis F (2018) Nanotechnology for environmental remediation: materials and applications. Molecules 23:1760

Gul MZ, Rupula K, Beedu SR (2022) Microbial consortia for pollution remediation—success stories. In: Microbial consortium and biotransformation for pollution decontamination. Elsevier

Gyawali S (2021) Enhanced digestate nitrification in sequencing batch reactors (SBR). University of South-Eastern Norway

Gzar HA, Al-Rekabi WS, Shuhaieb ZK (2021) Applicaion of moving bed biofilm reactor (MBBR) for treatment of industrial wastewater: a mini review. J Phys: Conf Ser, 012024

Han X, Xu K, Taratula O, Farsad K (2019) Applications of nanoparticles in biomedical imaging. Nanoscale 11:799–819

Han J-C, Ahmad M, Yousaf M, Rahman SU, Sharif HMA, Zhou Y, Yang B, Huang Y (2023) Strategic analysis on development of simultaneous adsorption and catalytic biodegradation over advanced bio-carriers for zero-liquid discharge of industrial wastewater. Chemosphere 332:138871

He Q, Yan X, Fu Z, Zhang Y, Bi P, Mo X, Xu P, Ma J (2022) Rapid start-up and stable operation of an aerobic/oxic/anoxic simultaneous nitrification, denitrification, and phosphorus removal reactor with no sludge discharge. Biores Technol 362:127777

He Y, Liu Y, Li X, Zhu T, Liu Y (2023) Unveiling the roles of biofilm in reducing N_2O emission in a nitrifying integrated fixed-film activated sludge (IFAS) system. Water Res 243:120326

Hernandez-Martinez GR, Youssef CB, Espinoza JER, Zepeda A (2021) Complete and simultaneous removal of aromatic compounds and ammonium in the nitrifying process: current knowledge of physiology behavior and dynamic response of microbial communities. In: Microbial ecology of wastewater treatment plants, pp 101–126

Hussain A, Kumari R, Sachan SG, Sachan A (2021) Biological wastewater treatment technology: advancement and drawbacks. In: Microbial ecology of wastewater treatment plants. Elsevier

Iannacone F (2021) Simultaneous removal of carbon, nitrogen and phosphorus from municipal wastewater through continuous-flow oxygen-controlled moving bed biofilm reactors

Ibrahim RS, Sabeen AH, Zainon Noor Z, Ahmad Mutamim NS (2020) Comparative study of suspended and attached growth in membrane bioreactors for wastewater treatment. Water Environ J 34:273–289

Jagaba AH, Kutty SRM, Noor A, Birniwa AH, Affam AC, Lawal IM, Kankia MU, Kilaco AU (2021) A systematic literature review of biocarriers: central elements for biofilm formation, organic and nutrients removal in sequencing batch biofilm reactor. J Water Process Eng 42:102178

Jaramillo F, Orchard M, Munoz C, Zamorano M, Antileo C (2018) Advanced strategies to improve nitrification process in sequencing batch reactors—a review. J Environ Manage 218:154–164

Jiang Y, Wu R, Zhang W, Xin F, Jiang M (2023) Construction of stable microbial consortia for effective biochemical synthesis. Trends Biotechnol 41:1430–1441

Jones MW, Peters GP, Gasser T, Andrew RM, Schwingshackl C, Gütschow J, Houghton RA, Friedlingstein P, Pongratz J, Le Quéré C (2023) National contributions to climate change due to historical emissions of carbon dioxide, methane, and nitrous oxide since 1850. Sci Data 10:155

Jürgens K, Taylor GT (2018) Microbial ecology and biogeochemistry of oxygen-deficient water columns. In: Gasol JM, Kirchman DL (eds) Microbial ecology of the oceans, 3rd edn. Wiley, NJ, USA, pp 231–288

Kamilya T, Majumder A, Yadav MK, Ayoob S, Tripathy S, Gupta AK (2022) Nutrient pollution and its remediation using constructed wetlands: insights into removal and recovery mechanisms, modifications and sustainable aspects. J Environ Chem Eng 10:107444

Karlikanovaite Balıkçı A (2019) Nitrogen removal and microbial community shift in oxic-settling-anoxic sludge reduction process

Kaur S, Diwan F, Reddersen B (2022) Microbial based 'cradle-to-cradle' approach to attain sustainable development goals of food and water security, good health and well being without crossing the planetary boundaries

Khan AH, Rudayni HA, Chaudhary AA, Imran M, Vambol S (2022) Modern use of modified sequencing batch reactor in wastewater treatment. Ecol Quest 33:99–110

Kharangate-Lad A, D'Souza NC (2021) Remediation of toxic environmental pollutants using nanoparticles and integrated nano-bio systems. In: Rhizobiont in bioremediation of hazardous waste, pp 443–482

Khatkar SK, Dudi K, Lonkar SA, Sidhu KS, Khatkar AB, Chandla NK, Panghal A (2023) An overview of membrane technology in dairy and food industry. In: Novel technologies in food science, pp 65–108

Khumalo SM (2018) Biological nutrient removal from industrial wastewater using a sequencing batch reactor

Kulkarni SJ (2024) An overview on recent developments in biological and hybrid wastewater treatment technology in India. In: Biological and hybrid wastewater treatment technology: recent developments in India, pp 25–55

Kumar S, Kumar R, Pandey A (2021) Current developments in biotechnology and bioengineering: strategic perspectives in solid waste and wastewater management. Elsevier

Kundu D, Dutta D, Samanta P, Dey S, Sherpa KC, Kumar S, Dubey BK (2022) Valorization of wastewater: a paradigm shift towards circular bioeconomy and sustainability. Sci Total Environ 848:157709

Kuśnierz M, Domańska M, Hamal K, Pera A (2022) Application of integrated fixed-film activated sludge in a conventional wastewater treatment plant. Int J Environ Res Public Health 19:5985

Landa-Acuña D, Hualpa-Cutipa E, Polo AG, Acosta RAS, Vargas-De-La-Cruz CB (2021) Biological wastewater treatment technology: microalgae. In: Wastewater treatment. Elsevier

Lee E, Rout PR, Bae J (2021) The applicability of anaerobically treated domestic wastewater as a nutrient medium in hydroponic lettuce cultivation: nitrogen toxicity and health risk assessment. Sci Total Environ 780:146482

Li Z-H, Ma Z-B, Yu H-Q (2018) Respiration adaptation of activated sludge under dissolved oxygen and hypochlorite stressed conditions. Biores Technol 248:171–178

Li X, Wu S, Dong Y, Fan H, Bai Z, Zhuang X (2021a) Engineering microbial consortia towards bioremediation. Water 13:2928

Li Y, Zhang M, Xu D, Shan X, Zheng P (2021b) Potential of anammox process towards high-efficient nitrogen removal in wastewater treatment: theoretical analysis and practical case with a SBR. Chemosphere 281:130729

Li J, Wang Z, Wang Y (2023) Integrating membrane aerated biofilm reactors with biological nitrogen removal processes: a new paradigm for achieving sustainable wastewater treatment plants. Chem Eng J, 146025

Liu T, Hu S, Guo J (2019) Enhancing mainstream nitrogen removal by employing nitrate/nitrite-dependent anaerobic methane oxidation processes. Crit Rev Biotechnol 39:732–745

Liu T, Lu Y, Zheng M, Hu S, Yuan Z, Guo J (2021) Efficient nitrogen removal from mainstream wastewater through coupling partial nitritation, anammox and methane-dependent nitrite/nitrate reduction (PNAM). Water Res 206:117723

Liu N, Sun Z, Zhang H, Klausen LH, Moonhee R, Kang S (2023) Emerging high-ammonia-nitrogen wastewater remediation by biological treatment and photocatalysis techniques. Sci Total Environ 875:162603

Lundberg JO, Weitzberg E (2022) Nitric oxide signaling in health and disease. Cell 185:2853–2878

Lyu X, Nuhu M, Candry P, Wolfanger J, Betenbaugh M, Saldivar A, Zuniga C, Wang Y, Shrestha S (2024) Top-down and bottom-up microbiome engineering approaches to enable biomanufacturing from waste biomass. J Ind Microbiol Biotechnol, kuae025

Malik D, Narayanasamy N, Pratyusha V, Thakur J, Sinha N (2023) Inorganic nutrients: macrominerals. In: Textbook of nutritional biochemistry. Springer

McInnes EF, Papineni S, Rinke M, Schorsch F, Marxfeld HA (2023) Agrochemicals. In: Haschek and Rousseaux's handbook of toxicologic pathology. Elsevier

Morgan-Sagastume F (2018) Biofilm development, activity and the modification of carrier material surface properties in moving-bed biofilm reactors (MBBRs) for wastewater treatment. Crit Rev Environ Sci Technol 48:439–470

Mushtaq N, Singh DV, Bhat RA, Dervash MA, Hameed OB (2020) Freshwater contamination: sources and hazards to aquatic biota. In: Fresh water pollution dynamics and remediation, pp 27–50
N'Goran KM, Yao KM, Trokourey A (2019) Phosphorus and nitrogen speciation in waters and sediments highly contaminated by an illicit urban landfill: the Akouedo landfill, Côte d'Ivoire. Reg Stud Mar Sci 31:100805
Ni M, Pan Y, Chen Y, Zhang X, Huang Y, Song Z (2021) Effects of seasonal temperature variations on phosphorus removal, recovery, and key metabolic pathways in the suspended biofilm. Biochem Eng J 176:108187
Ni M, Chen Y, Pan Y, Huang Y, Li D-p, Li L, Huang B, Song Z (2022) Study on community structure and metabolic mechanism of dominant polyphosphate-accumulating organisms (PAOs) and glycogen-accumulating organisms (GAOs) in suspended biofilm based on phosphate recovery. Sci Total Environ 815:152678
Ni B-J, Pan Y, Guo J, Virdis B, Hu S, Chen X, Yuan Z (2016) Denitrification processes for wastewater treatment
Nikolova AS, Belichovska D (2024) The Effect of nitrite use in meat products on human health. KNOWLEDGE-Int J 63:271–276
Nowrouzi M, Abyar H (2021) A framework for the design and optimization of integrated fixed-film activated sludge-membrane bioreactor configuration by focusing on cost-coupled life cycle assessment. J Clean Prod 296:126557
Osman AI, Chen Z, Elgarahy AM, Farghali M, Mohamed IM, Priya A, Hawash HB, Yap PS (2024) Membrane technology for energy saving: principles, techniques, applications, challenges, and prospects. Adv Energy Sustain Res 5:2400011
Palani G, Arputhalatha A, Kannan K, Lakkaboyana SK, Hanafiah MM, Kumar V, Marella RK (2021) Current trends in the application of nanomaterials for the removal of pollutants from industrial wastewater treatment—a review. Molecules 26:2799
Pandey G, Tharmavaram M, Rawtani D (2020) Functionalized halloysite nanotubes: an "ecofriendly" nanomaterial in environmental industry. In: Handbook of functionalized nanomaterials for industrial applications. Elsevier
Parde D, Ghosh R, Rajpurohit P, Bhaduri S, Behera M (2024) Nutrient retrieval techniques in wastewater treatment. In: Biological and hybrid wastewater treatment technology: recent developments in India. Springer
Park K-Y, Yu Y-J, Yun S-J, Kweon J-H (2019) Natural organic matter removal from algal-rich water and disinfection by products formation potential reduction by powdered activated carbon adsorption. J Environ Manage 235:310–318
Pattnaik R, Ray L, Saha K, Pal S, Singh H (2023) Microbial nitrogen transformation and recovery in wastewater: current strategies and applications
Paul A, Dasgupta D, Hazra S, Chakraborty A, Haghighi M, Chakraborty N (2023) Membrane bioreactor: a potential stratagem for wastewater treatment. In: Membranes for water treatment and remediation. Springer
Paull N (2020) Green wall technology for sustainably improving environmental quality: investigations into green wall plant health and particulate deposition. University of Technology Sydney (Australia)
Perdomo SA, Marmolejo-Tejada JM, Jaramillo-Botero A (2021) Bio-nanosensors: fundamentals and recent applications. J Electrochem Soc 168:107506
Perumal G, Doble M (2021) Fabrication of nanostructured scaffolds for tissue engineering applications. In: Nanomaterials and their biomedical applications, pp 317–334
Peter O (2019) Optimisation of the design and operation of chemically dosed activated sludge plants
Pickett M (2018) The ICARUS floating membrane photobioreactor for microalgae cultivation in wastewater: advancing technology from lab to field prototype. University of South Florida
Poikane S, Kelly MG, Herrero FS, Pitt J-A, Jarvie HP, Claussen U, Leujak W, Solheim AL, Teixeira H, Phillips G (2019) Nutrient criteria for surface waters under the European Water Framework Directive: current state-of-the-art, challenges and future outlook. Sci Total Environ 695:133888

Ponce-Jahen SJ, Cercado B, Estrada-Arriaga EB, Rangel-Mendez JR, Cervantes FJ (2024) Anammox with alternative electron acceptors: perspectives for nitrogen removal from wastewaters. Biodegradation 35:47–70

Qrenawi LI, Rabah FK (2023) Membrane bioreactor (MBR) as a reliable technology for wastewater treatment. J Membr Sci Res 9

Rahman TU, Roy H, Islam MR, Tahmid M, Fariha A, Mazumder A, Tasnim N, Pervez MN, Cai Y, Naddeo V (2023) The advancement in membrane bioreactor (MBR) technology toward sustainable industrial wastewater management. Membranes 13:181

Rajesh Banu J, Merrylin J, Kavitha S, Yukesh Kannah R, Selvakumar P, Gopikumar S, Sivashanmugam P, Do K-U, Kumar G (2021) Trends in biological nutrient removal for the treatment of low strength organic wastewaters. Curr Pollut Rep 7:1–30

Robles Á, Aguado D, Barat R, Borrás L, Bouzas A, Giménez JB, Martí N, Ribes J, Ruano MV, Serralta J (2020) New frontiers from removal to recycling of nitrogen and phosphorus from wastewater in the circular economy. Biores Technol 300:122673

Rodríguez-de la Garza JA, Pérez-Rodríguez P, González-Méndez LM, Martínez-Amador SY (2023) Recent advances in biological wastewater treatment. In: Current status of fresh water microbiology, pp 205–240

Rong Y, Liu X, Wen L, Jin X, Shi X, Jin P (2020) Advanced nutrient removal in a continuous A2/O process based on partial nitrification-anammox and denitrifying phosphorus removal. J Water Process Eng 36:101245

Rout PR, Shahid MK, Dash RR, Bhunia P, Liu D, Varjani S, Zhang TC, Surampalli RY (2021) Nutrient removal from domestic wastewater: a comprehensive review on conventional and advanced technologies. J Environ Manage 296:113246

San León D, Nogales J (2022) Toward merging bottom–up and top–down model-based designing of synthetic microbial communities. Curr Opin Microbiol 69:102169

Saravanan A, Kumar PS, Jeevanantham S, Karishma S, Tajsabreen B, Yaashikaa P, Reshma B (2021) Effective water/wastewater treatment methodologies for toxic pollutants removal: processes and applications towards sustainable development. Chemosphere 280:130595

Shabbirahmed AM, Kumaravel M, Yadav KK, Mohanty SS, Somu P (2022) Recent advances in biological nitrogen removal from wastewater: special focus on reactor configuration and nano-mediated microbial nitro-transformation. In: Development in wastewater treatment research and processes, pp 111–132

Shortle JS, Mihelcic JR, Zhang Q, Arabi M (2020) Nutrient control in water bodies: a systems approach. J Environ Qual 49:517–533

Shukla AK, Alam J, Alhoshan M (2022) Recent advancements in polyphenylsulfone membrane modification methods for separation applications. Membranes 12:247

Siciliano A, Limonti C, Curcio GM, Molinari R (2020) Advances in struvite precipitation technologies for nutrients removal and recovery from aqueous waste and wastewater. Sustainability 12:7538

Simm R, Lens P, Pakshirajan K (2021) Biological treatment technologies. IWA Publishing

Somu P, Paul S (2021) Recent advancements and challenges in the field of nanotechnology for wastewater treatment, recycle, and reuse. In: The future of effluent treatment plants. Elsevier

Song T, Zhang X, Li J, Wu X, Feng H, Dong W (2021) A review of research progress of heterotrophic nitrification and aerobic denitrification microorganisms (HNADMs). Sci Total Environ 801:149319

Srivastava P, Marquez GP, Gupta S, Mittal Y, Soda S, Dwivedi S, Odedishemi AF, Freguia S (2024) Diversity of anaerobic ammonium oxidation processes in nature. Chem Eng J, 149257

Staroń A, Długosz O (2021) Antimicrobial properties of nanoparticles in the context of advantages and potential risks of their use. J Environ Sci Health, Part A 56:680–693

Taghipour S, Hosseini SM, Ataie-Ashtiani B (2019) Engineering nanomaterials for water and wastewater treatment: review of classifications, properties and applications. New J Chem 43:7902–7927

Talan A, Tyagi RD, Drogui P (2021) Critical review on insight into the impacts of different inhibitors and performance inhibition of anammox process with control strategies. Environ Technol Innov 23:101553
Toledo NAB, Contreras CH, Corredor CAA, Pérez CAR (2021) Effect of biodegradable detergents on water quality. NVEO-Nat Volatiles Essent Oils J NVEO, 12080–12095
Tràn NTT (2023) The sequencing batch reactor (SBR) technology-evaluation of a full-scale SBR plant for the landfill leachate treatment
Trojanowicz K, Trela J, Plaza E (2021) Possible mechanism of efficient mainstream partial nitritation/anammox (PN/A) in hybrid bioreactors (IFAS). Environ Technol 42:1023–1037
Tyagi S, Rawtani D, Khatri N, Tharmavaram M (2018) Strategies for nitrate removal from aqueous environment using nanotechnology: a review. J Water Process Eng 21:84–95
US Environmental Protection Agency (EPA) (2009) National recommended water quality criteria. Office of Water, EPA 822-R-08-002, Washington, DC. https://www.epa.gov/wqc/national-recommended-water-quality-criteria
Varjani S, Pandey A, Gnansounou E, Khanal SK, Raveendran S (2020) Current developments in biotechnology and bioengineering: resource recovery from wastes. Elsevier
Verma AK, Rout PR, Lee E, Bhunia P, Bae J, Surampalli RY, Zhang TC, Tyagi RD, Lin P, Chen Y (2020) Biodiversity and sustainability. In: Sustainability: fundamentals and applications, pp 255–275
Wang J, Ying X, Huang Y, Chen Y, Shen D, Zhang X, Feng HJ (2021a) Numerical study of hydrodynamic characteristics in a moving bed biofilm reactor. Environ Res 194:110614
Wang LK, Wang M-HS, Shammas NK, Holtorff MS (2021b) Independent Physicochemical wastewater treatment system consisting of primary flotation clarification, secondary flotation clarification, and tertiary treatment. Environmental Flotation Engineering, pp 189–227
Waqas S, Bilad MR, Man Z, Wibisono Y, Jaafar J, Mahlia TMI, Khan AL, Aslam M (2020) Recent progress in integrated fixed-film activated sludge process for wastewater treatment: a review. J Environ Manage 268:110718
Warsinger DM, Chakraborty S, Tow EW, Plumlee MH, Bellona C, Loutatidou S, Karimi L, Mikelonis AM, Achilli A, Ghassemi A (2018) A review of polymeric membranes and processes for potable water reuse. Prog Polym Sci 81:209–237
Weissbrodt DG, Laureni M, van Loosdrecht MC, Comeau Y, Chen G, Ekama G, van Loosdrecht M, Brdjanovic D (2020). Basic microbiology and metabolism. Biological wastewater treatment: principles, modelling and design, 2nd edn. IWA Publishing, London, pp 11–76
World Health Organization (WHO) (2011) Guidelines for drinking-water quality, 4th ed. WHO Press, Geneva, Switzerland. https://www.who.int/publications/i/item/9789241548151
Wu P, Chen J, Garlapati VK, Zhang X, Jenario FWV, Li X, Liu W, Chen C, Aminabhavi TM, Zhang X (2022a) Novel insights into anammox-based processes: a critical review. Chem Eng J 444:136534
Wu Y, Zhang X, Chen J, Wang C, Zhang X, Chen J, Cai T, Liu W, Li X, Wu P (2022b) Effective utilization of refractory dissolved organic matters in domestic sewage allows to enhanced nitrogen removal by integrated fermentation, nitrification, denitratation and anammox process. Biores Technol 354:127227
Wu N-p, Tan B, Li M, Al-Dhabi NA, Lin B, He Q, Shen H-n, Cheng J-j, Tang W, Zhang Q (2024) Deciphering the role of hydraulic retention time (HRT) in integrated fixed-film activated sludge (IFAS) system under aniline stress: effects on microbial succession and communication. Chem Eng J 482:148834
Yang X (2021). Study of a hybrid system: moving bed biofilm reactor-membrane bioreactor (MBBR-MBR) in the treatment and reuse of textile industrial effluents
Yu C, Li L, Hu P, Yang Y, Wei W, Deng X, Wang L, Tay FR, Ma J (2021) Recent advances in stimulus-responsive nanocarriers for gene therapy. Adv Sci 8:2100540
Yusuf A, Almotairy ARZ, Henidi H, Alshehri OY, Aldughaim MS (2023) Nanoparticles as drug delivery systems: a review of the implication of nanoparticles' physicochemical properties on responses in biological systems. Polymers 15:1596

Zaman M (2020) Biological nutrient removal: minimizing carbon and oxygen requirements. The University of Western Ontario, Canada

Zhang L, Okabe S (2020) Ecological niche differentiation among anammox bacteria. Water Res 171:115468

Zhang W, Tooker NB, Mueller AV (2020a) Enabling wastewater treatment process automation: leveraging innovations in real-time sensing, data analysis, and online controls. Environ Sci: Water Res Technol 6:2973–2992

Zhang Y, Luo P, Zhao S, Kang S, Wang P, Zhou M, Lyu J (2020b) Control and remediation methods for eutrophic lakes in the past 30 years. Water Sci Technol 81:1099–1113

Zhang Y, Qiu X, Luo J, Li H, How S-W, Wu D, He J, Cheng Z, Gao Y, Lu H (2023) A review of the phosphorus removal of polyphosphate-accumulating organisms in natural and engineered systems. Sci Total Environ, 169103

Zhou Q, Sun H, Jia L, Wu W, Wang J (2022) Simultaneous biological removal of nitrogen and phosphorus from secondary effluent of wastewater treatment plants by advanced treatment: a review. Chemosphere 296:134054

Chapter 13
Catalytic Ceramic Membranes for Emerging Contaminant Treatment in Wastewater

Yen-Yi Lee, I.-Cheng Li, Bo-Wun Huang, Ya-Fen Wang, Sheng-Jie You, Minh-Ky Nguyen, Minh-Thuan Pham, and Guo-Ping Chang-Chien

Abstract Utilizing catalyst support for ceramic membranes (CMs) in wastewater filtration represents a promising strategy for enhancing filtration efficiency. Integrating catalysts into CMs has many advantages, including heightened chemical reactions and reaction sites, increased reaction activity, expanded surface area, enhanced thermal stability, improved mechanical strength, and augmented interaction between wastewater and the membrane. Nevertheless, this method has certain drawbacks, such as high costs, limited chemical resistance, and catalyst contamination and recovery challenges. This chapter also provides a comprehensive assessment of catalytic ceramic membranes (CCMs) in removing organic pollutants from wastewater, emphasizing the pivotal role of this technology in tackling environmental challenges. Furthermore, the research investigates the influence of various factors on

Y.-Y. Lee · M.-T. Pham (✉) · G.-P. Chang-Chien
Institute of Environmental Toxin and Emerging-Contaminant Research, Cheng Shiu University, Kaohsiung City, Taiwan
e-mail: 1905@gcloud.csu.edu.tw

Super Micro Mass Research and Technology Center, Cheng Shiu University, Kaohsiung City, Taiwan

Center for Environmental Toxin and Emerging-Contaminant Research, Cheng Shiu University, Kaohsiung City, Taiwan

I.-C. Li
Conservation and Research Center, Cheng Shiu University, Kaohsiung City, Taiwan

Department of Visual Communication Design, Cheng Shiu University, Kaohsiung City, Taiwan

B.-W. Huang
Department of Mechanical and Institute of Mechatronic Engineering, Cheng Shiu University, Kaohsiung City, Taiwan

Y.-F. Wang · S.-J. You
Department of Environmental Engineering, Chung Yuan Christian University, Taoyuan, Taiwan

M.-K. Nguyen
Faculty of Environment and Natural Resources, Nong Lam University, Ho Chi Minh City, Vietnam

A. M. Yatoo et al. (eds.), *Wastewater Monitoring and Management by Advanced Approaches*, Advances in Wastewater Research,
https://doi.org/10.1007/978-981-95-2601-7_13

membrane performance. It proposes prospects and market trends for further exploration, underscoring the significance of CCMs in offering potential solutions for wastewater treatment and environmental preservation.

Keywords Advanced oxidation processes · Catalytic ceramic membranes · Emerging contaminants · Market prospects · Wastewater treatment

Abbreviations

APT	Attapulgite
APAP	Acetaminophen
ATZ	Atrazine
BZA	Benzophenone
BPA	Bisphenol A
CBZ	Carbamazepine
CCMs	Catalytic ceramic membranes
CMs	Ceramic membranes
CNMs	Ceramic nanofiber membranes
EOCs	Emerging organic contaminants
H_2O_2	Hydrogen peroxide
$\bullet OH$	Hydroxyl radical
O_3	Ozone
O_2	Oxygen
$^1O_2$	Singlet oxygen
$O_2^{\cdot -}$	Superoxide radicals
O_{latt}	Oxygen in the lattice
O_{vac}	Oxygen vacancies
PBSA	P-Benzosulfonic acid
PMS	Peroxymonosulfate
ROS	Reactive oxygen species
$SO_4^{\cdot -}$	Sulfate radical
SMX	Sulfamethoxazole
TMP	Transmembrane pressure
UV	Ultraviolet

13.1 Introduction

The rapid increase in global population, along with industrialization and modernization, has led to a significant rise in daily wastewater, severely polluting water sources. If not properly managed, the increasing levels of pollutants will have severe and detrimental effects on ecosystems and living environments. According to the World Health Organization (WHO) (2024) (report in 2024), 1.4 million people die annually due to inadequate drinking water, sanitation, and hygiene, mainly in low- and middle-income countries. Unsafe sanitation causes 564,000 deaths, primarily from diarrheal diseases, and is a significant factor in several neglected tropical diseases like intestinal worms, schistosomiasis, and trachoma. Poor sanitation also contributes to malnutrition. According to a recent report published by the United Nations Educational, Scientific and Cultural Organization (UNESCO) (2024) on behalf of the United Nations (UN)-Water, 2.2 billion people currently do not have access to safe drinking water, and 3.5 billion people are without adequate sanitation. The UN's goal of ensuring universal access to these essential services by 2030 remains out of reach, raising concerns that these disparities may continue to grow. The impact of climate change on human health is also associated with water pollution and shortages (Chowdhary et al. 2020). Beyond traditional pollutants like heavy metals, organic compounds, and pathogenic microorganisms, emerging contaminants are becoming a significant concern (Yadav et al. 2021; Parida et al. 2021). These chemicals or microorganisms are less frequently monitored but can harm human health and the environment. Their presence in water bodies presents a significant challenge in wastewater management and requires advanced monitoring and treatment methods (Shah et al. 2020; Rout et al. 2021). Furthermore, pollution from these compounds arises from various sources, such as urban development, modern agricultural practices, industrialization, and human and animal treatment (Morin-Crini et al. 2022). These sources can lead to the emergence of new contaminants. According to the UN, emerging pollutants, such as antibiotics, drugs, steroids, endocrine disruptors, hormones, industrial additives, chemicals, and microplastics, are increasingly hazardous (programme). Municipal, industrial, and domestic wastewater are the primary pathways for their widespread distribution in aquatic environments. While more research is needed, it is widely recognized that these emerging pollutants are becoming a significant hazard (Mishra et al. 2023).

Water stress is a pressing issue as numerous cities worldwide grapple with rapidly depleting potable water supplies. This predicament underscores the need to reuse treated wastewater, thereby reducing the reliance on dwindling natural reserves for drinking purposes (Wang et al. 2021; Drangert 2021). To achieve effective wastewater remediation, the primary challenge lies in optimizing the overall efficiency of wastewater treatment technologies (Sangamnere et al. 2023). According to a report by Srujana Kathi, sustainable development necessitates innovative approaches to remediate contaminants from their source (Kathi and El Din Mahmoud 2024). Over the past decade, extensive research has focused on improving traditional wastewater treatment technologies like adsorption, advanced oxidation, biodegradation, and membrane filtration. However, no single technology can eliminate all contaminants, making hybrid systems more promising and more effective than single-treatment systems in enhancing treatment efficiency. Among these, catalytic ceramic membranes (CCMs) have emerged as a promising solution. Ceramic membranes (CMs) offer high chemical stability, thermal resistance, lightweight, flexible, and mechanical strength, making them suitable for harsh industrial environments (Mao et al. 2021; Ayode Otitoju et al. 2020). Advanced oxidation processes (AOPs), utilizing potent oxidizing radicals like hydroxyl radicals (•OH), superoxide anion radical ($\bullet O_2^-$), sulfate radicals (SO_4^-), and other reactive oxygen species (ROS), efficiently degrade pollutants without generating secondary waste (Lee et al. 2019; He et al. 2023). Integrating CMs with AOPs in CCMs enhances pollutant removal and membrane fouling control through synergistic effects, improving mass transfer rates and reaction efficiencies (Li et al. 2020; Zhou et al. 2023). This combination makes CCMs particularly effective for treating emerging contaminants in wastewater, offering significant potential for sustainable development and improved water security. Due to its cost-effectiveness, compactness, and scalability, membrane-based wastewater treatment has become a promising strategy to remove micropollutants (Dong et al. 2022; Hubadillah et al. 2022). CCMs are particularly valuable in treating industrial wastewater from sectors such as pharmaceuticals (Sawunyama et al. 2023), textiles (Chen et al. 2021), and petrochemicals (Bai et al. 2024). Their resilience and capacity to manage high concentrations of pollutants underscore their effectiveness and reliability in these demanding applications. Recent advancements include coating CCMs with nanomaterials such as manganese manganite oxide (MnMe oxide) (He et al. 2022), cobalt (II, III) oxide (Co_3O_4) (Cheng et al. 2023), titanium dioxide (TiO_2) (Kirk et al. 2024), manganese dioxide (MnO_2) (Duan et al. 2022), and iron oxychloride (FeOCl) (Zhao et al. 2022). Additionally, the use of activation agents like peroxymonosulfate (PMS) (Liangdy et al. 2023), ozone (O_3) (Xie et al. 2024), and hydrogen peroxide (H_2O_2) (Chen et al. 2023) has been explored to improve antifouling, pollutant removal, and disinfection performance. Integrating CCMs into hybrid treatment systems offers significant potential for sustainable development and improved water security worldwide. Their enhanced durability, catalytic efficiency, and multifunctional capabilities make them a promising solution for addressing complex water pollution challenges, inspiring new possibilities in wastewater management. A recent report by Muhammad Bilal Asif (Asif and Zhang 2021) highlights the novelty and industrial applications of CCMs in the water and wastewater treatment market. CCMs are praised for their chemical and thermal stability, low fouling tendency, and long lifespan, making them attractive for industrial use. The market for ceramic membranes is expected to grow at a compound annual rate of 12%. Full-scale CCM water treatment plants have been installed in countries like Japan, the USA, Singapore, and the UK, demonstrating their broad applicability and promising future in industrial applications. CMs are increasingly favored in water and wastewater treatment due to their durability, chemical and thermal stability, low fouling propensity, and extended lifespan. The market for CMs is expected to grow as environmental regulations tighten and water scarcity issues intensify. While the high initial cost has been a barrier, ongoing research is focused on using low-cost raw materials and innovative fabrication techniques to make them more affordable. Their applications span various industries, including petrochemical, pharmaceutical, food and beverage, and textiles, where they effectively handle complex contaminants. Future research aims to enhance membrane properties, minimize fouling, and integrate ceramic membranes with other treatment technologies to improve efficiency. With these advancements, ceramic membranes are poised to become crucial to sustainable water management strategies, offering significant market potential and contributing to a water-secure future.

This chapter delves into utilizing CCMs for wastewater treatment. The focus includes the inherent advantages of CCMs, such as chemical and thermal stability, low fouling propensity, and long lifespan, which make them highly effective for industrial applications. The characteristics and impacts of emerging contaminants in wastewater are explored, along with how CCMs can address these environmental challenges. The application of CCMs in wastewater treatment is examined, particularly regarding mechanisms, performance, and fabrication methods. The use of activation agents is also a crucial factor to consider. The market outlook and prospects for future research in this field are also analyzed. These insights aim to highlight the potential of CCMs in offering innovative solutions for wastewater treatment and environmental sustainability.

13.2 Characteristic and Impact of Emerging Contaminants in Wastewater

Emerging contaminants (ECs) in wastewater significantly challenge environmental sustainability and public health. These contaminants, which include pharmaceuticals, personal care products, pesticides, industrial chemicals, and nanomaterials, are not typically monitored in the environment but can cause significant adverse effects (Kumar et al. 2022; Prajapati et al. 2023). They often enter water systems in minute concentrations through various sources such as human excretion, improper disposal of medications, agricultural runoff, industrial discharges, household waste, and healthcare facilities (Nishat et al. 2023; Bobde et al. 2023). The complexity and low concentrations of ECs make them particularly challenging to remove using conventional water treatment processes. Ecologically, these contaminants can disrupt aquatic ecosystems, affecting marine organisms' reproductive and developmental processes (Prajapati et al. 2023). Endocrine-disrupting compounds (EDCs), a subset of ECs, are particularly harmful as they can cause hormonal imbalances in wildlife, reducing fertility and abnormal growth patterns. For human health, continuous exposure to low levels of ECs can lead to long-term effects such as antibiotic resistance, hormonal disruptions, and an increased risk of cancer. Pharmaceuticals and personal care products, even in trace amounts, can accumulate in the human body over time, posing significant health risks (Prajapati et al. 2023). ECs also degrade water quality, making it unsuitable for drinking and other uses, thereby increasing the burden on water treatment facilities to ensure safety standards are met. This degradation impacts not only public health but also economic sectors that depend on clean water, such as fisheries and tourism. The financial implications of EC contamination are substantial, as developing and implementing advanced treatment technologies to remove these contaminants can be costly. Additionally, industries reliant on clean water may suffer financial losses due to water quality issues. Addressing the challenge of ECs in wastewater requires innovative and effective removal strategies. Advanced treatment processes, such as membrane filtration, adsorption onto solid substrates, advanced oxidation processes, biological treatments, adsorption on granular activated carbon, and combined ozonation (O_3) and Ultraviolet (UV) treatment, have shown promise in removing ECs from water sources (Kumar et al. 2022; Bobde et al. 2023). Hybrid-constructed wetlands have emerged as a sustainable solution, demonstrating enhanced effectiveness in ECs removal.

Furthermore, integrating circular economy principles into water treatment—maximizing resource efficiency and minimizing waste output—offers a promising approach. This includes recovering valuable resources from wastewater, such as energy and nutrients, and integrating water treatment systems with other industrial processes (Das et al. 2023). Continued research and innovation in water treatment technologies and applying circular economy principles are essential to developing cost-effective and scalable solutions. By consolidating insights from diverse research and advancements, stakeholders can work towards ensuring the safety and sustainability of water resources for future generations.

13.3 Potential of Catalytic Ceramic Membrane Addressing Environmental Challenges

CMs are widely recognized for their thermal and chemical stability, effective abrasion and corrosion resistance, and high mechanical strength (Dong et al. 2022; Omar et al. 2023). These superior characteristics make CMs an optimal choice for applications in harsh conditions and complex chemical environments, significantly improving water and air quality. However, CMs production incurs high initial investment costs because it requires many advanced technologies and expensive raw materials (Hube et al. 2023; Asif and Zhang 2021). The porous structure of CMs is prone to clogging by small particles, necessitating frequent cleaning and maintenance, which increases operational and maintenance costs (Asif and Zhang 2021). Furthermore, CMs are only effective in removing larger particles and are insufficient for certain specific pollutants (Cai et al. 2023; Gu et al. 2022).

Integrating oxidation processes has been proposed to eliminate pollutants more thoroughly and control fouling better (Liangdy et al. 2023; Qiao et al. 2024). This combination not only enhances pollutant removal efficiency, cost-effectiveness, high permeability, stability, flexibility, and adaptability but also allows ceramic membranes to act as catalytic supports, providing both filtration and catalytic functions (Cai et al. 2023; Omar et al. 2024). CCMs significantly advance membrane technology by incorporating catalytic processes into traditional CMs. This integration enables CCMs to perform physical separation and chemical degradation of pollutants simultaneously, optimizing treatment efficiency (Liangdy et al. 2023; He et al. 2023; Chen et al. 2023).

Incorporating AOPs into CCMs has demonstrated high efficiency in treating emerging wastewater pollutants. A prominent study conducted by Jiaxuan Yang et al. in 2024 (Yang et al. 2024) on cobalt single-atom catalytic ceramic membranes (Co_1-NCNT-CM) combined with a peroxymonosulfate (PMS) system showed exceptional treatment performance for various wastewater pollutants. The combined processes of catalytic ozonation and sulfate radical oxidation provide robust solutions for pollutant degradation, improving removal rates and reducing fouling by forming ROS within membrane pores. Specifically, the treatment efficiency for phenol reached 100% within 15 min, while for other pollutants such as sulfamethoxazole (SMX), carbamazepine (CBZ), and atrazine (ATZ), the efficiencies were 93.24%, 69.83%, and 41.81%, respectively. This study also highlighted several advantages of the Co_1-NCNT-CM material with PMS, including high activity, stability over multiple cycles, environmental friendliness, catalytic activation capability with single-atom co-catalysts, multilayer functional regions, improved mass transfer, and enhanced surface roughness.

Another study by Qilin Gu et al. in 2021 (Gu et al. 2022) reported that CCMs are an up-and-coming and effective method for the future. CCMs not only provide an effective solution to environmental pollution issues but also open new research directions in catalytic materials and membrane technology. These studies have confirmed that integrating advanced oxidation processes into CMs enhances treatment efficiency and brings numerous economic and environmental benefits. This is particularly important in the context of increasing wastewater treatment demands and the need for advanced technologies to ensure environmental safety and sustainability.

13.4 Application of Catalytic Ceramic Membrane in Wastewater

CCMs without catalytic agents have been primarily studied to remove organic pollutants such as SMX, Benzophenone (BP-3), emerging organic contaminants (EOCs), etc. In Table 13.1, CMs with immobilized laccase from the fungi *Tinea versicolor* and *Pycnoporus sanguineus* (CS43) achieved 91% and 76% Bisphenol A (BPA) removal efficiencies at pH 5 after 30 min (Barrios-Estrada et al. 2018). This system demonstrates high degradation efficiency, easy enzyme immobilization, and industrial application potential, highlighting the advantages of using enzyme-functionalized CCMs for treating emerging pollutants.

Table 13.1 Performance and advantages of various catalytic ceramic membranes in wastewater treatment

Material	Functionalization	Fabrication	Optimal experimental parameter	Optimal performances	Advantages and characteristics	References
Ceramic membranes with immobilized laccase	Ceramic membranes with immobilized *P. sanguineus* (CS43) and *T. Versicolor*	Laccase immobilization by covalent binding on the ceramic	• Optimal pH for immobilization: pH 5 • BPA degradation activities were tested at various pH levels (3, 5, and 7) • Optimal pH for BPA degradation: pH 5	• *T. versicolor*: 6.16% (0.5727 U/cm^2) • *P. sanguineus* (CS43): 15.28% (1.4139 U/cm^2) • BPA degradation at 30 min with immobilized laccase from *T. versicolor*: 91% at pH 5 • BPA degradation at 30 min with immobilized laccase from *P. sanguineus* (CS43): 76%	• High degradation efficiency of BPA in less than 24 h • Laccase from *T. versicolor* exhibited higher activity compared to *P. sanguineus* (CS43) • Immobilized laccase system as a novel tool for tackling emerging contaminants degradation issues • The system shows potential for industrial applications due to the ease of immobilization and significant degradation capabilities	Barrios-Estrada et al. (2018)

(continued)

Table 13.1 (continued)

Material	Functionalization	Fabrication	Optimal experimental parameter	Optimal performances	Advantages and characteristics	References
Ceramic membrane modified with $CuMn_2O_4$ particles	$CuMn_2O_4$ particles are coated on a tubular ceramic membrane with UV/O_3 activation	Sol-gel technique	• Initial ozone dose: 1.0 mg/L • pH of water: 7.11 ± 0.19 • Soaking time for $CuMn_2O_4$ precursor solution: 120 min	• BP-3 removal efficiency: 76.6% after 120 min • Removal efficiencies of UV absorbers (BP-3, BZA, PBSA) were significantly improved, with complete degradation of BP-3 and BZA and 89.5% removal of PBSA	• High activity in catalytic ozonation for degrading UV absorbers and reducing membrane fouling • Enhanced ozone decomposition to generate hydroxyl radicals and hydrogen peroxide • Increased normalized permeate flux and reduced membrane fouling compared to a virgin ceramic membrane without coating	Guo et al. (2018)

(continued)

Table 13.1 (continued)

Material	Functionalization	Fabrication	Optimal experimental parameter	Optimal performances	Advantages and characteristics	References
Ceramic nanofiber membranes with Mn_2O_3	Ceramic nanofiber membranes anchored with Mn_2O_3 (Mn@CNMs) for O_3 activation	Mn_2O_3 nanoparticles were anchored on the pits of attapulgite (APT) nanofibers via the impregnation and in situ precipitation method with urea	• Sintering temperature: 500 °C • Precursor concentration: 0.05 mol/L $Mn(NO_3)_2$ • Ozone concentration: 10.5 mg/L, pH: 6.8 ± 0.2	• SMX removal: 81.3% during 7-h continuous filtration • Total removal of various emerging contaminants in real hospital wastewater: 98.5% • Ozone decomposition rate: 89.1%, high catalytic stability with less than 0.02 kPa increase in transmembrane pressure (TMP) during 7-h continuous filtration	• Nanosized pits of APT effectively prevented the aggregation and growth of Mn_2O_3 nanoparticles • Enhancing the catalytic performance • Interlaced APT nanofibers formed nano-sized network structures • Refining ROS and SMX in the close vicinity and increasing the interaction efficiency • Enhanced Mn (III)/Mn (IV) redox cycles contributed to the high catalytic stability • Effective in controlling membrane fouling with self-cleaning properties	Yang et al. (2022)

(continued)

Table 13.1 (continued)

Material	Functionalization	Fabrication	Optimal experimental parameter	Optimal performances	Advantages and characteristics	References
Ceramic membrane with manganese (III) oxide (Mn_2O_3)	Catalytic ceramic membrane integrated with Mn_2O_3 nanoparticles for PMS activation (MnCM/PMS)	A solid-state sintering method	• PMS dose: 0.1 mM • Membrane flux: 60 LMH, pH: 7.0 ± 0.2	• Removal efficiency: 81–97% for various EOCs; 99% for APAP at 40 LMH • Catalytic stability: TOC removal ~ 65%; manganese leaching 1.3–1.5 μg/L • Stability: sustained performance for 24-h operation with minimal ΔTMP increase (0.3 kPa) • pH range: effective from pH 4.0 to 9.0, best at neutral pH	• The membrane efficiently removes contaminants, is stable, works across various pH levels, and avoids toxic metals • Made with Mn_2O_3 nanoparticles, it operates well in both filtration modes, using singlet oxygen for degradation	Chen et al. (2022)

(continued)

Table 13.1 (continued)

Material	Functionalization	Fabrication	Optimal experimental parameter	Optimal performances	Advantages and characteristics	References
Attapulgite (APT) nanofibers with Mn_2O_3 nanoparticles	Ceramic nanofiber membrane coupled with ozonation active	Impregnation and in situ precipitation with urea as a precipitant, followed by calcination at 500 °C for 5 h	• Membrane layer • Thickness: 10 μm • Flux: 20 LMH • Initial SMX concentration: 20 mg/L • O_3 concentration: 10.5 mg/L • Initial pH: 6.8 ± 0.2 • Membrane layer • Thickness: 10 μm	• SMX removal efficiency: Up to 89% during 1 h catalytic continuous filtration	• High removal efficiency for SMX and other emerging organic • Contaminants • Enhanced hydrophilicity • Superior antifouling properties • Stable catalytic performance • Effective pollutant mineralization • Wide applicability • Enhanced ROS generation	Sang et al. (2022)
Ceramic membrane (CM) with manganese and cobalt oxides (MnO_2 $–Co_3O_4$)	Ceramic membrane coated with $MnO_2–Co_3O_4$ nanoparticles for ozonation active	Coating CM via dip-coating method multiple times (30, 40, 60, 80 coatings) followed by drying and calcination	• Ozone concentration: 2.0 mg/L • Initial BP-3 concentration: 0.5–3.0 mg/L • Evaluated different numbers of coatings and their impact on performance	• Highest BP-3 removal efficiency: 74.8% with CM coated 80 times with MnO_2 $–Co_3O_4$ Initial ozone concentration: 92.0% at 2.0 mg/L • Pore size: 76.0% with 50 nm pore size • pH level: 2–10	• Decreased roughness • Reduced fouling • Higher performance with more coatings • Improved catalytic activity • Stability in ozonation processes • Effective pollutant removal	Guo et al. (2016)

(continued)

Table 13.1 (continued)

Material	Functionalization	Fabrication	Optimal experimental parameter	Optimal performances	Advantages and characteristics	References
Cobalt single-atom catalyst-tailored ceramic membrane (Co_1-NCNT-CM)	N-doped CNT-supported Co single-atom modified CM (Co_1-NCNT-CM) with PMS	Coating method (Co_1-NCNT-CM on the ceramic membrane)	• Co atom doping: 0.5 wt% • Catalyst loading: 0.2 mg/cm^2 • PMS dosage: 0.3 mM • Reaction conditions • Contaminants: 10 mg/L (Phenol, 2,4-DCP, BPA, BPF, APAP, SMX, SMZ, SDZ, STZ), 5 mg/L (CBZ, ATZ) • pH: 7 • Temperature: 20 °C	• Specific flux decline: 84.73% of initial • Fouling resistance reduction: reversible: 87.18%; irreversible: 72.67% • Contaminant removal: Phenol: ~ 100% (15 min); SMX: 93.24%; CBZ: 69.83%; ATZ: 41.81% • SE matrix degradation: Phenol: 84.34%; SMX: 75.74%; CBZ: 35.01%; ATZ: 32.42% • Phenols and sulfonamides: > 97% • Stability and reusability: Phenol removal: ~ 80% (six cycles), Co ion leaching: < 0.001 mg/L	• High reactivity • Selective removal of phenols and sulfonamides • Antifouling properties • Stability over multiple cycles • Environmentally friendly • Catalytic activation with single-atom Co catalysts • Multilayer functional zones • Improved mass transfer • Enhanced surface roughness	Yang et al. (2024)

Activation agents such as O_3 and UV have shown significant improvements in the efficiency of CCMs for pollutants like BP-3 and SMX. Specifically, in Table 13.1, CMs coated with $CuMn_2O_4$ particles and activated by UV/O_3 achieved BP-3 removal efficiencies of up to 76.6%, higher than CMs coated with MnO_2–Co_3O_4 nanoparticles for O_3 active, which had BP-3 removal efficiencies of 74.8% (Barrios-Estrada et al. 2018; Guo et al. 2016). Although the difference between these two materials is only about 1.8%, CMs with $CuMn_2O_4$ activated by UV/ O_3 demonstrated higher removal efficiency for a broader range of pollutants (Guo et al. 2018). For UV-absorbing substances such as BP-3, benzophenone (BZA), and p-Benzosulfonic acid (PBSA), CMs with $CuMn_2O_4$ achieved complete degradation of BP-3 and BZA and 89.5% removal of PBSA. Similarly, for ceramic nanofiber membranes (CNMs) with Mn_2O_3 (Mn@CNMs/O_3) and Attapulgite nanofiber membranes (APT) with Mn_2O_3 nanoparticles, the SMX removal efficiency differed by 7.7%, with removal efficiencies of 81.3% and 89%, respectively (Sang et al. 2022; Yang et al. 2022). Mn@CNMs/O_3 membranes also achieved up to 98.5% removal efficiencies for various emerging pollutants in hospital wastewater (Yang et al. 2022).

For CCMs activated by PMS, high pollutant removal efficiencies were also demonstrated. CCMs integrating Mn_2O_3 particles to activate PMS (MnCM/PMS) removed between 81 and 97% of prominent organic pollutants (EOCs), achieving acetaminophen (APAP) removal efficiencies of up to 99% at a flow rate of 40 L per square meter per hour (LMH) (Chen et al. 2022). Additionally, Co_1-NCNT-CM achieved nearly 100% phenol removal after 15 min, and other compounds like SMX and CBZ reached removal efficiencies of 93.24% and 69.83%, respectively (Yang et al. 2024). This indicates that using PMS can optimize the pollutant removal capability of CMs while maintaining high stability and reusability.

In summary, studies have shown that the efficiency of CMs can be significantly improved when using activation agents such as O_3, UV, and PMS. In experiments without activation agents, pollutant removal efficiencies are promising but not as high as those with activation agents. Activators like O_3 and UV not only enhance removal efficiency but also improve the anti-fouling properties of the membranes. PMS also demonstrates high removal efficiencies and long-term stability. These results indicate the great potential of activated CMs in wastewater treatment and organic pollutant removal, contributing significantly to environmental protection and water quality improvement.

CCMs activated by PMS are used in AOPs to degrade various pollutants. The reaction mechanism involves activating PMS to generate sulfate radicals ($SO_4^{\cdot-}$), which are potent oxidizing agents. Below is a specific example of the reaction mechanism of the MnCM/PMS system (Chen et al. 2022), detailing the process steps and chemical equations:

The equations representing the formation of $O_2^{\cdot-}$, which can act as a precursor of $^1O_2$ are as follows:

$$\equiv Mn(III) - OH + HSO_5^- \rightarrow \equiv Mn(III) - (O)OSO_3^- + H_2O \rightarrow \\ \equiv Mn(IV) - OH + O_2^{\cdot-}$$

$$2O_2^{\cdot-} + 2H_2O \rightarrow {}^1O_2 + H_2O_2 + 2OH^-$$

$$O_2^{\cdot-} + H_2O_2 \rightarrow {}^1O_2 + OH^- + OH^-$$

$$O_2^{\cdot-} + OH^- \rightarrow {}^1O_2 + OH^-$$

The reduction of oxygen in the lattice (O_{latt}) can create oxygen vacancies (O_{vac}), combined with O_{vac} in MnCM, which can not only transform into active oxygen (O*) but also continue to react with PMS to generate ${}^1O_2$:

$$O_{latt}/O_{vac} \rightarrow O^*$$

$$O^* + HSO_5^- \rightarrow HSO_4^- + {}^1O_2$$

Besides, ${}^1O_2$ can also be formed through the natural decomposition of PMS:

$$HSO_5^- + SO_5^{2-} \rightarrow HSO_4^- + SO_4^{2-} + {}^1O_2$$

In summary, the general equation when the MnCM/PMS system removes pollutants is as follows:

$$HSO_5^- + Mn \rightarrow SO_4^{\cdot-} + Mn^{n+1}$$

$$SO_4^{\cdot-} + \text{Pollutant} \rightarrow \text{Product}$$

Furthermore, CCMs are utilized to activate O_3, thereby boosting efficiency through the generation and interaction of free radicals like (OH^-), ($O_2^{\cdot-}$), (${}^1O_2$). These radicals are essential in breaking down pollutants. Here is an example of the reaction mechanism for the 0.05 Mn@CNM catalytic ozonation system (Yang et al. 2022):

$$0.05\ \text{Mn@CNM} - \text{Active sites} + O_3 \rightarrow 0.05\ \text{Mn@CNM} - O^*\text{ad} + O_2^*$$

$$0.05\ \text{Mn@CNM} - O^*\text{ad} + O_3 + H_2O \rightarrow 0.05\ \text{Mn@CNM} - OH + O_2 + OH$$

$$O_2^* + O_3 \rightarrow O_2^{\cdot-} + O_2$$

$$O_2^* + O_3 \rightarrow 1O_2 + O_2$$

13.5 Market Outlook and Prospects for Future Research

A recent report by Muhammad Bilal Asif (Asif and Zhang 2021) highlights the novelty and industrial applications of CCMs in the water and wastewater treatment market. CCMs are praised for their chemical and thermal stability, low fouling tendency, and long lifespan, making them attractive for industrial use. The market for CMs is expected to grow at a compound annual rate of 12%. Full-scale CCMs water treatment plants have been installed in countries like Japan, the USA, Singapore, and the UK, demonstrating their broad applicability and promising future in industrial applications.

According to another report by Hubadillah et al. (2022), the global membrane separation technology market was projected to reach $28.10 billion by 2022, growing from $6.8 billion in 2005 to $9 billion in 2008 with a compound annual growth rate (CAGR) of 7.2%. The global membrane technology market is expected to expand at a CAGR of 10.5% from 2019 to 2027, increasing from $17.70 billion to $43.14 billion (Hubadillah et al. 2022). CCMs are also being extensively researched for new applications such as wastewater treatment in the pulp and paper industry, textile dyeing wastewater treatment, and membrane bioreactors (MBRs) (Jarrar et al. 2024).

Figure 13.1 provides a detailed visual representation of the various applications and mechanisms of CCMs in advanced emerging wastewater treatment. Although CCMs have many advantages, high production costs and complex fabrication processes remain significant barriers to large-scale deployment (Jarrar et al. 2024; Hubadillah et al. 2022). Therefore, current research focuses on reducing production costs by using low-cost alternative materials such as clay, agricultural waste, industrial waste, and animal bones, along with optimizing fabrication processes to enhance performance and reduce costs (Hubadillah et al. 2022; Kumar et al. 2016). With technological advancements and ongoing research efforts, the future of CCMs in environmental applications looks very bright, providing sustainable and efficient solutions to urgent environmental issues.

13.6 Conclusion and Future Prospects

CCMs combine the robustness of ceramics with enhanced catalytic properties, making them vital for emerging wastewater treatment. These membranes are chemically and thermally stable, resistant to fouling, and have a long lifespan, making them suitable for industrial applications. The global market for CCMs is growing at a compound annual rate of 12%, driven by stricter environmental regulations and water scarcity. CCMs are used in countries like Japan, the USA, Singapore, and the UK, showcasing their wide applicability. They effectively degrade contaminants; for example, immobilized laccase enzymes from fungi like *P. sanguineus* achieve up to 91% BPA removal. CCMs modified with $CuMn_2O_4$ particles have high removal efficiencies for BP-3, BZA, and PBSA. Advanced designs, such as Mn@CNMs, show

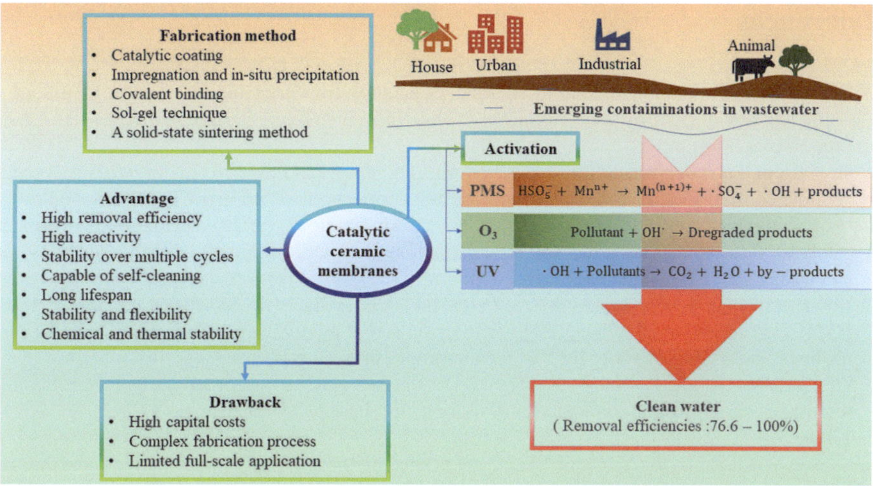

Fig. 13.1 A comprehensive overview of the potential of CCMs in emerging wastewater treatment

impressive removal rates, including 81.3% for SMX and 98.5% total contaminant removal in hospital wastewater. Although initial costs are high, research is focused on reducing expenses through innovative techniques and low-cost materials. However, to promote broader commercialization, future studies should focus on reducing production costs while improving the operational lifespan of CCMs. Strategies such as catalyst doping to enhance treatment efficiency and increase membrane durability are crucial, especially as fouling remains a major challenge. Research into self-cleaning ceramic membranes, which could significantly mitigate fouling, presents a promising avenue for future application. Moreover, reducing operational costs should be prioritized by focusing on scalable manufacturing methods and optimizing membrane regeneration processes. Such innovations would reinforce the position of CCMs as a key technology in sustainable water management, enhancing their market potential and contributing to long-term environmental solutions. In summary, CCMs provide robust and efficient wastewater treatment solutions, leveraging their structural and catalytic strengths to effectively tackle various contaminants.

Acknowledgements **This work was funded by the Higher Education Sprout Project** (2023~2028), Ministry of Education, Taiwan (R.O.C). Institute of Environmental Toxin and Emerging-Contaminant Research, Cheng Shiu University, Kaohsiung, Taiwan (R.O.C). Super Micro Mass Research and Technology Center, Cheng Shiu University, Kaohsiung, Taiwan (R.O.C). And Center for Environmental Toxin and Emerging-Contaminant Research, Cheng Shiu University, Kaohsiung, Taiwan (R.O.C).

References

Asif MB, Zhang Z (2021) Ceramic membrane technology for water and wastewater treatment: a critical review of performance, full-scale applications, membrane fouling and prospects. Chem Eng J 418:129481

Ayode Otitoju T, Ugochukwu Okoye P, Chen G, Li Y, Onyeka Okoye M, Li S (2020) Advanced ceramic components: materials, fabrication, and applications. J Ind Eng Chem 85:34–65

Bai H, Liang L, Cao P, Zhang H, Chen S, Yu H, Quan X (2024) Mgal2o4 incorporated catalytic ceramic membrane for catalytic ozonation of organic pollutants. Appl Catal B Environ 343:123527

Barrios-Estrada C, Rostro-Alanis MDJ, Parra AL, Belleville M-P, Sanchez-Marcano J, Iqbal HMN, Parra-Saldívar R (2018) Potentialities of active membranes with immobilized laccase for bisphenol A degradation. Int J Biol Macromol 108:837–844

Bobde PV, Sharma AK, Kumar R, Pandey JK, Wadhwa S (2023) Recent advances on sustainable removal of emerging contaminants from water by bio-based adsorbents. New J Chem 47:17626–17644

Cai C, Sun W, He S, Zhang Y, Wang X (2023) Ceramic membrane fouling mechanisms and control for water treatment. Front Environ Sci Eng 17:126

Chen M, Heijman SGJ, Rietveld LC (2021) State-of-the-art ceramic membranes for Oily wastewater treatment: modification and application. Membranes 11:888

Chen L, Maqbool T, Fu W, Yang Y, Hou C, Guo J, Zhang X (2022) Highly efficient manganese (Iii) oxide submerged catalytic ceramic membrane for nonradical degradation of emerging organic compounds. Sep Purif Technol 295:121272

Chen C, Lu L, Fei L, Xu J, Wang B, Li B, Shen L, Lin H (2023) Membrane-catalysis integrated system for contaminants degradation and membrane fouling mitigation: a review. Sci Total Environ 904:166220

Cheng X, Zhang Y, Fan Q, Wang L, Shi S, Luo X, Zhu X, Wu D, Liang H (2023) Preparation Of Co_3o_4@carbon nanotubes modified ceramic membrane for simultaneous catalytic oxidation and filtration of secondary effluent. Chem Eng J 454:140450

Chowdhary P, Bharagava RN, Mishra S, Khan N (2020) Role of industries in water scarcity and its adverse effects on environment and human health. In: Shukla V, Kumar N (eds) Environmental concerns and sustainable development: volume 1: air, water and energy resources. Springer Singapore, Singapore

Das S, Singh CK, Sodhi KK, Singh VK (2023) Circular economy approaches for water reuse and emerging contaminant mitigation: innovations in water treatment. Environ Dev Sustain 27:5794

Dong Y, Wu H, Yang F, Gray S (2022) Cost and efficiency perspectives of ceramic membranes for water treatment. Water Res 220:118629

Drangert JO (2021) Urban water and food security in this century and beyond: resource-smart cities and residents. Ambio 50:679–692

Duan Y, Yu J, Zhang R, Han P, Ren P, Liu M, Wong NH, Sunarso J (2022) Integrated Mno_2 nanosheet ultrafiltration ceramic membrane with micro-nano bubbles for catalytic treatment of dye wastewater. Sep Purif Technol 300:121786

Gu Q, Ng TCA, Bao Y, Ng HY, Tan SC, Wang J (2022) Developing better ceramic membranes for water and wastewater treatment: where microstructure integrates with chemistry and functionalities. Chem Eng J 428:130456

Guo Y, Xu B, Qi F (2016) A novel ceramic membrane coated with Mno_2–Co_3o_4 nanoparticles catalytic ozonation for benzophenone-3 degradation in aqueous solution: fabrication, characterization and performance. Chem Eng J 287:381–389

Guo Y, Song Z, Xu B, Li Y, Qi F, Croue JP, Yuan D (2018) A novel catalytic ceramic membrane fabricated with $CuMn_2O_4$ particles for emerging UV absorbers degradation from aqueous and membrane fouling elimination. J Hazard Mater 344:1229–1239

He Y, Wang L, Chen Z, Huang X, Wang X, Zhang X, Wen X (2022) Novel catalytic ceramic membranes anchored with mnme oxide and their catalytic ozonation performance towards atrazine degradation. J Membr Sci 648:120362

He Z, Ong JH, Bao Y, Hu X (2023) Chemocatalytic ceramic membranes for removing organic pollutants in wastewater: a review. J Environ Chem Eng 11:109548

Hubadillah SK, Jamalludin MR, Dzarfan Othman MH, Iwamoto Y (2022) Recent progress on low-cost ceramic membrane for water and wastewater treatment. Ceram Int 48:24157–24191

Hube S, Hauser F, Burkhardt M, Brynjólfsson S, Wu B (2023) Ultrasonication-assisted fouling control during ceramic membrane filtration of primary wastewater under gravity-driven and constant flux conditions. Sep Purif Technol 310:123083

Jarrar R, Abbas MKG, Al-Ejji M (2024) Environmental remediation and the efficacy of ceramic membranes in wastewater treatment—a review. Emergent Mater 7:1295–1327

Kathi S, El Din Mahmoud A (2024) Trends in effective removal of emerging contaminants from wastewater: a comprehensive review. Desalination Water Treat 317:100258

Kirk CH, Wang P, Chong CYD, Zhao Q, Sun J, Wang J (2024) Tio_2 photocatalytic ceramic membranes for water and wastewater treatment: technical readiness and pathway ahead. J Mater Sci Technol 183:152–164

Kumar RV, Goswami L, Pakshirajan K, Pugazhenthi G (2016) Dairy wastewater treatment using a novel low cost tubular ceramic membrane and membrane fouling mechanism using pore blocking models. J Water Process Eng 13:168–175

Kumar R, Qureshi M, Vishwakarma DK, Al-Ansari N, Kuriqi A, Elbeltagi A, Saraswat A (2022) A review on emerging water contaminants and the application of sustainable removal technologies. Case Stud Chem Environ Eng 6:100219

Lee WJ, Bao Y, Hu X, Lim T-T (2019) Hybrid catalytic ozonation-membrane filtration process with ceox and mnox impregnated catalytic ceramic membranes for micropollutants degradation. Chem Eng J 378:121670

Li C, Sun W, Lu Z, Ao X, Li S (2020) Ceramic nanocomposite membranes and membrane fouling: a review. Water Res 175:115674

Liangdy A, Lee WJ, Bao Y, Oh W-D, Lim T-T (2023) Hybrid process of persulfate-based advanced oxidation with meox-functionalized catalytic ceramic membrane for synergistic removal of micropollutants: recent developments, new insights, and prospects. Chem Eng J 466:143280

Mao X, Hong J, Wu YX, Zhang Q, Liu J, Zhao L, Li HH, Wang YY, Zhang K (2021) An efficient strategy for reinforcing flexible ceramic membranes. Nano Lett 21:9419–9425

Mishra RK, Mentha SS, Misra Y, Dwivedi N (2023) Emerging pollutants of severe environmental concern in water and wastewater: a comprehensive review on current developments and future research. Water Energy Nexus 6:74–95

Morin-Crini N, Lichtfouse E, Liu G, Balaram V, Ribeiro ARL, Lu Z, Stock F, Carmona E, Teixeira MR, Picos-Corrales LA, Moreno-Piraján JC, Giraldo L, Li C, Pandey A, Hocquet D, Torri G, Crini G (2022) Worldwide cases of water pollution by emerging contaminants: a review. Environ Chem Lett 20:2311–2338

Nishat A, Yusuf M, Qadir A, Ezaier Y, Vambol V, Ijaz Khan M, Ben Moussa S, Kamyab H, Sehgal SS, Prakash C, Yang H-H, Ibrahim H, Eldin SM (2023) Wastewater treatment: a short assessment on available techniques. Alex Eng J 76:505–516

Omar NMA, Othman MHD, Tai ZS, Milad M, Sokri MNM, Puteh MH (2023) Recent progress and technical improvement strategies for mitigating ceramic membrane bottlenecks in water purification processes: a review. Int J Appl Ceram Technol 20:3327–3356

Omar NMA, Othman MHD, Tai ZS, Kurniawan TA, Puteh MH, Jaafar J, Rahman MA, Ismail AF, Rajamohan N, Abdullah H, Wong KY (2024) Recent strategies for enhancing the performance and lifespan of low-cost ceramic membranes in water filtration and treatment processes: a review. J Water Process Eng 62:105399

Parida VK, Saidulu D, Majumder A, Srivastava A, Gupta B, Gupta AK (2021) Emerging contaminants in wastewater: a critical review on occurrence, existing legislations, risk assessment, and sustainable treatment alternatives. J Environ Chem Eng 9:105966

Prajapati D, Shah M, Yadav A, Panchal J (2023) A critical review on emerging contaminants: origin, discernment, and remedies. Sustain Water Resour Manag 9:69

Qiao Y, Han X, Chen F, Sun C, Xu L, Yao J, Wu Y, Qi Z, Liu R, Shen X (2024) Fouling control of different pretreatments on ceramic fouling ultrafiltration: a review. Environ Sci Water Res Technol 10:2273–2281

Rout PR, Zhang TC, Bhunia P, Surampalli RY (2021) Treatment technologies for emerging contaminants in wastewater treatment plants: a review. Sci Total Environ 753:141990

Sang J, Yang Y, Fu W, Chen X, Tang T, Sun X, Yang C, Zhang X (2022) Catalytic ceramic nanofiber membrane coupled with ozonation for degradation of sulfamethoxazole: critical parameters, mechanisms and applicability. Sep Purif Technol 302:122092

Sangamnere R, Misra T, Bherwani H, Kapley A, Kumar R (2023) A critical review of conventional and emerging wastewater treatment technologies. Sustain Water Resour Manag 9:58

Sawunyama L, Oyewo OA, Seheri N, Onjefu SA, Onwudiwe DC (2023) Metal oxide functionalized ceramic membranes for the removal of pharmaceuticals in wastewater. Surf Interfaces 38:102787

Shah AI, Din Dar MU, Bhat RA, Singh JP, Singh K, Bhat SA (2020) Prospectives and challenges of wastewater treatment technologies to combat contaminants of emerging concerns. Ecol Eng 152:105882

UN Enivironment Programme. Emerging pollutants in wastewater: an increasing threat. https://www.unep.org/events/un-environment-event/emerging-pollutants-wastewater-increasing-threat-0

UNESCO (2024) Water crises threaten world peace. https://www.un.org/Sustainabledevelopment/Blog/2024/03/Un-World-Water-Development-Report/

Wang D, Hubacek K, Shan Y, Gerbens-Leenes W, Liu J (2021) A review of water stress and water footprint accounting. Water 13:201

World Health Organization (2024) Sanitation. https://www.who.int/News-Room/Fact-Sheets/Detail/Sanitation

Xie X, Wang L, Wei J, He H, Guo Z, Wang C, Wen X, Song Y (2024) Progress in ceramic membrane coupling ozonation process for water and wastewater treatment: a critical review. Chem Eng J 493:152475

Yadav D, Rangabhashiyam S, Verma P, Singh P, Devi P, Kumar P, Hussain CM, Gaurav GK, Kumar KS (2021) Environmental and health impacts of contaminants of emerging concerns: recent treatment challenges and approaches. Chemosphere 272:129492

Yang Y, Fu W, Chen X, Chen L, Hou C, Tang T, Zhang X (2022) Ceramic nanofiber membrane anchoring nanosized Mn_2o_3 catalytic ozonation of sulfamethoxazole in water. J Hazard Mater 436:129168

Yang J, Zhao J, Wang H, Liu Y, Ding J, Wang T, Wang J, Zhang H, Bai L, Liang H (2024) Cobalt single-atom catalyst tailored ceramic membrane for selective removal of emerging organic contaminants. Environ Sci Ecotechnol 21:100416

Zhao Y, Zhao Y, Yu X, Kong D, Fan X, Wang R, Luo S, Lu D, Nan J, Ma J (2022) Peracetic acid integrated catalytic ceramic membrane filtration for enhanced membrane fouling control: performance evaluation and mechanism analysis. Water Res 220:118710

Zhou M, Chen J, Yu S, Chen B, Chen C, Shen L, Li B, Lin H (2023) The coupling of persulfate activation and membrane separation for the effective pollutant degradation and membrane fouling alleviation. Chem Eng J 451:139009

Chapter 14
Applications of Nanostructured Materials for Efficient Removal of Emerging Contaminants from Wastewater

Pooja Dhiman, Anita Rani, and Ranjana Chauhan

Abstract In the past few decades, water pollution has become a matter of concern not only for the scientific community but also for the common man. For the sake of a more comfortable lifestyle, the advancement of technology has led to rapid anthropogenic revolution in the industries, resulting in the discharge of effluents from industries into the water bodies. The discharge of untreated water from several industries, including textile, leather, pharmaceutical, agricultural, and mining, has been one of the sources of numerous persistent organic and inorganic pollutants, including heavy metals such as Cd^{2+}, Pb^{2+}, As(V), Hg^{2+}, Co^{2+}, Cr (VI) etc. in aquatic ecosystem consequently leading to severe environmental and public health issue. The rationale behind this is that the bio-magnification of these persistent and hazardous pollutants in the living body is responsible for severe health concerns. The present literature systematically organizes the utilization of adsorbent-supported advanced metallic Nanocomposites for the mitigation of these anthropogenic pollutants from wastewater. The agenda of this chapter is to elaborate on polymer-based Nanocomposites, carbon nanotubes/metal/metal oxide nanocomposites, graphene/graphene oxide, and chitosan-based nanocomposites, the most recent and advanced nanostructured materials for efficient removal of emerging contaminants from wastewater.

Keywords Nanocomposite · Photocatalyst · Pollutants · Removal · Treatment · Adsorbent

P. Dhiman (✉)
Department of Chemistry, School of Basic and Applied Sciences, Maharaja Agrasen University, Solan, Himachal Pradesh, India
e-mail: pooja.dhimanmanshi@gmail.com

A. Rani · R. Chauhan
Department of Chemistry, School of Basic and Applied Sciences, IEC University, Solan, Himachal Pradesh, India

A. M. Yatoo et al. (eds.), *Wastewater Monitoring and Management by Advanced Approaches*, Advances in Wastewater Research,
https://doi.org/10.1007/978-981-95-2601-7_14

14.1 Introduction

Expansion of industrial development, increase in population, and urban areas have remarkably raised the requirement for food, agricultural resources, water, and other natural products. Changes in climate, depletion and over-utilization of water resources, and anthropogenic water pollutants have exacerbated issues related to the provision of clean and usable water throughout the globe. Concerns related to water pollution, caused by various sources (Fig. 14.1), have put public health, ecosystems, and humanity in danger. Effluent from municipal sources, agricultural, and industrial led to over nutrients and depletion of oxygen in the water matrix, which can potentially endanger human health, wildlife, and aquatic life (Tariq and Mushtaq 2023).

The advanced water treatment urgently needs to be applied to reutilize the polluted water or dispose of the effluents without a detrimental impact on human health as well as on the environment. Currently, numerous technologies and strategies are in practice to reactivate and regenerate sources of water pollution. Application of conventional effluent water treatment assemblies, as sequential steps to acquire a required scale of physical, chemical, and biological treatment processes, and a combination of different techniques collectively to eliminate organic substances, nutrients, solid particles, chemicals, and toxic ions from the contaminated water, is

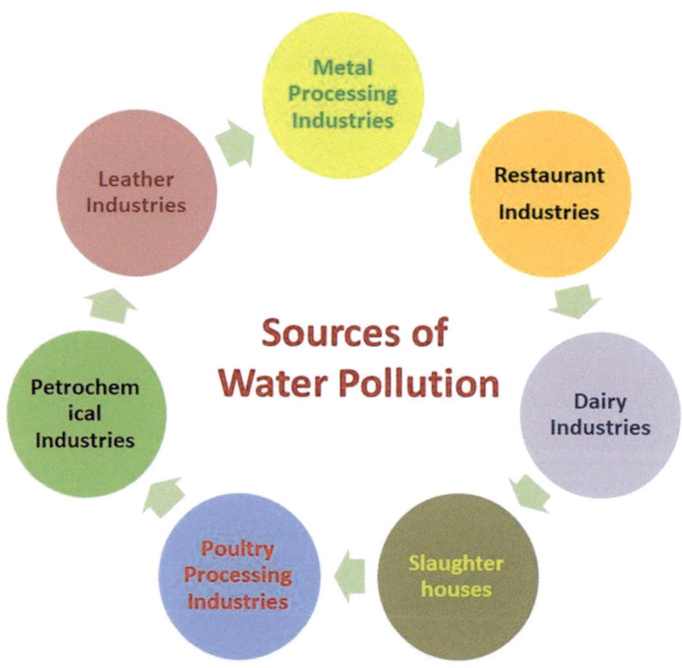

Fig. 14.1 Different sources of water pollutants

commonly practiced (Alsafran et al. 2023). Although conventional technologies are not efficient enough to meet the required expectations in purifying industrial effluent, high-strength landfill leachate, because of several associated disadvantages and limitations. Economic constraints, expanding regulations, water and energy loss are some of the challenges that persuade researchers and scientists, and available technologists to plan and design other available innovative industrial effluent treatment techniques for a sustainable future (Singh et al. 2023).

In the past few years, results of numerous novel strategies have been investigated in detail after treatment of polluted water by available water treatment techniques. A lot of attempts have been made to sort out the limitations and disadvantages pertaining to the efficiency of nanostructured materials deployed for the treatment and purification of contaminated water. Collaborative efforts of the researchers and scientists of diversified research domains, including chemistry, environmental science, ecology, biology, physics, engineering, and material science, have been engaged in the development of novel nanocomposite materials and design of advanced treatment systems to either separate or transform the contaminants from polluted water (Xia et al. 2023). A wide variety of unique, robust, and potential nanocomposite materials have recently been fabricated to be deployed as adsorbents, membrane filters, ion exchangers, photocatalysts, etc., for the purification of water under environmental remediation applications (Kumar et al. 2023). Previously, nano-materials and polymers have been deployed by researchers as potentially emerging composites for the treatment of polluted water in a more efficient way. To treat wastewater, the functions of these novel and newly developed polymer-based nanomaterials can be tuned as per the needs and the nature and composition of the pollutant into consideration. These kinds of nanocomposites belong to a more upgraded category of materials that present more advanced output with improved efficiencies because of the integration of the benevolent features of both robust photocatalyst and nanomaterial support (Das and Adak 2022). No doubt, the logical research and development, along with the definite applications, such kind of functionally active nanophotocatalytic materials, demand a deeper comprehension of their fundamental features and the science behind it. The agenda of the present study is to offer an apprehensive summary of the latest advances in this concern with respect to production, performance, utilization, and scope and future challenges of newly developed nanostructured composites for the treatment of water (Aydin et al. 2024). The present literature elaborates a recent report on more advanced technologies for treatment of water.

14.2 Conventional Water and Wastewater Treatment Approaches

Generally, the quality and quantity of wastewater treatment determine the number of techniques and processing stages required for its treatment. Practically, processes involved in the treatment of wastewater include preliminary, primary (1°), secondary

(2°), and tertiary (3°) treatments, which are a combination of biological, physical, and chemical methods (Huang et al. 2022). The prime motive of preliminary treatment is to eliminate all the coarse and suspended impurities from polluted water to overcome other associated issues, including pumps and pipes choking and resulting destruction, which may be used for water treatment (Shende 2024).

To assess the quality of water, several water quality parameters, including physical, chemical, microbial, and biological, should to be taken into consideration, which have been depicted in Fig. 14.2. The agenda of primary treatment or primary settling is the elimination of floating pollutants and the removal of such inorganic pollutants as well as organic contaminants that are settleable, which is why they are mitigated by either sedimentation or skimming, respectively (Dadebo et al. 2023). By the primary (1°) treatment, approximately 50–70% of the suspended solid contaminants and about 65% of the oily as well as greasy contaminants are eliminated from polluted water, and biochemical oxygen demand (BOD is quantity of oxygen in mgs required by involved microorganisms for degradation of organic matter can be declined by about 25–50% (Al-zou'by et al. 2023). The types of oily and greasy impurities, such as free, emulsified, or dissolved, govern the outcome of the water treatment from polluted water. As per previously published reports, the primary (1°) treatment is only able to partially mitigate organic pollutants, including organic nitrogen and phosphorus, as well as heavy metals from the contaminated aqueous matrix. Contrary to this, it is difficult to eliminate dissolved and colloidal impurities (Li et al. 2023a). The primary agenda of secondary (2°) water treatment techniques is focused on decreasing the extent of organic pollutants, including both dissolved and colloidal impurities. On the other hand, suspended pollutants in the contaminated water are mitigated by anoxic, aerobic, and anaerobic biological treatment techniques. The secondary (2°) water treatment techniques include trickling biofilters, rotating biological contactors, waste stabilization ponds, and activated sludge systems and related facilities (Sangamnere et al. 2023). Tertiary (3°) water treatment, generally referred to as more advanced water treatment techniques, is implemented when a specific water contaminant has to be eliminated after treatment by secondary methods (Zagklis and Bampos 2022). In practical situations, the existing forces of the physical water treatment methods, including electromagnetic forces (EMF) and gravitational forces, contribute to the separation of existing contaminants of the polluted water. Water filters, membranes, and screens are some of the commonly used barriers in physical methods of wastewater treatment (Ghazal et al. 2022). Sedimentation, coagulation/flocculation, flotation, and adsorption are a few other operable physicochemical processes deployed for the treatment of water (Gunawardhana et al. 2023). For removing definite pollutants from contaminated water, chemical methods generally deploy appropriate chemical reagents by carrying out practical reactions in wastewater. For the purification of wastewater, chemical oxidation, ion exchange, and chemical precipitation are some other chemical processes to purify wastewater (Shamsan et al. 2023). The purification of the wastewater bodies by conventional physicochemical processes has been challenging, with several other environmental concerns pertaining to cost and sustainability. In some practical scenarios, these

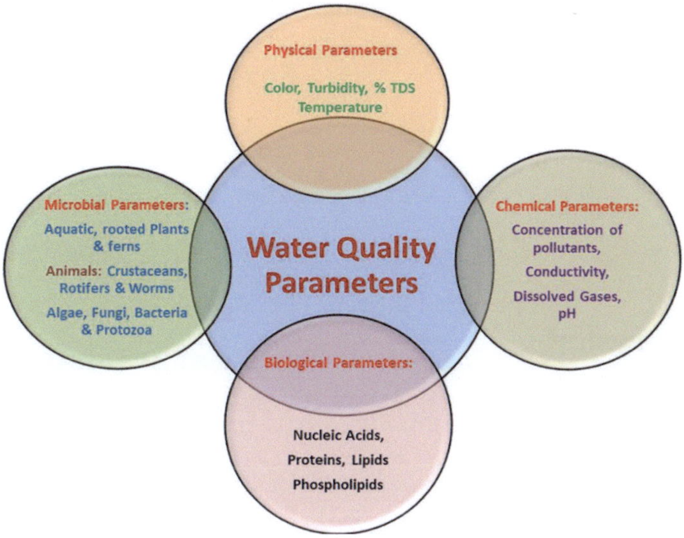

Fig. 14.2 Several water quality parameters to assess the quality of water

methods are quite partial, expensive, delayed, and account for additional secondary (2°) pollution (Khongkliang et al. 2024).

In the present era, the objective of the research community is not only restricted to the reusability of water but also aiming to increase the efficiency of wastewater treatment. Researchers made better attempts to meet the demands and regulations by transforming the concept to purify wastewater as a valuable source of energy. With the evolving technologies, there is an increased demand for designing and developing sustainable water treatment techniques, deploying reusable, eco-friendly nanostructured composites with greater cost-effectiveness and efficiency (Ranjan Mishra et al. 2022). Figure 14.3 represents the conventional water treatment techniques. More advanced functionally active nanostructured composites generally explore new dimensions for economically weaker nations for water treatment for matching their technological and financial constraints.

14.3 Advanced Water and Wastewater Treatment Technologies

The development of highly efficient water treatment techniques for polluted water is essential due to the aggravated water crises. Investigations in terms of water treatment and sanitation at several stages in different sectors are being carried out. Simplicity of the process, efficiency, cost involved, water quality, as well as health and environmental effects, and several associated considerable parameters required to be kept into

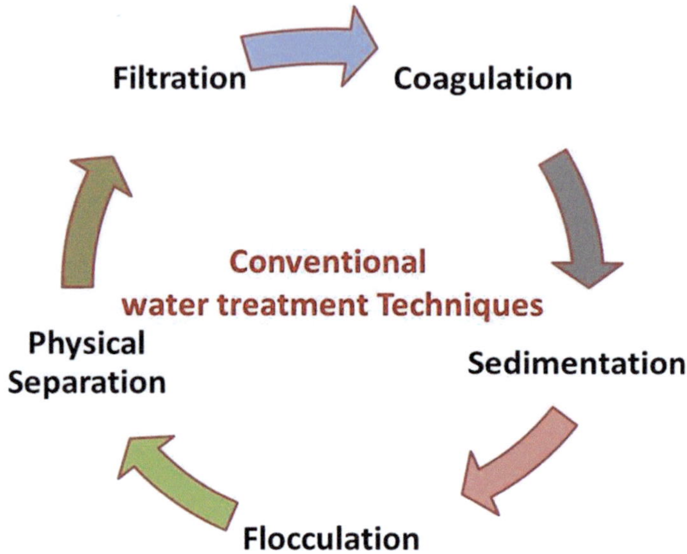

Fig. 14.3 Conventional water treatment techniques

consideration in recently developed water treatment techniques. Biofilm-supported systems and microbial-based fuel cells have recently been observed for utilization in the treatment of wastewater (Srivastava et al. 2022). The innovative technology is capable of generating electrical energy from the organic matter in wastewater without conducting any significant segregation. The present technique can broadly be deployed in the target areas without the availability of sufficient electrical infrastructures (Schwetschenau et al. 2022). Despite several associated advantages, this technology generally offers economic, environmental, and operational challenges; therefore, more research is needed to scale it up for industrial utilization. Microbial fuel cell supported systems offer a logical opportunity for treatment of municipal polluted water owing to the rationale, waste is free, and the energy content of municipal polluted water is approximately up to 9.3 folds the energy required by the polluted water in waste wastewater treatment plant (Toczyłowska-Mamińska and Mamiński 2022). Microbial supported fuel cells are better alternatives for the present high energy-consuming biological water treatment systems, although the involved costs, difficulty in technological upgradation, and challenges pertaining to materials of the electrode are the restrictions that limit the broader utilization of the present system in the treatment of polluted water (Zuo et al. 2023). Industrial commercialization and the overall performance of several analyzed nanostructured materials cells are governed by the generation of electrons and efficiency of proton transfer, evidently dependent on the engaged electrodes' performance (Khorram et al. 2023). That is why, for the upgradation of performance and efficiency of either cathode or anode or both electrodes (cathodes and anodes) with novel materials is important to design, develop, and investigate the nanostructured photocatalysts as pollutants remediation strategies for wastewater treatment approaches. Numerous investigations on carbonaceous materials, biomaterials, and polymer nano-composites and nano-materials (Mamidi et al. 2022) have been carried out for up-gradation of electrode performance, including surface area, stability, and conductance.

14.3.1 Ultrasound Technology

Ultrasound is one of the latest approaches being deployed for both detoxification of organic pollutants and as an additional tool for avoiding resultant fouling of filters and membranes involved in water treatment and purification processes (Das et al. 2022). Ultrasound energy has been recorded to degrade pollutants by providing heat, generating hydroxyl radicals (°OH), and subsequently oxidizing them. The ultrasound approach is a more advanced, efficient, and economically viable approach when deployed in conjunction with other relevant water treatment and purification techniques, because photocatalytic degradation and mineralization of water pollutants by the ultrasound technique are little bit challenging to take place at appropriate time-scale and dissipate energy (Ju et al. 2022). So far, it has also been observed that pharmaceutical and personal care commodities concerned pollutants are difficult to eliminate by conventional biological-based approaches, but these contaminants can easily be mineralized by ultrasound technology (Hassani et al. 2022). Generally, the ultrasound technique has been deployed in conjunction with several other water purification techniques, including biological strategies (Pryce et al. 2022), flocculation, coagulation, membrane filtration, electrochemical methods, and adsorption (Kolya and Kang 2023).

14.3.2 Microwave Technology

Microwave is yet another kind of technology that has demonstrated the capability to be broadly utilized in diversified pollutants segregation and water treatment assemblies (Fajal et al. 2023). It has expressed its ability to assist in the treatment of water by chemical methods for the objective of improvement of its kinetics, boosting efficiency and heat selectivity, permitting either selective activation or suppression of target reaction progress, and therefore reducing the dimension (size) of the engaged equipment (Yang et al. 2022). Additionally, the microwave approach had also been investigated in conjunction with several other commonly deployed oxidants in order to accelerate the generation of free radicals such as (°OH) from the employed oxidant, making degradation as well as mineralization of the water pollutants quicker and more efficient. Deployment of microwave processes along with nano-structured photocatalysts has also been documented to accelerate the elimination of water pollutants (Katwal and Pathania 2022). Several microwave-adsorbent nanostructured materials, including metal oxides (MO) and activated carbon, have so far been explored for boosting the adsorption-led photodegradation and mineralization of water pollutants from aqueous matrix in the water treatment process (Oladoye et al. 2023). Other relevant associated water treatment approaches, including photolysis and Fenton processes, can also be associated in conjunction with microwave technology to upgrade the performance and efficiency with respect to mineralization of the existing water contaminants (Xia et al. 2022).

14.3.3 Membrane Technology

Membrane technology is yet another one to be considered as one of the contributing ways to effectively and efficiently treat water on a commercial scale. The remarkable contribution of more advanced and modern membrane technologies, including microfiltration, ultrafiltration, and reverse osmosis in water treatment and subsequent purification, is primarily because of the associated advantageous opportunities which these water treatment techniques offer for greater sustainability with task-specific designs (Giri et al. 2022). In membrane technology, the consumption of lower energy with limited space is demanded in comparison to other conventional water treatment approaches. Recently, membrane bioreactors have been accepted as more trustworthy technology for treatment of polluted water, in which biological reactors are generally conjunct with filters for the filtration process to overcome the limitations of conventional membranes engaged in water treatment assemblies. Even after that, fouling is the prime drawback of membrane technology, restricting its broader utilization as bioreactors on a commercial scale (Muduli et al. 2022). So far, to sort out this issue of fouling, antifouling agents as self-cleaning agents are being deployed. Advanced water pollutants remediation techniques have been demonstrated in Fig. 14.4.

14.3.4 Membrane Bioreactors

Addressing the problem of membrane fouling, the deployment of novel nanostructured composites can enable diversified utilization of membrane bioreactors for wastewater treatment and consequently reduce the cost of treatment. The novel nanostructured composite materials can be effective in upgrading the efficiency of connected growth treatment assemblies, including coagulation and moving bed membrane bioreactors, in the treatment of wastewater (Asante-Sackey et al. 2022). Fabrication and deployment of nanostructured composite materials for membrane fabrication is a domain in which extensive research is being done by researchers. A wide range of materials, including ceramic, polymeric materials, and bioreactors, are being explored to upgrade the characteristics of membranes, including their porosity, wetting behavior, membrane lifetime, and chemical and mechanical stability (Xue et al. 2024). Nanostructured composites, which are the prime agenda of writing

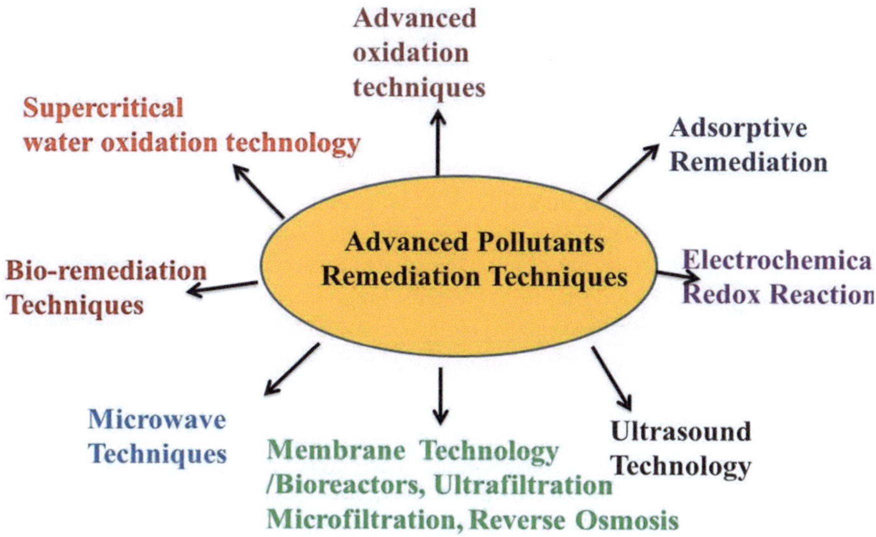

Fig. 14.4 Advanced pollutants remediation techniques

the present literature, are yet other novel adsorptive materials, which recently have become reasonably common and popular for membrane fabrication. More economic and flexible designs are generally associated with these materials, which are also capable of enhancing the physical, chemical, and functional characteristics of the deployed membranes (Bhattacharyya et al. 2022).

14.3.5 Electrochemical Redox Reaction

Electrochemical redox reactions have numerous advantageous characteristics over already operating water treatment strategies. These advantageous features of electrochemical redox reactions include (i) their potential to treat multiple categories of water pollutants simultaneously; (ii) in situ generation of participating radicals (°OH and $°O^{2-}$) and chemicals; (iii) modular construction of the cell; and (iv) recovery of energy (Ganiyu et al. 2021). Treatment of wastewater by potential electrochemical cells includes electrochemical reduction of oxyanions, halogenation of organic molecules, electrochemical oxidation of organic molecules, electro-deionization of inorganic salts, and inactivation of viruses, bacteria, and toxic metals removal via the electrocoagulation, electrodeposition, and electrosorption processes (Mousset et al. 2023). The widespread potential of electrochemical technologies has generated countless opportunities for the replacement of multiple established traditional water treatment technologies. The deployment of electrochemical advanced oxidation processes (EAOPs) at the place of electrochemical water disinfection instead of disinfection by chlorine (Cl_2) products and traditional advanced oxidation processes (AOPs) (de Sousa Filho et al. 2022). Although so far, electrochemical technologies have been deployed for a sole agenda, including electrochemical disinfection of wastewater, there are several other opportunities also, regarding the deployment of a single electrochemical cell to achieve several wastewater treatment targets. For instance, simultaneous oxidation of organic pollutants, electrochemical disinfection, removal of several metallic impurities, oxidation and reduction of contaminants, and electrocoagulation, oxidation of water pollutants, and electrochemical disinfection are a few examples (Zeng et al. 2023). Latest outcomes of the research in this context have also converted traditional water segregation membranes into potentially active surfaces, referred to as reactive electrochemical membranes (REMs). Further scientific developments and upgradation of this technology have expanded its applications for simultaneous electrochemical oxidation/reduction and filtration of water pollutants (Fitch et al. 2022).

14.3.6 Supercritical Water Oxidation Technology

Keeping the novel characteristic features of water under supercritical conditions, supercritical water oxidation (SCWO) technology has been reported as an emerging technology with potential to achieve greater efficiency in a short span of time, with elimination of the requirement of dewatering for wastewater, including sewage sludge. Contrary to several other established supercritical fluids, water has been observed to undergo remarkable challenges with respect to solvent behavior between normal and supercritical operating conditions. For the same reason, the workability of such a system is more challenging. From the outcomes of the investigation, the reported disadvantages of SCWO operating systems are (i) corrosion of the apparatus while its working at both greater temperature and greater pressure; (ii) incompatibility for wastewater with greater inorganic content; (iii) corrosion of the assembly because of in situ acid generation consequently results in decrease in density at elevated temperature; (iv) greater consumption of energy and (v) high cost of operation. To sort out such problems, there are several remarkable ongoing and completed investigations. But, from all the commercially constructed and operated water treatment plants, only two water treatment plants had been closed because of technical issues such as choking, corrosion, and economic reasons (Tang et al. 2021). That is why efforts are being made to reduce the existence of such issues by emphasizing more on oxidation by subcritical water rather than SCWO, catalytic operations, or partial oxidation. More investigations in this regard are needed to be carried out to look for more resistant and reasonably durable materials for SCWO technology. To run and upgrade this innovative technology efficiently, appropriate technical solutions, reactor design, and water-resistant materials with justified durability need to be identified and deployed. Otherwise, keeping the characteristic features of inorganic, organic compounds, and solid content into consideration, this technology can only be deployed at the commercial level for treatment of wastewater, else it shall be restricted to laboratory scale for research and development (R&D) related investigations (Adar et al. 2019). Oxidation investigations about subcritical water oxidation are also being upgraded with more advanced methods for biomass transformation. As in terms of the consumption of energy, oxidation by subcritical water is more economical owing to the reason that it runs at comparatively lower pressure and temperature in comparison to SCWO (Tang et al. 2021).

14.3.7 Advanced Oxidation Techniques

Advanced oxidation techniques (AOTs) are most commonly used techniques to oxidize organic impurities through the generation of highly reactive hydroxyl and superoxide radicals. AOTs mediated by hydroxyl radical deploy hydroxyl radicals (HO°), the most powerful oxidizing species after fluorine, with an $E^0 = 2.80$ V (Xiao et al. 2020). Such superactive and unselective radicals have remarkable potential to degrade the pollutant organic compounds without any secondary waste generation, yielding inorganic ions, CO_2, and H_2O as final discharged products (Ribeiro et al. 2017). AOTs can be categorized into photochemical, chemical, hydrochemical, sonochemical, and electrochemical processes, which are governed by the generation of HO• radicals, or as homogeneous and heterogeneous processes, determined by the number of phases involved during the transport of reaction species. Homogeneous AOTs comprise processes deploying hydrogen peroxide (H_2O_2), UV, dissolved O_3, Fenton reagent, wet peroxide oxidation (WPO), as well as wet oxidation (WO). On the other hand, heterogeneous AOTs include heterogeneous Fenton-like processes, heterogeneous photocatalysis, photocatalytic zonation, catalytic wet oxidation (CWO), as well as catalytic wet peroxide oxidation (CWPO) and catalytic processes. UV/H_2O_2, ozonation, and UV processes have been successfully executed on full-scale urban wastewater treatment plants (UWWTPs) to mineralize micropollutants (MPs). AOTs have yet not been implemented at a full level, including UV/O_3, UV/H_2O_2/Fe (mild photo-Fenton), UV/O_3/H_2O_2, and other advanced processes, including heterogeneous photo-catalysis, heterogeneous photo-Fenton, photo-catalytic ozonization, photocatalytic membrane processes, electrochemical oxidation, and sonolysis (Lama et al. 2022). Sulphate radical-mediated AOTs have recently attracted remarkable attention because the high redox potential of SO_4^{2-}, which can be compared with the HO° free radical (Thulasisingh et al. 2022). The in situ generated HO° free radical has been observed to interact with microplastic through either of these three competing routes: (i) addition (ii) hydrogen abstraction, and (iii) electron abstraction. SO_4^{2-} reacts preferably via an electron transfer mechanism. Both radicals can interact with inorganic ions, including Cl^- as well as CO_3^{2-}, leading to the generation of secondary reactive species, including chloride (Cl^-) and carbonates (CO_3^{2-}), with relatively greater reactivity with pollutants (Li et al. 2022). Furthermore, chlorine radical-mediated AOTs, including either UV or chlorine, are supported on selective oxidants generated by photolysis of chlorine: Chloride (Cl^-) selective as well as with relatively greater reactivity than HO°

for specific organic pollutants, including phenol, chloro-benzene, and benzoic acid (Ribeiro et al. 2019). The use of AOTs as hybrid processes has also been investigated via integration with membrane technologies and through the coupling of more than two AOTs. This hybrid strategy can result in: (i) a synergistic effect because of the positive interactions among the individual processes; (ii) a cumulative impact generated by elevated production of reactive oxygen species (ROS); (iii) in a rare antagonistic effect, because of excessive formation of ROS, which potentially can act as self-scavengers (Xie et al. 2022). The overall remediation of water contaminants by AOTs is greatly influenced by the quality of the water matrix requiring treatment, as well as by the nature of dissolved pollutants, which can have inhibitory, neutral, or promotional effects. The composition of the wastewater matrix has been reported to have a strong impact on the inactivation of pathogens in disinfection processes of water (Duan et al. 2021). The presence of scavengers in real water matrix also interrupts elimination of dissolved organic matter (DOM), expressed as the prime proportion of organic matter in effluent occurring in biologically treated urban wastewater (Du et al. 2023).

14.3.8 Bioremediation Techniques

Bio-surfactants are yet another class of novel nanostructured composite materials deployed in the remediation of water pollutants. Investigations have revealed that nanoscale composite materials are capable of raising the bioactivity of bio-surfactants and bio-catalysts, and due to their quantum size effects, surface morphological features of nanostructured composites improve their microbial metabolic performance. Magnetic biochar is a bio-nanocomposite commonly deployed for bioremediation of wastewater. Biochar consists of residues of biomasses, including domestic, agricultural, forestry, animal husbandry, and industrial is considered a supplementary transporter for nanomaterials owing to its greater porosity and enormous exposed surface area facilitated with numerous oxygen-rich functional groups, including hydroxyl, carbonyl, carboxyl, and epoxy groups (Valizadeh et al. 2021). The biochar was synthesized from *Pinus massoniana* pine needles by grinding, pyrolyzing, and heating at elevated temperature (>300 °C) before its reaction with ferrous sulfate ($FeSO_4.7H_2O$) and ferrous nitrate nonahydrate ($Fe(NO_3)_3.9H_2O$) to impart magnetic properties in the synthesized biochar using $NH_3.H_2O$ as solvent (Liang et al. 2020). Manz et al. (2023) fabricated this nanocomposite to decompose the 1, 4-dioxane molecules via per-sulfate activation. His investigations revealed that per-sulfate can also be activated by oxygenated functional groups and Fe (II). Also, if transformed into quinone and ketone, the carbon–carbon ($C = C$) double bond can be activated by per-sulfate. From all eco-friendly composite materials supposed to be deployed in nanocomposites, polymers have attracted special attention. Polymer-supported nanocomposite materials have been synthesized and investigated for more than five decades for the fabrication of industrially applicable high-performance composite materials, while fibrous protein-based polymer nanocomposites have prevailed in the animal and human body for millions of years (Ahmad et al. 2023). These biodegradable natural polymers have been reported to offer several potential advantages, including matrix components, including lipids, polysaccharides, and proteins, as natural polymers with the capability of generating biocompatible nanocomposites (Yang et al. 2023). In the past few years, intensive collaborative research attempts have been made to emphasize polymer nanocomposites to upgrade their functional, chemical, and physical characteristics for their widespread applications.

14.3.9 Adsorptive Remediation

A plethora of strategies are presently available in the market to remove heavy metal ions from wastewater, including solvent extraction, ion exchange, physical and chemical precipitation, reverse osmosis, membrane filtration, cyanide treatment, electrochemical precipitation, as well as adsorption processes (Alkhadra et al. 2022). The magnetic nanocomposites deployed for the remediation of heavy metals have been tabulated in Table 14.1. Even though reverse osmosis had been among the most efficient as well as effective methods to decrease heavy metal ions, this technique had been limited because of its high operational cost, in addition to a limited pH range (Najid et al. 2022). Similarly, most of the available techniques for the removal of heavy metals are reportedly imperfect because of their relatively higher cost and complicated procedure.

It is required to develop new technologies using widely available low-cost materials for the removal of heavy metal ions from the polluted matrix. Among various currently available water treatment techniques to remove heavy metal ions from wastewater, adsorption has been recorded to be one of the most effective as well as feasible techniques to treat water (Alam et al. 2022). Adsorption has been recorded to be far improved over other strategies because of its initial setup cost, simplicity of design, insensitivity to hazardous pollutants, in addition to its ease of operation. The magnetic nanocomposite structures for the remediation of pharmaceutical pollutants have been summarized in Table 14.2. Moreover, this method does not get engaged in the generation of toxic substances.

The utilization of nanostructured novel composite materials is not only limited to this, but the different nanocomposites for the removal of industrial dyes from wastewater have been tabulated in Table 14.3.

Keeping the several widespread utilizations and the mode of action of the nanostructured photocatalysts into consideration, it has been revealed that remediation of water pollutants from different sources is sequestrated either by one dominant or multiple operating forces in the real-time practical scenarios. So far, investigated working mechanisms for the elimination of water pollutants from effluent include pore filling, electrostatic attractions, and adsorbed species exchange, hydrophobic interactions, H-bonding, adsorption, partition uncarbonized fraction, and pi–pi interactions. The brief demonstration of the operating mechanism for the remediation of water pollutants from real discharged effluent has been depicted in Fig. 14.5.

Table 14.1 Magnetic nanocomposite deployed for heavy metal elimination

Magnetic nanosorbent	Target pollutant	Maximum adsorption capacity ($mg.g^{-1}$)	Remarks	References
Fe_3O_4-SiO_2 Fe_3O_4-SiO_2–NH_2	Cr(III)	36.9[a] 108.2[a]	Time: 2 h pH: 5.4 Temp.: 25 °C Concentration: 104 $mg.L^{-1}$ Isotherm model: Freundlich	Liu et al. (2023)
Chitosan–clinoptilolite–Fe_3O_4	Pb(II)	137.0	Time: 2 h pH: 6.0 Temp.: 60 °C Concentration: 10–90 $mg.L^{-1}$ Isotherm model: Langmuir Kinetic order: Pseudo second	Mohammed (2024)
Fe_3O_4–activated carbon	Cd(II), Pb(II)	49.8 86.2	Time:1 h pH: 6.0 and pH: 5.0 Temp.: 25 °C Concentration: 10–200 $mg.L^{-1}$	Dogari et al. (2023)
Fe_3O_4@C	Cr(VI)	61.7	Time: 2.5 h pH:4.0 Temperature: 25 °C Concentration: 20–100 $mg.L^{-1}$ Isotherm model: Langmuir Kinetic order: Pseudo-second	Li et al. (2023b)

(continued)

Table 14.1 (continued)

Magnetic nanosorbent	Target pollutant	Maximum adsorption capacity ($mg.g^{-1}$)	Remarks	References
Fe_3O_4–multi-walled carbon nanotubes	Pb(II) Zn(II)	67.3 3.3	Time: 6 h pH: 5.0 Temp.: 25 °C Concentration: 200 $mg.L^{-1}$ 30 $mg.L^{-1}$ Isotherm model: Langmuir Kinetic order: Pseudo second	Ozdemir et al. (2023)
Fe_3O_4–grapheme Oxide–NH_2	Cr(VI)	123.4	Time:12 h pH: 2.0 Temp.: 25 °C Isotherm model: Langmuir Kinetic order: Pseudo Second	Bagheri et al. (2024)
GO–Fe_3O_4	As(III) As(V)	147.0 113.0	Time: 24 h pH: 7.0, pH:3.0, Temp.: 25 °C Concentration: 0.1–1200 $mg.L^{-1}$ Isotherm model: Langmuir Kinetic order: Pseudo second	Dung et al. (2022)
Fe_3O_4-polyaniline	Hg(II)	307.0	Time: 5.17 h pH: 5.4 Temp.: 20 °C Concentration: 20–100 $mg.L^{-1}$ Isotherm model: Langmuir Kinetic order: Pseudo second	Kumara et al. (2023)

(continued)

Table 14.1 (continued)

Magnetic nanosorbent	Target pollutant	Maximum adsorption capacity ($mg.g^{-1}$)	Remarks	References
Fe_3O_4-NH_2–poly-dopamine	Cr(VI)	284.1	Time: 4 h pH: 2.0 Temp.: 25 °C Concentration: 50–1000 $mg.L^{-1}$ Isotherm model: Langmuir Kinetic order: Pseudo second	Qiu et al. (2023)
Fe_3O_4–chitosan–poly-ethylene-imine	Pb (II)	124.0	Time: 105 min pH: 5.0 Temp.: 50 °C Concentration: 30–90 $m.L^{-1}$ Isotherm model: Langmuir Kinetic order: Pseudo second	Shi et al. (2023)
Fe_3O_4@C	As(V)	20.1	Time: 2 h pH: 1.0–2.0 Temp.: 25 °C Concentration: 10–50 $mg.L^{-1}$ Isotherm model: Langmuir Kinetic order: Pseudo second	Peng et al. (2022)
Chitosan–Fe_3O_4–GO	Cu(II)	217.4	Time:12 h pH: 7.0 Temp.: 25 °C Concentration: 20–300 $mg.L^{-1}$ Isotherm model: Langmuir Kinetic order: Pseudo second	Bulin et al. (2024)

(continued)

Table 14.1 (continued)

Magnetic nanosorbent	Target pollutant	Maximum adsorption capacity ($mg.g^{-1}$)	Remarks	References
Fe_3O_4-Z-NCNT/PC nanocomposite	Pb(II)	739.87 $mg.g^{-1}$	pH: 2–9 Higher exposed surface area (SA): 1358 $m^2.g^{-1}$ Larger pore volume: 1.61 $cm^3.g^{-1}$ Adsorption capability even after ten cycles, greater selectivity, elevated capacity 20 min	Liosis et al. (2021)
GO functionalized tryptophan (MGO-Trp) nanocomposite	Cd(II) Pb (II) Cr(VI)	5400, 4800, and 3400 $\mu mol.g^{-1}$ for Pb(II), Cd(II), Cr(VI) ions respectively	pH: 7 for Pb (II) and Cd(II), 2 for Cr(VI), Environment-friendly, surface area (SA) 10.157 $m^2.g^{-1}$ with pore diameter 13.229 nm, exothermic reaction, nano-sized structure (50–75 nm)	Mahmoud et al. (2020)
Magnetic GO nano-composites	Cu(II) Pb (II) Ni(II) Zn(II) Cr(III)	62 89 $mg.g^{-1}$ in case of Cu(II), 200 $mg.g^{-1}$ for Pb(II), 51.02 $mg.g^{-1}$ in case of Ni(II), 63.69 $mg.g^{-1}$ in case of Zn(II) and 24.33 $mg.g^{-1}$ in case of Cr(III)	Time: 30 min pH: 5 for Pb(II), 6 in case of Cu(II), 7 in case of Zn(II), 6 in case of Cr(III), 8 in case of Ni(II) Effective elimination for four cycles	Guo et al. (2014)

(continued)

Table 14.1 (continued)

Magnetic nanosorbent	Target pollutant	Maximum adsorption capacity ($mg.g^{-1}$)	Remarks	References
Silica-coated magnetic GO nanocomposite	Ni(II) Pb(II) Cd(II) Cr(VI) Ag(I) Co(II) Cu(II)	182.98 $mg.g^{-1}$ in case of Cr(VI), 116.35 $mg.g^{-1}$ in case of Co(II), 226.08 $mg.g^{-1}$ for Ni(II), 149.59 $mg.g^{-1}$ in case of Cu(II), 100.81 $mg.g^{-1}$ for Cd(II), 168.55 $mg.g^{-1}$ in case of Pb(II) as well as 14.09 $mg.g^{-1}$ for Ag(I)	Time: 5 min pH: 6 In a batch experiment, elimination was quicker as well as the equilibrium has been attained in 35 min, chemically adsorptive behavior	Jamasbi et al. (2022)
ZIF-67-Fe_3O_4 eggshell membrane nanocomposites	Cu(II)	344.82 $mg.g^{-1}$	pH: 5 Higher exposed surface area (SA) 1263.9 $m^2.g^{-1}$	Katmış (2023)

[a]Experimental data

Table 14.2 Magnetic nanocomposite structures for the remediation of pharmaceutical pollutants

Magnetic nanosorbent	Target pesticide/ pharmaceutical pollutant	Maximum adsorption capacity ($mg.g^{-1}$)	Remarks	References
Fe_3O_4–NH_2–molecularly imprinted polymer	Chlorpyrifos, diazinon, phosalone	172.4 192.3 196.1	Time:1 h Temp.: 25 °C Concentration: 0.2–1.2 mg/L Isotherm model: Langmuir Kinetic order: Pseudo-second order	Werner et al. (2024)
GO–Fe_3O_4	2.4-dichlorophenoxyacetic	67.3[a]	pH: 3.0 Concentration: 50–750 mg/L Isotherm model: Freundlich Kinetic order: Pseudo-second order	Nguyen et al. (2024)
Fe_3O_4–reduced GO	Ametryn	54.8	Time:70 min pH: 5.0 Temp.: 25 °C Concentration: 2–54 mg/L Isotherm model: Langmuir Kinetic order: Pseudo-second	Sun et al. (2023)
Fe_3O_4–rGO–thio-urea dioxide composite	Tetracycline	1233	Time: 24 h pH: 4.0 Temp.: 40 °C Concentration: 5–100 mg/L Isotherm model: Langmuir Kinetic order: Pseudo second	Harikumar et al. (2024)

(continued)

Table 14.2 (continued)

Magnetic nanosorbent	Target pesticide/ pharmaceutical pollutant	Maximum adsorption capacity ($mg.g^{-1}$)	Remarks	References
Fe_3O_4–reduced GO–nitrile-tri-acetic acid	Tetracycline	212[a]	Time: 24 h pH: 4.0 Temp.: 25 °C Concentration: 50 mg/L Kinetic order: Pseudo second	Idham et al. (2022)
GO–Fe_3O_4	Methadone	87.2[a]	Time: 30 min pH: 6.2 Temp.: 22.5 °C Concentration: 30 mg/L Isotherm model: Langmuir Kinetic order: Pseudo second	Nodehi et al. (2024)
Fe_3O_4–AC–chitosan	Erythromycin Ciprofloxacin Amoxicillin	90.1 178.6 526.3	Time: 2 h pH: 7.0 Temp.: 25 °C Concentration: 5–60 mg/L Isotherm model: Langmuir Kinetic order: Pseudo second	Abd El-Monaem et al. (2022)
Fe_3O_4@SiO_2–Si–quaternary chitosan	Diclofenac	240.4	Time: 30 min pH: 6.0 Temp.: 25 °C Concentration: 40–670 mg/L Isotherm model: Langmuir	Vijayasri and Singh (2024)

(continued)

Table 14.2 (continued)

Magnetic nanosorbent	Target pesticide/ pharmaceutical pollutant	Maximum adsorption capacity ($mg.g^{-1}$)	Remarks	References
Hydroxy-apatite @C–Fe_3O_4	Ampicillin	3.5	Time: 24 h pH: 6 Temp.: 25 °C Concentration: 0–5[b] mg/L Isotherm model: Langmuir	Das et al. (2013)
Multi-walled carbon nanotubes–N–$CoFe_2O_4$	17-β-estradiol, sulfamethoxazole	7.4[a] 20.0[a]	Time: 100 min pH: 5.5 Temp.: 25 °C Concentration: 0.4–2.4 mg/L Isotherm model: Freundlich Kinetic order: Pseudo second	Prokić et al. (2022)

[a] Experimental value, [b] mmol/L

The requirements for these methods are the availability of the most appropriate adsorbent materials with their tunability in both model-based design as well as the standard of working in order to provide high-quality water treated by a highly sophisticated technique. In case of adsorption process in its reverse mode, the adsorbents ought to be regenerated in case of employment several times via an appropriate desorption technique at relatively lower cost for its maintenance, higher efficiency, in addition to ease of operation.

14.4 Conclusion and Future Prospects

The concern about the quality of water being supplied to the public has forced the researchers to carry out their research for highly sophisticated, advanced, cost-effective, as well as efficient wastewater treatment technologies. Lots of research had already been carried out on the synthesis as well as applications of nanomaterials in pollutants remediation from wastewater. Very recently, magnetic nanomaterials, specifically nanostructured iron oxides, have attracted special attention in the treatment of wastewater. The most commonly reported cost-effective and directly deployable method for the fabrication of magnetic ferrous nanoparticles has recently been recorded as the co-precipitation method of synthesis. However, the limitations of the co-precipitation method are challenges in regulating the particle size, at a later stage,

Table 14.3 Different nanocomposites for the removal of industrial dyes

Nanocomposite	Textile pollutant	Adsorption capacity	Remarks	References
Novel magnetic nanocomposite-MECNT as an adsorbent	Cationic dyes including methylene blue (MB), rhodamine blue (RB) and crystal violet (CV)	287 $mg.g^{-1}$ in case of crystal violet (CV), 231 $mg.g^{-1}$ in case of RhB, 302 $mg.g^{-1}$ in case of MB	pH: 8 Time: 0–6 h for crystal violet, three times reusable Model of isotherm: Langmuir Model of kinetic: pseudo-second-order Composite expressed high potential for endothermic, spontaneous adsorption, after five cycles of adsorption, the crystal violet (CV), methylene blue (MB), and rhodamine B (RB) were only 11.0, 9.0, and 10.0% adsorbent left from experimental matrix	Rostami et al. (2024)
Magnetic beads (AO-γ-Fe_2O_3) OMWCNTs nanocomposite	Methylene blue	905.5 $mg.g^{-1}$	pH: 4–10 Time: 48 h Process led by intraparticle diffusion Isotherm model: Freundlich Langmuir isotherm	Dehbi et al. (2020)

(continued)

Table 14.3 (continued)

Nanocomposite	Textile pollutant	Adsorption capacity	Remarks	References
Laccase–CNT nanocomposites	Congo red	CNTs had an adsorption potential of 625 $mg.g^{-1}$, while laccase-loaded CNTs exhibited an adsorption of ~ 769.23 $mg.g^{-1}$	pH: ~ 7 Time:180 min A robust technology of direct immobilization on laccases over carbon nanotubes regenerative adsorption potential, eco-friendly nature, smaller energy cost, efficient substrate conversion Isotherm model: Langmuir model The nanocomposite removed 96% of the dye, laccase-loaded carbon nanotube demonstrated improved adsorptive elimination in comparison to bare carbon nanotubes (CN)	Ambika and Singh (2020)
mCS/GO-QAS (Magnetic graphene oxide derivative) ammonium salt, chitosan	Basic brown 4	650 $mg.g^{-1}$	pH: 6 Time: 30 min Reusability up to three cycles, hydrophobic, Temkin models, as well as Kinetic model: pseudo-second Order nanocomposite had an adsorptive elimination efficiency of 95% but reusability efficiency declined to 64%; process is exothermic and spontaneous	de Figueiredo Neves et al. (2020)

(continued)

Table 14.3 (continued)

Nanocomposite	Textile pollutant	Adsorption capacity	Remarks	References
Meso-porous carbon-supported cobalt ferrite nanocomposites	Congo red (CR), methylene blue (MB)	Methylene blue (MB) expressed adsorptive potential of 43.05 $mg.g^{-1}$, whereas Congo red (CR) demonstrated an adsorptive capacity of 270.64 $mg.g^{-1}$	pH: 3–11 Time: 3 min Composite can adsorb both cationic and anionic dyes Kinetic order: Pseudo-second-order Isotherm model: Langmuir The adsorptive process is reported to be spontaneous and exothermic for 97.6% for Congo red (CR) and methylene blue (MB) expressed 75.3% adsorption efficiency. The adsorptive capability declined after two cycles	Wu et al. (2020)
Magnetic chitosan (MC) nanocomposite	Reactive red (RR) 141, reactive yellow (RY) 14	Absorbent capability of 89.7 $mg.g^{-1}$ in the case of reactive red (RR) 141 and 98.8 $mg.g^{-1}$ in the case of s (RY) 14	pH: 5.6 Reusable for five cycles Isotherm model: Freundlich Kinetic order: Pseudo-first-order Elimination capability in case of reactive red (RR)141 has been ~ 99.5% as well as in case of reactive yellow (RY) 14 was ~ 92.7%, investigated up to five cycles; reduced adsorption potential has been ascribed to the availability of only a few adsorptive locations	Jaafari et al. (2020)

(continued)

Table 14.3 (continued)

Nanocomposite	Textile pollutant	Adsorption capacity	Remarks	References
Magnetic oxidized multi-walled carbon nanotube (OMWCNT)–κ-carrageenan–Fe_3-O_4 nanocomposite	Anionic reactive black (RB) 5 as well as crystal violet (CV)	–	Time: 300 min Adaption by crystal violet (CV) is linearly proportionate to pH, whereas inversely in the case of reactive black (RB) 5, the process demonstrated super-paramagnetic features Kinetic order: Pseudo-second-order Isotherm model: Langmuir The adsorptive process is spontaneous and endothermic	Homaeigohar (2020)
Magnetic NH_2–CNT-Fe_2O_3-ZIF-8 nanocomposites	Rhodamine B, malachite green	–	pH: 6 Time: 145 min NH_2–CNT/Fe_2O_3 had a surface area of 1659 $m^2.g^{-1}$ while ZIF-8 has a surface area of 1485 $m^2.g^{-1}$ Kinetic order: Pseudo-second-order Isotherm model: Langmuir Elimination efficiency of 90.93% for NH_2–CNT/Fe_2O_3 while 19.38% in the case of ZIF-8. After five cycles, the reported elimination efficiency was 91%	Ambika and Singh (2021)

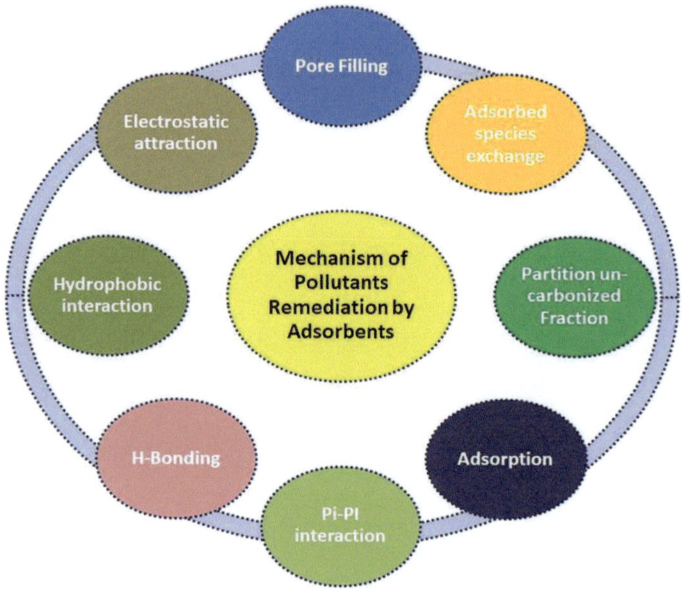

Fig. 14.5 Mechanism of pollutants remediation by adsorbents

interfere with the material's chemical, physical, and magnetic features. Contrary to this, hydrothermal and solvothermal methods used to synthesize Nanoparticles facilitate better control over the morphology, particle size, and favor primarily, the generation of core–shell structures. On the other hand, the thermal technique demands a supply of greater temperature in addition to applied pressure.

The present study concludes that magnetic particles can efficiently be merged with other materials, constituting nanostructured composites, to be evaluated for their adsorptive performance for the elimination of widely spread contaminants from wastewater. The adsorptive performance efficiency of nanostructured composites acting as adsorbents to eliminate inorganic pharmaceutical-derivative compounds as well as pesticides to a considerable extent is dependent on parameters such as characteristics features of novel nanostructured composite, their affinity between the nanosorbents and the contaminants, real-time experimental conditions, and so on. While evaluating the viability of the novel nanostructured composite, several factors, including pH, contact time for adsorption, concentration of contaminant, and adsorbent dosage, must also be regulated as per need for obtaining desired outcomes. Another aspect of research about nanomaterials is their regeneration and reuse. To deploy nanomaterials for the purification of volumes of water in bulk, the adsorbent ought to be reused, in case of both economic as well as environmental concerns. Outcomes of the deployment of nanostructured composite material in the treatment of natural water are significant because such investigations have revealed that the co-existence of other pollutants might also interfere with the adsorptive remediation of the target contaminant.

Additionally, fabrication of nanostructured composite materials on a large scale is also challenging, demanding optimization of the ideal conditions, number of reagents used, number of steps adopted, product yield, and simultaneously, a material with the required characteristic features of an adsorbent. In case of industrial-scale commercialization, more detailed investigations need to be carried out to evaluate the impacts of such novel nanostructured composite materials on human health and the environment. Furthermore, toxicological investigations regarding magnetic nanoparticles are restricted, and their outcomes are still not at a conclusive stage. In brief, investigations with several novel magnetic nanostructured composite materials are being carried out by eminent researchers, focusing on the deployment as an adsorbent in the treatment of wastewater for the elimination of diversified contaminants.

References

Abd El-Monaem EM, Eltaweil AS, Elshishini HM et al (2022) Sustainable adsorptive removal of antibiotic residues by chitosan composites: an insight into current developments and future recommendations. Arab J Chem 15:103743

Adar E, İnce M, Bilgili MS (2019) Evaluation of development in supercritical water oxidation technology. Desalin Water Treat 161:243–253

Ahmad A, Zahra M, Alam F et al (2023) A sustainable approach for the multi-dimensional exploitation of mixed biochar based nano-composites. Fuel 336:126930

Alam G, Ihsanullah I, Naushad M, Sillanpää M (2022) Applications of artificial intelligence in water treatment for optimization and automation of adsorption processes: recent advances and prospects. Chem Eng J 427:130011

Alkhadra MA, Su X, Suss ME et al (2022) Electrochemical methods for water purification, ion separations, and energy conversion. Chem Rev 122:13547–13635

Alsafran M, Saleem MH, Al Jabri H et al (2023) Principles and applicability of integrated remediation strategies for heavy metal removal/recovery from contaminated environments. J Plant Growth Regul 42:3419–3440

Al-zou'by JY, Alzoubi FY, Migdadi AB, Al-Zboon K (2023) Evaluating the impacts of manufactured silver nanoparticles dispersed in various wastewaters on biochemical oxygen demand kinetics of the resulting wastewaters. Nanotechnol Environ Eng 8:119–129

Ambika, Singh PP (2020) Advances in carbon nanomaterial-based green nanocomposites. In: Emerging carbon-based nanocomposites for environmental applications, pp 175–201

Ambika, Singh PP (2021) Functionalized magnetic carbon nanomaterials for environmental remediation. In: Environmental applications of carbon nanomaterials-based devices, pp 227–249

Asante-Sackey D, Rathilal S, Tetteh EK, Armah EK (2022) Membrane bioreactors for produced water treatment: a mini-review. Membranes 12:275

Aydin D, Gübbük İH, Ersöz M (2024) Recent advances and applications of nanostructured membranes in water purification. Turk J Chem 48:1–20

Bagheri V, Naseri A, Sajedi-Amin S et al (2024) Using Fe_3O_4-graphene oxide-modified chitosan with melamine magnetic nanocomposite in the removal and magnetic dispersive solid-phase microextraction of Cr (VI) ion in aquatic samples. Chem Pap 78:381–396

Bhattacharyya A, Liu L, Lee K, Miao J (2022) Review of biological processes in a membrane bioreactor (MBR): effects of wastewater characteristics and operational parameters on biodegradation efficiency when treating industrial oily wastewater. J Mar Sci Eng 10:1229

Bulin C, Guo T, Ma Y (2024) Spectroscopic and statistical physics elucidation for Cu (II) remediation using magnetic bio adsorbent based on Fe_3O_4-chitosan-graphene oxide. Int J Biol Macromol 276:133895

Dadebo D, Nasr M, Fujii M, Ibrahim MG (2023) Bench-scale continuous-flow system for coagulation/flocculation/sedimentation treatment of washing bay effluents: a comparison between natural and chemical coagulants. Eng Proc 37:79

Das A, Adak MK (2022) Photo-catalyst for wastewater treatment: a review of modified Fenton, and their reaction kinetics. Appl Surf Sci Adv 11:100282

Das B, Mandal M, Upadhyay A et al (2013) Bio-based hyperbranched polyurethane/Fe_3O_4 nanocomposites: smart antibacterial biomaterials for biomedical devices and implants. Biomed Mater 8:35003

Das S, O'Connell MG, Xu H et al (2022) Assessing advances in anti-fouling membranes to improve process economics and sustainability of water treatment. ACS ES&T Eng 2:2159–2173

de Figueiredo Neves T, Dalarme NB, da Silva PMM et al (2020) Novel magnetic chitosan/ quaternary ammonium salt graphene oxide composite applied to dye removal. J Environ Chem Eng 8:103820

de Sousa Filho JW, Lenza GA, Tonhela MA et al (2022) Full-scale application of an electrochemical disinfection solution in a municipal drinking water treatment plant. Water Environ J 36:86–95

Dehbi A, Dehmani Y, Omari H et al (2020) Comparative study of malachite green and phenol adsorption on synthetic hematite iron oxide nanoparticles (α-Fe_2O_3). Surf Interfaces 21:100637

Dogari H, Salehi MM, Hassanzadeh-Afruzi F et al (2023) Magnetic polyacrylonitrile-melamine nanoadsorbent (PAN-Mel@ Fe_3O_4) for effective adsorption of Cd (II) and Pb (II) from aquatic area. Mater Sci Eng B 298:116871

Du Z, Li G, Ding S et al (2023) Effects of UV-based oxidation processes on the degradation of microplastic: fragmentation, organic matter release, toxicity and disinfection byproduct formation. Water Res 237:119983

Duan X, Zhou X, Wang R et al (2021) Advanced oxidation processes for water disinfection: features, mechanisms and prospects. Chem Eng J 409:128207

Dung MD, Nga TTV, Lan NT, Thanh NK (2022) Adsorption behavior and mechanism of As (V) on magnetic Fe_3O_4–graphene oxide (GO) nanohybrid composite material. Anal Sci 38:427–436

Fajal S, Hassan A, Mandal W et al (2023) Ordered macro/microporous ionic organic framework for efficient separation of toxic pollutants from water. Angew Chem 135:e202214095

Fitch A, Balderas-Hernandez P, Ibanez JG (2022) Electrochemical technologies combined with physical, biological, and chemical processes for the treatment of pollutants and wastes: a review. J Environ Chem Eng 10:107810

Ganiyu SO, Martínez-Huitle CA, Oturan MA (2021) Electrochemical advanced oxidation processes for wastewater treatment: advances in formation and detection of reactive species and mechanisms. Curr Opin Electrochem 27:100678

Ghazal H, Koumaki E, Hoslett J et al (2022) Insights into current physical, chemical and hybrid technologies used for the treatment of wastewater contaminated with pharmaceuticals. J Clean Prod 361:132079

Giri A, Khakre Y, Shreeraj G et al (2022) The order–disorder conundrum: a trade-off between crystalline and amorphous porous organic polymers for task-specific applications. J Mater Chem A 10:17077–17121

Gunawardhana T, Hong JG, Choi Y et al (2023) Water quality characteristics and reuse potential using adsorption as a post-treatment option for a full-scale hydrocyclone, coagulation, flocculation, and dissolved air flotation system. Environ Geochem Health 45:8585–8598

Guo X, Du B, Wei Q et al (2014) Synthesis of amino functionalized magnetic graphenes composite material and its application to remove Cr (VI), Pb (II), Hg (II), Cd (II) and Ni (II) from contaminated water. J Hazard Mater 278:211–220

Harikumar B, Subhiksha V, Okla MK et al (2024) Magnetic Fe_3O_4 nanospheres supported N/ S-SnO_2 nanorod for highly effective visible light photocatalytic degradation of tetracycline. J Alloy Compd 986:174017

Hassani A, Malhotra M, Karim AV et al (2022) Recent progress on ultrasound-assisted electrochemical processes: a review on mechanism, reactor strategies, and applications for wastewater treatment. Environ Res 205:112463

Homaeigohar S (2020) The nanosized dye adsorbents for water treatment. Nanomaterials 10:295

Huang S, Yu J, Li C et al (2022) The effect review of various biological, physical and chemical methods on the removal of antibiotics. Water 14:3138

Idham MF, Falyouna O, Eljamal R et al (2022) Antibiotics removal from aqueous environments: a mini review on graphene oxide-based nanomaterials application

Jaafari J, Barzanouni H, Mazloomi S et al (2020) Effective adsorptive removal of reactive dyes by magnetic chitosan nanoparticles: kinetic, isothermal studies and response surface methodology. Int J Biol Macromol 164:344–355

Jamasbi N, Ziarani GM, Mohajer F et al (2022) Silica-coated modified magnetic nanoparticles (Fe_3O_4@ SiO_2@$(BuSO_3H)_3$) as an efficient adsorbent for Pd^{2+} removal. Chemosphere 307:135622

Ju L, Lu X, Yang S et al (2022) A multi-time scale dispatching optimal model for rural biomass waste energy conversion system-based micro-energy grid considering multi-energy demand response. Appl Energy 327:120155

Katmış A (2023) Fabrication and characterization of eggshell membrane biotemplated metal-organic framework nanoclusters

Katwal R, Pathania D (2022) Nanostructured photocatalysts in the degradation of toxic pollutants. Heterog Nanocatalysis Energy Environ Sustain 2:73–93

Khongkliang P, Nuchdang S, Rattanaphra D et al (2024) Efficiency enhancement of electrocoagulation, ion-exchange resin and reverse osmosis (RO) membrane filtration by prior organic precipitation for treatment of anaerobically-treated palm oil mill effluent. Chemosphere 363:142899

Khorram AG, Fallah N, Nasernejad B et al (2023) Electrochemical-based processes for produced water and oily wastewater treatment: a review. Chemosphere 338:139565

Kolya H, Kang C-W (2023) Next-generation water treatment: exploring the potential of biopolymer-based nanocomposites in adsorption and membrane filtration. Polymers 15:3421

Kumar N, Gusain R, Pandey S, Ray SS (2023) Hydrogel nanocomposite adsorbents and photocatalysts for sustainable water purification. Adv Mater Interfaces 10:2201375

Kumara KSM, Shivakumar P, Ganesh V et al (2023) Hydrogels of PANI doped with Fe_3O_4 and GO for highly stable sensor for sensitive and selective determination of heavy metal ions. Inorg Chem Commun 158:111553

Lama G, Meijide J, Sanromán A, Pazos M (2022) Heterogeneous advanced oxidation processes: current approaches for wastewater treatment. Catalysts 12:344

Li D, Feng Z, Zhou B et al (2022) Impact of water matrices on oxidation effects and mechanisms of pharmaceuticals by ultraviolet-based advanced oxidation technologies: a review. Sci Total Environ 844:157162

Li J, How ZT, Benally C et al (2023a) Removal of colloidal impurities by thermal softening-coagulation-flocculation-sedimentation in steam assisted gravity drainage (SAGD) produced water: performance, interaction effects and mechanism study. Sep Purif Technol 313:123484

Li J, Su J, Yang Q, Yang Z (2023b) Hydrothermal synthesis of Zr-doped chitosan carbon-shell protected magnetic composites (Zr–Fe_3O_4@C) for stable removal of Cr (VI) from water: Enhanced adsorption and pH adaptability. Mater Chem Phys 306:128057

Liang Y, Li M, Pan F et al (2020) Alkaline phosphomonoesterase-harboring microorganisms mediate soil phosphorus transformation with stand age in Chinese Pinus massoniana plantations. Front Microbiol 11:571209

Liosis C, Papadopoulou A, Karvelas E et al (2021) Heavy metal adsorption using magnetic nanoparticles for water purification: a critical review. Materials 14:7500

Liu Z, Lei M, Zeng W et al (2023) Synthesis of magnetic Fe_3O_4@ SiO_2-(-NH_2/-COOH) nanoparticles and their application for the removal of heavy metals from wastewater. Ceram Int 49:20470–20479

Mahmoud ME, Osman MM, Abdel-Aal H, Nabil GM (2020) Microwave-assisted adsorption of Cr (VI), Cd (II) and Pb (II) in presence of magnetic graphene oxide-covalently functionalized-tryptophan nanocomposite. J Alloy Compd 823:153855

Mamidi N, Delgadillo RMV, Barrera EV et al (2022) Carbonaceous nanomaterials incorporated biomaterials: the present and future of the flourishing field. Compos B Eng 243:110150

Manz KE, Kulaots I, Greenley CA et al (2023) Low-temperature persulfate activation by powdered activated carbon for simultaneous destruction of perfluorinated carboxylic acids and 1, 4-dioxane. J Hazard Mater 442:129966

Mohammed AN (2024) Adsorption efficiency of chitosan/clinoptilolite (CS/CZ) composite for effective removal of Cd^{+2} and Cr^{+6} ions from wastewater effluents of dairy cattle farms. Environ Monit Assess 196:611

Mousset E, Fournier M, Su X (2023) Recent advances of reactive electroseparation systems for water treatment and selective resource recovery. Curr Opin Electrochem 42:101384

Muduli M, Chanchpara A, Choudhary M et al (2022) Critical review on sustainable bioreactors for wastewater treatment and water reuse. Sustain Water Resour Manag 8:159

Najid N, Kouzbour S, Gourich B (2022) Improvement of permeation flux by evaluating the effects of coagulated-humic acids flocs and pH during coagulation-microfiltration hybrid process for water treatment: Performance, fouling modeling, specific energy consumption, and processes comparison. Chem Eng Process Process Intensif 179:109081

Nguyen THT, Nguyen KT, Le BH et al (2024) Synthesis of magnetic Fe_3O_4/graphene aerogel for the removal of 2, 4-dichlorophenoxyacetic acid herbicide from water. RSC Adv 14:22304–22311

Nodehi M, Baghayeri M, Hosseini F, Behazin R (2024) Development of an ultra-sensitive electrochemical sensor: AgNPs deposition on Fe_3O_4NPs-GO surface for accurate quantification of morphine and pethidine in biological samples. Top Catal 68:1–14

Oladoye PO, Kadhom M, Khan I et al (2023) Advancements in adsorption and photodegradation technologies for rhodamine B dye wastewater treatment: fundamentals, applications, and future directions. Green Chem Eng 5:440–460

Ozdemir S, Dündar A, Dizge N et al (2023) Preconcentrations of Pb (II), Ni (II) and Zn (II) by solid phase bio-extractor using thermophilic Bacillus subtilis loaded multiwalled carbon nanotube biosorbent. Chemosphere 317:137840

Peng Y, Li Y, Tang S et al (2022) Dynamic adsorption of As (V) onto the porous α-Fe_2O_3/Fe_3O_4/C composite prepared with bamboo bio-template. Water 14:1848

Prokić D, Vukčević M, Mitrović A et al (2022) Adsorption of estrone, 17β-estradiol, and 17α-ethinylestradiol from water onto modified multi-walled carbon nanotubes, carbon cryogel, and carbonized hydrothermal carbon. Environ Sci Pollut Res 29:4431–4445

Pryce D, Kapelan Z, Memon FA (2022) A comparative evaluation of the sustainability of alternative aeration strategies in biological wastewater treatment to support net-zero future. J Clean Prod 374:134005

Qiu X, Wang B, Zhao X et al (2023) Green and sustainable imprinting technology for removal of heavy metal ions from water via selective adsorption. Sustainability 16:339

Ranjan Mishra S, Gadore V, Ahmaruzzaman M (2022) Nanostructured composite materials for treatment of dye contaminated water. In: Nanohybrid materials for water purification. Springer, pp 97–120

Ribeiro MCM, Starling MCVM, Leão MMD, de Amorim CC (2017) Textile wastewater reuse after additional treatment by Fenton's reagent. Environ Sci Pollut Res 24:6165–6175. https://doi.org/10.1007/S11356-016-6921-9

Ribeiro ARL, Moreira NFF, Puma GL, Silva AMT (2019) Impact of water matrix on the removal of micropollutants by advanced oxidation technologies. Chem Eng J 363:155–173

Rostami MS, Khodaei MM, Benassi E (2024) Surface modified of chitosan by TiO_2@MWCNT nanohybrid for the efficient removal of organic dyes and antibiotics. Int J Biol Macromol 274:133382

Sangamnere R, Misra T, Bherwani H et al (2023) A critical review of conventional and emerging wastewater treatment technologies. Sustain Water Resour Manag 9:58

Schwetschenau SE, Kovankaya Y, Elliott MA et al (2022) Optimizing scale for decentralized wastewater treatment: a tool to address failing wastewater infrastructure in the United States. ACS ES&T Eng 3:1–14

Shamsan A, Fouad M, Yacoob W et al (2023) Performance of a variety of treatment processes to purify wastewater in the food industry. Curr Chem Lett 12:431–438

Shende KM (2024) Mitigating operation, quality and regulatory challenges in the packaged drinking water industry: a failure mode effect analysis (FMEA) approach

Shi J, Teng X, Suo Q et al (2023) Novel magnetic polyethersulfone-polyethyleneimine-based microbeads for removal of lead ions from water: kinetics and thermodynamics. J Mol Liq 387:122632

Singh BJ, Chakraborty A, Sehgal R (2023) A systematic review of industrial wastewater management: evaluating challenges and enablers. J Environ Manage 348:119230

Srivastava RK, Sarangi PK, Vivekanand V et al (2022) Microbial fuel cells for waste nutrients minimization: recent process technologies and inputs of electrochemical active microbial system. Microbiol Res 265:127216

Sun S, Meng X, Lv Z et al (2023) Research progress on the removal of pesticides in water by Fe_3O_4-based adsorbents in the past decade: a review. Arab J Chem 17:105405

Tang X, Zheng Y, Liao Z et al (2021) A review of developments in process flow for supercritical water oxidation. Chem Eng Commun 208:1494–1510

Tariq A, Mushtaq A (2023) Untreated wastewater reasons and causes: a review of most affected areas and cities. Int J Chem Biochem Sci 23:121–143

Thulasisingh A, Nagapushnam P, Balakrishnan Y, Kannaiyan S (2022) Different methods and technologies of advanced oxidation processes adopted in industrial wastewater treatment. In: Wastewater treatment. CRC Press, pp 35–46

Toczyłowska-Mamińska R, Mamiński MŁ (2022) Wastewater as a renewable energy source—utilisation of microbial fuel cell technology. Energies 15:6928

Valizadeh S, Lee SS, Baek K et al (2021) Bioremediation strategies with biochar for polychlorinated biphenyls (PCBs)-contaminated soils: a review. Environ Res 200:111757

Vijayasri K, Singh AK (2024) Magnetic adsorbents for the removal of pharmaceutical contaminants. In: Advanced materials for pharmaceutical wastewater treatment. CRC Press, pp 63–97

Werner J, Zgoła-Grześkowiak A, Grześkowiak T, Frankowski R (2024) Biopolymers-based sorbents as a future green direction for solid phase (micro) extraction techniques. TrAC Trends Anal Chem 173:117659

Wu C, Dong D, Yu X et al (2020) Mesoporous carbon/cobalt ferrite nanocomposite: a charge and pH independent magnetic adsorbent for dye pollutant treatment. Diam Relat Mater 105:107796

Xia H, Li C, Yang G et al (2022) A review of microwave-assisted advanced oxidation processes for wastewater treatment. Chemosphere 287:131981

Xia C, Li X, Wu Y et al (2023) A review on pollutants remediation competence of nanocomposites on contaminated water. Environ Res 222:115318

Xiao J, Xie Y, Rabeah J et al (2020) Visible-light photocatalytic ozonation using graphitic C_3N_4 catalysts: a hydroxyl radical manufacturer for wastewater treatment. Acc Chem Res 53:1024–1033

Xie Z-H, He C-S, Zhou H-Y et al (2022) Effects of molecular structure on organic contaminants' degradation efficiency and dominant ROS in the advanced oxidation process with multiple ROS. Environ Sci Technol 56:8784–8795

Xue Y, Kamali M, Al-Salem SM et al (2024) Naturally derived materials to enhance the membrane properties in (waste) water treatment applications-mechanisms, scale-up challenges and economic considerations. J Water Process Eng 57:104647

Yang F, Du M, Yin K et al (2022) Applications of metal-organic frameworks in water treatment: a review. Small 18:2105715

Yang X, Yu C, Hassan B et al (2023) Pyrolytic mechanisms of typical organic components of sewage sludge in the presence of CaO: polysaccharides, proteins, and lipids. Sci Total Environ 901:166020

Zagklis DP, Bampos G (2022) Tertiary wastewater treatment technologies: a review of technical, economic, and life cycle aspects. Processes 10:2304

Zeng Q, Zhang Y, Chen P et al (2023) Electrocoagulation coupled with electrooxidation for the simultaneous treatment of multiple pollutants in contaminated sediments. J Environ Sci 124:89–97

Zuo K, Garcia-Segura S, Cerrón-Calle GA et al (2023) Electrified water treatment: fundamentals and roles of electrode materials. Nat Rev Mater 8:472–490

Chapter 15
Highly Active Metal-Oxide Electrodes: Emphasizing Their Efficiency and Cost-Effectiveness in Electrochemical Remediation of Organic Contaminants from Wastewater

Dusmant Maharana, Saismrutiranjan Mohanty, Sanjoy Kumar Maji, Sibani Sahu, and Biswajit Mishra

Abstract The rapid growth in population and industrialization has significantly improved human living standards, but it has also led to considerable environmental degradation. A major concern resulting from this expansion is the increasing presence of organic contaminants in ecosystems. These toxic pollutants cause long-lasting damage, including biomagnification, environmental persistence, endocrine disruption, and carcinogenic effects, all of which pose significant threats to both ecosystems and human health. In response to these challenges, there has been increasing attention toward innovative methods for efficiently removing organic pollutants from water and wastewater. Typically, toxic organic compounds undergo physical, chemical, or biological treatments to break them down or neutralize their toxicity before being released into the environment. This chapter delves into advanced oxidation processes

D. Maharana
Department of Environmental Science, Parul Institute of Applied Sciences, Parul University, Vadodara, Gujarat, India

State Key Laboratory of Water Environment Simulation, School of Environment, Beijing Normal University, Beijing, China

S. Mohanty (✉) · B. Mishra
Department of Environmental Sciences, School of Applied Sciences, Centurion University of Technology and Management, Bhubaneswar, Odisha, India
e-mail: saismruti1994@gmail.com

S. K. Maji
Department of Chemistry, School of Applied Sciences, KIIT Deemed to Be University, Bhubaneswar, Odisha, India

S. Sahu
Department of Chemistry, School of Applied Sciences, Centurion University of Technology and Management, Bhubaneswar, Odisha, India

A. M. Yatoo et al. (eds.), *Wastewater Monitoring and Management by Advanced Approaches*, Advances in Wastewater Research,
https://doi.org/10.1007/978-981-95-2601-7_15

(AOPs), a physicochemical approach designed to reduce environmental pollution, with a particular focus on the electrochemical oxidation process using highly active metal-oxide electrodes. Both early and recent studies have evaluated the effectiveness of electrochemical methods in decomposing organic contaminants in wastewater. Modern treatment techniques often involve advanced nanocrystalline electrodes such as PbO_2, SnO_2, boron-doped diamond (BDD), Pt/carbon felt, $ZnWO_4$ film electrodes, and various Ti-based electrodes, including SnO_2–Sb, BDD, RuO_2, and Pt. The chapter also examines promising alternative electrode materials for electrochemical oxidation processes in treating organic wastewater. This method offers numerous advantages, including cost-effectiveness, environmental compatibility, energy efficiency, flexibility, and automation potential, positioning it as a highly promising solution for tackling pollution-related issues.

Keywords Metal-oxide electrodes · Electrochemical process · Organic contaminants · Intermediate products · Energy efficiency

15.1 Introduction

Environmental pollution primarily arises from natural processes, such as chemical or thermal contamination, rather than solely from human activities. However, with the continual growth of the global population and intensified industrialization, ensuring environmental safety has become an imperative necessity. Water contamination has emerged as a major environmental issue in recent years, raising serious concerns. Water can become contaminated with a range of harmful organic wastes, such as chlorophenols, herbicides, dyes, pesticides, and pharmaceuticals, all of which can be extremely toxic even at low concentrations. This issue has captured the attention of numerous researchers (Santos et al. 2023; Xu et al. 2024). When organic pollutants build up directly in the environment, both terrestrially and aquatically, they are the main cause of many environmental problems (Relyea 2005; Tripathi et al. 2014a, b). Several factors converge to exacerbate water pollution, including the world's rapidly growing population, swift economic development, poor water management practices, accidental chemical spills like oil, and improper wastewater discharge from industries, agriculture, and pharmaceuticals. Recently, several environmental experts have expressed concern about groundwater pollution, which receives a large volume of organic sludge. As a result, this crucial situation has prompted environmental authorities across the world to monitor both water usage and illness. Furthermore, the global expansion of agricultural products, such as crops, fibers, and vegetables, has increased the demand for agrochemicals to ensure high productivity. As a result, the use of agrochemicals has increased dramatically. Because organic chemicals including 2,4-dichlorophenol (2,4-DCP), triclosan, rhodamine B, and 2,4,5-trichlorophenoxyacetic acid (2,4,5-T) have detrimental effects on aquatic and human health, their extensive use has created environmental concerns in recent years (Shang et al. 2011). The U.S. Environmental Protection Agency (US-EPA) and the Institute of Medicine (IOM) have designated a few of them as priority-controlled pollutants (US-EPA 2012; IOM 2012).

Modern applications of advanced oxidation processes (AOPs) represent a promising global technology, playing a critical role in environmental restoration (Anipsitakis and Dionysiou 2003). AOPs such as UV/Fe^{3+} oxalate or H_2O_2, Fenton processes, photocatalysis, photoelectrocatalysis, ozone water systems, metal–oxalic acid, and O_3/UV have all been developed and studied by research groups in order to remove organic impurities from water (Ma et al. 2010; Ojani et al. 2012). These are the traditional methods, mostly supported by phase separation techniques capable of oxidizing pollutants chemically. The concepts of reactive oxygen species (ROS) and free radicals have long been recognized in relation to the degradation of organic materials (Huang et al. 1993; Bolton 2001). AOPs are increasingly acknowledged for their remarkable efficacy in both physical and chemical realms. Their ability to induce similar alterations in pollutant chemical structures via active free radicals and their thermodynamic feasibility underscore their effectiveness (Domenech et al. 2004). Free radicals, particularly hydroxyl radicals (•OH), are of special importance due to their strong oxidation capacity and resilience against organic oxidation (Niu et al. 2013a, b; Maharana et al. 2015a, b, c). Nevertheless, additional research indicates that AOPs may produce other oxidizing species in addition to •OH (Anipsitakis and Dionysiou 2003). Reactive oxygen species, such as the •OH radical produced during the oxidation process, may usually oxidize organic contaminants.

Over the past decade, electrolysis and electrochemical oxidation have demonstrated greater effectiveness in degrading a variety of organic contaminants (Martínez-Huitle et al. 2023; Gaber et al. 2024). The electrochemical process, a widely recognized treatment technology, has the ability to generate efficient oxidation, leading to the complete removal of many hazardous contaminants from the environment. Electrode materials are essential for the electro-oxidation of pollutants, as they can significantly influence the evolution of the oxidation process, potentially leading to the formation of various hazardous chemical intermediates (Niu et al. 2013a, b; Lin et al. 2013; Maharana et al. 2015a). Electrode materials are typically categorized into two main groups based on their oxidation mechanism: active and non-active. In both types of anodes, the initial step in the degradation of carbon-based compounds involves the oxidation of water (H_2O) molecules, yielding the •OH radical. This process can be represented by the equation: A (•OH): A + H_2O. The chemical reactivity of heterogeneous A (•OH) is significantly affected by the electrode's characteristics. Intense interactions between active anodes and •OH radicals may prompt the rapid formation of larger oxide or superoxide (AO) radicals (Quiroz et al. 2005, 2006). This reaction can occur when an electrode coated with metal oxide has access to higher oxidation states than the typical oxygen evolution overpotential, which is A (•OH) $\rightarrow$ AO + H^+ + e^-. Rather than oxygen evolution occurring as a side reaction through the chemical degradation of higher oxide species (AO $\rightarrow$ A + ½ O_2), the redox pair AO/A serves as a mediator in the oxidation of organic contaminants. This is achieved through the reaction mechanism: AO + R $\rightarrow$ A + RO. Moreover, organic molecules possess the ability to directly interact with A (•OH), leading to the production of fully oxidized reaction products like CO_2 and

H_2O. This phenomenon arises due to the limited interaction between free radicals such as •OH and the surface of a non-active electrode. An alternative way to express the equation is: [A (•OH)] + R → A + mCO_2 + nH_2O + H^+ + e-, where “R” represents a carbon-based molecule, “m” signifies the number of carbon (C) atoms, and “n” indicates the number of hydrogen atoms. Achieving complete oxidation to CO_2 and H_2O in this scenario, without any heteroatoms, necessitates (2 m + n) oxygen atoms. In addition to oxidation to O_2 [A (•OH) → A + ½ O_2 + H^+ + e-], the described reaction pathway can also face competition from side reactions involving A (•OH), such as the indirect dimerization to hydrogen peroxide (H_2O_2) through the following process: H_2O_2 + 2A → 2A (•OH).

The non-active electrode primarily serves as a passive substrate for electron removal, engaging solely in the anodic reaction of organic molecules without providing a catalytically active site for their adsorption (Martinez-Huitle and Ferro 2006). This type of anodized material is often subjected to mechanical operations like water oxidation and outer-sphere electron transfer procedures. Additionally, the oxidation of carbon-based chemicals is facilitated by the •OH radical generated through water discharge. The correlation between the rate of organic oxidation of physiosorbed A (•OH) and the electrochemical activity, valued in terms of oxygen evolution overpotential, is anticipated to be strongly influenced by the efficacy of the A (•OH) interaction. If this connection weakens, the anode’s effectiveness in degrading organic compounds with A (•OH) decreases. For instance, boron-doped diamond (BDD) anodes have been studied as efficient electrodes for the electrochemical oxidation of organic contaminants (Quiroz et al. 2006; Lin et al. 2013). These researchers suggest that BDD anodes are the best non-active electrodes with the aforementioned qualities. In contrast to active electrodes like IrO_2 and RuO_2, which only partially oxidize organic contaminants (Da Pozzo et al. 2005; Martinez-Huitle and Ferro 2006), non-active oxides with doped chemicals, such as Ti/SnO_2 and Pb/PbO_2, are more effective at converting oxidized organics into CO_2 (Quiroz et al. 2005; Martinez-Huitle and Ferro 2006). Recent advancements in electrocatalysis and electrochemical methodologies have demonstrated remarkable effectiveness in degrading environmental pollutants, owing to distinctive attributes (Gaber et al. 2024).

This study investigates the application of various metal-oxide-coated Ti electrodes for the removal of diverse organic chemicals from water. The emphasis is placed on assessing how operational parameters such as preliminary concentration of organic solution, initial pH value, electrode gap, and current density impact the oxidation kinetics of organic substances. Verification of organic compound oxidation across different electrodes is achieved by measuring the decrease in total organic carbon (TOC). Furthermore, comprehensive analysis and discussion are conducted on the breakdown by-products of organic contaminants. Several key characteristics of metal-oxide electrodes and their degradation techniques have been emphasized for their considerable potential and attractiveness. These include environmental compatibility, cost-effectiveness, versatility, energy efficiency, and the potential for future integration with artificial intelligence (AI) technologies.

15.2 Role of Electrode Materials in Electro-Oxidation of Organic Contaminants

Electrode materials play a crucial role in the effectiveness of advanced oxidation processes (AOPs), which are used for the degradation of organic pollutants in wastewater. AOPs rely on the generation of highly reactive species, particularly hydroxyl radicals (•OH), which can degrade a wide range of contaminants. The choice of electrode material directly impacts the efficiency, selectivity, and energy consumption of the process. Various types of metal-oxide electrodes are examined and evaluated in terms of their efficacy in the electro-oxidation of organic contaminants. Electrodes like SnO_2–Sb, SnO_2–Sb/PbO_2 coated on Ti plate (Lin et al. 2012; Niu et al. 2013a, b; Maharana et al. 2015a, b), RuO_2 (Comninellis 1994), Pt–Ir (Szpyrkowicz et al. 2000), BDD coated on Ti plate (Chen et al. 2003a, b; Lin et al. 2015), $ZnWO_4$ film electrode (Zhao and Zhu 2006), and Pt/carbon felt (Sires et al. 2007) have been widely explored in recent wastewater treatment processes. There has been a growing interest in developing highly efficient and stable catalysts for the electro-oxidation of water and wastewater. In addition to water electrolysis, these catalysts could also provide an effective alternative for removing organic pollutants from water or wastewater through electrodegradation (Mirehbar et al. 2023).

This chapter reviews a compilation of data from 1991 to 2024 on the degradation of organic pollutants using various anodes or electrodes. It outlines the resultant by-products under different conditions, as presented in Table 15.1, and gathers this information as key findings aligned with the theme's objectives. Pioneering researchers in the 1970s delved into the potential of electrochemical oxidation for effectively addressing toxic phenols on a pilot-scale plant. Building upon this groundwork, in the early 1980s, researchers introduced synthetic wastewater solutions to deepen the exploration in this field (Mieluch et al. 1975; Smith de Sucre and Watkinson 1981; Chettiar and Watkinson 1983). Smith and Watkinson explored phenol oxidation for wastewater treatment applications using PbO_2 anodes in both split and undivided electrochemical reactors. Their findings revealed that the ratios of oxidized phenols varied between 53 and 77%, contingent upon the type of cell and operational parameters, such as current intensity ranging from 10 to 20 A and differing pH values, respectively (Smith de Sucre and Watkinson 1981). However, the technique has not found a commercial use due to its poor response rate and efficiency (Chettiar and Watkinson 1983). As demonstrated in Table 15.1, SnO_2 and PbO_2 are the most often studied anodic materials for electro-oxidation processes. In research study, PbO_2 is often recognized as a prominent material for oxygen evolution in electrolysis. However, there has been notable exploration into utilizing a Ti-based PbO_2 electrode for efficiently oxidizing phenolic compounds such as benzoquinone, phenol, and chlorinated phenols within aqueous solutions (Comninellis 1992; Houk et al. 1998; Rodgers et al. 1999). Based on the composition of the Ti/PbO_2 electrode, possible removal efficiency was obtained in all cases of conditional parameters. Although PbO_2 electrodes are very inexpensive, stable, and effective in degrading organic pollutants, they are susceptible to electrochemical corrosion, and the release of Pb^{2+} ions can result in severe water pollution (Lin et al. 2015; Maharana et al. 2015a). To circumvent these issues, Ti can be utilized as a support for thermally deposited PbO_2 sheets. On the other hand, Ti-supported Pt anode was employed for the effective degradation of colors, tannery effluents, benzoic acid, and bisphenol-A (Szpyrkowicz et al. 2000; Tanaka et al. 2002). In our investigation, textile dyes such as rhodamine B were shown to degrade effectively on Ti-supported SnO_2–Sb electrodes. Yet, significant advancements in the oxidation of organic pollutants have been noted with the SnO_2–Sb_2O_5 anode, exhibiting a current efficiency five times higher than that observed with the Pt anode (Kötz et al. 1991; Stucki et al. 1991). In addition, Comninellis (1992) reported that the current efficiency at the PbO_2 and RuO_2 anodes was 0.18 and 0.14, respectively. Meanwhile, the current efficiency reached 0.58 at SnO_2–Sb_2O_5 on Ti-based anode, achieving a 71% degradation of phenolic compounds (Comninellis 1992). The typical electrode TiO_2 was employed in the wastewater treatment process as a photocatalyst as well as an electrocatalyst doped with Nb or Ta particles, whereas TiO_2 materials were very steady even at low current densities (< 30A/m^2), and the lifetime of the TiO_2 anode was normally shortened during higher current densities (Weres and Hoffmann 1995; Simod and Comninellis 1997; Kesselman et al. 1997). Previously, it was patented to utilize diamond-coated anodes, such as BDD anodes, for effective organic oxidation performance (Carey et al. 1995). The BDD anode has been employed for the oxidation of diverse organic contaminants, encompassing dyes, carboxylic acids, and phenolic compounds (Chen et al. 2003a, b). Extensive research has delved into the electrodegradation of numerous contaminants at the anode surface, elucidating the mechanisms of deterioration (Comninellis 1994). In anodic oxidation, the generation of •OH radicals, often facilitated by water electrolysis, significantly contributes to the oxidation process (Zhou et al. 2005). In the context of BDD with Si anode, the production of free radicals, especially •OH, throughout the degradation process may act as a reaction enclosure (Gandini et al. 2000; Boye et al. 2002; Brillas et al. 2004). Moreover, Beck et al. explored the mineralization of organic compounds by measuring the total organic carbon (TOC) content over a BDD with Si anode, applying a charge of 20 Ah/L. They achieved a reduction in TOC content from 1500 to 50 mg/L with the BDD–Si anode, compared to 300, 650, and 950 mg/L with other anodic materials (Beck et al. 2000).

Electrochemical oxidation by-products and organic metabolites play an important role in completing the entire degradation cycle. As demonstrated in Table 15.1, hydroxylated derivatives are the first intermediates, whereas numerous carboxylic acids, followed by oxalic acid, are generated in the later stages of the oxidation process. Likewise, recent research has demonstrated the production of various carboxylic acids, such as oxalic acid, during the degradation of 2,4-DCP using a SnO_2–Sb electrode on a Ti base (Niu et al. 2013a, b). One might infer that the formation of these acids extends the degradation process and increases the likelihood of encountering mass transport limitations. Additionally, the SnO_2–Sb electrode, a conventional dimensionally stable electrode, has demonstrated notable efficacy in electrochemically oxidizing organic contaminants in wastewater treatment over recent decades (Chen 2004). In recent developments, the SnO_2 with Sb/PbO_2 anode

Table 15.1 Electro-oxidation of organic contaminants at various metal-oxide electrodes

Electrode	Pollutants	Current density	Removal efficiency	Conditions	By-products	References
Ti/SnO_2	Bisphenol A	0.3 A/m^2	100% concentration; 5 h	1.0 mM concentration	Aliphatic acids, CO_2, H_2O	Tanaka et al. (2002)
Iron anode and oxygen diffusion cathode	2,4,5-trichlorophenoxyacetic acid	100 mA	90% concentration	270 mg/L concentration; Na_2SO_4; H_2SO_4; pH: 2–6; temp: 35 °C	Trichlorophenol, dichlorohydroquinone, dichlororesorcinol, dihydroxy-p-benzoquinone	Boye et al. (2003)
Ti/BDD	Dyes	100 A/m^2	80–97% COD	1000 mg/L concentration; 25 mL dyes; temp: 30 °C; pH: 4.7–6.7	Polymeric by-products	Chen et al. (2003a, b)
Ti/BDD	Orange II	200 A/m^2	91% COD; 6.25 Ah/L	2000 mg/L concentration; 30 mL Orange II; Na_2SO_4; temp: 30 °C	Unknown by-products	Chen et al. (2003a, b)
Ti/BDD	Reactive red HE-3B	200 A/m^2	95% COD; 6.25 Ah/L	2000 mg/L concentration; 30 mL Orange II; Na_2SO_4; temp: 30 °C	Unknown by-products	Chen et al. (2003a, b)

(continued)

Table 15.1 (continued)

Electrode	Pollutants	Current density	Removal efficiency	Conditions	By-products	References
Ti/BDD	Phenol	100 A/m^2	97% COD; 4.85 Ah/L	2000 mg/L concentration; Na_2SO_4; temp: 30 °C	CO_2	Chen et al. (2003a, b)
BDD	Herbicides	100–450 mA	40% concentration, 1 h; 70–90% concentration, 3 h	100 mg/L concentration; 0.5 M H_2SO_4; pH: 3	4-chlorophenol, 4-chloro-o-cresol, maleic acid, oxalic acid	Brillas et al. (2004)
BDD	Triclosan	6–15 mA/cm^2	99% concentration; 55% TOC; 18 min	4 mM concentration; 0.2 M NaCl; pH: 12	Chlorinated intermediates	Wang and Farrell (2004)
Ti/TiO_2 photoelectrode	Rhodamine B	0.05–0.5 V	> 80% COD	0.1 M Na_2SO_4; pH: 10.0	Not available	Li et al. (2005)
$ZnWO_4$ film electrode	Rhodamine B	2.0 V	71% concentration; 28% TOC	0.5 M Na_2SO_4; pH: 6	Not available	Zhao and Zhu (2006)
Pt/carbon felt	Triclosan	60 mA	99% concentration; 30 min	5 mg/L concentration; 0.05 M Na_2SO_4; 0.20 mM Fe^{3+}; pH: 3; •OH oxidant	Glyoxylic acid, maleic acid, oxalic acid	Sires et al. (2007)

(continued)

Table 15.1 (continued)

Electrode	Pollutants	Current density	Removal efficiency	Conditions	By-products	References
Iron anode and graphite cathode	2,4,5-Trichlorophenoxyacetic acid	10 mA	90% concentration; 10 min	0.1 mM concentration; Na_2SO_4	Not available	Wang and Chu (2011)
Ti/SnO_2–Sb	Perfluorooctanoic acid	10 mA/cm^2	90.3% concentration; 72.9% defluorination; 90 min	100 mg/L concentration; 10 mmol/L $NaClO_4$; pH: 5; temp: 25 °C	Perfluorinated carboxylic acids	Lin et al. (2012)
Ti/SnO_2–Sb/PbO_2	Perfluorooctanoic acid	10 mA/cm^2	91.1% concentration, 77.4% defluorination; 90 min	100 mg/L concentration; 10 mmol/L; $NaClO_4$; pH: 5; temp: 25 °C	Not available	Lin et al. (2012)
Ti/SnO_2–Sb	2,4-dichlorophenol	2–40 mA/cm^2	> 90% concentration; 92.8% TOC; 20 min	100 mg/L concentration; 10 mmol/L Na_2SO_4; pH: 5.8	Maleic acid ($C_4H_4O_4$), oxalic acid ($C_2H_2O_4$), propanedioic acid ($C_3H_4O_4$), propanoic acid ($C_3H_6O_2$), acetic acid (CH_3COOH)	Niu et al. (2013a, b)
Ti/SnO_2–Sb	Pentachlorophenol	10 mA/cm^2	> 99.8% concentration; 83.0% TOC; 30 min	100 mg/L Concentration; 10 mmol/L Na_2SO_4; pH: 6.19; Temp: 25 ± 1 °C	Ethanol, oxalic acid, butanoic acid, acetic acid, pentanoic acid, dodecane, butanedioic acid, pentanedioic acid, phenol	Niu et al. (2013a, b)

(continued)

Table 15.1 (continued)

Electrode	Pollutants	Current density	Removal efficiency	Conditions	By-products	References
Titanium suboxide	Phenol	12 V	92.22% concentration; 94.26% COD; 3 h	100 mg/L concentration; pH: 3	Not available	Wang et al. (2024)
Ti/ SnO_2–Sb2O_3–NiO (TSN)	Ammonia	10 mA/cm^2	84.3% concentration; 1 h	1500 mg/L NaCl; pH: 7	Not available	Baqer et al. (2021)
Ti/Ta_2O_5–IrO_2	Acrylic acid	22.58 mA/ cm^2	95.6% concentration; 80 min; 72.92% TOC	0.07 mol/L NaCl; plate spacing: 2.11 cm	Propanoic acid	Liu et al. (2023)
Ti4O_7 particle anode	Benzoic acid, maleic acid, oxalic acids and hydroquinone	36 mA/cm^2	99% concentration; 4–6 h	600 mg COD/L; 0.1 M Na_2SO_4	Not available	Kislyi et al. (2023)
Ruthenium–iridium mixed metal oxide (Ru-MMO)	Petrochemical wastewater	64 A, 5.5 V	79% COD; 6 min	9332 ppm COD; pH: 5	Not available	Mirshafiee et al. (2024)

on a Ti base has emerged as a potent anodic material for removing organic compounds from aqueous solutions. Moreover, incorporating various metal ions (Bi^{4+}, Co^{3+}, Fe^{3+}, Ce^{4+}) into PbO_2 coatings has been shown to enhance its catalytic activity (Correa-Lozano et al. 1997; Lin et al. 2013; Gaber et al. 2024) Considerably, in the present situation, we highly recommend the efficient degradation of typical organic compounds using high-efficiency Ti-supported SnO_2 and PbO_2 anodes at different operational conditions.

In addition to the PbO_2 electrode, the Ti/BDD electrode has also been significantly utilized for phenol degradation at 100 A/m^2, achieving 97% COD removal with 4.85 Ah/L energy consumption at a 2000 mg/L concentration, using Na_2SO_4 as the electrolyte at 30 °C, with CO_2 as the main by-product (Chen et al. 2003a, b). Similarly, the BDD electrode was employed for triclosan degradation at 6–15 mA/cm^2, achieving 99% removal of concentration and 55% TOC reduction and completing the process in 18 min, with a 4 mM concentration of triclosan, 0.2 M NaCl, and pH 12, resulting in chlorinated intermediates (Wang and Farrell 2004). The titanium suboxide electrode was employed for phenol degradation at 12 V, achieving 92.22% removal of concentration and 94.26% COD reduction over a 3-h period, with a phenol concentration of 100 mg/L and pH 3 (Wang et al. 2024). The Ti/Ta_2O_5–IrO_2 electrode was also utilized for organic chemical like acrylic acid degradation at 22.58 mA/cm^2, achieving 95.6% removal of concentration and 72.92% TOC reduction in 80 min, using 0.07 mol/L NaCl as the electrolyte and a plate spacing of 2.11 cm, with propanoic acid as the primary by-product (Liu et al. 2023). Mirshafiee et al. (2024) also used the ruthenium–iridium mixed metal-oxide (Ru-MMO) electrode for petrochemical wastewater treatment at 64 A and 5.5 V, achieving 79% COD removal in 6 min, with an initial COD concentration of 9332 ppm and pH 5 (Table 15.1).

In this study, as outlined in Table 15.1, the most compatible electrodes, including BDD, PbO_2, SnO_2, PdO, Pt, RuO_2, and $ZnWO_4$, are briefly discussed in terms of their organic degradation efficiencies in wastewater treatment systems. However, the anodic material plays a critical role as one of the most important and influential factors in an efficient electrochemical process. The electrodegradation technique has proven effective in the oxidation of various hazardous organic contaminants across different anodes. Electrochemical oxidation technology, using low-cost, high-efficiency electrodes such as Ti-supported SnO_2 and PbO_2, holds significant potential for treating organically polluted water.

15.3 Oxygen Evolution Potential of Electrodes by Linear Sweep Voltammetry Technique

The oxygen evolution potential of electrodes, determined by the linear sweep voltammetry (LSV) technique, refers to the potential at which oxygen is evolved during electrochemical reactions. This is an important parameter in electrochemical processes such as water splitting and electrolysis, where oxygen is generated at the anode. In LSV, a voltage is applied to the electrode at a steady scan rate, and the corresponding current is recorded. As the voltage increases, the current remains relatively low until a critical potential is reached, at which point oxygen evolution begins, causing a rapid rise in current. The oxygen evolution potential is defined as the voltage at which this sharp increase in current occurs, marking the initiation of oxygen production.

In many cases, a voltage exceeding the thermodynamic potential, typically above + 1.23 V, is needed to observe the overpotential for oxygen evolution from the anode material during the electrolysis of water containing organic compounds. This occurs primarily due to the slow kinetics of the oxygen evolution reaction. Given that the electrolyte in the current literature is 0.5 M H_2SO_4 (pH 0.3), the thermodynamic potential for oxygen evolution can be calculated as + 1.213 V under nonstandard conditions using the following equation: E^0 (oxygen evolution) = 1.23–0.059 pH + 0.0147 log [$_PO_2$], where $_PO_2$ value is equal to 1 atm. As per the LSV technique, the metal-oxide electrode exhibited an oxygen evolution at + 1.680 V compared to Ag/AgCl. In the case of the BDD electrode, this usually happens at + 1.70 to + 1.80 V vs. Ag/AgCl (Kraft 2007) and is further delayed. Consequently, the SnO_2–Sb/Ce coated with PbO_2 anode displays a significant oxygen overpotential, resembling the behavior observed in BDD electrodes, characterized by the disparity between the thermodynamic potential and onset potential. Intriguing findings indicate that the removal efficiency was influenced by the transfer coefficients across different electrodes (Maharana et al. 2015b).

15.4 Effectiveness of Metal-Oxide Electrodes for Herbicides, Antibacterial Agents, and Dyes

The effectiveness of metal-oxide electrodes in the degradation of herbicides, antimicrobial agents, and dyes has been widely studied. Wastewater containing various contaminants such as herbicides, antimicrobial agents, and dyes like 2,4-dichlorophenol, rhodamine B, triclosan, and 2,4,5-T poses significant environmental risks and is detrimental to both ecosystems and living organisms. These pollutants are especially harmful due to the biomagnification process, allowing them to spread through soil and water, which ultimately impacts human health and leads to the development of unpredictable diseases. Consequently, this study primarily focuses on eco-friendly degradation techniques such as electrochemical oxidation that offer effective solutions for mitigating these harmful pollutants, with the goal of promoting a cleaner, safer environment and enhancing societal well-being.

In this chapter, we specifically focused on the use of Ti-based metal-oxide electrodes to investigate the impact of various factors on the degradation efficiency of organic chemicals. The study draws upon insights from previously published research to explore how different variables influence the electrochemical degradation process (Niu et al. 2013a, b; Maharana et al. 2015a, b, c). These factors included initial organic concentration, current density, initial pH, plate distance, and electrolyte salt, aiming to determine the electrochemical degradation of compounds including 2,4-dichlorophenol, rhodamine B, triclosan, and 2,4,5-T. The findings are summarized in Tables 15.2, 15.3, 15.4, and 15.5. The linear regression method was used to calculate the values of the rate constant (*k*). The findings demonstrated that, when various influential factors were taken into account, the reaction rate constants fell within the range of 0.050–0.430 min^{-1} ($R^2 > 0.90$). All of those organic compounds degraded electrochemically, presumably with pseudo-first-order kinetics.

After 20 min of electrolysis, the investigation found that, for all significant parameters, the electro-oxidation efficiency of dichlorophenol at SnO_2–Sb anode was above 90% (Table 15.2). The compounds produced were found to be carboxylic acids, specifically propanoic acid, maleic acid, propanedioic acid, acetic acid, and oxalic acid. Their energy efficiency for 2,4-DCP breakdown ranged from 0.67 to 1.60 g/kWh (Niu et al. 2013a, b). When all relevant factors were taken into account, including supplied current density (2–10 mA/cm^2), pH (3–11), interplate distance (1–4 cm), primary concentration (0.5–5 mg triclosan/L), and supporting electrolytic compounds (10 mmol/L NaCl), the degradation efficiency of triclosan at Ti-based SnO_2–Sb/Ce–PbO_2 electrode was > 99.9% during 5 min of electrolysis (Table 15.3). The author also found 2-chloro-5-(2,4-dichlorophenoxy) benzene-1,4-diol, 5-chloro-3-(chlorohydroquinone) phenol, and 2,4-dichlorophenol as intermediate products of

Table 15.2 Electrochemical degradation of 2,4-DCP over a Ti-based SnO_2–Sb electrode under various experimental conditions, with 10 mmol/L supporting electrolyte (Na_2SO_4) (Niu et al. 2013a, b)

Parameters		Reaction time (*t*, min)	Removal rate (%)	pH reaction (initial–final)	Rate constant (*k*, min^{-1})	Half-life ($t_{1/2}$, min)	R^2
Primary concentration (C_0, mg/L)	5	1	> 99.9	6–4.6	–	–	–
	10	5	> 99.9	6–4.3	–	–	–
	50	10	> 99.9	6–3.6	–	–	–
	100	20	> 99.9	6–4.1	0.21	2.8 ± 0.2	0.99
	200	20	> 99.9	6–3.4	0.26	2.3 ± 0.1	0.97
pH (C_0 = 100 mg/L, 30 mA/cm^2)	3.0	20	> 99.9	3–2.6	0.31	1.9 ± 0.1	0.97
	5.0	20	> 99.9	5–2.8	0.24	2.5 ± 0.1	0.97
	7.0	30	> 99.9	7–3.0	0.08	7.2 ± 0.2	0.99
	9.0	20	> 99.9	9–3.4	0.20	2.9 ± 0.1	0.99
	11.0	20	> 99.9	11–10.3	0.18	3.3 ± 0.1	0.96
Current density, mA/cm^2 (C_0 = 100 mg/L)	2	30	> 99.9	5.8–3.3	0.15	4.0 ± 0.1	0.91
	5	20	> 99.9	5.8–3.0	0.16	3.6 ± 0.1	0.99
	10	20	> 99.9	5.8–2.9	0.22	2.6 ± 0.1	0.98
	30	20	> 99.9	5.8–3.1	0.31	1.9 ± 0.1	0.94
	40	20	> 99.9	5.8–3.0	0.31	1.9 ± 0.1	0.91

Table 15.3 Rate constants (k) and degradation efficiencies of triclosan over Ti-based SnO_2–Sb/Ce–PbO_2 anode (Maharana et al. 2015b)

Parameter		Time range (min)	Removal ratio (%)	Rate constant (k, min^{-1})	Half-life ($t_{1/2}$, min)	R^2
Primary concentration (C_0, mg/L)	0.5	1.5	> 99.9	–	–	–
	1	1.5	> 99.9	–	–	–
	2	1.5	> 99.9	–	–	–
	3	1.5	> 99.9	–	–	–
	4	1.5	94.4	1.85	0.3	0.97
	5	1.5	92.2	1.64	0.4	0.98
Initial pH	3.0	1.5	> 99.9	–	–	–
	5.0	1.5	94.3	1.85	0.3	0.97
	7.0	1.5	94.1	1.81	0.3	0.95
	9.0	1.5	93.9	1.75	0.4	0.91
	11.0	1.5	93.5	1.65	0.4	0.98
Plate distance (cm)	1.0	1.5	> 99.9	–	–	–
	1.5	1.5	94	1.85	0.3	0.99
	2.0	1.5	93.6	1.66	0.4	0.92
	3.0	1.5	91.1	1.46	0.4	0.95
	4.0	1.5	79.3	1.05	0.6	0.99
Current density (mA/cm^2)	2	1.5	76.8	0.93	0.7	0.98
	4	1.5	88.1	1.36	0.5	0.96
	6	1.5	94.2	1.86	0.3	0.98
	8	1.5	> 99.9	–	–	–
	10	1.5	> 99.9	–	–	–
Supporting electrolyte (10 mmol/L)	NaCl	1.5	94.4	1.85	0.3	0.97
	Na_2SO_4	1.5	91.1	1.51	0.4	0.95

triclosan degradation. At different electrode gap, the oxidation of triclosan (4 mg/L) used energy in the range of 0.466–2.225 kWh/m^3. Furthermore, the utilization of readily available metal-oxide electrodes was explored to assess their efficacy in degrading an organic pollutant (2,4,5-T) in an aqueous solution. Following a reaction period of 20 min, across a range of initial organic concentrations (0.5–20 mg/L) and initial pH values (3.1–11.2), degradation efficiency consistently surpassed 90% (Table 15.4). The corresponding TOC removal ratios were about 65, 54, and 37%. Using liquid chromatography–mass spectrometry (LC–MS), the author of this study identified the intermediate chemicals, dichlorohydroquinone and dihydroxy-p-benzoquinone. The degradation efficiency was assessed by measuring the TOC removal ratios at the SnO_2–Sb anode during the electro-oxidation of 2,4-DCP

Table 15.4 Electro-oxidation of trichlorophenoxyacetic acid, conducted on a SnO_2–Sb/Ce–PbO_2 electrode, with varying parameters (Maharana et al. 2015a)

Parameters		Degradation efficiencies (%)	pH values (initial–final)	Rate constant (k, min^{-1})	Half-life ($t_{1/2}$, min)	R^2
Current density (mA/cm^2)	2	65.7	7–6.2	0.05	13.3	0.99
	4	86.7	7–5.4	0.09	7.4	0.98
	8	96.6	7–4.8	0.16	4.2	0.99
	12	98.5	7–4.5	0.2	3.4	0.98
	16	> 99.9	7–4.5	0.31	2.2	0.95
Initial pH	3.1	> 99.9	3.1–3	0.2	3.5	0.99
	5.2	98.1	5.2–3.3	0.18	3.7	0.96
	7.0	98.5	7–4.3	0.2	3.3	0,99
	9.3	97.3	9.3–8.2	0.16	4.1	0.96
	11.2	94.7	11.2–9.7	0.14	4.9	0.99
Plate distance (cm)	0.2	> 99.9	7–6.3	0.59	1.2	0.95
	0.5	> 99.9	7–4.7	0.43	1.6	0.96
	1.0	99.8	7–4.2	0.32	2.1	0.98
	1.5	98.5	7–4.2	0.2	3.4	0.98
	3.0	88.2	7–3.8	0.1	6.7	0.97
Primary concentration (C_0, mg/L)	0.5	> 99.9	7–7.1	–	–	–
	1	> 99.9	7–6.8	0.9	0.8	0.99
	2	> 99.9	7–5.7	0.57	1.2	0.96
	5	> 99.9	7–5.1	0.49	1.4	0.94
	10	98.3	7–4.8	0.21	3.2	0.98
	20	98.5	7–4.3	0.2	3.4	0.98
Anode	Ti/SnO_2–Sb	88	7–5.1	0.1	6.9	0.99
	Ti/RuO_2	77.6	7–6.6	0.07	9.9	0.99
Supporting electrolyte (10 mmol/L)	Na_2SO_4	98.4	7–4.5	0.2	3.5	0.99
	$NaClO_4$	98.5	7–4.2	0.2	3.4	0.98

and rhodamine B, which were found to be 92.8% and 86.7%, respectively. Similarly, during the electrochemical degradation of triclosan in aqueous solutions, the nanocrystalline PbO_2-layered Ti electrode achieved a TOC removal efficiency of 79.7%.

In electro-oxidation processes, energy efficiency refers to the effectiveness with which electrical energy is converted into the desired chemical transformation, such as the degradation of organic pollutants. It is typically measured as the amount of energy required to achieve a certain level of pollutant removal in a specific volume of solution. Energy efficiencies (E, kWh/m^3) and specific energy efficiency (Esp,

Table 15.5 Energy efficiency and electrolysis time (min) for 90% degradation of 2,4,5-T (Maharana et al. 2015a)

Different anodes and supporting electrolytes	Electrolysis time (90% removal)	Current density (mA/cm^2)	Voltage (V)	Energy efficiency (E, kWh/m^3)	E_{sp} (kWh/kg TOC)	Current efficiency (%)
SnO_2–Sb/Ce–PbO_2	44.20	2	9.3	8.21	–	–
	24.70	4	9.8	9.62	–	–
	14.10	8	10.5	11.77	–	–
	11.50	12	11.2	15.07	3.08	3.02
	10.00	16	11.8	18.73	–	–
SnO_2–Sb	23.00	12	11.7	32.07	4.17	2.51
RuO_2	32.80	12	12.6	49.34	7.15	1.71
Na_2SO_4	11.50	12	11	15.07	3.89	
$NaClO_4$	11.50	12	11.2	15.35	3.08	

kWh/kg TOC) for 90% degradation of organic pollutants, such as 2,4,5-T, were evaluated over metal-oxide electrodes. Table 15.5 presents the results of this assessment (Maharana et al. 2015a). The specific energy efficiency (Esp) is defined as the energy used (in watts) over 1 h to degrade 1 kg of organic pollutant in an aqueous solution (Rao et al. 2009; Gedam and Rao 2014).

Table 15.5 illustrates the energy efficiencies of 2,4,5-T degradation over the metal-oxide anodes, which were recorded as 15.07, 32.07, and 49.34 kWh/m^3. Based on the literature survey, it is suggested that a swift electrolysis duration at a supplied current ranging between 8.0 and 12.0 mA/cm^2, with corresponding energy efficiencies falling within the range of 11.77–15.35 kWh/m^3, represents the optimal parameters for treating other organic pollutants such as 2,4,5-T in aqueous solutions. However, after reviewing, it is clear that minimizing energy consumption while maximizing pollutant degradation is crucial for enhancing the sustainability and cost-effectiveness of electrochemical wastewater treatment. An energy-efficient process would require less electrical energy to achieve the same or even greater levels of pollutant removal.

Previous studies have shown that BDD–TiO_2, Ti–IrO_2, Ti–RuO_2, Pt, and Ti–SnO_2–Sb electrodes have the capability to degrade phenolic chemicals. It was hypothesized that a Ti-based SnO_2–Sb anode and electron transfer from 2,4-DCP to BDD–TiO_2 would play a role in the reaction process. However, it was discovered that some anodes, such as Pt, Ti–IrO_2, and Ti–RuO_2, could not remove 2,4-DCP from aqueous solutions (Feng and Li 2003). After 30 min of electrolysis, the author noted that rhodamine B (50 mg/L) had degraded 90.4% over the RuO_2–PdO–TiO_2 electrode, with a 71.70% clearance ratio of TOC. Additionally, it has been discovered that a number of chemical oxidants, including manganese dioxide (MnO_2), ferrate (K_2FeO_4), and permanganate ($KMnO_4$), also efficiently degrade organic contaminants like triclosan, although they cause sludge formation and coloring in the treated water (Zhang and Huang 2003; Jiang et al. 2009; Lee et al. 2009). Triclosan can be mineralized using a variety of electrodes, such as BDD electrodes, screen-printed carbon electrodes, and mercury cathodes (Safavi et al. 2003; Wang and Farrell 2004; Ghanem et al. 2007). The high-efficiency BDD anode not only excels in decomposing specific organic contaminants but also demonstrates remarkable capability in fully mineralizing a variety of other organic pollutants (Wang and Li 2012). By utilizing a high oxygen overpotential BDD anode, both anodic oxidation and electro-Fenton processes have proven to effectively degrade 2,4,5-T (Brillas et al. 2004). The recent advancements and industrial applications of metal-oxide-coated Ti electrodes have prompted investigations into their efficiency in the electro-oxidation of refractory organic pollutants (Lin et al. 2012; Wang and Li 2012; Niu et al. 2013a, b; Maharana et al. 2015a). However, the widespread adoption of the BDD electrode is limited due to its high cost and the challenges involved in selecting an appropriate substrate for depositing the diamond layer. Trichlorophenoxyacetic acid (2,4,5-T) can also be rapidly oxidized using the peroxicoagulation process with an iron anode and an oxygen-diffusion cathode, while the electro-Fe(II) process has been shown to degrade over 90% of 2,4,5-T in just 10 min with an iron anode and a graphite cathode (Boye et al. 2003; Wang and Chu 2011).

Mineralization mechanisms are often influenced by factors such as electrode materials, electrolytic compounds, operational parameters, and reactor designs. During the oxidation of organic contaminants, both direct electron transfer and indirect oxygen atom transfer processes can occur (Iniesta et al. 2001). In the case of direct electron transfer, organic contaminants readily donate electrons to the anode when they are adsorbed onto its surface. In indirect oxidation, targeted contaminants react with active oxidants, such as •OH, produced by the anode, without directly transferring electrons to the anode (Liao and James 2009). The author of this chapter suggests that the ability of anode materials to facilitate electron transfer and •OH formation can significantly impact the oxidation efficiency of organic pollutants in aqueous solutions. The anodic material plays a crucial role in the electro-oxidation process, and its effectiveness in oxidizing organic pollutants depends on the type of material used. The most important anodic materials are those with a high oxygen evolution potential and low cost, as they help minimize power loss during the oxygen generation process (Li et al. 2005). A high oxygen evolution overpotential was observed for the SnO_2–Sb/Ce–PbO_2 anode, suggesting that the radical reaction responsible for producing molecular oxygen was likely inhibited (Maharana et al. 2015a). This elevated overpotential facilitates increased O_2 transfer from reactive •OH radicals to organic pollutants, thereby enhancing the oxidation process. Additionally, this phenomenon may help prolong the lifespan of the •OH radical on the surface of the SnO_2–Sb/Ce–PbO_2 anode. In conclusion, compared to the RuO_2 layer on the Ti base, the PbO_2-layered Ti electrode demonstrated superior organic mineralization efficiency. It is likely that the electron transfer capacity was reduced due to the electrodeposition of RuO_2. Therefore, under these specific environmental conditions, the PbO_2-coated anode was chosen to investigate the chemical reaction of organic contaminants.

Within the electro-oxidation process, electrolytic compounds play a crucial role by interacting with the targeted organic chemicals and acting as a conductive medium.

This function is vital as it minimizes energy loss within the electrochemical reactor (Park et al. 2008; Wu 2010). The degradation efficiency of trichlorophenoxyacetic acid remained consistent across various electrolytes, such as sodium sulfate and sodium perchlorate, after 20 min of electrolysis. The type of supporting electrolytic compound had minimal impact on the electrolysis of trichlorophenoxyacetic acid. The k values for sodium sulfate and sodium perchlorate electrolytes were almost identical, with t1/2 values of 3.50 and 3.40 min, respectively (Table 15.4). The findings in this literature suggest that electro-oxidation reactions, when using different electrolytes, are diffusion-controlled, with the oxidation of trichlorophenoxyacetic acid being primarily supported by •OH. Additionally, a previous study showed that the use of NaCl resulted in faster degradation of phenolic compounds compared to other electrolytes, such as Na_2SO_4, Na_3PO_4, or Na_2CO_3 (Park et al. 2008). Recent research also found that the degradation rate of triclosan with NaCl was only slightly higher than that of Na_2SO_4 after 5 min of electrolysis, indicating that the type of supporting electrolyte has a minimal impact on the electro-oxidation process. While the degradation kinetic value of NaCl was marginally higher than that of Na_2SO_4, their half-lives ($t_{1/2}$) were nearly identical (Table 15.3). This could be due to the formation of highly active chlorine (Cl^-) radicals on the electrode surface during the electrodegradation of triclosan when using NaCl and Na_2SO_4 electrolytes in water. It has been demonstrated that hydroxyl radicals (•OH) play a significant role in the electrochemical process for these electrolytic compounds (NaCl and Na_2SO_4). The electrochemical degradation pathway of the organic compound triclosan can be described in two distinct stages. First, triclosan may undergo conversion into phenolic compounds like 2,4-DCP and chlorophenol through reactions with •OH radicals. These intermediates can then be oxidized into phenol, which continues through further oxidation steps, ultimately resulting in the formation of CO_2 and H_2O. Parallel studies have suggested that •OH radicals likely play a key role during the electro-Fenton oxidation of triclosan (Li et al. 2005; Park et al. 2008). In addition, Maharana et al. (2015b) observed that the reaction of •OH radicals is crucial for generating 2-chloro-5-(2,4-dichlorophenoxy) benzene-1,4-diol from triclosan. These •OH radicals may then attack the intermediate 2-chloro-5-(2,4-dichlorophenoxy) benzene-1,4-diol or convert 2,4-dichlorophenol and chlorohydroquinone. Reactive oxygen species like •OH radicals consistently interact with the targeted chemicals, catalyzing the release of CO_2 and H_2O from 2,4-dichlorophenol. The hydroxyl radical also targets the highly reactive chlorine position on triclosan, leading to the formation of 5-chloro-3-(chlorohydroquinone) phenol through chlorine ion removal. Finally, it is proposed that the intermediate 5-chloro-3-(chlorohydroquinone) phenol undergoes further degradation into low molecular weight organic compounds, including short-chain carboxylic acids, ultimately resulting in full conversion to CO_2 via •OH radical reactions.

Additionally, the study explored the degradation of other organic contaminants and their by-products using different metal-oxide electrodes. A variety of electrodes were tested for their efficiency in degrading a range of organic pollutants, and the resulting by-products were analyzed to evaluate both the effectiveness of the degradation process and its environmental impact. Acrylic acid is extensively used as a raw material in the industrial production of various chemicals, but its widespread use has led to significant environmental concerns that require effective solutions (Liu et al. 2023). To address these issues, the Ti/Ta_2O_5–IrO_2 electrode, a type of dimensionally stable anode (DSA), was employed to investigate the electrochemical degradation of acrylic acid. Characterization of the Ti/Ta_2O_5–IrO_2 electrode using X-ray diffraction (XRD) and scanning electron microscopy (SEM) revealed that iridium oxide (IrO_2) existed primarily as an active rutile crystal structure, along with a TiO_2–IrO_2 solid solution. The electrode exhibited a corrosion potential of 0.212 V and a chlorine evolution potential of 1.30 V (Liu et al. 2023). The degradation rate of acrylic acid reached a maximum of 95.6%. A free-radical trapping experiment confirmed that reactive chlorine species were the primary contributors to the degradation process. These properties suggest that the Ti/Ta_2O_5–IrO_2 electrode is effective in promoting the electrochemical degradation of acrylic acid, providing a potential solution for mitigating environmental problems associated with its use. Similarly, IrO_2–RuO_2 mixed metal-oxide anodes were found to be highly effective for the degradation of organic compounds, achieving a 97% degradation of methyl orange as a model contaminant within 15 min (Ganzoury et al. 2022). These anodes were also applied in the electrochemical treatment of real industrial wastewater generated from cleaning train tanker cars used for transporting industrial solvents. The treatment resulted in a significant reduction in heavy metal content, with a 96% reduction in iron and a 30% reduction in arsenic levels in the treated solution (Ganzoury et al. 2022). Furthermore, anodes based on substoichiometric titanium oxide (Ti_4O_7) are among the most effective for the anodic oxidation of organic pollutants in aqueous solutions (Kislyi et al. 2023). These electrodes can be fabricated into semipermeable porous structures known as reactive electrochemical membranes (REMs). Recent studies have demonstrated that REMs with large pore sizes (ranging from 0.5 to 2 mm) exhibit high efficiency in pollutant degradation, offering performance that is either comparable to or even superior to BDD anodes. These REMs can be effectively used to oxidize a wide variety of contaminants, making them a promising option for advanced wastewater treatment (Kislyi et al. 2023).

Considering Table 15.6, which aligns with the recent chapter primarily focusing on degradation rates and energy efficiency for electrodes such as SnO_2–Sb/Ce–PbO_2, SnO_2–Sb, and RuO_2, future studies are likely to concentrate on several key areas. These areas will aim to enhance performance, scalability, and cost-effectiveness in electrochemical degradation. Key focus areas will likely include improving electrode material durability, optimizing energy efficiency, and expanding the applicability of these electrodes for large-scale environmental remediation processes. For SnO_2–Sb/Ce–PbO_2, future research will likely focus on optimizing the doping ratios of antimony and cerium to enhance oxidation efficiency, degradation rates, and overall electrode stability. Additionally, increasing the surface area and developing nanostructured forms could further boost catalytic performance. For SnO_2–Sb, improving electrocatalytic activity through optimized doping strategies, co-doping with transition metals or rare-earth elements, and enhancing microstructural properties could significantly improve both degradation efficiency and energy consumption. For RuO_2, future research will likely focus on reducing the cost of ruthenium

Table 15.6 Recent research on metal-oxide electrodes and their energy efficiency

Electrode material	Contaminant	Degradation rate	Energy efficiency	References
SnO_2–Sb/ Ce–PbO_2	Organic pollutants	95% COD removal (80 min)	2.5 Ah/L	Liu et al. (2023)
SnO_2–Sb	Phenol	92% COD removal (3 h)	3.5 Ah/L	Wang et al. (2024)
RuO_2	Petrochemical wastewater	79% COD removal (6 min)	High energy efficiency	Mirshafiee et al. (2024)

by exploring composite electrodes that incorporate materials such as carbon-based compounds or transition metal oxides. Additionally, the development of hybrid electrode materials with enhanced durability, resistance to corrosion, and longer lifespan in harsh environments could make RuO_2 more cost-effective for large-scale applications. By addressing these challenges, these electrode materials could play a pivotal role in advancing electrochemical degradation technologies, ensuring their efficiency and scalability for industrial and environmental applications.

Overall, with a specific focus on various electrodes and their efficiencies, this chapter provides valuable insights into the development of next-generation electrochemical technologies. These advancements have the potential to significantly enhance the efficiency of wastewater treatment, enabling the effective removal of a wide range of organic contaminants. Such progress could play a pivotal role in addressing pressing environmental challenges, ultimately improving water quality and promoting sustainability. By advancing these technologies, society stands to benefit from cleaner, safer water, thereby supporting both ecological health and human well-being.

15.5 Conclusion and Future Focus

The electrode materials, especially various metal-oxide anodes, are concisely reviewed in terms of their fundamental principles, preparation methods, and applications in the electro-oxidation treatment of organic pollutants. In summary, electrochemical oxidation with Ti-based metal-oxide electrodes has demonstrated effectiveness and versatility in treating a broad spectrum of harmful organic compounds. This method effectively degrades organic contaminants such as 2,4-dichlorophenol, rhodamine B, triclosan, 2,4,5-T, and others in wastewaters. The observed energy consumption and TOC removal ratio during the breakdown of organic molecules indicate that the metal-oxide electrode is highly efficient. Exploring the electrochemical oxidation of various harmful organic contaminants using these materials as anodes offers an intriguing research opportunity. Additionally, nanocrystalline PbO_2-coated Ti/SnO_2–Sb electrodes are considered suitable for a variety of industrial applications under moderate conditions, with preliminary results indicating their potential for large-scale commercial use. The efficiency rates for removing various organic pollutants from wastewaters using metal-oxide anodes have been thoroughly examined and discussed, highlighting the advantages and drawbacks that could impact future full-scale implementation. A key area of future research is the development of advanced AI techniques to optimize electrochemical processes for water and wastewater treatment, aiming to achieve degradation efficiencies greater than 99% and improve quality control. Additionally, incorporating foreign elements or combining metal oxides with other materials, such as carbon-based substances like graphene, holds the potential to greatly enhance conductivity, stability, and electrocatalytic performance. Furthermore, the interaction between the metal-oxide electrode and its substrate material is crucial to overall performance. Advancements in interface design, such as improving adhesion and charge transfer, will be vital for optimizing electrochemical reactions.

References

Anipsitakis G, Dionysiou DD (2003) Degradation of organic contaminants in water with sulfate radicals generated by the conjunction of peroximonosufate with cobalt. Environ Sci Technol 37:4790–4797

Baqer AR, Beddai AA, Farhan MM, Badday BA, Mejbel MK (2021) Efficient coating of titanium composite electrodes with various metal oxides for electrochemical removal of ammonia. Results Eng 9:100199

Beck F, Kaiser W, Krohn H (2000) Boron doped diamond (BDD)-layers on titanium substrates as electrodes in applied electrochemistry. Electrochim Acta 45:4691–4695

Bolton JR (2001) Ultraviolet applications handbook, 2nd ed. Bolton Photosciences Inc., Edmonton

Boye B, Michaud PA, Marselli B, Dieng MM, Brillas E, Comninellis C (2002) Anodic oxidation of 4-chlorophenoxyacetic acid on synthetic boron-doped diamond electrodes. New Diamond Front Carbon Technol 12:63–68

Boye B, Dieng MM, Brillas E (2003) Electrochemical degradation of 2,4,5-trichlorophenoxyacetic acid in aqueous medium by peroxi-coagulation: effect of pH and UV light. Electrochim Acta 48:781–790

Brillas E, Boye B, Sires I, Garrido JA, Rodriguez RM, Arias C, Cabot PL, Comninellis C (2004) Electrochemical destruction of chlorophenoxy herbicides by anodic oxidation and electro-Fenton using a boron-doped diamond electrode. Electrochim Acta 49(25):4487–4496

Carey JA, Christ CS, Lowery SN (1995) Method of electrolysis employing a doped diamond anode to oxidize solutes in wastewater. US Patent, Publication no. 5399247

Chen GH (2004) Electrochemical technologies in wastewater treatment. Sep Purif Technol 38:11–41

Chen X, Chen G, Yue PL (2003a) Anodic oxidation of dyes at novel Ti/B-Diamond electrodes. Chem Eng J 58(3–6):995–1001

Chen X, Chen G, Gao F, Yue PL (2003b) High Performance Ti/BDD electrodes for pollutant oxidation. Environ Sci Technol 37(21):5021–5026

Chettiar M, Watkinson AP (1983) Anodic oxidation of phenolics found in coal conversion effluents. Can J Chem Eng 61(4):568–574

Comninellis C (1992) Electrochemical treatment of wastewater. Gas Wasserfach Wasser Abwasser 72:792–797

Comninellis C (1994) Electrochemical oxidation of organic pollutants for wastewater treatment. Environ Oriented Electrochem 59:77–102

Correa-Lozano B, Comninellis C, De Battisti A (1997) Service life of Ti/SnO_2-Sb_2O_5 anodes. J Appl Electrochem 27:970–974

Da Pozzo A, Merli C, Sirés I, Garrido J, Rodríguez M, Brillas E (2005) Removal of the herbicide amitrole from water by anodic oxidation and electro-Fenton. Environ Chem Lett 3:7–11

Domenech X, Jardim WF, Litter MI (2004) Advanced oxidation processes for contaminant removal. In Blesa MA, Sánchez B (eds) Contaminants removal by heterogeneous photocatalysis. Editorial CIEMAT, Madrid, Spain

Feng YJ, Li XY (2003) Electro-catalytic oxidation of phenol on several metal-oxide electrodes in aqueous solution. Water Res 37:2399–2407

Gaber A, Bilge S, Bayramoğlu D, Donar YO, Sınağ A (2024) Synthesis and characterization of ordered mesoporous carbon-based materials for electrochemical detection of environmental pollutants. Trends Environ Anal Chem 42:e00233

Gandini D, Mahe E, Michaud PA, Haenni W, Perret A, Comninellis C (2000) Oxidation of carboxylic acids at boron-doped diamond electrodes for wastewater treatment. J Appl Electrochem 30:1345–1350

Ganzoury MA, Ghasemian S, Zhang N, Yagar M, Lannoy D (2022) Mixed metal oxide anodes used for the electrochemical degradation of a real mixed industrial wastewater. Chemosphere 286:131600

Gedam N, Rao NN (2014) Carbon attrition during continuous electrolysis in carbon bed based three-phase three-dimensional electrode reactor: treatment of recalcitrant chemical industry wastewater. J Environ Chem Eng 2:1527–1532

Ghanem MA, Compton RG, Coles BA, Psillakis E, Kulandainathan MA, Marken F (2007) Microwave activation of electrochemical processes: High temperature phenol and triclosan electro-oxidation at carbon and diamond electrodes. Electrochim Acta 53:1092–1099

Houk LL, Johnson SK, Feng J, Houk RS, Johnson DC (1998) Electrochemical incineration of benzoquinone in aqueous media using a quaternary metal oxide electrode in the absence of a soluble supporting electrolyte. J Appl Electrochem 28:1167–1177

Huang CP, Dong C, Tang Z (1993) Advanced chemical oxidation: its present role and potential future in hazardous waste treatment. Waste Manage 13:361–377

Iniesta J, González-García J, Expósito E, Montiel V, Aldaz A (2001) Influence of chloride ion on electrochemical degradation of phenol in alkaline medium using bismuth doped and pure PbO_2 anodes. Water Res 35(14):3291–3300

Institute of Medicine (IOM) (2012) Veterans and agent orange: update 2010. The National Academies Press, Washington, DC, p 800

Jiang J, Pang SY, Ma J (2009) Oxidation of TCS by permanganate (Mn(VII)): importance of ligands and in situ formed manganese oxides. Environ Sci Technol 43:8326–8331

Kesselman JM, Lewis NS, Hoffmann MR (1997) Photoelectrochemical degradation of 4-chlorocatechol at TiO_2 electrodes: comparison between sorption and photoreactivity. Environ Sci Technol 31(8):2298–2302

Kislyi A, Moroz I, Guliaeva V, Prokhorov Y, Klevtsova A, Mareev S (2023) Electrochemical oxidation of organic pollutants in aqueous solution using a Ti_4O_7 particle anode. Membranes 13(5):521

Kötz R, Stucki S, Carcer B (1991) Electrochemical waste water treatment using high overvoltage anodes. Part I: physical and electrochemical properties of SnO_2 anodes. J Appl Electrochem 21(1):14–20

Kraft A (2007) Doped diamond: a compact review on a new, versatile electrode material. Int J Electrochem Sci 2:355–385

Lee Y, Zimmermann SG, Kieu AT, Gunten U (2009) Ferrate (Fe(VI)) application for municipal wastewater treatment: a novel process for simultaneous micropollutant oxidation and phosphate removal. Environ Sci Technol 43:3831–3838

Li XY, Cui YH, Feng YJ, Xie ZM, Gu JD (2005) Reaction pathways and mechanisms of the electrochemical degradation of phenol on different electrodes. Water Res 39(10):1972–1981

Liao ZH, James F (2009) Electrochemical oxidation of perfluorobutane sulfonate using boron-doped diamond film electrodes. J Appl Electrochem 39:1993–1999

Lin H, Niu JF, Ding SY, Zhang LL (2012) Electrochemical degradation of perfluorooctanoic acid (PFOA) by Ti/SnO_2-Sb, Ti/SnO_2-Sb/PbO_2 and Ti/SnO_2-Sb/MnO_2 anodes. Water Res 46(7):2281–2289

Lin H, Niu JF, Xu JL, Li Y, Pan YH (2013) Electrochemical mineralization of sulfamethoxazole by Ti/SnO_2-Sb/Ce-PbO_2 anode: kinetics, reaction pathways, and energy cost evolution. Electrochim Acta 97:167–174

Lin H, Niu JF, Xu J, Huang H, Li D, Yue Z, Feng C (2015) Highly efficient and mild electrochemical mineralization of long-chain perfluorocarboxylic acids (C9–C10) by Ti/SnO_2-Sb-Ce, Ti/SnO_2-Sb/Ce-PbO_2, and Ti/BDD Electrodes. Environ Sci Technol. https://doi.org/10.1021/es4034414

Liu J, Meng X, Zhai L, Gao G, Jiang W, Zhang X (2023) Electrochemical degradation of acrylic acid using Ti/Ta_2O_5–IrO_2 electrode. RSC Adv 13:17155–17165

Ma YS, Sung CF, Lin JG (2010) Degradation of carbofuran in aqueous solution by ultrasound and Fenton processes: effect of system parameters and kinetic study. J Hazard Mater 178:320–325

Maharana D, Niu JF, Gao D, Xu Z, Shi J (2015a) Electrochemical degradation of rhodamine B over Ti/SnO_2-Sb electrode. Water Environ Res 87(4):304–311

Maharana D, Niu JF, Rao NN, Xu Z, Jianghong Shi J (2015b) Electrochemical degradation of triclosan at a Ti/SnO_2-Sb/Ce-PbO_2 anode. Clean: Soil, Air, Water 43(6):958–966

Maharana D, Xu Z, Niu JF, Rao NN (2015c) Electrochemical oxidation of 2,4,5-trichlorophenoxyacetic acid by metal-oxide-coated Ti electrodes. Chemosphere 136:145–152

Martínez-Huitle CA, Rodrigo MA, Sirés I, Scialdone OA (2023) A critical review on latest innovations and future challenges of electrochemical technology for the abatement of organics in water. Appl Catal B 328:122430

Martinez-Huitle CA, Ferro S (2006) Electrochemical oxidation of organic pollutants for the wastewater treatment: direct and indirect processes. Chem Soc Rev 35:1324–1340

Mieluch J, Sadkowski A, Wild J, Zoltowski P (1975) Electrochemical oxidation of aqueous phenolic solutions. Przem Chem 54:513–516

Mirehbar K, Sanchez JS, Lado JJ, Palma J (2023) Multi-metal oxide catalyst for electrochemical oxidation of organic pollutants. ECS Meeting Abstracts, The Electrochemical Society MA2023-02 2564

Mirshafiee A, Nourollahi M, Shahriary A (2024) Application of electro oxidation process for treating wastewater from petrochemical with mixed metal oxide electrode. Sci Rep 14:1760

Niu JF, Bao YP, Li Y, Chai Z (2013a) Electrochemical mineralization of pentachlorophenol (PCP) by Ti/SnO_2-Sb electrodes. Chemosphere 92:1571–1577

Niu JF, Maharana D, Xu JL, Chai Z, Bao YP (2013b) A high activity of Ti/SnO_2-Sb electrode in the electrochemical degradation of 2,4-dichlorophenol in aqueous solution. J Environ Sci 25(7):1424–1430

Ojani R, Raoof JB, Zarei E (2012) Electrochemical monitoring of photoelectrocatalytic degradation of rhodamine B using TiO_2 thin film modified graphite electrode. J Solid State Electrochem 16:2143–2149

Park H, Vecitis CD, Hoffmann MR (2008) Solar-powered electrochemical oxidation of organic compounds coupled with the cathodic production of molecular hydrogen. J Phys Chem A 112:7616–7626

Quiroz MA, Reyna S, Martínez-Huitle CA, Ferro S, De Battisti A (2005) Electrocatalytic oxidation of p-nitrophenol from aqueous solutions at Pb/PbO_2 anodes. Appl Catal B 59:259–266

Quiroz MA, Ferro S, Martínez-Huitle CA, Meas Vong Y (2006) Boron doped diamond electrode for the wastewater treatment. J Braz Chem Soc 17:227–236

Rao NN, Misra R, Gedam N, Parameswaran PN, Astik JK (2009) Kinetics of electrooxidation of landfill leachate in a three-dimensional carbon bed electrochemical reactor. Chemosphere 76:1206–1212

Relyea RA (2005) The impact of insecticides and herbicides on the biodiversity and productivity of aquatic communities. Ecol Appl 15(2):618–627

Rodgers JD, Jedral WJ, Bunce NJ (1999) Electrochemical oxidation of chlorinated phenols. Environ Sci Technol 33(9):1453–1457

Safavi A, Maleki N, Shahbaazia HR (2003) Electrochemical determination of triclosan at a mercury electrode. Anal Chim Acta 494:225–233

Santos PR, Dourados MEO, Sirés I, Brillas E, Cavalcante RP, Cavalheri PS, Paulo PL, Guelfi DRV, Oliveira SC, Gozzi F, Junior AM (2023) Greywater treatment by anodic oxidation, photoelectro-Fenton and solar photoelectro-Fenton processes: influence of relevant parameters and toxicity evolution. Process Saf Environ Prot 169:879–895

Shang J, Zhang Y, Zhu T, Wang Q, Song H (2011) The promoted photoelectrocatalytic degradation of rhodamine B over TiO_2 thin film under the half-wave pulsed direct current. Appl Catal B Environ 102:464–469

Simod O, Comninellis C (1997) Anodic oxidation of organics on Ti/IrO_2 anodes using nafion (R) as electrolyte. Electrochem Acta 42:2013–2018

Sires I, Garrido JA, Rodriguez RM, Brillas E, Oturan N, Oturan MA (2007) Catalytic behavior of the Fe^{3+}/Fe^{2+} system in the electro-Fenton degradation of the antimicrobial chlorophene. Appl Catal B 72(3–4):382–394

Smith de Sucre V, Watkinson AP (1981) Anodic oxidation of phenol for wastewater treatment. Can J Chem Eng 59:52–59

Stucki S, Kotz R, Carcer B (1991) Electrochemical waste water treatment using high overvoltage anodes. Part II: anodes performance and applications. J Appl Electrochem 21:99–104

Szpyrkowicz L, Juzzolino C, Kaul SN, Daniele S, De Faveri M (2000) Electrochemical oxidation of dyeing baths bearing disperse dyes. Ind Eng Chem Res 39(9):3241–3248

Tanaka S, Nakata Y, Kimura T, Yustiawati, Kawasaki M, Kuramitz H (2002) Electrochemical decomposition of bisphenol A using Pt/Ti and SnO_2/Ti anodes. J Appl Electrochem 32(2):197–201

Tripathi PK, Liu M, Gan L (2014a) Tuned surface area and mesopore diameter of ordered mesoporous carbon: ultrahigh decontamination of di(2-ethylhexyl)phthalate. RCS Adv 4(45):23853–23860

Tripathi PK, Liu M, Zhao Y, Ma X, Gan L, Noonan O, Yub C (2014b) Enlargement of uniform micropores in hierarchically ordered micro-mesoporous carbon for high level decontamination of bisphenol A. J Mater Chem A 2(22):8534–8544

US-EPA (2012) EPA's reanalysis of key issues related to dioxin toxicity and response to NAS comments, vol. 1. U.S. Environmental Protection Agency, Washington DC (CAS No 1746-01-6)

Wang YR, Chu W (2011) Degradation of 2,4,5-trichlorophenoxyacetic acid by a novel Electro Fe(II)/Oxone process using iron sheet as the sacrificial anode. Water Res 45:3883–3889

Wang JK, Farrell J (2004) Electrochemical inactivation of triclosan with boron doped diamond film electrodes. Environ Sci Technol 38:5232–5237

Wang JG, Li XM (2012) Electrochemical treatment of wastewater containing chlorophenols using boron-doped diamond film electrodes. J Cent South Univ Technol 19:1946–1952

Wang B, Liao Y, Wang T (2024) Metal oxide particle electrodes for degradation of high concentration phenol wastewater via electrocatalytic advanced oxidation. Chin J Chem Eng 75:204–213

Weres O, Hoffmann MR (1995) Electrochemical device for generating hydroxyl free radicals and oxidizing chemical substances dissolved in water. US Patent, Publication no. 5439577

Wu TN (2010) Electrochemical removal of pentachlorophenol in a lab-scale platinum electrolyzer. Water Sci Technol 62:2313–2320

Xu HX, Sun X, Yang H (2024) Research progress of electrolytic treatment technology for organic wastewater. J Solid State Electrochem. https://doi.org/10.1007/s10008-024-05871-0

Zhang H, Huang CH (2003) Oxidative transformation of triclosan and chlorine by manganese oxides. Environ Sci Technol 37:1430–2421

Zhao X, Zhu Y (2006) Synergetic degradation of rhodamine B at a porous $ZnWO_4$ film electrode by combined electro-oxidation and photocatalysis. Environ Sci Technol 40:3367–3372

Zhou MH, Dai QZ, Lei LC, Ma CA, Wang DH (2005) Long life modified lead dioxide anode for organic wastewater treatment: electrochemical characteristics and degradation mechanism. Environ Sci Technol 39(1):363–370

Chapter 16
Membrane Filtration Techniques: Purifying Wastewater from Emerging Toxicants

Argyris Panagopoulos and Panagiotis A. Michailidis

Abstract This chapter examines the urgent issue of emerging toxicants in the environment and their impacts. Emerging toxicants, including pharmaceuticals and personal care products (PPCPs), pesticides, fire retardants, plasticizers, and various industrial chemicals, are increasingly being detected in aquatic ecosystems due to widespread industrial, agricultural, and urban activities. Their presence poses threats to both the environment and human health. Key sources of these contaminants are identified, which include domestic and hospital effluent discharge, industrial wastewater discharge, and agricultural runoff. Membrane filtration techniques, including microfiltration (MF), reverse osmosis (RO), ultrafiltration (UF), nanofiltration (NF), and forward osmosis (FO), are explored as effective solutions for removing emerging contaminants from water. These techniques utilize semipermeable membranes to filter substances based on several properties, like size, charge, and others. While emerging toxicants pose challenges, advances in membrane technology coupled with improved pollution prevention efforts could help curb their spread and protect ecosystems and public health on a global scale. Overall, this chapter highlights the urgent need for concerted action to curb the spread of emerging toxicants and safeguard the environment and public health.

Keywords Membrane filtration · Wastewater treatment · Toxicant removal · Water purification · Emerging contaminants · Filtration techniques · Emerging toxicants

16.1 Introduction

The escalation of anthropogenic endeavors has concurrently led to an augmentation in waste generation, subsequently yielding a significant elevation in the concentration of pollutants within aquatic environments. Human activities continuously generate

---

A. Panagopoulos (✉) · P. A. Michailidis
School of Chemical Engineering, National Technical University of Athens, Athens, Greece
e-mail: argyrispan@chemeng.ntua.gr

A. M. Yatoo et al. (eds.), *Wastewater Monitoring and Management by Advanced Approaches*, Advances in Wastewater Research,
https://doi.org/10.1007/978-981-95-2601-7_16

contaminants of emerging concern (CECs) that infiltrate the environment, posing threats to the planet's ecological balance (Zhang et al. 2021; Ahmad et al. 2022). The imperative to address these contaminants assumes paramount importance, as their elimination is crucial to ensuring the viability of life on Earth (Morin-Crini et al. 2022; Pradhan et al. 2023).

The relentless pursuit of development and technological innovation has led to accelerated progress in various spheres, accompanied by a corresponding increase in the complexity of methods and substances employed. Nevertheless, the historical record highlights instances where chemical compounds intended to enhance the standard of living have inadvertently had detrimental effects on the environment/humans. Contamination of water, air, and soil is a long-standing issue, with long-established pollutants such as arsenic and lead posing ongoing challenges (Samal et al. 2022; Osuoha et al. 2023; Pradhan et al. 2023).

The inexorable drive for economic growth has precipitated a significant expansion of industrial production, resulting in every manufacturing facility generating wastewater composed of chemicals developed/used in the pursuit of technological advancement. This wastewater encompasses a diverse array of hydrocarbons, heavy metals, and pathogenic microorganisms, all of which pose significant threats to the environment/humans. These pollutants are often characterized by their persistence, with natural degradation processes either ineffective or unable to cope with the sheer volume of substances involved. Moreover, the presence of micropollutants, substances present in trace amounts, has proven particularly challenging to monitor and analyze. These can be of synthetic origin, such as halogenated molecules, or natural compounds, such as antibiotics or estrogens. While their subtle effects were previously difficult to discern, advancements in research have subsequently highlighted their potential negative impacts, prompting heightened concern among stakeholders regarding these low-level compounds (Hale et al. 2020; Panagopoulos 2022a; Panagopoulos and Giannika 2022a).

The proliferation of human activities across various sectors, including agricultural, industrial, and urban development, has led to a multitude of pressing issues affecting global growth metrics. As consumer demands continue to rise and living standards improve, the environmental consequences of this expansion are starkly evident in the form of heightened pollution. Furthermore, the cumulative presence of various pollutants, such as industrial wastewater effluents, chemicals, spills, and soil contamination stemming from the use of pesticides and sludge, as well as the disposal of hazardous waste, has been identified in aquatic ecosystems. In addition, the increased detection of low-concentration yet ecologically significant contaminants, including antibiotics and endocrine disruptors, as environmental requirements intensify, has sparked concern. Specifically, a class of pollutants known as "emerging toxicants," "emerging contaminants," or "emerging pollutants" is highlighted as a potential threat to human well-being and ecological harmony of the environment (Pradhan et al. 2023). These complex compounds are typically identified in aqueous environments, with far-reaching implications for both human health and environmental sustainability (Morin-Crini et al. 2022; Billings et al. 2024).

Environmental toxicants infiltrate the ecosystem through various anthropogenic pathways and subsequently permeate environmental matrices. According to recent studies, environmental toxicants primarily originate from direct and indirect pollution sources that are dispersed throughout the aquatic system. These chemical compounds, whether synthetic or naturally occurring, exhibit properties that enable them to disrupt the ecological balance and pose a threat to the integrity of the environment and human well-being (Petrie 2021). Notably, environmental toxicants are capable of persisting and mobilizing in various environmental matrices, including water, air, and soil sediments, thereby contributing to their far-reaching and potentially harmful impacts (Wilkinson et al. 2016; Carrizo et al. 2022).

Water contamination has become a pressing concern due to the consequences of anthropogenic activities. Biological contaminants, including viruses and bacteria, have infiltrated water sources, necessitating their categorization as emerging or reemerging pathogens. Their virulence has sparked increased concern as they continue to drive mortality rates globally. The scope of waterborne illnesses has expanded, incorporating an array of diseases, including those previously thought to be under control. Currently, the most prevalent pollutants include pharmaceutical products, personal care products (PCPs), and flame retardants. Notably, a plethora of emerging toxicants have been identified, with a profound impact on the endocrine system. To date, over 200 distinct compounds have been detected, with some already being monitored (Wilkinson et al. 2016; Sharma et al. 2021; Osuoha et al. 2023).

The sustenance of life on our planet relies heavily on the availability of water. Unfortunately, various urban pursuits, industrial manufacturing, agricultural practices, and healthcare facilities generate substantial amounts of micropollutants, thereby contaminating the aquatic environment. These anthropogenic activities exacerbate the already pressing issue of freshwater scarcity. A plethora of substances, including pharmaceutical active compounds, PCPs, and endocrine-disrupting compounds (EDCs), have been increasingly detected in water bodies over the past few decades. These compounds, collectively referred to as “emerging toxicants/contaminants,” pose a significant threat due to the lack of comprehensive toxicological data and established guidelines. Ubiquitous and pervasive, these hazardous substances can enter aquatic environments through various wastewater pathways, including wastewater from hospitals, agricultural runoff, industrial effluent discharge, and effluent from wastewater treatment plants.

In confronting the escalating challenges posed by emerging toxicants in wastewater, the implementation of advanced membrane filtration techniques emerges as a beacon of hope. By harnessing the selective permeability of membranes based on properties (e.g., pore size, charge), technologies such as nanofiltration (NF), microfiltration (MF), reverse osmosis (RO), ultrafiltration (UF), and forward osmosis (FO) offer promising avenues for efficiently removing these contaminants from water sources. These membrane filtration methods not only hold the potential to safeguard ecosystems and public health but also signify a proactive step toward mitigating the spread of emerging toxicants globally. Through the synergy of cutting-edge membrane technology and enhanced pollution prevention strategies, a sustainable solution beckons to mitigate the detrimental impacts of these contaminants, underscoring the critical role of membrane filtration in the preservation of the environment and the well-being of future generations.

16.2 Emerging Toxicants

Adverse ecological and human health consequences can be attributed to chemicals of both synthetic and natural origin, which are capable of inducing known or suspected harm. A distinct category of toxicants comprises those from a chemical class that has hitherto received limited scientific investigation, with stakeholders expressing concern regarding potential environmental or health impacts. Furthermore, concerns have been raised regarding the inadequacy of existing environmental assessment frameworks for accommodating these emerging contaminants (Wilkinson et al. 2017; Petrie 2021; Kumar et al. 2023).

The traditional methods used to indicate environmental contaminants have limitations when it comes to classifying the intricate system of emerging toxicants. The conventional characterization of a compound by its persistence, lipophilicity, and toxicity has proven insufficient in capturing the intricacies of emerging contaminants. Notably, some emerging contaminants exhibit pseudo-persistence, wherein they persist in the environment due to ongoing releases from sewage treatment facilities. Lipophilicity, traditionally utilized to predict the ease with which a contaminant crosses cell membranes or enters tissues, is an incomplete characterization tool, as many emerging contaminants possess pharmacological properties that facilitate active transport into cells and tissues. Additional characteristics of emerging contaminants have hindered our capacity for reliable risk assessment vis-à-vis human and environmental health. Current analytical capabilities surpass our understanding of the toxic effects of these compounds on humans and wildlife. Establishing a cause–effect relationship necessitates the investigation of uptake, mode of action, and biological end points for each emerging contaminant, rather than merely attributing adverse effects to co-occurrence (Sivaranjanee and Kumar 2021; Morin-Crini et al. 2022; Pradhan et al. 2023; Billings et al. 2024).

The implications of anthropogenic activities on the environment are far-reaching, as pollutants have seeped into nearly every terrestrial and aquatic sphere. The term "emerging toxicants" encompasses not only newly detected pollutants but also previously known substances whose hazardous effects have only recently been elucidated due to advancements in scientific research. It is essential to recognize that the majority of emerging toxicants are well-established pollutants whose toxic potential has only recently been acknowledged. The label "emerging" applies not merely to the substance itself but also to the growing concern surrounding its impact. As such, emerging toxicants are frequently denoted as "contaminants of emerging concern" or "CECs." The paucity of toxicological data for the vast majority of ubiquitous chemicals, particularly in the context of long-term, low-level exposure, suggests that numerous chemicals with a lengthy history of use may contribute to the burgeoning problem of emerging contaminants in the future (Wilkinson et al. 2016; Sivaranjanee and Kumar 2021; Carrizo et al. 2022; Rohani 2023; Billings et al. 2024).

16.3 Categories of Emerging Toxicants

The categories of emerging toxicants are presented in Fig. 16.1, while categories and examples of emerging toxicants are presented in Table 16.1.

16.3.1 Pharmaceuticals and Drugs

Many medicinal products, broadly classified as pharmaceuticals or drugs, have been developed to diagnose, alleviate, or cure various ailments. These remedial substances are formulated to interact with the human body's biological processes, yielding a range of effects that can mitigate the progression of disease or restore normal physiological functions. The diversity of pharmaceuticals encompasses a vast array of compounds, including synthetic and natural substances, which can be dispensed in various forms, such as tablets, capsules, liquids, or injectables (Zakari-Jiya et al. 2022; Osuoha et al. 2023).

It is imperative to note that pharmaceuticals play a crucial role in restoring wellness, preventing diseases, and maintaining the efficiency of both humans and animals. Over the past 20 years, pharmaceuticals have gained considerable attention as potential bioactive substances in the environment. Unregulated or undergoing regulatory cycles, they pose a significant risk to water bodies since they can persist at tiny concentrations, potentially jeopardizing drinking water supplies, ecosystems, and human health. Notwithstanding their presence in water for an extended period, the

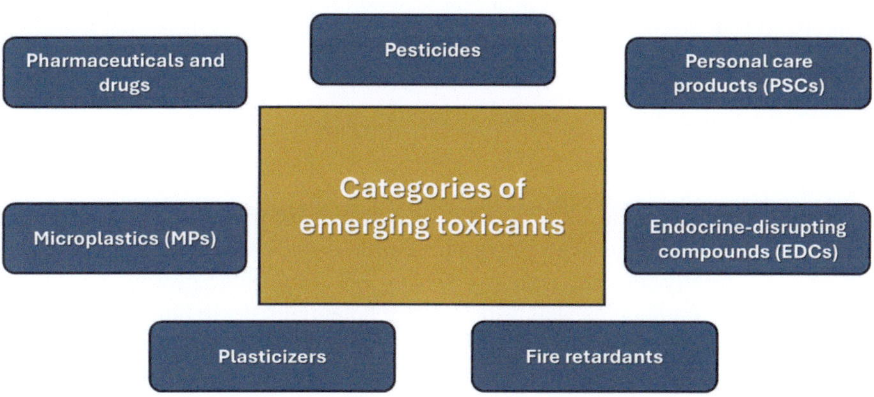

Fig. 16.1 Categories of emerging toxicants

Table 16.1 Categories and examples of emerging toxicants

Category	Examples
Pharmaceuticals	Synthetic and natural substances like antibiotics, antidepressants, and anti-inflammatories
Pesticides	Synthetic and natural substances used to kill unwanted plants and pests
Personal care products (PCPs)	Ultraviolet (UV) filters, fragrances, and disinfectants used in soaps, shampoos, and cosmetics
Endocrine-disrupting compounds (EDCs)	Estrogens, androgens, and glucocorticoids that interfere with hormone functions
Fire retardants	Brominated compounds like polybrominated diphenyl ethers used in furniture and electronics
Plasticizers	Phthalates and other additives that increase plastic flexibility and durability
Microplastics	Particles less than 5 mm from plastic product degradation and synthetic fiber shedding

levels of pharmaceuticals in the environment have only recently been measured and identified as potentially hazardous to the environment (Llompart et al. 2019; Freitas and Radis-Baptista 2021; Hejna et al. 2022).

Pharmaceuticals originate from veterinary/human medications, as well as food additives, and are initially released into the environment through animal or human excretion with minimal modification to their chemical structure. Surface water, groundwater, and, to a lesser extent, treated water all contain pharmaceutical residues at levels typically less than 100 ng/l, while processed water generally exhibits concentrations below 50 ng/l. Among various categories of emerging contaminants, nonsteroidal anti-inflammatory drugs, benzothiazoles, synthetic endocrine disruptors, perfluorinated compounds, and benzotriazoles warrant particular attention, characterized by distinct physicochemical properties and unique bioaccumulation profiles (Wilkinson et al. 2016; Carrizo et al. 2022; Zhou et al. 2023).

16.3.2 Pesticides

The application of pesticides has been a crucial aspect of modern agricultural practices, seeking to mitigate the effects of pests and maximize crop yields. These synthetic or naturally derived substances are capable of exerting a profound impact on the external environment, as well as presenting potential risks to human health. The widespread deployment of pesticides has been accompanied by growing concerns regarding their potential to contaminate soil, water, and air, thereby compromising ecosystem stability and posing threats to biodiversity. Furthermore, cumulative exposure to pesticides can have deleterious effects on human neurological and reproductive functions, prompting a heightened sense of urgency to elucidate the mechanisms underlying their toxicity and to develop novel, safer alternatives for pest management (Baldissarelli et al. 2019; Bilal et al. 2019; Rohani 2023).

The designation "pesticide" encompasses a diverse array of natural formulations designed to prevent, eliminate, or regulate the growth of pests, weeds, fungi, insects, and molds. Pesticides have been categorized into distinct classes based on their mode of application, definition, source of origin, mechanism of action, and toxicity. The examination and classification of pesticides have been a long-standing issue in ecological contexts, with recent findings revealing the widespread presence of chemically active ingredients, such as lipid-regulating agents, analgesics, antimicrobial agents, cleaning products, chemicals, and actively formulated mixtures from PCPs, including parabens and ultraviolet (UV) filters, across various channels globally. The excessive use of pesticides and chemical compounding agents in contemporary horticultural practices has led to pollution of diverse environmental matrices, encompassing air, land, and water (Ryu et al. 2022; Zheng et al. 2022; Rajan et al. 2023).

In various sectors, including agriculture, plant care, and environmental conservation, numerous pesticides are employed, with some exhibiting the potential to contaminate groundwater via leaching, spills, and soil runoff. Noteworthy is the case of herbicides like atrazine, which, as a selective herbicide, inhibits photosynthesis in vulnerable plants, leading to cardiovascular complications, retinal degenerative disorders, muscle degeneration, and cancer in humans. Oxyfluorfen, an herbicide that impedes protoporphyrinogen oxidase activity, causes irreparable cell membrane damage. Human exposure to oxyfluorfen can result in liver dysfunction and blood biomarker disturbances.

Furthermore, these pesticides pose a threat to marine life, inducing severe illnesses in both humans and animals, as well as disrupting the soil microbiota. To mitigate these ecological concerns, various physicochemical methods, including composting, incineration, and landfill remediation, have been employed to remediate pesticide contamination in the environment (Warren et al. 2003; de Araújo et al. 2022; Okoye et al. 2022; Rajan et al. 2023).

16.3.3 Personal Care Products (PCPs)

The corpus of PCPs encompasses a diverse array of items that individuals utilize to maintain their physical appearance and well-being. A significant proportion of these products are designed to clean and moisturize the skin, including soaps and shampoos. Additionally, PCPs may also comprise cosmetics, such as lip balms, eye shadows, and foundation, aimed at beautifying and enhancing one's appearance. Furthermore, the category of PCPs also includes oral hygiene products, such as toothpaste and mouthwash, aimed at promoting dental health and cleanliness. The widespread use of PCPs has led to a multifaceted industry, with a vast array of consumer choices and preferences driving market demand and innovation (Wilkinson et al. 2016; Kumar et al. 2023).

Adolescent sanitation products embody a diverse array of medicinal and cosmetic substances categorized as harmful substances. PCPs comprise a range of active compounds, including antioxidants, disinfectants, insect repellents, UV filters, and fragrances. UV filters are commonly utilized in sunscreens and are incorporated into several beauty care products, shampoos, etc. Fragrances are employed in beauty care products and moisturizers. Disinfectants are commonly employed to prevent contamination in hospitals, food processing industries, and private residences to prevent microbial contamination. Additionally, disinfectants are also used as antimicrobials in cleansers and skin creams, as well as in toothpastes. Antioxidants are widely employed as additives in foodstuffs, pharmaceuticals, and cosmetics (Samal et al. 2022; Zakari-Jiya et al. 2022).

The environmental impacts of pharmaceuticals, such as triclosan and UV filters, have been documented to have endocrine-disrupting effects on marine life (Nethathe et al. 2021; Nozaki et al. 2023). Exposure to ibuprofen has also been shown to disrupt endocrine cells, leading to reproductive and physical health issues, which are often irreversible. Furthermore, ibuprofen was found to impede the growth of beneficial microorganisms in an aquatic ecosystem. Notably, another potentially hazardous pharmaceutical, paracetamol, has been associated with numerous toxic effects, including genetic material interference and lipid oxidation in cell membranes, ultimately resulting in cell death or necrosis (Archer et al. 2017; Wilkinson et al. 2017; Arguello-Pérez et al. 2019; Jan-Roblero and Cruz-Maya 2023).

Several PCPs exhibit pronounced persistence or pseudoconstancy, posing a threat to both human health and biological ecosystems due to their capacity for bioaccumulation. Even at minute concentrations, they have been known to induce significant adverse effects. PCPs are distinguished from common natural toxins, like polycyclic aromatic hydrocarbons, by their extreme potency and unpredictable behavior. This allows them to migrate and disperse easily within aquatic environments. The majority of PCPs enter the environment through effluent releases from sewage treatment plants. The continued presence of PCPs in wastewater treatment facilities underlies concerns regarding their potential for natural dissemination. Their persistence has been shown to have negative implications for aquatic organisms (Llompart et al. 2019; Osuoha et al. 2023).

16.3.4 Endocrine-Disrupting Compounds (EDCs)

The potency of certain environmental pollutants has instigated considerable concern among scientists and public health professionals alike, as these compounds have been found to exert detrimental effects on the endocrine system. Notably, EDCs have been identified as a critical threat to human health, as they can disrupt the delicate balance of hormones within the body. These EDCs, often present in everyday products such as plasticware, personal care items, and pesticides, have been demonstrated to mimic or interfere with the activity of hormones, thereby precipitating a range of adverse health consequences, including altered development, growth, and reproductive function (Biemann et al. 2021; Pironti et al. 2021).

In addition to the accumulation of emerging contaminants, a number of endocrine-disrupting mixtures have been identified in freshwater environments. Endocrine-disrupting mixtures are pollutants that interfere with the typical functioning of the endocrine system, leading to hostile effects on reproduction, growth, and immune response. Comprising a diverse range of synthetic and natural chemicals, including pesticides, plasticizers, and organometallic compounds, EDCs are widely dispersed in water and other environments (Caporale et al. 2022; Bérubé et al. 2023). As a result of their structural similarity to natural chemicals, EDCs can have a detrimental impact on the endocrine systems of humans/animals. There are three primary categories of endocrine-disrupting substances: androgenic, thyroidal, and estrogenic (Zhang et al. 2016; Chmielewski et al. 2021; Gadupudi et al. 2021).

Various classes of mutagens and endocrine disruptors, including estrogens, androgens, glucocorticoids, and their metabolites, infiltrate the marine environment through urban or industrial wastewater, agricultural runoff, or airborne particulate matter resulting from combustion processes. Prolonged exposure to endocrine disruptors can elicit unforeseen consequences for the health and integrity of ecosystems, as documented by various studies (Gałązka and Jankiewicz 2022; Płotka-Wasylka et al. 2023; Singh et al. 2023).

16.3.5 Fire Retardants

Fire retardants comprise a class of substances that delay/inhibit the spread of fire. Characterized by their ability to reduce the flammability of a material, these substances are frequently employed in various industrial and domestic applications, including upholstered furniture, electrical wiring, and construction materials. Fire retardants can be grouped into two main types: additive and reaction-based. Additive fire retardants, which function by combining with the combustible material, have been shown to exhibit diminished efficacy over time due to degradation and migration. In contrast, reaction-based fire retardants undergo a chemical reaction upon exposure to heat, producing a flame-retarding environment. Recent studies have highlighted the potential benefits of combining multiple fire retardants in a synergistic manner, leading to enhanced fire-resistance properties. The utilization of fire retardants, substances designed to delay or extinguish combustion, is a widespread phenomenon (Mathews et al. 2022; Jafari et al. 2024). A substantial number of more than 150 distinct formulations are presently used as fire retardants, with brominated fire retardants, specifically, being the most commonly utilized type. These brominated fire retardants are widely utilized in various polymeric materials for their fire-retarding properties (Feng et al. 2023).

Polybrominated diphenyl ethers (PBDEs) and phosphate-based fire retardants, which are commonly found in mechanical and consumer products, have raised concerns regarding their potential environmental and health impacts. Furthermore, two surfactants, perfluorooctane sulfonate and perfluorooctanoic acid, have been extensively utilized over the past five decades, particularly in fire-fighting foams. PBDEs, comprising penta-, octa-, and deca-BDE, have been used since the 1970s and are regarded as the most prominent brominated fire retardants of the past few years. The use of PBDEs has been prohibited by the European Union (EU) Restriction of Hazardous Substances since 2006, leading to a shift toward the utilization of alternative options (Banks et al. 2020; Mastalska-Popławska et al. 2023; Pozo et al. 2023).

Due to the constraints imposed on the use of PBDEs, organophosphate fire retardants have emerged as a more suitable alternative to brominated fire retardants. These compounds are predominantly employed in a variety of applications, including plasticizers, floor cleaning solutions, grease, elastic materials, polyurethane foams, furniture, and electrical equipment. As a rapidly growing class of fire retardants, organophosphate compounds have been found to be effective in reducing the combustibility of numerous consumer and building materials, alongside other alternatives such as alumina trihydrate and halogenated and nitrogen-based fire retardants (Kung et al. 2022; Das et al. 2023).

16.3.6 Plasticizers

Plasticizers are a class of additives utilized to facilitate the widespread adoption of synthetic materials, particularly polymers. These substances are designed to enhance the flexibility, durability, and overall performance of plastics by promoting an increase in molecular mobility and reducing intermolecular interactions. The efficacy of plasticizers lies in their ability to modify the viscoelastic properties of polymers, rendering them more suitable for diverse applications. Furthermore, plasticizers can improve the processing characteristics of plastics, making them easier to mold, shape, and fabricate. Consequently, their incorporation has been instrumental in extending the range of feasible applications for plastics, which have become ubiquitous in modern society (Horn et al. 2004; Jia et al. 2018; Gili et al. 2022).

Sulfonamides, a distinct class of plasticizers, have received relatively limited scrutiny regarding their environmental presence, human exposure, and toxicity. Phthalates, a type of plasticizer utilized since the 1930s in various industrial and commercial applications, have been associated with adverse health effects, including carcinogenic potential, genetic mutation, and reproductive and developmental issues in both human and animal populations (Billings et al. 2021, 2024; Nagorka et al. 2022).

16.3.7 Microplastics (MPs)

The prevalence of microplastics (MPs) has garnered considerable attention in recent years, as these diminutive yet pervasive pollutants have been discovered to be ubiquitously distributed throughout the environment. Herein lies a concern, as MPs have been found to be ingested by a variety of organisms, from aquatic life forms to terrestrial animals, including humans. The consequences of MP ingestion are far-reaching, compromising the health and well-being of affected organisms as well as potentially disrupting the ecological balance of entire ecosystems (Andrady 2011; Hale et al. 2020; Pan et al. 2023). Furthermore, the sources of MP pollution are multifaceted, encompassing a broad array of human activities, including the release of synthetic fibers during laundry cycles, the breakdown of plastic products, and the inadequate disposal of plastic waste. As MPs continue to pose a significant threat to the planet's ecological integrity, it is imperative that stakeholders and policymakers collaborate to develop and apply effective strategies for mitigating the impact of these minuscule yet insidious pollutants (Guo et al. 2020; Kabir et al. 2023).

The minute yet pervasive properties of MPs have led to widespread concern regarding their pernicious nature. As a consequence of their diminutive size, MPs boast an extensive surface area, permitting them to adsorb a plethora of exogenous substances coinciding with them in marine ecosystems. Consequently, MPs can function as vectors, facilitating the interaction of xenobiotics with aquatic organisms. Following ingestion, MPs may liberate xenobiotics, thereby enhancing their bioavailability and potentially facilitating their biomagnification/bioaccumulation across the food chain, as well as modulating their detrimental impacts (Campanale et al. 2020; Kye et al. 2023).

16.4 Sources of Emerging Toxicants

The origins of emerging contaminant pollution are multifaceted and necessitate a comprehensive examination. Several sources contribute to the presence of these contaminants in the environment, with varying degrees of severity. First, municipal and industrial wastewater treatment plants (WWTPs) are critical points of contamination, as they often lack the infrastructure and regulatory frameworks to effectively remove or degrade emerging contaminants. Second, agricultural activities, including the use of pesticides and fertilizers, lead to the introduction of these pollutants into the environment. Additionally, the disposal of contaminated soil, sediments, and other waste products can also contribute to the problem. Furthermore, human activities such as the improper disposal of hazardous waste and the release of industrial effluents also pose significant threats to environmental and human health. It is essential to identify and address these sources of contamination to mitigate the impact of emerging contaminants on ecosystems and human populations alike (Altamira-Algarra et al. 2022; Morin-Crini et al. 2022; Billings et al. 2024).

Table 16.2 Sources of emerging toxicants

Source	Examples
Domestic and hospital effluents	Pharmaceuticals, PCPs, excess nutrients, pathogens
Industrial wastewater	1,4-dioxane, benzotriazoles, polybrominated diphenyl ethers (PBDEs)
Agricultural runoff	Pesticides, fertilizers, veterinary drugs from crop cultivation and livestock

Wastewater is the term used to describe water that has been rendered unsuitable for human consumption following its utilization in industrial processes and household activities. These anthropogenic activities aid the accumulation of toxic pollutants in natural water bodies. The aquatic environment is beset by pollution from diverse sources, including fertilizers, infiltration of septic and sewage tanks, agrochemical runoff, and seepage from landfill sites. Research indicates that even after undergoing treatment at conventional wastewater treatment facilities, quantities of emerging pollutants persist in untreated wastewaters. Minute levels of pharmaceuticals, EDCs, PCPs, and their by-products have been detected in effluents from such facilities, which collectively form a group of contaminants that bioaccumulate within the food chain. These contaminants traverse groundwater networks via surface seepage and groundwater recharge points. Moreover, groundwater is extensively contaminated due to the continuous integration of waste emanating from wastewater treatment facilities into the aquatic system, thereby containing a multitude of emerging pollutants (Al-Maadheed et al. 2019; Kung et al. 2022; Panagopoulos 2022a; Panagopoulos and Giannika 2022a).

Effluents originating from industrial and urban wastewater treatment facilities are directly released into aquatic bodies, including rivers, lakes, and ponds. Furthermore, concerns surrounding the environmental fate of these effluents arise when they percolate through the groundwater, adsorb sediment particles, and compromise soil fertility. The trajectory of emerging contaminants from their point of origin to their eventual sinking point in the aquatic system is heavily contingent upon the properties of these effluents, including volatility, polarity, adsorption characteristics, and persistence. The sources of emerging toxicants are summarized in Table 16.2.

16.4.1 Domestic and Hospital Effluents

The discharge of wastewater from domestic and hospital sources poses a significant environmental and public health concern. Domestic effluents, generated by households, commercial establishments, and agricultural activities, are a principal component of urban wastewater, often containing a range of pollutants, including chemical pollutants, pathogens, and excess nutrients. Hospital effluents, in contrast, are characterized by the presence of pharmaceutical residues, disinfectants, and other healthcare-related contaminants. The combination of these pollutants can pose a threat to aquatic ecosystems, human health, and the overall sustainability of urban water cycle management. Furthermore, the disposal of untreated or insufficiently treated hospital effluents has been linked to antimicrobial resistance, waterborne diseases, and environmental degradation. In freshwater ecosystems, domestic wastewater poses a significant threat due to its high concentration of pathogenic microorganisms, degradable organic matter, and nutrient-rich constituents, which collectively exacerbate eutrophication processes. Furthermore, the presence of nonbiodegradable substances such as petroleum-based compounds, synthetic detergents, pesticides, and heavy metals, which bioaccumulate in the food chain, has detrimental long-term consequences for biodiversity (Al-Maadheed et al. 2019; Panagopoulos and Haralambous 2020a; Morin-Crini et al. 2022; Panagopoulos and Giannika 2022a).

Over the past decade, the composition of domestic wastewater has undergone a marked shift, partly attributable to the increased use of PCPs and household chemicals. Domestic wastewater has been identified as a primary source of microbial pollutants, nitrogen, and phosphorus in natural water bodies, leading to eutrophication and the proliferation of toxin-producing cyanobacteria. As such, untreated domestic wastewater poses a significant risk to both human and aquatic health. The geographic distribution of wastewater contaminants varies in response to regional differences in domestic usage patterns. WWTPs have been monitored for the concentration levels of antibiotics in influent/effluent streams, revealing values within the range of a few tens of micrograms per liter. Macrolides, sulfonamides, trimethoprim, and fluoroquinolones have been identified as the most prevalent antibiotics detected globally in wastewater. Notably, β-lactam, a medication often prescribed in large quantities, exhibits negligible presence in wastewater due to its susceptibility to chemical hydrolysis/chemical transformation in alkaline or acidic environments (Gadupudi et al. 2021; Petrie 2021; Samal et al. 2022; Panagopoulos 2023).

Furthermore, a variety of antifungal and antimicrobial agents, including miconazole, thiabendazole, triclocarban, and triclosan, are widely utilized in household products such as shampoos, toothpastes, soaps, and dermal creams. It has been observed that concentrations of triclocarban and triclosan tend to be higher in wastewater, while miconazole and thiabendazole, which are employed as therapeutic treatments, enter the aquatic system at significantly lower concentrations. Various categories of therapeutic drugs, including beta-adrenoreceptor antagonists, anticonvulsants and antipsychotics, hypolipidemic agents, anti-inflammatory agents, glucocorticoids, and synthetic hormones such as estradiol, are employed as medication. Notably, a segment of these substances is excreted unchanged through the human body and eventually finds its way into aquatic ecosystems. In the long term, these pharmacological substances pose a significant threat to the well-being of aquatic organisms (Verlicchi et al. 2010; Vaudreuil et al. 2022; Ramírez-Moreno et al. 2023).

PCPs, specifically those designed to address the harmful effects of UV radiation, owe their efficacy to the presence of UV-blocking filters. These filters are integral components of various cosmetic formulations, including sunscreens, dyes, etc. Importantly, these substances can ingress aquatic ecosystems via two primary pathways. Initially, they may be introduced through recreational activities involving water bodies such as lakes/rivers. In addition, utilized domestic water, subsequently discharged through WWTPs, can also serve as a conduit for the deposition of these chemical agents in aquatic environments. In the realm of dietetics, artificial sweeteners are commonly employed as additives to low-calorie foods and beverages, serving as substitutes for natural sugars. These substances, being nonmetabolizable, are excreted from the human body in their original form. While some artificial sweeteners, such as cyclamate and saccharin, demonstrate biodegradable properties, others, including sucralose, a chlorinated compound, and acesulfame, exhibit persistent characteristics that may compromise the aquatic ecosystem (Zakari-Jiya et al. 2022; Kumar et al. 2023; Osuoha et al. 2023).

16.4.2 Industrial Wastewater

Industrial wastewater is a pervasive issue within the manufacturing sector, warranting concerted attention in the pursuit of environmental sustainability. The discharge of effluent streams from industrial processes has significant implications for the ecosystem, encompassing concerns about water quality, aquatic life, and human health. Notably, the industrial sector's contribution to global wastewater generation is substantial, with a significant proportion attributed to industrial production activities. Industrial facilities, by virtue of their operation, produce a diverse array of wastewater streams, characterized by varying levels of contaminant loads and physical properties. The intensity and composition of industrial wastewater are influenced by factors such as production processes, raw materials, and treatment technologies employed (Morin-Crini et al. 2022; Panagopoulos 2022a).

Furthermore, the heterogeneous nature of industrial wastewater necessitates tailored management strategies to ensure effective treatment and minimize environmental impacts. Efforts to mitigate the environmental footprint of industrial wastewater have focused on the applications of best management practices (BMPs), coupled with ongoing research and development of innovative treatment technologies. The deployment of cutting-edge wastewater treatment systems, such as brine treatment systems, membrane bioreactors and advanced oxidation processes, has the potential to substantially reduce pollutant loads and enhance the overall effectiveness of wastewater management. The release of industrial compounds into the environment can have detrimental consequences for the ecosystem. A plethora of effluent streams, emanating from various industries such as pharmaceuticals, tanneries, iron smelting, textiles, dishware manufacturing, and paper and pulp mills are known to contain a multitude of emerging contaminants. Emerging industrial contaminants can be identified as follows: The versatile solvent 1,4-dioxane plays a crucial role in the production of various chemicals/laboratory reagents. When copolymerized with 1,1,1-trichloroethane, this solvent exhibits striking resistance to natural biodegradable processes, thereby necessitating its persistence in groundwater and propensity for soil leaching. Moreover, 1,4-dioxane has been discovered to be present in certain pharmaceuticals and personal care products (PPCPs) (Verlicchi et al. 2012; Eberle et al. 2016; Freitas and Radis-Baptista 2021; Luo et al. 2021; Samal et al. 2022; Vaudreuil et al. 2022; Wang et al. 2023). Several important classes of pharmaceutical antibiotics contain benzotriazole derivatives, which exhibit antifungal, antibacterial, and anthelmintic properties. Notably, these compounds have been documented to persist in aquatic environments. PBDEs are synthetic additives employed to retard fire and flame hazards in a diverse range of household and commercial settings. In coastal and estuarine ecosystems, the presence of PBDEs has been convincingly detected. Through industrial manufacturing and household products, including furniture, electronics, and bedding, these chemicals are translocating into the environment. Exposure to PBDEs has been linked to the potential development of neurodevelopmental and thyroid dysfunctions. Furthermore, PBDEs are suspected to exhibit endocrine-disrupting properties (Blum et al. 2019; Vaudreuil et al. 2022; Osuoha et al. 2023; Pozo et al. 2023).

16.4.3 Agricultural Runoff

Soil and water resources are often compromised by the inadvertent release of pollutants and erosive sediments through agricultural runoff. Agricultural operations, encompassing practices such as crop cultivation and livestock rearing, generate considerable amounts of runoff that can carry detrimental substances into nearby waterways and ecosystems. Typically, the primary contaminants responsible for polluting these water bodies are pesticides, fertilizers, and animal waste, all of which can have adverse effects on aquatic life and human health. Moreover, the excessive amounts of suspended sediment introduced by agricultural runoff can significantly alter stream morphology and impair water quality, thereby compromising its usability for various purposes (Altamira-Algarra et al. 2022; Castellanos-Estupiñan et al. 2022).

Contamination of surface water emanating from agricultural sources is primarily attributed to the effluent of toxins, hormones, veterinary medicines, and man-made chemicals employed in agricultural practices. The influx of emerging toxicants in agricultural land soil, subsequently transported to aquatic systems through leaching, runoff, and drainage processes, poses significant environmental concerns. These contaminants are released directly into the agricultural environment via the application of pesticides/other solid waste and indirectly through animal biowaste emanating from veterinary medicines. The prospect of developing novel technologies for the efficient use of pesticides is promising, with existing pesticides being transformed into nanoparticulate forms (1–100 nm) to enhance their efficacy and bioavailability, including increased solubility, stability, absorption into plant tissue, and toxicity toward pests (Altamira-Algarra et al. 2022; Castellanos-Estupiñan et al. 2022).

While there are theoretical concerns about the potential transfer of these nanoparticles to surface water and subsequent contamination of aquatic ecosystems, many substances introduced into agricultural systems are susceptible to degradation through physical, chemical, and biological processes. Although these reactions may culminate in the complete breakdown of a chemical, often yielding carbon dioxide, water, and methane, they may also yield stable intermediate compounds, commonly referred to as degradation or transformation products. The conventional wastewater treatment process commences with a primary treatment stage, where particulate matter precipitates and is eliminated, subsequent to which a secondary treatment phase employs microbial breakdown to degrade organic pollutants, including chemical compounds. Activated sludge treatment is a common methodology employed in the secondary treatment stage, where potentially hazardous substances may undergo transformation yet can also adhere to sludge particles within the system. Ultimately, the excess sludge is extracted, characterized by high nutrient concentrations, and often utilized as a fertilizer in agricultural settings. Endocrine disruptors that become entrapped in sludge may inadvertently contaminate the environment upon application of biosolids from wastewater treatment facilities on arable land (Wang et al. 2018; Becker and Pinheiro 2019; Castellanos-Estupiñan et al. 2022).

16.5 Membrane Filtration Techniques

Membrane filtration techniques have garnered substantial consideration in recent years due to their versatility and effectiveness in tackling a wide range of industrial applications. The fundamental principle underlying these technologies is the utilization of semipermeable membranes, which selectively allow certain molecules or ions to pass through while rejecting others. These membranes, typically fabricated from materials such as polymeric materials or nanomaterials, are designed to operate at various pressure and temperature conditions, enabling their use in diverse environments. By exploiting the unique properties of membrane materials, membrane separation technologies have been employed to achieve pollutant removal, water treatment, and biosensing, among other objectives (Panagopoulos et al. 2019).

For instance, UF/MF membranes have been utilized to remove suspended solids and larger particles from contaminated water, whereas RO has been used to desalinate seawater and remove dissolved salts. Additionally, membrane bioreactors have been developed to integrate biological treatment with membrane technology, enhancing the efficiency of wastewater treatment processes. Overall, membrane separation technologies have emerged as a crucial tool for addressing various industrial, environmental, and biomedical challenges, and their applications continue to expand as researchers explore new materials, designs, and operating conditions to optimize their performance and effectiveness (Panagopoulos 2020a, b; Rosen and Farsi 2022).

While membrane technologies have exhibited significant efficacy in eliminating a wide range of emerging contaminants, including endocrine disruptors, a critical consideration is the substantial intrinsic cost associated with this process. Nevertheless, the membrane filtration method has emerged as a preferred treatment option owing to its pronounced selectivity across a broad spectrum, unhindered continuous operation, absence of by-product or metabolite formation, and remarkable adaptability to conventional water treatment processes (Panagopoulos 2020c; Panagopoulos et al. 2020).

Membrane separation methods employ a variety of materials, e.g., polymers and ceramics, which are subjected to hydrostatic pressure to facilitate the removal of high molecular weight solutes. Consequently, the filtrative characteristics, removal efficiency, and impurity rejection are contingent upon the selected membrane material, its constituent compounds, and the environmental conditions under which they are exposed. A multitude of complex factors influence the removal process, including the membrane material's pore size, hydrophobicity, functional groups, and the quality of the feed water, as well as other factors that are essential to consider (Panagopoulos and Haralambous 2020a). Membrane processes are typically categorized based on the pore diameters and molecular weight thresholds governing the transport of solutes across the phases, in addition to the driving force facilitating this transfer. Specifically, MF, UF, NF, and RO are the most broadly utilized pressure-driven membrane processes for segregating endocrine disruptors. Furthermore, novel membrane technologies, including FO, have shown considerable promise (Zeman and Zydney 2017; Saji et al. 2020; Panagopoulos 2021a, b).

16.5.1 Microfiltration (MF)

The process of microporous membrane separation, commonly referred to as MF, involves the use of a semipermeable membrane with small pores to separate particles of varying sizes from a fluid or gas stream. This technique is frequently employed in industrial applications, such as wastewater treatment and bioprocessing, to remove impurities and contaminants of microbial origin from the medium. Through the selective permeation of the membrane, MF enables the efficient separation of suspended solids, bacteria, and viruses, thereby enhancing the quality and purity of the resulting effluent. The benefits of MF lie in its ability to conserve water/energy, reduce waste generation, and minimize the risk of contamination in pharmaceutical and biotechnological processes (Zeman and Zydney 2017; Panagopoulos 2021c). While MF has been utilized successfully for the removal of the aforementioned contaminants, its application for eliminating endocrine disruptors is limited due to the size of its pores (Gadupudi et al. 2021; Panagopoulos 2021d; Pironti et al. 2021).

As a standalone process, MF proves insufficient for effluent treatment and is often combined with other removal technologies. In certain instances, MF is applied as a pretreatment step to remove natural organic matter and colloids and to prevent membrane fouling, while in other cases, it is employed as a posttreatment step to enhance biological treatment outcomes and maintain membrane quality, thereby extending its lifespan. The integration of MF with NF and RO has been shown to be a successful strategy for wastewater treatment (Panagopoulos and Giannika 2022a).

16.5.2 Ultrafiltration (UF)

The process of UF constitutes a crucial step in modern membrane separation technology. This methodology, which is fundamentally distinct from conventional filtration procedures, involves the selective removal of minute solutes, colloidal particles, and dissolved macromolecules from a liquid matrix. The primary objective of UF is to partition the liquid into a purified stream, free from impurities, and a concentrated stream, rich in suspended solids. By virtue of its unique membrane structure and controlled operating parameters, UF enables the efficient separation of complex mixtures, encompassing both chemical and biological substances. Notably, this technique has garnered significant attention across various fields, encompassing biotechnology, environmental science, and the pharmaceutical industry, due to its versatility and potential to alleviate environmental concerns and enhance product quality (Zeman and Zydney 2017; Panagopoulos 2022b, c).

UF, a low-pressure separation technique grounded in the sieving principle, surpasses MF in the efficacy of removing emerging contaminants due to its miniature pore sizes. Additionally, UF surpasses MF in the removal of proteins. The membrane–compound relationship plays a vital role in the removal procedure, particularly for endocrine disruptors. Their low molecular mass renders size exclusion ineffective, thereby necessitating adsorption onto the membrane, an interaction reliant on the adsorbate–adsorbent affinity. Furthermore, the membrane–compound interaction's significance is evident in the marked variation in removal rates of the same endocrine disruptor when diverse membrane types are employed (Panagopoulos and Haralambous 2020b; Panagopoulos and Giannika 2022b). The amalgamation of UF with alternative technologies, including ultrasound and advanced oxidative processes, or pretreatment methods such as chlorination/ozonation, significantly enhances the removal efficacy of highly water-soluble endocrine disruptors, e.g., organic acid compounds and estrogens. The efficacy of these hybrid processes lies in their ability to not only produce reusable permeate but also prevent membrane fouling. Moreover, the combined deployment of UF and ozonation has been found to be particularly effective in eliminating endocrine disruptors (Zeman and Zydney 2017; Panagopoulos 2022d).

16.5.3 Nanofiltration (NF)

NF technology exploits membranes characterized by compact, high-charged structures, conferring a removal efficiency nestled between that of UF and RO. The NF membranes possess minute pores and a significant charge density, predominantly negative, owing to the presence of ionizable groups. These membranes can effectively preclude both monovalent/multivalent ions, with the latter being more effectively removed (>90%) due to electrostatic interactions between the charged groups.

Consequently, NF allows the separation of substances based on their charge properties and size. Ordinarily, the NF process offers the most economical option, as it operates under lower pressure than RO, thereby reducing operational costs. Moreover, it exhibits a lower molecular weight cutoff than MF and UF, thereby prohibiting a broader range of compounds (Zeman and Zydney 2017; Schäfer and Fane 2021; Panagopoulos 2022e).

The efficacy of NF is significantly contingent upon the membrane material, nature of the contaminant, and composition of the solution. NF exhibits high rejection rates for all hormones, regardless of the water matrix, primarily attributed to size exclusion and hydrophobic interactions. Studies have revealed that the weakest NF performance is observed in groundwater, characterized by high alkalinity, total hardness, and conductivity, which may have reduced electrostatic interactions and, subsequently, the rejection of compounds (Schäfer and Fane 2021; Panagopoulos 2022a).

16.5.4 Reverse Osmosis (RO) and Forward Osmosis (FO)

The processes of RO and FO, which employ semipermeable membranes with diminutive pore sizes smaller than those of NF membranes, hinge on the distinct driving forces utilized to filter contaminated aqueous media (Panagopoulos 2023). In the case of RO, a mechanism is triggered by the application of hydraulic pressure (i.e., external pressure), surpassing the osmotic pressure, thereby inducing water flow from a more concentrated flux (e.g., seawater, brackish water) toward the permeate. Conversely, FO exploits the osmotic pressure differential as the driving force for separation. A pressure gradient is generated by the addition of a concentrated solution (i.e., draw solution), compelling the water flux. Although FO is not a pressure-driven membrane technology, its separation mechanisms exhibit similarities to those of RO and NF (Panagopoulos and Giannika 2023). Several research studies have demonstrated that both RO and FO are commonly highly effective in eliminating endocrine disruptors, with a few compounds exhibiting removal rates lower than 50% due to the peculiarity of their chemical structures. Notably, RO tends to exhibit superior separation performance compared to FO, despite the latter's benefits (Panagopoulos and Giannika 2024a).

The employment of RO has emerged as a preeminent technique for the treatment of heterogeneous water matrices, owing to its remarkable capability to eliminate a broad spectrum of soluble bioactive contaminants present at minute concentrations. Similarly, the performance of other filtration processes hinges on the specificity of the membrane's physicochemical structure, which influences both removal efficiency and optimal operating conditions. The fabrication of RO membranes involves the utilization of various materials, including polyamide, polyether, and cellulose (Panagopoulos and Giannika 2024b). The categories of membrane filtration techniques are summarized in Table 16.3.

Table 16.3 Categories of membrane filtration techniques (Panagopoulos and Giannika 2022a, 2024b)

Membrane-based processes	Abbr	Driving force	Separation size range
Nanofiltration	NF	Pressure gradient	1–10 nm
Microfiltration	MF	Pressure gradient	10–0.1 μm
Ultrafiltration	UF	Pressure gradient	< 0.1 μm–5 nm
Reverse osmosis	RO	Pressure gradient	< 5 nm
Forward osmosis	FO	Osmotic pressure gradient	< 5 nm

16.6 Conclusions and Prospects

This chapter provides a comprehensive overview of emerging toxicants, their sources and impacts, and membrane filtration techniques for their removal. Emerging contaminants such as pharmaceuticals, PCPs, pesticides, plasticizers, and EDCs now pervade our environment due to widespread industrial and agricultural activities. Their ubiquity in aquatic ecosystems poses major threats to both environmental and human health.

Key sources of emerging contaminant pollution were identified, including domestic and industrial wastewater, agricultural runoff, and hazardous waste disposal. These pathways introduce an array of problematic substances into the biosphere. While wastewater treatment aims to reduce contaminant loads, conventional facilities are often inadequate for completely removing emerging contaminants. Their complex nature challenges conventional risk assessment and treatment approaches.

Membrane filtration is highlighted as an effective strategy for purifying water and eliminating emerging contaminants. A variety of membrane processes, such as MF, RO, UF, NF, and FO, were reviewed based on their unique driving forces and separation mechanisms. These techniques leverage semipermeable membranes to selectively filter contaminants based on size, charge, and other properties. When optimized through material selection and hybridization with other treatment steps, membrane filtration can achieve high removal efficacies for diverse emerging contaminants.

While emerging contaminants pose formidable threats, advancements in membrane technology offer promising solutions. Continued research on membrane development alongside source control measures could substantially curb emerging contaminant pollution. An integrated approach considering all pathways and a contamination mitigation strategy considering the limitations of conventional treatment are necessary to safeguard ecosystems and public health on a global scale. Sustainable management of industrial activities and agricultural practices must also be prioritized to curb contamination at its source.

References

Ahmad A, Kurniawan SB, Abdullah SRS et al (2022) Contaminants of emerging concern (CECs) in aquaculture effluent: Insight into breeding and rearing activities, alarming impacts, regulations, performance of wastewater treatment unit and future approaches. Chemosphere 290:133319

Al-Maadheed S, Goktepe I, Latiff ABA, Shomar B (2019) Antibiotics in hospital effluent and domestic wastewater treatment plants in Doha, Qatar. J Water Process Eng 28:60–68. https://doi.org/10.1016/j.jwpe.2019.01.005

Altamira-Algarra B, Puigagut J, Day JW et al (2022) A review of technologies for closing the P loop in agriculture runoff: Contributing to the transition towards a circular economy. Ecol Eng 177:106571

Andrady AL (2011) Microplastics in the marine environment. Mar Pollut Bull 62:1596–1605

Archer E, Petrie B, Kasprzyk-Hordern B, Wolfaardt GM (2017) The fate of pharmaceuticals and personal care products (PPCPs), endocrine disrupting contaminants (EDCs), metabolites and illicit drugs in a WWTW and environmental waters. Chemosphere 174:437–446. https://doi.org/10.1016/j.chemosphere.2017.01.101

Arguello-Pérez MA, Mendoza-Pérez JA, Tintos-Gómez A et al (2019) Ecotoxicological analysis of emerging contaminants from wastewater discharges in the Coastal Zone of Cihuatlán (Jalisco, Mexico). Water 11:1386. https://doi.org/10.3390/w11071386

Baldissarelli DP, Vargas GDLP, Korf EP et al (2019) Remediation of soils contaminated by pesticides using physicochemical processes: a brief review. Planta Daninha 37:e019184975

Banks APW, Engelsman M, He C et al (2020) The occurrence of PAHs and flame-retardants in air and dust from Australian fire stations. J Occup Environ Hyg 17:1699246. https://doi.org/10.1080/15459624.2019.1699246

Becker N, Pinheiro IG (2019) Potentiality of permeable pavements in the improvement of surface runoff quality: a review. Urbe 11:e20180009. https://doi.org/10.1590/2175-3369.011.002.AO07

Bérubé R, LeFauve MK, Heldman S et al (2023) Adipogenic and endocrine disrupting mixture effects of organic and inorganic pollutant mixtures. Sci Total Environ 876:162587. https://doi.org/10.1016/j.scitotenv.2023.162587

Biemann R, Blüher M, Isermann B (2021) Exposure to endocrine-disrupting compounds such as phthalates and bisphenol A is associated with an increased risk for obesity. Best Pract Res Clin Endocrinol Metab 35:101546

Bilal M, Iqbal HMN, Barceló D (2019) Persistence of pesticides-based contaminants in the environment and their effective degradation using laccase-assisted biocatalytic systems. Sci Total Environ 695:133896

Billings A, Jones KC, Pereira MG, Spurgeon DJ (2021) Plasticisers in the terrestrial environment: Sources, occurrence and fate. Environ Chem 18:111–130

Billings A, Jones KC, Pereira MG, Spurgeon DJ (2024) Emerging and legacy plasticisers in coastal and estuarine environments: a review. Sci Total Environ 908:168462

Blum A, Behl M, Birnbaum LS et al (2019) Organophosphate ester flame retardants: are they a regrettable substitution for polybrominated diphenyl ethers? Environ Sci Technol Lett 6:638–649

Campanale C, Massarelli C, Savino I et al (2020) A detailed review study on potential effects of microplastics and additives of concern on human health. Int J Environ Res Public Health 17:1212

Caporale N, Leemans M, Birgersson L et al (2022) From cohorts to molecules: adverse impacts of endocrine disrupting mixtures. Science 375:8244. https://doi.org/10.1126/science.abe8244

Carrizo JC, Vo Duy S, Munoz G et al (2022) Suspect screening of pharmaceuticals, illicit drugs, pesticides, and other emerging contaminants in Argentinean Piaractus mesopotamicus, a fish species used for local consumption and export. Chemosphere 309:136769. https://doi.org/10.1016/j.chemosphere.2022.136769

Castellanos-Estupiñan MA, Carrillo-Botello AM, Rozo-Granados LS et al (2022) Removal of nutrients and pesticides from agricultural runoff using microalgae and cyanobacteria. Water 14:558. https://doi.org/10.3390/w14040558

Chmielewski J, Czarny-Działak M, Wróblewska I et al (2021) The challenge of endocrine disrupting compounds (EDCs) in the current environment for public health. J Elem 26:893–903
Das O, Mensah RA, Balasubramanian KBN et al (2023) Functionalised biochar in biocomposites: the effect of fire retardants, bioplastics and processing methods. Compos Part C Open Access 11:100368. https://doi.org/10.1016/j.jcomc.2023.100368
de Araújo EP, Caldas ED, Oliveira-Filho EC (2022) Pesticides in surface freshwater: a critical review. Environ Monit Assess 194:452
Eberle D, Ball R, Boving TB (2016) Peroxone activated persulfate treatment of 1,4-dioxane in the presence of chlorinated solvent co-contaminants. Chemosphere 144:728–735. https://doi.org/10.1016/j.chemosphere.2015.08.063
Feng J, Liu L, Zhang Y et al (2023) Rethinking the pathway to sustainable fire retardants. Exploration 3:20220088. https://doi.org/10.1002/EXP.20220088
Freitas LDAA, Radis-Baptista G (2021) Pharmaceutical pollution and disposal of expired, unused, and unwanted medicines in the Brazilian context. J Xenobiot 11:61–76
Gadupudi CK, Rice L, Xiao L, Kantamaneni K (2021) Endocrine disrupting compounds removal methods from wastewater in the United Kingdom: a review. Science 3:11
Gałązka A, Jankiewicz U (2022) Endocrine disrupting compounds (nonylphenol and bisphenol A)–sources, harmfulness and laccase-assisted degradation in the aquatic environment. Microorganisms 10:2236
Gili A, del Gaudio I, King R et al (2022) Modelling plasticiser loss inside closed environments. Polym Degrad Stab 206:110204. https://doi.org/10.1016/j.polymdegradstab.2022.110204
Guo JJ, Huang XP, Xiang L et al (2020) Source, migration and toxicology of microplastics in soil. Environ Int 137:105263
Hale RC, Seeley ME, La Guardia MJ et al (2020) A global perspective on microplastics. J Geophys Res Oceans 125:e2018JC014719
Hejna M, Kapuścińska D, Aksmann A (2022) Pharmaceuticals in the aquatic environment: a review on eco-toxicology and the remediation potential of algae. Int J Environ Res Public Health 19:7717
Horn O, Nalli S, Cooper D, Nicell J (2004) Plasticizer metabolites in the environment. Water Res 38:3693–3698. https://doi.org/10.1016/j.watres.2004.06.012
Jafari P, Zhang R, Huo S et al (2024) Machine learning for expediting next-generation of fire-retardant polymer composites. Compos Commun 45:101806
Jan-Roblero J, Cruz-Maya JA (2023) Ibuprofen: toxicology and biodegradation of an emerging contaminant. Molecules 28:2907
Jia P, Xia H, Tang K, Zhou Y (2018) Plasticizers derived from biomass resources: a short review. Polymers 10:1303
Kabir MS, Wang H, Luster-Teasley S et al (2023) Microplastics in landfill leachate: sources, detection, occurrence, and removal. Environ Sci Ecotechnology 16:100256
Kumar M, Sridharan S, Sawarkar AD et al (2023) Current research trends on emerging contaminants pharmaceutical and personal care products (PPCPs): a comprehensive review. Sci Total Environ 859:160031
Kung HC, Hsieh YK, Huang BW et al (2022) An overview: organophosphate flame retardants in the atmosphere. Aerosol Air Qual Res 22:220148
Kye H, Kim J, Ju S et al (2023) Microplastics in water systems: a review of their impacts on the environment and their potential hazards. Heliyon 9:e14359
Llompart M, Celeiro M, Dagnac T (2019) Microwave-assisted extraction of pharmaceuticals, personal care products and industrial contaminants in the environment. TrAC Trends Anal Chem 116:136–150
Luo YH, Long X, Wang B et al (2021) A synergistic platform for continuous co-removal of 1,1,1-trichloroethane, trichloroethene, and 1,4-dioxane via catalytic dechlorination followed by biodegradation. Environ Sci Technol 55:00542. https://doi.org/10.1021/acs.est.1c00542
Mastalska-Popławska J, Wójcik Ł, Izak P (2023) Applications of hydrogels with fire retardant properties—a review. J Solgel Sci Technol 105:608–624

Mathews LD, Capricho JC, Peerzada M et al (2022) Recent progress and multifunctional applications of fire-retardant epoxy resins. Mater Today Commun 33:104702
Morin-Crini N, Lichtfouse E, Liu G et al (2022) Worldwide cases of water pollution by emerging contaminants: a review. Environ Chem Lett 20:2311–2338
Nagorka R, Birmili W, Schulze J, Koschorreck J (2022) Diverging trends of plasticizers (phthalates and non-phthalates) in indoor and freshwater environments—why? Environ Sci Eur 34:46. https://doi.org/10.1186/s12302-022-00620-4
Nethathe B, Chipangura J, Hassan IZ et al (2021) Diclofenac toxicity in susceptible bird species results from a combination of reduced glomerular filtration and plasma flow with subsequent renal tubular necrosis. PeerJ 9:12002. https://doi.org/10.7717/peerj.12002
Nozaki K, Tanoue R, Kunisue T et al (2023) Pharmaceuticals and personal care products (PPCPs) in surface water and fish from three Asian countries: species-specific bioaccumulation and potential ecological risks. Sci Total Environ 866:161258. https://doi.org/10.1016/j.scitotenv.2022.161258
Okoye CO, Okeke ES, Okoye KC et al (2022) Occurrence and fate of pharmaceuticals, personal care products (PPCPs) and pesticides in African water systems: a need for timely intervention. Heliyon 8:e09143
Osuoha JO, Anyanwu BO, Ejileugha C (2023) Pharmaceuticals and personal care products as emerging contaminants: need for combined treatment strategy. J Hazard Mater Adv 9:100206
Pan Y, Gao SH, Ge C et al (2023) Removing microplastics from aquatic environments: A critical review. Environmental Science and Ecotechnology 13:100222
Panagopoulos A (2020a) Process simulation and techno-economic assessment of a zero liquid discharge/multi-effect desalination/thermal vapor compression (ZLD/MED/TVC) system. Int J Energy Res 44:4948. https://doi.org/10.1002/er.4948
Panagopoulos A (2020b) Techno-economic evaluation of a solar multi-effect distillation/thermal vapor compression hybrid system for brine treatment and salt recovery. Chem Eng Process Process Intensif 152:107934. https://doi.org/10.1016/j.cep.2020.107934
Panagopoulos A (2020c) A comparative study on minimum and actual energy consumption for the treatment of desalination brine. Energy 212:118733. https://doi.org/10.1016/j.energy.2020.118733
Panagopoulos A (2021a) Beneficiation of saline effluents from seawater desalination plants: fostering the zero liquid discharge (ZLD) approach—a techno-economic evaluation. J Environ Chem Eng 9:105338. https://doi.org/10.1016/j.jece.2021.105338
Panagopoulos A (2021b) Techno-economic assessment of minimal liquid discharge (MLD) treatment systems for saline wastewater (brine) management and treatment. Process Saf Environ Prot 146:007. https://doi.org/10.1016/j.psep.2020.12.007
Panagopoulos A (2021c) Energetic, economic and environmental assessment of zero liquid discharge (ZLD) brackish water and seawater desalination systems. Energy Convers Manag 235:113957. https://doi.org/10.1016/j.enconman.2021.113957
Panagopoulos A (2021d) Water-energy nexus: desalination technologies and renewable energy sources. Environ Sci Pollut Res 28:21009–21022
Panagopoulos A (2022a) Brine management (saline water & wastewater effluents): sustainable utilization and resource recovery strategy through minimal and zero liquid discharge (MLD & ZLD) desalination systems. Chem Eng Process Process Intensif 176:108944
Panagopoulos A (2022b) Study and evaluation of the characteristics of saline wastewater (brine) produced by desalination and industrial plants. Environ Sci Pollut Res 29:23736–23749. https://doi.org/10.1007/s11356-021-17694-x
Panagopoulos A (2022c) Process simulation and analysis of high-pressure reverse osmosis (HPRO) in the treatment and utilization of desalination brine (saline wastewater). Int J Energy Res 46:23083. https://doi.org/10.1002/er.8607
Panagopoulos A (2022d) Techno-economic assessment and feasibility study of a zero liquid discharge (ZLD) desalination hybrid system in the Eastern Mediterranean. Chem Eng Process Process Intensif 178:109029. https://doi.org/10.1016/j.cep.2022.109029

Panagopoulos A (2022e) Techno-economic assessment of zero liquid discharge (ZLD) systems for sustainable treatment, minimization and valorization of seawater brine. J Environ Manage 306:114488. https://doi.org/10.1016/j.jenvman.2022.114488
Panagopoulos A, Giannika V (2022a) Decarbonized and circular brine management/valorization for water & valuable resource recovery via minimal/zero liquid discharge (MLD/ZLD) strategies. J Environ Manage 324:116239
Panagopoulos A, Giannika V (2022b) Comparative techno-economic and environmental analysis of minimal liquid discharge (MLD) and zero liquid discharge (ZLD) desalination systems for seawater brine treatment and valorization. Sustainable Energy Technol Assess 53:102477. https://doi.org/10.1016/j.seta.2022.102477
Panagopoulos A, Giannika V (2023) Study on the water resources and the opportunities for sustainable desalination & minimal/zero liquid discharge (MLD/ZLD) practices in Greece (Eastern Mediterranean). Sustain Water Resour Manag 9:106. https://doi.org/10.1007/s40899-023-00884-5
Panagopoulos A, Giannika V (2024a) Techno-economic analysis (TEA) of zero liquid discharge (ZLD) systems for treatment and utilization of brine via resource recovery. Chem Eng Process Process Intensif 200:109773. https://doi.org/10.1016/j.cep.2024.109773
Panagopoulos A, Giannika V (2024b) A comprehensive assessment of the economic and technical viability of a zero liquid discharge (ZLD) hybrid desalination system for water and salt recovery. J Environ Manage 359:121057. https://doi.org/10.1016/j.jenvman.2024.121057
Panagopoulos A, Haralambous KJ (2020a) Environmental impacts of desalination and brine treatment—challenges and mitigation measures. Mar Pollut Bull 161:111773
Panagopoulos A, Haralambous K-J (2020b) Minimal liquid discharge (MLD) and zero liquid discharge (ZLD) strategies for wastewater management and resource recovery—analysis, challenges and prospects. J Environ Chem Eng 8:104418. https://doi.org/10.1016/j.jece.2020.104418
Panagopoulos A, Haralambous KJ, Loizidou M (2019) Desalination brine disposal methods and treatment technologies—a review. Sci Total Environ 693:133545
Panagopoulos A, Loizidou M, Haralambous KJ (2020) Stainless steel in thermal desalination and brine treatment: current status and prospects. Metals Mater Int 26:1463–1482
Panagopoulos A (2023) Zero liquid discharge and minimal liquid discharge strategies for sustainable saline wastewater (brine) management and valorization. In: Resource recovery in industrial waste waters
Petrie B (2021) A review of combined sewer overflows as a source of wastewater-derived emerging contaminants in the environment and their management. Environ Sci Pollut Res 28:32095–32110
Pironti C, Ricciardi M, Proto A et al (2021) Endocrine-disrupting compounds: an overview on their occurrence in the aquatic environment and human exposure. Water 13:1347
Płotka-Wasylka J, Mulkiewicz E, Lis H et al (2023) Endocrine disrupting compounds in the baby's world—a harmful environment to the health of babies. Sci Total Environ 881:163350
Pozo K, Oyola G, Jorquera H et al (2023) Environmental signature and health risk assessment of polybrominated diphenyl ethers (PBDEs) emitted from a landfill fire in Santiago de Chile. Environ Pollut 330:121648. https://doi.org/10.1016/j.envpol.2023.121648
Pradhan B, Chand S, Chand S et al (2023) Emerging groundwater contaminants: a comprehensive review on their health hazards and remediation technologies. Groundw Sustain Dev 20:100868
Rajan S, Parween M, Raju NJ (2023) Pesticides in the hydrogeo-environment: a review of contaminant prevalence, source and mobilisation in India. Environ Geochem Health 45:5481–5513
Ramírez-Moreno FJ, Gómez-Oliván LM, García-Medina S et al (2023) Oxidative stress as regulator of neuronal impairment after exposure to hospital effluents in Danio rerio. Sci Total Environ 893:164906. https://doi.org/10.1016/j.scitotenv.2023.164906
Rohani MF (2023) Pesticides toxicity in fish: histopathological and hemato-biochemical aspects—a review. Emerg Contam 9:100234
Rosen M, Farsi A (2022) Sustainable energy technologies for seawater desalination

Ryu HD, Han H, Park JH, Kim YS (2022) New insights into the occurrence and removal of 36 pesticides in pesticide wastewater treatment plants in Korea. Chemosphere 309:136717. https://doi.org/10.1016/j.chemosphere.2022.136717

Saji VS, Meroufel AA, Sorour AA (2020) Corrosion and fouling control in desalination industry

Samal K, Mahapatra S, Hibzur Ali M (2022) Pharmaceutical wastewater as emerging contaminants (EC): treatment technologies, impact on environment and human health. Energy Nexus 6:100076

Schäfer AI, Fane AG (2021) Nanofiltration: principles, applications, and new materials, vols 1 and 2

Sharma R, Agrawal PR, Kumar R, et al (2021) Current scenario of heavy metal contamination in water. In: Contamination of water: health risk assessment and treatment strategies

Singh A, Singh G, Singh P, Mishra VK (2023) Overview of sources, fate, and impact of endocrine disrupting compounds in environment and assessment of their regulatory policies across different continents. Total Environ Res Themes 7:100071. https://doi.org/10.1016/j.totert.2023.100071

Sivaranjanee R, Kumar PS (2021) A review on remedial measures for effective separation of emerging contaminants from wastewater. Environ Technol Innov 23:101741. https://doi.org/10.1016/j.eti.2021.101741

Vaudreuil MA, Vo Duy S, Munoz G, Sauvé S (2022) Pharmaceutical pollution of hospital effluents and municipal wastewaters of Eastern Canada. Sci Total Environ 846:157353. https://doi.org/10.1016/j.scitotenv.2022.157353

Verlicchi P, Galletti A, Petrovic M, BarcelÓ D (2010) Hospital effluents as a source of emerging pollutants: an overview of micropollutants and sustainable treatment options. J Hydrol 389:416–428

Verlicchi P, Al Aukidy M, Galletti A et al (2012) Hospital effluent: investigation of the concentrations and distribution of pharmaceuticals and environmental risk assessment. Sci Total Environ 430:109–118. https://doi.org/10.1016/j.scitotenv.2012.04.055

Wang M, Zhang D, Dong J, Tan SK (2018) Application of constructed wetlands for treating agricultural runoff and agro-industrial wastewater: a review. Hydrobiologia 805:1–31

Wang CX, Qian Y, Zhang ZJ et al (2023) Current status and prospects of research on 1,4-dioxane pollution and treatment technologies in the water environment. J Groundw Sci Eng 11:158–170. https://doi.org/10.26599/JGSE.2023.9280014

Warren N, Allan IJ, Carter JE et al (2003) Pesticides and other micro-organic contaminants in freshwater sedimentary environments—a review. Appl Geochem 18:159–194

Wilkinson JL, Hooda PS, Barker J et al (2016) Ecotoxic pharmaceuticals, personal care products, and other emerging contaminants: a review of environmental, receptor-mediated, developmental, and epigenetic toxicity with discussion of proposed toxicity to humans. Crit Rev Environ Sci Technol 46:336–381

Wilkinson J, Hooda PS, Barker J et al (2017) Occurrence, fate and transformation of emerging contaminants in water: an overarching review of the field. Environ Pollut 231:954–970

Zakari-Jiya A, Frazzoli C, Obasi CN et al (2022) Pharmaceutical and personal care products as emerging environmental contaminants in Nigeria: a systematic review. Environ Toxicol Pharmacol 94:103914

Zeman LJ, Zydney AL (2017) Microfiltration and ultrafiltration: principles and applications

Zhang C, Li Y, Wang C et al (2016) Occurrence of endocrine disrupting compounds in aqueous environment and their bacterial degradation: a review. Crit Rev Environ Sci Technol 46:1061881. https://doi.org/10.1080/10643389.2015.1061881

Zhang S, Gitungo S, Dyksen JE et al (2021) Indicator compounds representative of contaminants of emerging concern (CECs) found in the water cycle in the United States. Int J Environ Res Public Health 18:1288

Zheng LL, Zhang J, Liu XZ et al (2022) Degradation of pesticide wastewater with simultaneous resource recovery via ozonation coupled with anaerobic biochemical technology. Chemosphere 300:134520. https://doi.org/10.1016/j.chemosphere.2022.134520

Zhou T, Zhang Z, Liu H et al (2023) A review on microalgae-mediated biotechnology for removing pharmaceutical contaminants in aqueous environments: occurrence, fate, and removal mechanism. J Hazard Mater 443:130213

Chapter 17
Smart Wastewater Management Techniques: The Role of Artificial Intelligence and Machine Learning

Siddharth Swami and Surindra Suthar

Abstract Water is a vital resource for various sectors, but poor management can lead to higher operational costs and regulatory risks. The global water shortage is currently affecting over 2 billion people, and this shortage has increased demand for effective water management. Industries are developing innovative methods for waste treatment, including biological treatment and water recycling. Artificial intelligence (AI) models are being used to optimize the dosage of photocatalysts for wastewater treatment, achieving a 30–40% reduction in water use and higher efficiency in pollutant removal. AI systems can also predict potential environmental damage and hazards, allowing industries to take preventive measures. Smart sensors and Internet of Things (IoT) devices are improving real-time tracking of water quality and usage across manufacturing processes. This chapter explores the latest advancements in wastewater management through advanced technologies such as deep learning, machine learning, and the Internet of Things (IoT). This technological integration results in enhanced and efficient wastewater management systems.

Keywords Wastewater · Artificial intelligence · Efficiency · Machine learning · Automation · Technology

17.1 Introduction

The availability of an important natural resource in the form of water directly impacts the basic functionality of every sector. According to the United Nations Environment Programme (UNEP) annual report, approximately 70% of the freshwater is utilized in industries, which points to this sector's dependency on freshwater (Zhao et al. 2020). Different manufacturing activities like oil and gas extraction, agricultural activities, mining, and power generation are all reliant on both water quantity and

S. Swami · S. Suthar (✉)
School of Environment and Natural Resources, Doon University, Dehradun, Uttarakhand, India
e-mail: ssuthar@doonuniversity.ac.in

A. M. Yatoo et al. (eds.), *Wastewater Monitoring and Management by Advanced Approaches*, Advances in Wastewater Research,
https://doi.org/10.1007/978-981-95-2601-7_17

water quality. Poor water management in all its various forms could also expose the organization to higher operational costs and regulatory risks, which would be detrimental to company profits (Wang et al. 2023). This situation is exacerbated by the fact that there is increased demand for effective water management due to the global shortage of water supply, which, according to the World Bank, already affecting over 2 billion people (Wang et al. 2023; Baskar et al. 2024). There are practices that, if adopted by industries, would help to address these concerns. Among these issues is the disposal of wastewater (framework represented in Fig. 17.1), which, if not properly done, can have grave environmental impacts. The release of industrial effluents without treatment can lead to or contribute to the pollution of surface water, kill aquatic organisms, and even be ingested by humans through the food chain. It has also been illustrated by the World Economic Forum that pollution of water has a direct link to the destruction of biological diversity necessary for the stability of ecosystems (Manu and Thalla 2017). In recent years, these problems have made some industries come up with more innovative and progressive methods of treating waste. Chemically, physically, and biologically, such measures have also been put in place to ensure that water is reused and environmental vices are minimized. Some of these methods include: biological treatment methods such as the activated sludge processes, ensuring about 90% removal of biochemical oxygen demand (BOD) and suspended solids as well (Zhang et al. 2023). Water recycling in industrial processes can reduce the amount of water withdrawals by even 50%, and this has been supported by the International Energy Agency (IEA) (Baarimah et al. 2024; Alprol et al. 2024). It is also worth mentioning that policies and regulations play a fundamental role in the effective management of water resources. Integrated water resource management (IWRM) is promoted by the United Nations Environment Programme, which can harness stakeholder participation for effective management of water resources (Benson et al. 2020; Cairone et al. 2024). It would be necessary to take into account the European Union's Water Framework Directive, which allows for the preservation of water sources whilst encouraging a circular economy within the manufacturing sector (Abdallah et al. 2020). The application of artificial intelligence (AI) in dealing with new technologies is set to change the operations of industrial sectors that rely heavily on water. This synergy can assist with real-time assessments and individual-centered options, thereby improving productivity and engaging stakeholders (Deepnarain et al. 2020). The use of artificial intelligence in innovations and technologies is gradually becoming an important component in resource management, especially now that emerging startups are making great efforts toward reducing water usage and treatment. There is a rising global concern about water shortage. The United Nations in 2015 projected that by the year 2025, about two-thirds of the global population will be living in a water-stressed environment (Singh et al. 2023). Industries such as textiles, which have extensive dyeing, printing, and finishing processes, are faced with resource constraint problems.

According to the report, even a single plain cotton t-shirt requires over 2500 L of water for its production, and the need for better methods is very high. To prevent this, industries are now adopting water-saving techniques (Baarimah et al. 2024; Zhang et al. 2023). The application of AI models such as radial basis function

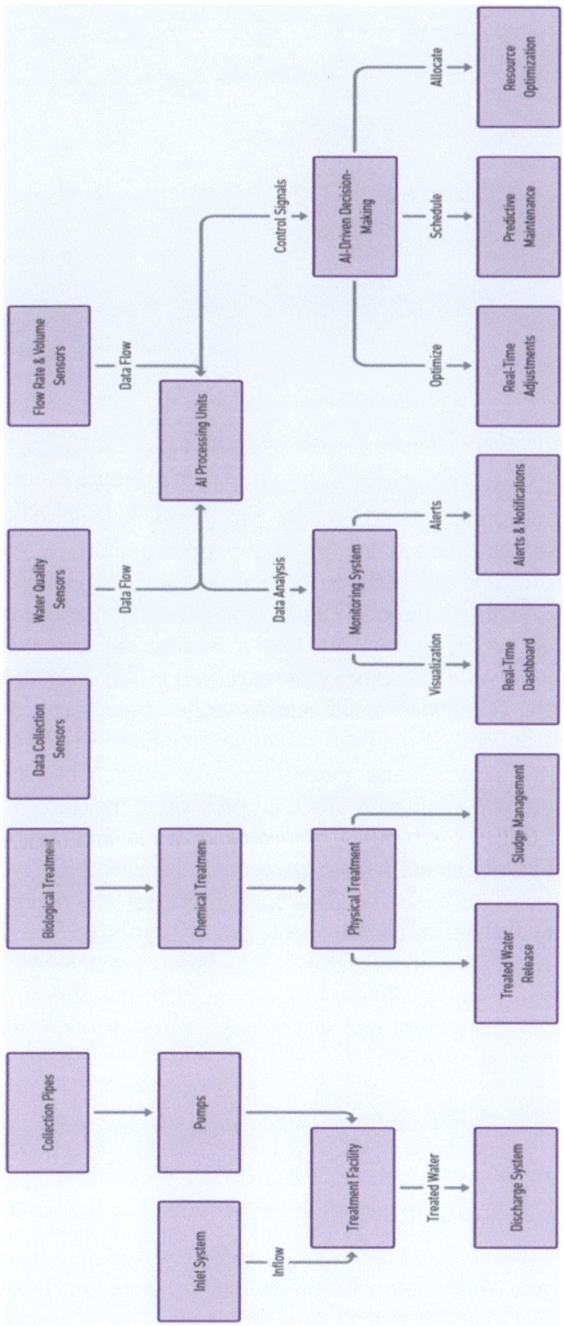

Fig. 17.1 Wastewater management framework through data analytics by integrating data-driven wastewater management

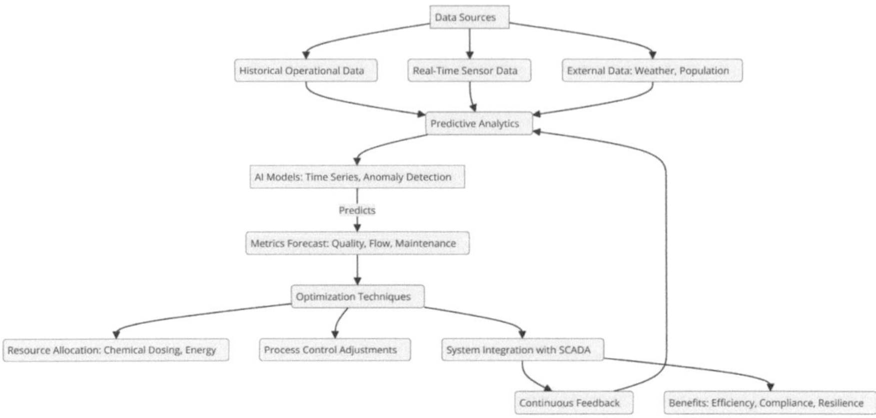

Fig. 17.2 Predictive analytics to apply optimization techniques

(RBF) artificial neural networks (ANN), and adaptive neuro-fuzzy inference systems (ANFIS) for photocatalyst dosage optimization further illustrates the contribution of AI in wastewater treatment (Abdallah et al. 2020; Aghdam et al. 2023). These AI-based models can determine the best appropriate dosages of chemicals that could be required in water treatment, thereby cutting down the amount of water and cost required in the operation. Attaining a 30–40% reduction in water use and higher efficiency in pollutant removal are some of the expected results from the application of RBF ANN in textile wastewater treatment (Sharghi et al. 2019). It is not only in water management where AI operates effective predictive strategies (as shown in Fig. 17.2), but AI systems can also predict potential damage and hazards to the environment, and therefore, industries can take preventive measures (Gulati et al. 2024). AI applications can view archives of water indicators, including their quality and amount used in a particular activity, to estimate times when there would be high or low water use (Awolusi et al. 2016). This sort of strategy and planning not only protects and promotes the health of the people but also goes further in protecting the environment by preventing the excessive extraction of water resources. Smart sensors and Internet of Things (IoT) devices are necessarily going to improve the real-time tracking of water quality and water usage across several manufacturing processes (Pérez-Beltrán et al. 2024).

17.2 Wastewater Management Through Effective Advanced Technological Interventions

The International Water Association estimates that the deployment of intelligent water management systems would lead to a 50% reduction in nonrevenue water losses and enhance the operational efficiency of industrial implementations. Various industries, such as paper industries, chemical industries, and the renewable energy generation industry, all share the common concern of industrial water consumption (Aghdam et al. 2023; Li et al. 2024). The paper industry alone uses about 4% of all global withdrawals of fresh water due to the requirements for such great quantities for pulping and bleaching (Chang et al. 2024). Many water-intensive biological reactions are followed by solvent extractions during the production of pharmaceuticals. Some estimates go as high as an astonishing 5000 L of water for every kilogram of the drug produced during the manufacturing processes. Water is essential in the manufacturing process of solar panels as it is used for the washing of materials, cooling, and refining silicon (Ganthavee and Trzcinski 2024; Wang et al. 2023). The International Energy Agency estimates that the solar photovoltaic (PV) sector can use approximately 1500–3000 L of water for every megawatt-hour (MWh) of electricity produced, with the variation depending on the cooling approaches and technology applied (Li et al. 2024; Singh et al. 2023; Ganthavee and Trzcinski 2024).

Some industrial activities and industries do not strictly adhere to environmental policies, which increases the challenges related to the contamination of water resources. A large proportion of this pollution comes from the failure of industries to properly treat wastewater before it is discharged (Blumensaat et al. 2019).

In dealing with these scenarios, some researchers used predictive technologies such as the boosted regression tree (BRT) method, a supervised machine learning approach (process represented in Fig. 17.3) that enables prediction and classification of water quality parameters with ease. It has been proven that BRT improves the correct attribution of the causes of pollution, and hence the accurate definition of management activities (Bahramian et al. 2023).

17.2.1 Water Pollution and Wastewater Treatment Facilities

Parameters such as composition or flow levels of a particular type of waste may be of varying significance during tail-end processing, which in many industries is termed industrial waste treatment, but there are specific qualities of wastewater treatment facilities (WWTPs) that invalidate this. Major areas of concern in WWTPS are: chemical composition variability, inflow rate variability, lack of performance data of equipment, skilled manpower requirement, and lack of computer-based sophisticated mechanisms for operational improvement (Sahu et al. 2023). On the other hand, manufacturing processes treat effluents within specific water quality parameters to maintain efficiency in operations, like steady state systems; WWTPs need to respond appropriately all the time to the provided inflow parameters as they vary all the time due to multiple factors (Fang et al. 2023; Blumensaat et al. 2019). This makes it complex since operators are constantly expected to make control changes to achieve target parameters. Such challenges encountered due to organic load, contaminants, and other wastewater characteristics require adequate management to avoid undue difficulties. A free stream turbulence likely flows to the sequester site in the WWTP in a wide variety of ranges; there are references to the values shifting from merely

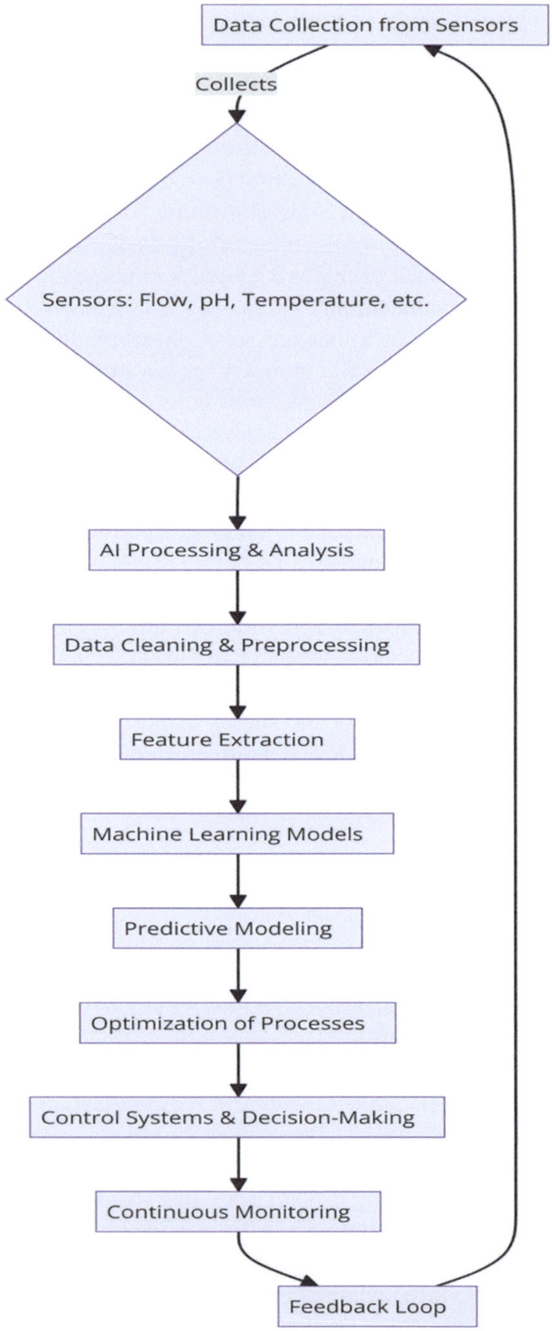

Fig. 17.3 Machine learning process through data processing obtained from sensors

10 to 500 m^3/h because of the storm surge or torrential rain at peak level (Mahmoud et al. 2020; Gulati et al. 2024; Saddiqi et al. 2023). Such variations and much higher values compel an operational range that is entirely different from that used in industries, with a clear distinction in the level of input noise. WWTPs are designed with other harsh seasons in mind, especially during the midsummer months, and are characterized by high flows and temperatures that may seemingly increase operational stress. This raises concerns for designing capacity for WWTPs since they need to consider average flow and also peak conditions that may be experienced during dry weather (Saddiqi et al. 2023; Zhang et al. 2023; Mahmoud et al. 2023). At such times, high solid concentrations can lead to decreased treatment efficiencies and increased treatment and operational costs. The other complex phenomenon is that of biological nature in wastewater treatment which is particularly in the activated sludge systems, where relationships of microorganisms' language complexity regulate the treatment effectiveness (Fu et al. 2023; Bahramian et al. 2023). Developing treatment process models is complicated by variability in microbial populations, as well as variability in the condition of the influent coming to the treatment system. Existing models often do not accurately model microbial dynamics, resulting in uncertainty in treatment efficiency and stability (Elsayed et al. 2024). These dynamic conditions also pose maintenance and operational issues. The necessity for continued adaptation to meet the demands of varying influent characteristics increases the operational and mechanical costs and difficulty in controlling chemical and energy demand (Benson et al. 2020; Senthil Rathi et al. 2024). Dissatisfactory management of such variations may lead to ineffective operations, increased operational costs, or create risks of noncompliance with regulatory requirements. Variations in nitrogen, phosphorus, and organic compounds and their concentrations in the wastewater, which need treatment processes to be adapted to suit them differently, can significantly influence the effectiveness of the treatment (Karam et al. 2020). For example, variations in nutrient inputs may result in difficulties in the operations needed to meet discharge permit conditions that are restrictive in the level of nutrient pollution that can be discharged. There is a lack of information on WWTP performance, the condition of its equipment, and the level of advancement of its technologies. Such ignorance can be detrimental to maintenance and likely causes wastage (Granata et al. 2017; Alhajeri et al. 2024). An estimated 40% of existing WWTPs are equipped with obsolete technologies, leading to higher operational costs, as well as lower treatment efficiency. Qualified operators constitute a further constraint. Dealing with complicated and often complex treatment processes and equipment, they are required to be skilled. Also the joint statement, the National Environmental Training Centre and Utility Workers Union states that to address the changing conditions of wastewater, a skilled active operator is required for effective treatment operations (Fang et al. 2023; Sahu et al. 2023; Nasir and Li 2024). Wastewater treatment plants (WWTPs) have begun to adopt new technologies such as artificial neural networks and expert systems. The first progress in the development of aid expert systems was a significant milestone in the implementation of AI in wastewater management. These systems focus on developing dynamic models of WWTPs that can emulate various treatment processes (Karam et al. 2020;

Essahlaoui et al. 2024). Over the past years, there has been progress in the understanding of a complete mixed-activated sludge system, leading to the development of a set of mathematical equations representing all processes taking place within the system. These models usually include components such as primary sedimentation tanks, aeration tanks, and secondary clarifiers. In practice, the performance predictions generated by these AI systems can be verified with real-life operational data, which may be useful for improving decision-making and treatment processes (Sen et al. 2018; Nasir and Li 2024). Such models can result in savings in energy use and chemical consumption of up to 15%, thereby increasing operational efficiency. One of the most appealing responses to integrated food waste (FW) leachate in WWTPs is regarded as a method of improving the efficiency of water treatment (Granata et al. 2017; Ding et al. 2024). Several investigations show that the acid fermentation liquid (AFL) concentration in FW leachate could be considered high and could act as an appropriate organic carbon supplement in advanced wastewater treatment processes for nitrogen and phosphorus removal. Research indicates that sewage effluent is usually characterized by very low carbon-to-nitrogen C:N ratios, which typically range between 5:1 and 15:1 ratios, and such conditions impede nutrient retrieval effectively (Qi et al. 2020; Benson et al. 2020). The case of leachate from food waste has been found to have C: N ratio, in most cases more than 25:1. This offers a better substrate for microorganisms during co-digestion of different materials. Co-digestion of FW with sewage sludge not only enhances nutrient retrieval but also boosts biogas production (Kumari 2024). One study noted that using anaerobic processes in conjunction with FW leachates could raise methane generation by as much as 30%, greatly enhancing the energy recovery potential of wastewater treatment facilities. This upsurge is because leachate contains more biodegradable organic materials, which enable microorganisms to work more efficiently in the digestion phase. The challenges posed by water quality forecasting have been a topic of immense focus in the last few decades, leading to the establishment of advanced control systems (Zhang et al. 2024; Alprol et al. 2024).

One such instrument is the water quality index (WQI), which incorporates a variety of factors that affect the water, such as turbidity, total suspended solid (TSS), biochemical oxygen demand (BOD), dissolved oxygen (DO), temperature, pH, calcium, and chemical oxygen demand (COD), thus allowing a multi-indicator assessment of overall water quality under single index (Ye et al. 2020; Cairone et al. 2024). WQI's effective coping can assist in developing better monitoring and early warning systems in developing urban areas, which are under development and ecological pressure. Some problems arise due to the presence of uncertainty in water quality data, caused by human activities and also variability in the environment (Sakti et al. 2023; Wani et al. 2024). However, AI models are becoming a feasible option, using machine learning techniques to manage artificial intelligence models, which could tackle the nonlinear and intricate structure of water quality data. AI models could predict the parameters of water quality at an accuracy level of more than 90%, which would help in faster decision-making processes (Goralski and Tan 2020). The use of different artificial intelligence (AI) models for the prediction and management of water quality indicators (as represented in Fig. 17.4) has been a subject of interest in recent works. Multiple linear regression (MLR), radial basis function (RBF), particle swarm optimization (PSO), artificial neural networks genetic algorithm (ANNGA), and support vector machines (SVM), among others, are effective. Nonlinear relationships that are common in complex datasets can also be analyzed to improve water quality evaluations (Ismail et al. 2024). Neural networks (NNs) have become the main tool used to monitor surface water quality. Artificial neural networks (ANN) and multiple linear regression (MLR) techniques were applied to estimate groundwater quality in the Shivganga River basin. For instance, in 2015, samples of 34 groundwater points were taken during the monsoon and premonsoon seasons from various locations. The model proposed was able to accurately reproduce the behavior of the Levenberg–Marquardt three-layer back propagation algorithm (BPA) in ANN model construction, so that water quality index predictions made using the model were consistent (Yusoff et al. 2024; Ghosh et al. 2024). In addition to groundwater quality estimation, hazardous dye treatment with ANN was reported as well. A study is reported in which the author overcame the challenges associated with the implementation of ANN–RSM optimization for dye removal sequences from wastewater (Nguyen et al. 2024). This work highlighted ANN's ability to understand and represent complex and nonlinear processes arising due to interactions between different treatment parameter combinations, resulting in higher removal efficiencies (Ding et al. 2024). More so, there has been the development and integration of advanced techniques that predict the water quality of hybrid models. The wavelet denoising technique combined with the adaptive neuro-fuzzy inference system (WDT-ANFIS) showed remarkable accuracy levels compared to other models. Various studies revealed that the WDT-ANFIS model managed to obtain R^2 values equal to, or higher than, 0.9, concerning several water quality parameters (Entezari et al. 2023). These results show great potential for using wavelet transforms combined with neuro-fuzzy approaches to water quality control and forecasting, therefore enhancing decisiveness in the management of water resources. It has been reported that AI-based ensemble methods have proven to be effective when modelling DO levels in the future and using both one-step and multistep approaches (Wani et al. 2024; Alvi et al. 2023). One of the remarkable advantages of neural network ensemble (NNE) is its problem of asymmetrical cycles, hence it can be useful in the multistep modelling of dissolved oxygen. Auto-regressive integrated moving average (ARIMA) is a conventional statistical technique that was used with the support of three models: ANFIS, SVM, and back propagation neural network (BPNN) (Sakti et al. 2023). The goal of utilizing such different methodologies was to improve the forecasting abilities of the separate models. Conservative ensemble techniques such as weighted average NNE (WAE) and simple average ensemble (SAE) were also used to improve prediction performance (Chen and Chen 2020). The results of the application of these ensembles showed considerable gains in model performance (Alhajeri et al. 2024). In the calibration phase, this approach demonstrated that the use of NNE could improve the average performance of the standard models by up to 14%, which was evidence of the success of the strategy of merging models to utilize the advantages and compensate for the weaknesses (Al Aani et al. 2019). Besides the prediction of DO, even waste effluent quality indicators such as BOD, COD, and TN have been targeted by applying ensemble modelling. The methods used were multivariate modelling, including Feedforward neural network, ANFIS, SVM, and MLR techniques. To this end, simple ensemble strategies were utilized, such as WAE, SAE, and NNE, which aimed to improve the predictive performance of these models (Huang et al. 2024). Although all ensemble methods delivered strong results, NNE was the most accurate method among the WAE and SAE. This shows how NNE can improve the prediction of turbulent water quality parameters and hence assist in better water resource management and water quality regulations since these parameters are often complex (Bisaria et al. 2024). The implementation of AI-based integrated model approaches in predicting water quality illustrates the power of advanced modelling techniques' ability to enhance the quality of forecasting, hence being useful to the management of water quality in various environments. The successful use of science-based models in controlling and optimizing wastewater treatment processes depends on the availability of reliable data (Najafzadeh and Zeinolabedini 2019). The changing conditions in wastewater treatment, coupled with IoT sensor networks, often result in misrepresented or erroneous data on several occasions. For example, research shows that environmental issues may damage as much as 30% of the data collected by IoT sensors, thus preventing further obstacles to treatment effectiveness through already high data voids (How et al. 2020).

17.3 Optimization of Wastewater Management with Enhanced Efficiency by Artificial Intelligence

There has been a growing body of literature that has sought to understand the effect of information loss concerning the calibration of WWTP models. It is in such regions, where turbidity levels often surpass 100 nephelometric turbidity units (NTU), that coagulation processes prove most effective as they can reduce these levels to less than 5 NTU in the effluent (Kumari 2024; Lim et al. 2023; Sengorur et al. 2015). The issues of data availability and quality must be resolved to ensure the seamless adoption of scientific models during the treatment of wastewater. With the help of modern analytical methods and effective coagulation processes, WWTPs can improve operational efficiency while at the same time increasing the chances for a more accurate assessment of the water quality and enhancing environmental protection (Qi et al. 2020; Alsabt et al. 2024). Flocculation is a vital step in the process of wastewater treatment, where tiny agglomerating particles known as flocs are formed from the accumulation of colloidal particles. These flocs assist in contaminant removal through sedimentation, thus enhancing the propellant's efficiency, especially methylparaben, to about 85%. Other recent studies have looked for new approaches to remove toxic heavy metals, some of them using new technologies. In an experiment, this methodology involved both sonication and a custom-designed nanocomposite material to extract heavy metals from water (Bagheri et al. 2023). One of the ways of recovering metals, including lead and cadmium, with an efficiency of more than 90%, proved to be a

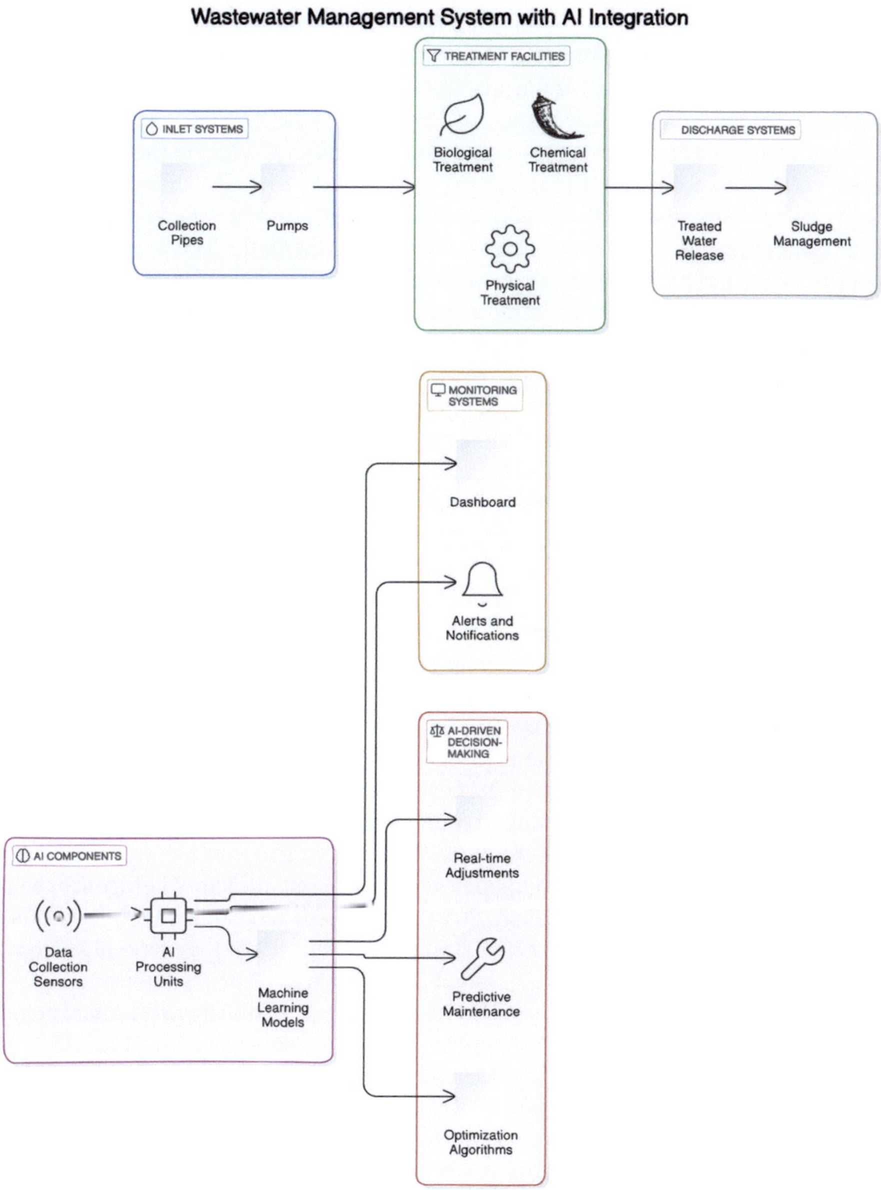

Fig. 17.4 Technological integration in the wastewater management process

successful combination of sonication technology and synthetic nanocomposite materials. Highly sophisticated optimization combinations have also been carried out to produce specific adsorbents for the specific color combinations that are complex in wastewater (Duarte et al. 2023; Manu and Thalla 2017). Research studies have shown the potential of a new biosorbent sourced from agricultural waste in the treatment of wastewater, with color intensity reduction to more than 70%, which implies its suitability as an environmentally friendly option in color removal. There is, however, a significant impact of wastewater treatment technologies on the heavy metal removal practices appropriate for sustainable water management due to the development of new and advanced flocculation removal techniques for contaminants (Rezk et al. 2023a, b).

The application of artificial neural networks (ANN) and response surface methodology (RSM) is quite effective in the optimization of water treatment processes. RSM can model relationships between numerous factors, whereas ANN specializes in attributes that are complex and nonlinear and coexist with these factors (Lim et al. 2023; Singh et al. 2023). This allows a more comprehensive view of how various parameters work together to influence processes such as treatment efficiency. The laboratory jar test is one of the most frequently employed techniques to estimate the optimum coagulant dose for treatment processes. This remains an inconvenient method since it is involved in the typical and traditional processes that are lengthy and use many chemicals (Alprol et al. 2024; Geetha et al. 2023; Zhao et al. 2020). The jar test requires tests of processes such as dissolved oxygen (DO), turbidity, conductivity, pH, temperature measurements, and many more, and such parameters can vary highly between samples of raw water. One of the problems is that the values of turbidity range from 1 NTU (Nephelometric Turbidity Units) in clean water to more than 100 NTU in polluted sources, which affects the coagulant dose that needs to be used for the treatment to work. To confront these problems, the studies' authors have developed predictive models using different artificial intelligence techniques (Baskar et al. 2024; Guo et al. 2024). It includes techniques such as particle swarm optimization (PSO), extreme learning machine (ELM), genetic algorithms (GA), multilayer perceptron (MLP), radial basis function (RBF) networks, adaptive ANFIS, and self-organizing maps (SOM) for an accurate estimate of the ideal coagulant dose at any point in water treatment module as shown in Fig. 17.5. Recent studies suggest that models such as general regression neural networks (GRNN), ANFIS, and multilayer perceptron (MLP) can greatly enhance the doers' ability to ascertain the suitable coagulant dosage (Rezk et al. 2023a, b; Shah et al. 2024). A case study, for instance, revealed that ANN models for optimal chemical dose prediction could increase current accuracy by more than 30%, enabling a more economical use of chemicals while limiting the costs of operations. Such shifts in AI modelling fast-track the determination of coagulant dosage in a chemical and also enable better treatment effects (Morain et al. 2024) (Table 17.1).

Water treatment facilities can operate more efficiently and effectively in the removal of contaminants as they do not have to depend on laboratory tests, which are time-consuming when there are changes in water quality. The combination of ANN with RSM and AI tools forms a comprehensive model for the optimal dosages

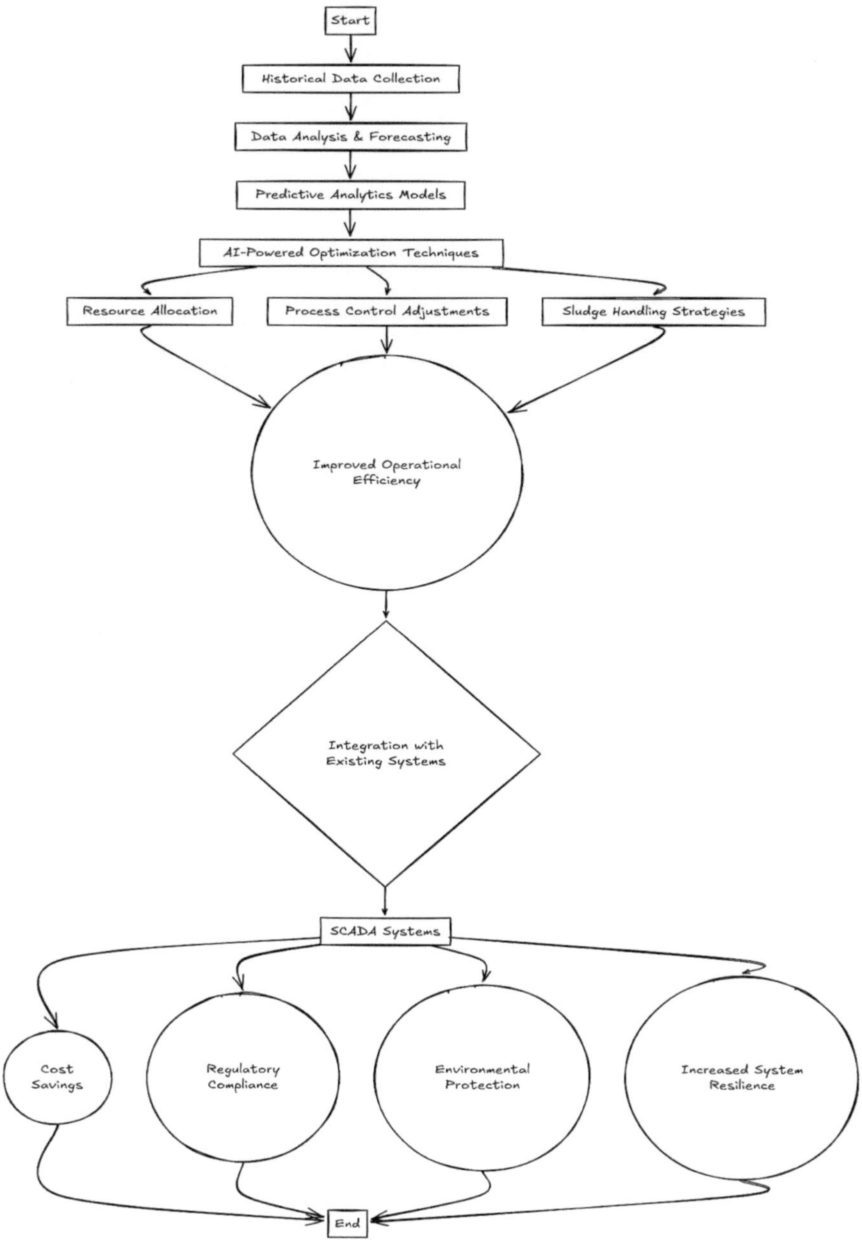

Fig. 17.5 Data-based decision-making system to obtain improved operational efficiency

Table 17.1 Requirements and advantages of different advanced technologies used in wastewater management

Technology	Specifications	Input variables	Advantages	Challenges	References
Internet of Things (IoT)	Smart meters, flow sensors, and water-quality sensors Real-time streaming, Latency, system uptime, Packet loss, and stable network	Turbidity, temperature, real-time pH, and flow rate	Provides real-time monitoring and control with enhanced system responsiveness and efficiency	Security risks with data loss and maintenance challenges for various sensor nodes	Wani et al. (2024), Sharghi et al. (2019), Bagheri et al. (2023)
Artificial intelligence (AI)	Particle swarm optimization (PSO), principal component analysis (PCA), convolutional neural network (CNN), and feedforward neural network (FNN) Time series, image, extensive datasets, mean square error (MSE), R^2, and precision	Contaminant concentration, conductivity, temperature, turbidity, and pH	Contamination detection with high precision, adaptability to real-time data, with improved responses and control	The requirement of detailed large datasets for model training to achieve the required accuracy and high computational cost	Manu and Thalla (2017), Rezk et al. (2023a, b)
Machine learning (ML)	Support vector machine (SVM), K-nearest neighbor (KNN), Decision tree, and random forest Structured, clean and labelled data, accuracy, F1 score, receiver operating characteristic area under the curve (ROC-AUC)	Nutrient levels, dissolved oxygen, flow rate, and pH	Effective for regression and classification in treatment and contamination effectiveness	The technology is limited by data quality and volume, hyperparameter tuning is resource-intensive	Baskar et al. (2024)
Deep learning (DL)	Long- short-term memory (LSTM), recurrent neural network (RNN), and generative adversarial network (GAN) High-quality sequential data, time series, root mean squared error (RMSE), and accuracy	Seasonal changes, historic flow data, and pollutant levels	Useful for anomaly detection and predictive maintenance, best performance for time-series modelling	Requirement of extensive training data and large resources, process prone to overfitting	Senthil Rathi et al. (2024), Aghdam et al. (2023), Qi et al. (2020)

(continued)

Table 17.1 (continued)

Technology	Specifications	Input variables	Advantages	Challenges	References
Sensors	pH sensors, turbidity sensors, and conductivity sensors Analogue and digital, regular calibration, sensitivity, accuracy, and response time	Turbidity, conductivity, pH, and temperature	Accuracy is ensured by obtaining real-time data in dosing and operational adjustments	Requirement of frequent calibration, variations due to environmental factors, and high installation cost	Zhao et al. (2020), Singh et al. (2023)
Data processing techniques	Principal component analysis (PCA), normalization, and feature selection Clean labelled data, variance explained, and execution time	Raw data from sensors (e.g., pH, turbidity, conductivity, and temperature)	Enhanced model performance, reduced data complexity, improved computational efficiency	The danger of data loss is a time-consuming process with large datasets because of data processing complexities	Awolusi et al. (2016), Alprol et al. (2024)
Cloud computing	Cloud processing, data storage Structured and semi-structured, stable internet, latency, and data transfer speed	Aggregated data from all the IoT devices and sensor outputs	Remote accessibility, high storage capacity, and scalability for growing data requirements	Uninterrupted connectivity, necessity, security, and data privacy issues	Alvi et al. (2023), Kumari (2024)
Predictive maintenance	Anomaly detection, predictive machine learning models Time series, historical data, mean time to failure (MTTF)	Energy usage, flow rate, and equipment status	Extending equipment lifespan, reducing downtime, and early fault detection	Requires consistent data input, complex in legacy systems	Bagheri et al. (2023), Najafzadeh and Zeinolabedini (2019)

of coagulants in water treatment and is more efficient and sustainable (Alvi et al. 2023; Ngulube et al. 2024). Multilayer perceptron neural networks (MLP NNs) are one of the most widely used methods for the estimation of the optimal coagulant dose in water treatment processes. A study applied the data-driven method in the plant assessing the effectiveness of coagulation-flocculation with the purpose of copper (Cu) removal and the volume of sludge produced (Senthil Rathi et al. 2024; Wani et al. 2024; Cairone et al. 2024). Genetic algorithms (GA) performed better than particle swarm optimization (PSO) in terms of higher compliance rates for Cu removal and sludge level. These results show that certain optimization methods should accompany the neural networks to improve the treatment process. In the case of another application, an ANN-MLP model was trained to determine the efficiency of using pumice, zeolite, and sepiolite adsorbents to treat wastewater resulting from andesite marble production (Awolusi et al. 2016; Oruganti et al. 2023). The study has proven that, despite MLP NNs being beneficial, the extreme learning machine (ELM) in conjunction with radial basis function (RBF) neural networks was a more effective architectural model for such an application. The ELM-RBF model has not only shown better prediction accuracy but also less computational time, thus demonstrating the advantages of hybrid modelling in complex wastewater situations (Mafat et al. 2024; Karam et al. 2020). It is possible to put a number on the effectiveness of those models of artificial intelligence; for instance, research indicates that effective training of MLP models makes it possible to predict the MP of various and up to 85% accuracies of several water quality parameters and remain within the set ranges. Such a depth of accuracy is imperative for some industries that make use of such water to not incur high penalties and interruptions in operations due to pollution (Shah et al. 2024). The implementation of these models can reduce the dosing of coagulants necessary, and thus, fewer chemicals will be used in treatments while simultaneously achieving better results. This will help facilities avoid pollution by correctly estimating the amounts of coagulants that are required for each process, thus reducing waste (Alzahrani et al. 2023; Bagheri et al. 2023).

17.4 Conclusion and Future Prospects

The integration of advanced technologies in wastewater treatment reveals a significant advantage in optimizing coagulant dosing and enhancing water quality predictions. Models such as multilayer perceptron (MLP) and response surface methodology (RSM) demonstrate the potential for capturing nonlinear and complex relationships in treatment processes, resulting in improved accuracy and efficiency of the output. The study concludes that hybrid approaches, particularly those combining extreme learning machine (ELM) with radial basis function (RBF) networks, provide better results as compared to traditional methods and hence present a robust framework for addressing the challenge of wastewater management. Artificial intelligence-driven solutions streamline processes like regulation and improve environmental compliance, ecological sustainability, and public health scenarios. There remains a need for exploration of machine learning algorithms for predictive maintenance and IoT technologies for real-time monitoring. These innovative approaches are essential for developing sustainable and effective wastewater treatment practices, aligning with global efforts to protect water resources and promote environmental sustainability.

References

Abdallah M, Talib MA, Feroz S, Nasir Q, Abdalla H, Mahfood B (2020) Artificial intelligence applications in solid waste management: a systematic research review. Waste Manage 109:231–246

Aghdam E, Mohandes SR, Manu P, Cheung C, Yunusa-Kaltungo A, Zayed T (2023) Predicting quality parameters of wastewater treatment plants using artificial intelligence techniques. J Clean Prod 405:137019

Al Aani S, Bonny T, Hasan SW, Hilal N (2019) Can machine language and artificial intelligence revolutionize process automation for water treatment and desalination? Desalination 458:84–96

Alhajeri NS, Tawfik A, Nasr M, Osman AI (2024) Artificial intelligence-enabled optimization of Fe/Zn@ biochar photocatalyst for 2, 6-dichlorophenol removal from petrochemical wastewater: a techno-economic perspective. Chemosphere 352:141476

Alprol AE, Mansour AT, Ibrahim MEED, Ashour M (2024) Artificial intelligence technologies revolutionizing wastewater treatment: current trends and future prospective. Water 16(2):314

Alsabt R, Alkhaldi W, Adenle YA, Alshuwaikhat HM (2024) Optimizing waste management strategies through artificial intelligence and machine learning-an economic and environmental impact study. Clean Waste Syst 8:100158

Alvi M, Batstone D, Mbamba CK, Keymer P, French T, Ward A et al (2023) Deep learning in wastewater treatment: a critical review. Water Res 245:120518

Alzahrani AI, Chauhdary SH, Alshdadi AA (2023) Internet of things (IoT)-based wastewater management in smart cities. Electronics 12(12):2590

Awolusi OO, Nasr M, Kumari S, Bux F (2016) Artificial intelligence for the evaluation of operational parameters influencing nitrification and nitrifiers in an activated sludge process. Microb Ecol 72:49–63

Baarimah AO, Bazel MA, Alaloul WS, Alazaiza MY, Al-Zghoul TM, Almuhaya B et al (2024) Artificial intelligence in wastewater treatment: research trends and future perspectives through bibliometric analysis. Case Stud Chem Environ Eng 10:100926

Bagheri M, Farshforoush N, Bagheri K, Shemirani AI (2023) Applications of artificial intelligence technologies in water environments: from basic techniques to novel tiny machine learning systems. Process Saf Environ Prot 180:10–22

Bahramian M, Dereli RK, Zhao W, Giberti M, Casey E (2023) Data to intelligence: the role of data-driven models in wastewater treatment. Expert Syst Appl 217:119453

Baskar G, Omer SN, Saravanan P, Rajeshkannan R, Saravanan V, Rajasimman M, Shanmugam V (2024) Status and future trends in wastewater management strategies using artificial intelligence and machine learning techniques. Chemosphere 362:142477

Benson D, Gain AK, Giupponi C (2020) Moving beyond water centricity? Conceptualizing integrated water resources management for implementing sustainable development goals. Sustain Sci 15(2):671–681

Bisaria K, Singh R, Gupta M, Mathur A, Dixit A (2024) Novel acoustic-activated alkali-functionalized Trapa bispinosa peel biochar for green immobilization of chlorpyrifos from wastewater: artificial intelligence modelling and experimental validation. Biomass Convers Biorefinery 14(6):7763–7782

Blumensaat F, Leitão JP, Ort C, Rieckermann J, Scheidegger A, Vanrolleghem PA, Villez K (2019) How urban storm-and wastewater management prepares for emerging opportunities and threats:

digital transformation, ubiquitous sensing, new data sources, and beyond-a horizon scan. Environ Sci Technol 53(15):8488–8498

Cairone S, Hasan SW, Choo KH, Li CW, Zarra T, Belgiorno V, Naddeo V (2024) Integrating artificial intelligence modeling and membrane technologies for advanced wastewater treatment: research progress and future perspectives. Sci Total Environ 362:173999

Chang HL, Liu YL, Keng CJ, Jiang HL, Hu J (2024) Challenges of industrial wastewater treatment: utilizing Membrane bioreactors (MBRs) in conjunction with artificial intelligence (AI) technology. J Ind Prod Eng 41(5):422–427

Chen T, Chen CYJ (2020) Hierarchical intelligence platform designed for wastewater management systems: information technology integration. Desalin Water Treat 183:167–172

Deepnarain N, Nasr M, Kumari S, Stenström TA, Reddy P, Pillay K, Bux F (2020) Artificial intelligence and multivariate statistics for comprehensive assessment of filamentous bacteria in wastewater treatment plants experiencing sludge bulking. Environ Technol Innov 19:100853

Ding Y, Sun Q, Lin Y, Ping Q, Peng N, Wang L, Li Y (2024) Application of artificial intelligence in (waste) water disinfection: emphasizing the regulation of disinfection by-products formation and residues prediction. Water Res 253:121267

Duarte MS, Martins G, Oliveira P, Fernandes B, Ferreira EC, Alves MM et al (2023) A review of computational modeling in wastewater treatment processes. ACS ES&T Water 4(3):784–804

Elsayed A, Ghaith M, Yosri A, Li Z, El-Dakhakhni W (2024) Genetic programming expressions for effluent quality prediction: towards AI-driven monitoring and management of wastewater treatment plants. J Environ Manage 356:120510

Entezari A, Aslani A, Zahedi R, Noorollahi Y (2023) Artificial intelligence and machine learning in energy systems: a bibliographic perspective. Energ Strat Rev 45:101017

Essahlaoui F, Elhajrat N, Halimi M, Azrour M, Kılıç Z, Abbassi AE (2024) Intelligent real-time monitoring system for wastewater management using artificial neural network. In: Sustainable and green technologies for water and environmental management. Springer Nature Switzerland, Cham, pp 15–24

Fang B, Yu J, Chen Z, Osman AI, Farghali M, Ihara I et al (2023) Artificial intelligence for waste management in smart cities: a review. Environ Chem Lett 21(4):1959–1989

Fu G, Sun S, Hoang L, Yuan Z, Butler D (2023) Artificial intelligence underpins urban water infrastructure of the future: a holistic perspective. Camb Prism Water 1:e14

Ganthavee V, Trzcinski AP (2024) Artificial intelligence and machine learning for the optimization of pharmaceutical wastewater treatment systems: a review. Environ Chem Lett 22:1–26

Geetha M, Bonthula S, Al-Maadeed S, Al-Lohedan H, Rajabathar JR, Arokiyaraj S, Sadasivuni KK (2023) Research trends in smart cost-effective water quality monitoring and modeling: special focus on artificial intelligence. Water 15(18):3293

Ghosh D, Adhikary S, Sau S (2024) AI-based smart water quality monitoring and wastewater management: an integrated bio-computational approach. In: Innovations in machine learning and IoT for water management. IGI Global, pp 127–151

Goralski MA, Tan TK (2020) Artificial intelligence and sustainable development. Int J Manag Educ 18(1):100330

Granata F, Papirio S, Esposito G, Gargano R, De Marinis G (2017) Machine learning algorithms for the forecasting of wastewater quality indicators. Water 9(2):105

Gulati S, Chhabra L, Tomar K, Nagvani S, Jacob MK, Samarjeet (2024) Artificial intelligence in wastewater treatment. In: Application of artificial intelligence in wastewater treatment. Springer Nature Switzerland, Cham, pp 85–105

Guo H, Qin Q, Hu M, Chang JS, Lee DJ (2024) Treatment of refinery wastewater: current status and prospects. J Environ Chem Eng 12:112508

How ML, Cheah SM, Chan YJ, Khor AC, Say EMP (2020) Artificial intelligence-enhanced decision support for informing global sustainable development: a human-centric AI-thinking approach. Information 11(1):39

Huang H, Ma R, Ren H (2024) Scientific and technological innovations of wastewater treatment in China. Front Environ Sci Eng 18(6):72

Ismail R, Alsadi J, Hatamleh R, Telfah DA, Jaradat A, Aljamal M et al (2024) Employing CNN and black widow optimization for sustainable wastewater management in an environmental engineering context. Asian J Civ Eng 25:1–16

Karam A, Zaher K, Mahmoud AS (2020) Comparative studies of using nano zerovalent iron, activated carbon, and green synthesized nano zerovalent iron for textile wastewater color removal using artificial intelligence, regression analysis, adsorption isotherm, and kinetic studies. Air Soil Water Res 13:1178622120908273

Kumari S (2024) Exploring emerging technologies for AI-integrated advancements in wastewater treatment: harnessing emerging technologies for AI in wastewater treatment. In: The AI cleanse: transforming wastewater treatment through artificial intelligence: harnessing data-driven solutions, pp 223–244

Li X, Su J, Wang H, Boczkaj G, Mahlknecht J, Singh SV, Wang C (2024) Bibliometric analysis of artificial intelligence in wastewater treatment: Current status, research progress, and future prospects. J Environ Chem Eng 12:113152

Lim YM, Swamy V, Ramakrishnan N, Chan ES, Kesuma HP (2023) Volatile organic compounds (VOCs) in wastewater: recent advances in detection and quantification. Microchem J 195:109537

Mafat IH, Palla S, Surya DV (2024) Machine learning and artificial intelligence for algal cultivation, harvesting techniques, wastewater treatment, nutrient recovery, and biofuel production and optimization. In: Value added products from bioalgae based biorefineries: opportunities and challenges, pp 463–487. Springer Nature Singapore, Singapore

Mahmoud AS, Farag RS, Elshfai MM (2020) Reduction of organic matter from municipal wastewater at low cost using green synthesis nano iron extracted from black tea: artificial intelligence with regression analysis. Egypt J Pet 29(1):9–20

Mahmoud AED, Khan NA, Hung YT (2023) Advances in artificial intelligence applications in sustainable water remediation. In: Artificial intelligence and modeling for water sustainability. CRC Press, pp 53–69

Manu DS, Thalla AK (2017) Artificial intelligence models for predicting the performance of biological wastewater treatment plant in the removal of Kjeldahl Nitrogen from wastewater. Appl Water Sci 7:3783–3791

Morain A, Ilangovan N, Delhom C, Anandhi A (2024) Artificial intelligence for water consumption assessment: state of the art review. Water Resour Manag 38:1–22

Najafzadeh M, Zeinolabedini M (2019) Prognostication of waste water treatment plant performance using efficient soft computing models: an environmental evaluation. Measurement 138:690–701

Nasir FB, Li J (2024) Understanding machine learning predictions of wastewater treatment plant sludge with explainable artificial intelligence. Water Environ Res 96(10):e11136

Ngulube KF, Abdelhaleem A, Fujii M, Nasr M (2024) Synergism of artificial intelligence and techno-economic for sustainable treatment of methylene blue dye-containing wastewater by photocatalysis. Sustainability 16(2):529

Nguyen DV, Park J, Lee H, Han T, Wu D (2024) Assessing industrial wastewater effluent toxicity using boosting algorithms in machine learning: a case study on ecotoxicity prediction and control strategy development. Environ Pollut 341:123017

Oruganti RK, Biji AP, Lanuyanger T, Show PL, Sriariyanun M, Upadhyayula VK et al (2023) Artificial intelligence and machine learning tools for high-performance microalgal wastewater treatment and algal biorefinery: a critical review. Sci Total Environ 876:162797

Pérez-Beltrán CH, Robles AD, Rodriguez NA, Ortega-Gavilán F, Jiménez-Carvelo AM (2024) Artificial intelligence and water quality: from drinking water to wastewater. TrAC Trends Anal Chem 172:117597

Qi J, Hou Y, Hu J, Ruan W, Xiang Y, Wei X (2020) Decontamination of methylene Blue from simulated wastewater by the mesoporous rGO/Fe/Co nanohybrids: artificial intelligence modeling and optimization. Mater Today Commun 24:100709

Rezk H, Olabi AG, Abdelkareem MA, Sayed ET (2023a) Artificial intelligence as a novel tool for enhancing the performance of urine fed microbial fuel cell as an emerging approach for simultaneous power generation and wastewater treatment. J Taiwan Inst Chem Eng 148:104726

Rezk H, Olabi AG, Sayed ET, Alshathri SI, Abdelkareem MA (2023b) Optimized artificial intelligent model to boost the efficiency of saline wastewater treatment based on hunger games search algorithm and ANFIS. Sustainability 15(5):4413

Saddiqi MM, Zhao W, Cotterill S, Dereli RK (2023) Smart management of combined sewer overflows: from an ancient technology to artificial intelligence. Wiley Interdiscip Rev Water 10(3):e1635

Sahu S, Kaur A, Singh G, Arya SK (2023) Harnessing the potential of microalgae-bacteria interaction for eco-friendly wastewater treatment: a review on new strategies involving machine learning and artificial intelligence. J Environ Manage 346:119004

Sakti AD, Mahdani JN, Santoso C, Ihsan KTN, Nastiti A, Shabrina Z et al (2023) Optimizing city-level centralized wastewater management system using machine learning and spatial network analysis. Environ Technol Innov 32:103360

Sen D, Fashokun A, Angelotti R, Brooks M, Bhaumik H, Card C et al (2018) An artificial intelligence platform for asset management contributes to better decision-making tools for operations, maintenance, and utility management. Water Environ Res 90(4):355–375

Sengorur B, Koklu R, Ates A (2015) Water quality assessment using artificial intelligence techniques: SOM and ANN—a case study of Melen River Turkey. Water Qual Expo Health 7:469–490

Senthil Rathi B, Senthil Kumar P, Sanjay S, Prem Kumar M, Rangasamy G (2024) Artificial intelligence integration in conventional wastewater treatment techniques: techno-economic evaluation, recent progress and its future direction. Int J Environ Sci Technol 22:1–26

Shah SMH, Abba SI, Yassin MA, Lawal DU, Aliyu F, Al-Qadami EHH et al (2024) New strategy based on Hammerstein-Wiener and supervised machine learning for identification of treated wastewater salinization in Al-Hassa region, Saudi Arabia. Environ Sci Eur 36(1):114

Sharghi E, Nourani V, AliAshrafi A, Gökçekuş H (2019) Monitoring effluent quality of wastewater treatment plant by clustering based artificial neural network method. Desalin Water Treat 164:86–97

Singh NK, Yadav M, Singh V, Padhiyar H, Kumar V, Bhatia SK, Show PL (2023) Artificial intelligence and machine learning-based monitoring and design of biological wastewater treatment systems. Biores Technol 369:128486

Wang Y, Cheng Y, Liu H, Guo Q, Dai C, Zhao M, Liu D (2023) A review on applications of artificial intelligence in wastewater treatment. Sustainability 15(18):13557

Wani AK, Rahayu F, Ben Amor I, Quadir M, Murianingrum M, Parnidi P et al (2024) Environmental resilience through artificial intelligence: innovations in monitoring and management. Environ Sci Pollut Res 31(12):18379–18395

Ye Z, Yang J, Zhong N, Tu X, Jia J, Wang J (2020) Tackling environmental challenges in pollution controls using artificial intelligence: a review. Sci Total Environ 699:134279

Yusoff NHM, Chew WJ, Chong CH, Wan YK (2024) Artificial intelligence in color classification of 3D-printed enhanced adsorbent in textile wastewater. J Water Process Eng 65:105776

Zhang S, Jin Y, Chen W, Wang J, Wang Y, Ren H (2023) Artificial intelligence in wastewater treatment: a data-driven analysis of status and trends. Chemosphere 336:139163

Zhang J, Yao X, Zhao Y, Li R, Chen X, Jin H et al (2024) Optimizing the UV-Fenton degradation of m-Cresol wastewater: an experimental and artificial intelligence modeling approach. Ind Eng Chem Res 63(2):921–929

Zhao L, Dai T, Qiao Z, Sun P, Hao J, Yang Y (2020) Application of artificial intelligence to wastewater treatment: a bibliometric analysis and systematic review of technology, economy, management, and wastewater reuse. Process Saf Environ Prot 133:169–182

MIX
Papier aus verantwortungsvollen Quellen
Paper from responsible sources
FSC® C105338

If you have any concerns about our products,
you can contact us on
ProductSafety@springernature.com

In case Publisher is established outside the EU,
the EU authorized representative is:
Springer Nature Customer Service Center GmbH
Europaplatz 3, 69115 Heidelberg, Germany

Printed by Libri Plureos GmbH
in Hamburg, Germany